SELECTED TOPICS IN PHYSICS, ASTROPHYSICS AND BIOPHYSICS

SELECTED TOPICS IN PHYSICS, ASTROPHYSICS AND BIOPHYSICS

PROCEEDINGS OF THE XIVTH
LATIN AMERICAN SCHOOL OF PHYSICS,
CARACAS 10–28 JULY 1972

Edited by

E. ABECASSIS DE LAREDO
Physics Center, Instituto Venezolano de Investigaciones Científicas (I.V.I.C.)

and

N. K. JURISIC
Facultad de Ciencias, Universidad Central de Venezuela

D. REIDEL PUBLISHING COMPANY
DORDRECHT-HOLLAND / BOSTON U.S.A.

First printing: December 1973

Library of Congress Catalog Card Number 73–83563

ISBN-13: 978-94-010-2635-2 e-ISBN-13: 978-94-010-2633-8
DOI: 10.1007/ 978-94-010-2633-8

Published by D. Reidel Publishing Company,
P.O. Box 17, Dordrecht, Holland

Sold and distributed in the U.S.A., Canada, and Mexico
by D. Reidel Publishing Company, Inc.
306 Dartmouth Street, Boston,
Mass. 02116, U.S.A.

Softcover reprint of the hardcover 1st edition 1973

TABLE OF CONTENTS

PREFACE

This volume contains the lecture notes of ten courses given at the XIV Latin American School of Physics (XIV LASP) which took place in Caracas, Venezuela, from the 10th to the 28th of July 1972. The LASP is held each year in a different Latin American country. Its purpose is to bring together young Latin American physicists at the doctorate level to attend lectures given by well known scientists. The participants are also invited to give seminars on their research work.

The topics of the courses given this year were chosen according to the existent fields of interest in Latin America. Two of these courses, namely those covering astrophysics and biophysics were given in such a way as to be accessible to all participants independently of their main field of interest.

The XIV LASP has received financial support from institutions in Venezuela and abroad, making possible a meeting of ninety-two Latin American physicists and ten distinguished lecturers. For this we are indebted to the following Institutions: Consejo Nacional de Investigaciones Científicas y Tecnológicas de Venezuela, Organization of American States, Instituto Venezolano de Investigaciones Científicas, and its physicists, Universidad Central de Venezuela, Consejo de Desarrollo Científico y Humanístico de la U.C.V., Universidad Simon Bolivar, Embassy of U.S.A. in Venezuela, Embassy of France in Venezuela, The British Council in Venezuela, Ministerio de Educación de Venezuela and the Latin-American Center of Physics.

Also we want to thank, deeply: Professor M. Moshinsky, Professor O. Rojo, Professor L. Chang, Dr J. Fernández and Dr M. Puma without whose collaboration this meeting would not have been possible.

The help of Mrs A. M. Gutierrez, Mrs E. de Garcia and Mr L. Hernández was invaluable and appreciated by all of us.

E. A. L.

N. K. J.

Caracas,
October 1973

PREFACE

PART A

SOLID STATE PHYSICS

CRYSTAL LATTICE DEFECTS AND MATTER TRANSPORT

A. R. ALLNATT

Dept. of Chemistry, University of Western Ontario, London 72, Ontario, Canada

1. Introduction

These lectures are about the movements of atoms from one part of a solid to another, over distances of many atomic spacings. Such motions manifest themselves in diffusion and electrical conductivity measurements and are among the most important crystal properties which can be traced to the presence of crystal lattice defects.

We shall be concerned with bulk migration of *point defects* such as vacant lattice sites, interstitial atoms and foreign atoms (substitutional or interstitial) in single crystals. We will say nothing about the effect of extended defects such as dislocations, grain boundaries and surfaces on atomic migration. The two basic properties of point defects in connection with the topic are their concentration and their mobility. The basic facts are long known and simply stated.

1.1. Concentration of intrinsic defects

Crystals at thermodynamic equilibrium contain vacant sites and interstitial atoms, the intrinsic defects. To take an atom from the surface and squeeze it into an interstitial position costs energy. Similarly to form a vacant site by taking an atom to the surface, requires a definite formation energy. The concentration of intrinsic lattice defects at thermal equilibrium is governed by a Boltzmann factor

$$c \sim e^{-E_f/kT} \tag{1.1}$$

where E_f is the energy to form one defect, a quantity which we shall not attempt to define closely at this point. Frequently one or another of vacancies or interstitials dominates. Some results for dominant defects are in the Table I. These defect structures are established from combined evidence of several kinds of measurement including transport properties. For reviews see References [1, 2]. Typical concen-

TABLE I

Systems	Intrinsic defects
Metals (f.c.c.) Rare gas solids	Vacancies
Alkali halides	Schottky defects (anion and cation vacancies)
AgCl, AgBr	Cationic Frenkel defects (cation interstitial and cation vacancies)
Fluorite structures (e.g. CaF_2, UO_2)	Anionic Frenkel defects

Abecassis de Laredo and Jurisic (eds.), Selected Topics in Phys. Astrophys. and Biophys. 3–43.

trations for substances, mainly metals and alkali halides, with which we will be concerned are in the range 10^{-5}–10^{-3} near the melting point.

1.2. Mobility and Migration Mechanisms

All the mechanisms by which atoms migrate are thermally activated processes. The migrating atom undergoes many ordinary sized vibrations before a suitable thermal fluctuation supplies the local concentration of energy and momentum necessary for the atom to pass through intermediate configurations of high energy between one lattice site and the next. The probability per unit time that a suitable fluctuation occurs (the jump frequency) is governed by a Boltzmann factor

$$w \sim e^{-E_m/kT}$$

where E_m is the energy of activation. The fluctuation of energy dissipates rapidly and successive jumps are dynamically independent. The principle migration mechanisms which have been proposed are as follows:

(i) *Vacancy mechanism.* A lattice atom can move into a adjacent vacancy if a suitable thermal fluctuation occurs (Figure 1a). As successive atoms move in this way the vacancy migrates through the crystal.

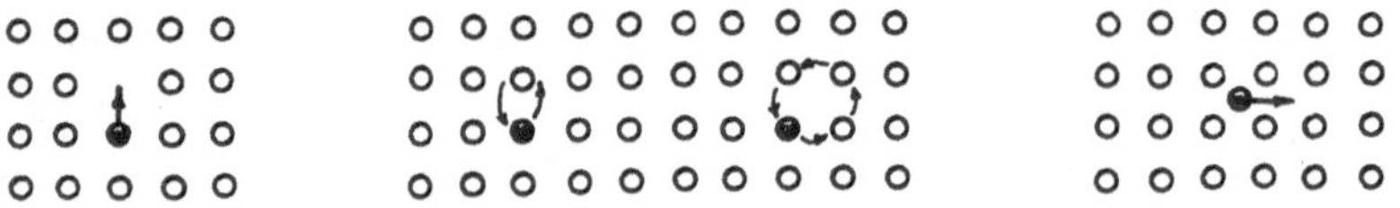

(a) Vacancy (b) exchange and four-atom ring (c) Interstitial

Fig. 1.

(ii) *Exchange and ring mechanisms.* A pair of neighbouring atoms may exchange places or a ring of neighbouring atoms may rotate so that each atom jumps into the one before it (Figure 1b). These mechanisms have so far not been identified in any of the systems we shall discuss.

(iii) *Interstitial mechanism.* An interstitial atom moves directly between adjacent interstitial sites (Figure 1c). It occurs for H, C, O, and N impurities in metallic systems and noble metals (Ag, Au) in Pb, Sn, In.

(iv) *Interstitialcy mechanism.* An interstitial atom pushes one of the neighbouring normal lattice atoms into an interstitial position and itself takes up the normal lattice site. The three possibilities for a f.c.c. lattice are shown in Figure 2. For the direct jump, 1, the two atoms move in a collinear manner whereas for the two indirect jumps, 2 and 3, the motion is non-collinear. Types 1 and 2 probably occur in AgCl.

(v) *Crowdion and split interstitial mechanisms.* In the crowdion configuration shown in Figure 3a for the bcc lattice the neighbours of the interstitial relax along a close-packed direction. In the split interstitial (Figure 3b) there are two atoms symmetrically placed about one normal site. As we shall see, these forms may be important in sodium.

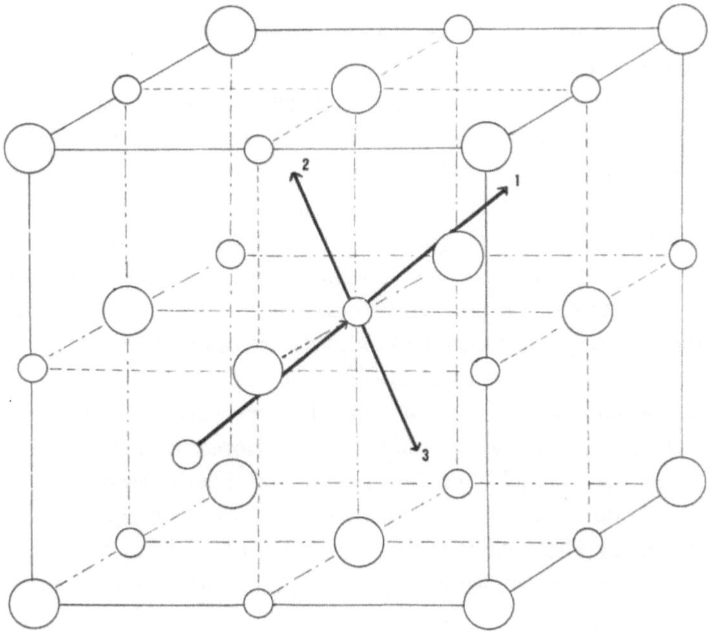

Fig. 2.

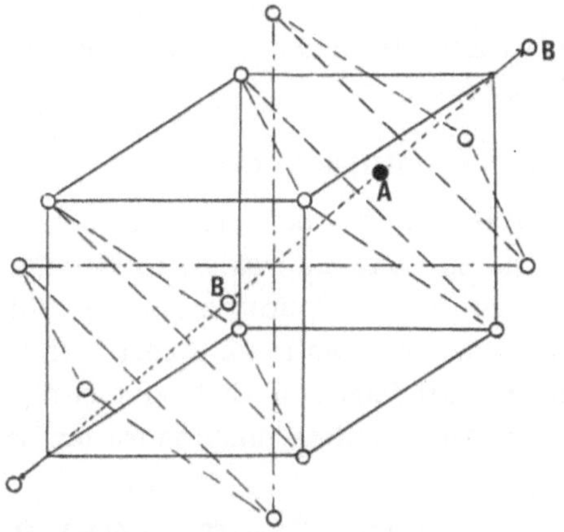

Fig. 3a. The crowdion configuration. A is the extra atom and its two neighbours along the ⟨111⟩ axis, denoted by B, have relaxed equal amounts from A along the axis.

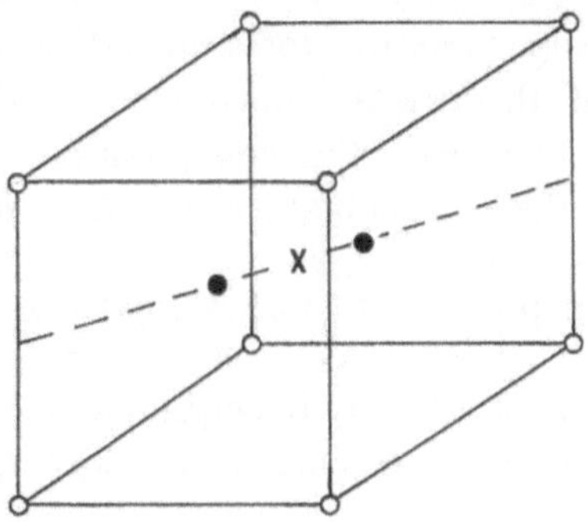

Fig. 3b. The split interstitial configuration.

(vi) *Divacancy and trivacancy mechanisms.* When a binding energy exists between nearest neighbour vacancies then an appreciable fraction of divacancies can occur. These can migrate without dissociating into single vacancies. They probably occur in Ag, and (as anion-cation pairs) in NaCl and KCl. Even higher aggregates have been proposed e.g. tri-vacancies in KCl but not positively identified. These mechanisms illustrate the importance of defect-defect interactions rather directly.

In recent work on matter transport [3] we note (1) the increasingly sophisticated combinations of theory and accurate experiment used to distinguish between migration mechanisms, and (2) the need to reconsider the earlier rather simple statistical mechanical theories of jump frequencies and defect interactions.

In the lectures we focus attention on the following:

(1) The isotope effect in diffusion. This is a powerful aid in assigning defect migration mechanisms, and has raised some questions about the correctness of the 'equilibrium' rate theories of jump frequencies. (2) Thermotransport. This provides another aspect of jump frequencies and the need for non-equilibrium theories of jump frequencies. (3) The theory of defect interactions, particularly in ionic crystals where the interactions are of long range and practically important in determining transport properties.

2. Some Background Results

In this section we outline without detailed derivations some well-known methods and results relevant to the three topics.

2.1. Defect Concentration and Interactions

The imperfect crystal can be imagined to be built up from a perfect one by (a) forming vacancies by removal of atoms (ions, molecules) to infinity, (b) forming vacancies and filling each with a substitutional impurity atom, (c) adding atoms to interstitial sites. Crystals of interest will not contain more than 1% of such point defects.

Let N_i denote the number point defects of kind i and let $N \equiv (N_1, N_2, \dots N_\sigma)$ denote the total defect composition. The partition function for temperature T and volume V is given by

$$\exp(-A\beta) = Q(N, T, V) = \sum_{\{N\}} \exp[-A(\{N\}, V, T)\beta], \tag{2.1}$$

A is the Helmholtz free energy and β is $(kT)^{-1}$. The sum is over all configurations $\{N\}$ of the defects, i.e. all assignments of atoms to lattice sites. $A(\{N\}, V, T)$ is the Helmholtz free energy of the crystal constrained to defect configuration $\{N\}$, calculated from the eigen-energies for that configuration.

If the defects all occupy the same sublattice then we can write

$$Q(N, T, V) = \exp\left[-\sum_{i=0}^{\sigma} N_i A_i \beta\right] \sum_{\{N\}} \exp[-A_{\text{int}}(\{N\})\beta], \tag{2.2}$$

where N_0 is the number of host atoms in the crystal and A_0 the free energy per atom in the perfect crystal. A_i for $i \neq 0$ is the change in energy when one defect of kind $_i$ is formed in the perfect crystal. With these definitions A_{int} must be the energy of defect-defect interactions.

(i) *Neglect defect interactions.* A_{int} is zero and so we have

$$\sum_{\{N\}} \exp\left[-A_{\mathrm{int}}(\{N\})\,\beta\right] = \Omega(N), \tag{2.3}$$

where $\Omega(N)$, the total number of configurations, is given by

$$\Omega(N) = \left(\sum_{i=0}^{\sigma} N_i\right)! \Big/ \prod_{i=0}^{\sigma} N_i! \tag{2.4}$$

Since we generally work at constant pressure we calculate the Gibbs free energy

$$G = A + pV = A + p\left[V_0 + \Delta V(N)\right], \tag{2.5}$$

where V_0 is the volume of the perfect crystal and $\Delta V(N)$ the change in volume when the N defects are formed. From Equation (2.2)–(2.5) the chemical potential of species i is

$$\mu_i \equiv \left(\frac{\partial G}{\partial N_i}\right)_{T,\,p,\,N_{j\neq i}} = g_i + kT\log c_i, \tag{2.6}$$

where

$$g_i = A_i + p\left(\frac{\partial V}{\partial N_i}\right)_{T,\,p,\,N_{j\neq i}}$$

and $c_i = N_i/(\sum_{j=0}^{\sigma} N_j)$ is the concentration of defects. We can show in the usual manner that

$$G = \sum_{i=0}^{\sigma} \mu_i N_i\,. \tag{2.7}$$

If we repeat this for the case that some defects and atoms (ions) are restricted to particular sublattices e.g. ionic crystals then (2.6) and (2.7) still hold but with $c_i = N_i/B_i$ where B_i the number of sites on the sublattice occupied by i, and with the sum in (2.7) over all species.

To find the number of intrinsic lattice defects at thermodynamic equilibrium we minimize G at constant T and P with respect to the number of each kind of intrinsic defect with the restrictions: (a) there is a constant number of each kind of atom, (b) electrical neutrality is maintained, (c) constant structure is maintained e.g. in an ionic crystal MX the number of M sites equals the number of X sites.

As an example consider an ionic crystal MX doped by impurity salt IX_2 and containing Schottky defects. Call the cation and anion vacancies + and − respectively. The Gibbs free energy is

$$G = N_M\mu_M + N_X\mu_X + N_I\mu_I + N_+\mu_+ + N_-\mu_- \tag{2.8}$$

and the constraints are

(a) N_M, N_X, N_I are each constant
(b) $N_M + 2N_I = N_X$
(c) $N_M + N_I + N_+ = N_X + N_-$.

Carrying out the variation we find

$$\mu_+ + \mu_- = 0, \quad \text{so} \quad c_+c_- = \exp(-g_s\beta) \equiv K_s, \tag{2.9}$$

where $g_s \equiv g_+ + g_-$ is the energy to form a Schottky pair. For other examples e.g. Frenkel defects, vacancies in metals and semiconductors see References [1, 2, 4, 5].

(ii) *Defect clusters.* When defect interactions cannot be neglected a simple extension of the proceeding method is generally made. Consider for example the doped ionic crystal again. It is well-known from ionic conductivity measurements that a nearest-neighbour pair divalent impurity-cation vacancy is stable by a binding energy of ~0.4 eV compared with infinite separation. The cation vacancies in such 'complexes' do not contribute to the conductivity since they are bound whereas the unpaired ones do contribute.

The method used to find the concentration of complexes is to treat them as a new species and continue to ignore interactions between defect species. Repeating the previous arguments we need Ω_c, the number of configurations on the cation sub-lattice for N_+ unpaired cation vacancies, N_I unpaired impurities, N_k complexes, and N_M host cations on the $B \equiv (N_M + N_I + N_+ + 2N_k)$ cation sites. If we consider first the number of ways of setting down complexes Ω_k, then unpaired impurities Ω_I, and then unpaired vacancies Ω_+ we get

$$\begin{aligned} \Omega &= \Omega_k \Omega_I \Omega_+ \\ \Omega_k &\simeq \left\{ z^{N_k} \prod_{s=0}^{N_k - 1} (B - 2s) \right\} / N_k! \\ \Omega_I &= (B - 2N_k)!/N_I!\,(B - N_I - 2N_k)! \\ \Omega_+ &= (N_M + N_+ - zN_I)!/N_+!\,(N_M - zN_I)! \end{aligned} \tag{2.10}$$

where z is the number of nearest neighbours per site. Equation (2.10) is approximate because it neglects interference terms coming from the hinderance of rotation of one complex by another close by. Notice that in Ω_+ we assigned the vacancies to sites not already occupied by k or I and which are not neighbours of the N_I free impurities. The chemical potentials are now

$$\begin{aligned} \mu_k &= g_k + kT \log z + kT \log c_k \\ \mu_+ &= g_+ + kT \log c_+ \gamma_+ \\ \log \gamma_+ &= -(zc_I + c_k) + 0(c^2). \end{aligned}$$

Thus the activity coefficient γ_+ is not unity any more but contains excluded volume corrections which are correct only to lowest order in c. Similar remarks apply to γ_I, γ_M. By minimising G with respect to N_k, N_+, N_- with the appropriate 3 constraints we get Equation (2.9) and also

$$\mu_k = \mu_I + \mu_+$$

which gives the mass-action equation

$$\frac{c_k}{c_+ c_I \gamma_+ \gamma_I} = z e^{\xi\beta} \equiv K_2 \tag{2.11}$$

$$-\xi \equiv g_k - g_+ - g_I . \tag{2.12}$$

For dilute solutions the activity coefficients can be taken as unity. The same method can be applied to larger defect clusters; further examples of this mass-action method can be found in References [1, 2, 4, 5].

Equation (2.11) can be applied to vacancy-impurity interactions in metals where interactions are short-ranged [1]. In ionic crystals there are long range (Coulombic) interactions between the unpaired defects and it is less satisfactory. It is then generally assumed that a correction can be made by approximating γ_+, γ_I by the Debye-Hückel activity coefficients from electrolyte theory. This is certainly correct at low enough concentrations but it is not known precisely under what conditions it breaks down and it seems difficult to investigate in the context of the method just described. With the increasingly detailed analysis of ionic conductivity results it is worthwhile to develope some other method which allows investigations of this kind. This will be done in Section 5.

2.2. Relation of Transport Coefficients to Defect Concentrations and Jump Rates

2.2.1. *Random Walk Method*

Let us fix attention on one particular species i.e. one kind of impurity atom, interstitial, vacancy, in a crystal which may have concentration gradients, electric fields, temperature gradients etc. We wish to calculate the flux $\mathbf{J}(\mathbf{r}, t)$, the number of atoms of the species crossing unit area in unit time in some direction. Let $n(x, t)$ be the number density of atoms at some position x on the x-axis at time t and let $K(x+\Delta x, t+\Delta t/x, t)$ be the conditional probability that an atom known to be at x at time t will be at $x+\Delta x$ at time $t+\Delta t$. For simplicity we restrict our gradients to the x-direction (generalization is simple). According to the definition we have

$$\int K(x, t+\Delta t/x-\Delta x, t)\,\mathrm{d}(\Delta x) = 1 .$$

Conservation of atoms implies

$$n(x, t+\Delta t) = \int K(x, t+\Delta t/x-\Delta x, t)\, n(x-\Delta x, t)\,\mathrm{d}(\Delta x) . \tag{2.13}$$

For not too large times and displacements we can Taylor expand in Δt on the left, and on the right we can Taylor expand the integrand considered as a function of $x-\Delta x$ about the value at x. We define moments of the displacement by

$$\langle(\Delta x)^n\rangle = \int (\Delta x)^n\, K(x+\Delta x, t+\Delta t/x, t)\,\mathrm{d}(\Delta x), \quad n = 1, 2, \ldots .$$

Equation (2.13) can then be written as

$$n(x,t) + \frac{\partial n(x,t)}{\partial t} + \cdots$$
$$= n(x,t) - \frac{\partial}{\partial x}\left[\langle \Delta x\rangle\, n(x,t)\right] + \frac{1}{2!}\frac{\partial^2}{\partial x^2}\left[\langle (\Delta x)^2\rangle\, n(x,t)\right]\ldots.$$

Comparing this with the equation of conservation of matter

$$\frac{\partial n(x,t)}{\partial t} = -\frac{\partial}{\partial x} J_x(x,t)$$

we find the flux of atoms in the x direction at (x, t) is

$$J_x(x,t) = n\frac{\langle \Delta x\rangle}{\Delta t} - n\frac{\partial}{\partial x}\frac{\langle (\Delta x)^2\rangle}{2\Delta t} - \frac{\partial n}{\partial x}\frac{\langle (\Delta x)^2\rangle}{2\Delta t}$$
$$+ \text{ terms in } \langle (\Delta x)^3\rangle \text{ and higher moments}. \qquad (2.14)$$

This holds for migration along any chosen direction x in any crystal isotropic or anisotropic. The terms in $\langle (\Delta x)^3\rangle$ and higher moments can be neglected for the applications to systems with small gradients which follow. For more extensive discussions of random walks see for example References [6, 7]. We next consider some common uses of this equation.

2.2.1.1. *Diffusion in chemically homogeneous systems with no external forces or temperature gradients*

There are no forces to bias the jumps and the system is uniform in its properties everywhere so $\langle \Delta x\rangle$ is zero and $\langle (\Delta x)^2\rangle$ is independent of x. Equation (2.14) then yields Fick's law

$$J_x = -D_{xx}\frac{\partial n}{\partial x} \qquad (2.15)$$

with D_{xx} given by the Einstein equation

$$D_{xx} = \frac{\langle (\Delta x)^2\rangle}{2\Delta t}. \qquad (2.16)$$

For isotropic crystals $\langle (\Delta x)^2\rangle$, $\langle (\Delta y)^2\rangle$, $\langle (\Delta z)^2\rangle$ will all be equal and we can write

$$D_{xx} = D_{yy} = D_{zz} = D = \langle (\Delta \mathbf{r})^2\rangle / 6\Delta t, \qquad (2.17)$$

where $\mathbf{r}$ denotes the position vector for one atom. The flux equation is then

$$\mathbf{J} = -D\nabla n \qquad (2.18)$$

and combining this with the equation for conservation of matter we obtain

$$\frac{\partial n}{\partial t} = D\nabla^2 n \qquad (2.19)$$

To determine D experimentally we have to pick a geometry for which we can solve the last equation with appropriate boundary conditions. The most common method is the thin layer method. A radioactive tracer diffuses from a layer on a plane surface of the crystal, perpendicular to the x-axis say. The solution is

$$n(x, t) = \frac{n_0}{\sqrt{(\pi D t)}} \exp(-x^2/4Dt), \tag{2.20}$$

where n_0 is the activity per unit area at $t=0$. We microtome the crystal into thin slices after a diffusion anneal of time t and measure the activity in each slice, the thickness being found by weighing. A plot of logarithm of activity vs. the square of the penetration thickness should give a straight line from which D can be found. A 'tail' in the penetration plot at large x generally indicates enhanced diffusion down dislocations, grain boundaries, etc.

Other methods are to diffuse activity into the solid from a constant composition in the vapor and measure the penetration profile, or to diffuse activity from the solid into the vapor and measure the build up in the vapor.

2.2.1.2. *Diffusion in inhomogeneous systems*

In the previous experiments because each tracer atom 'sees' only homogeneous material the detailed interpretation is simple. If we inter-diffuse two blocks of different materials then there are composition gradients and $\langle \Delta x \rangle \neq 0$. Since D depends on composition the interpretation of results is more difficult and detailed information at the molecular level hard to extract. We shall not attempt to discuss these experiments which are dealt with in References [6, 8]. In systems with temperature or electrical gradients $\langle \Delta x \rangle$ is again not zero, e.g. Section 4.

2.2.2. *Correlation Factors for Diffusion in Uniform Systems*

2.2.2.1. *Definition of the correlation factor*

In the Einstein equation for D_{xx}, Equation (2.16), we can write the total displacement of a typical atom as

$$\Delta x = \sum_{i=1}^{n} x_i,$$

where x_i is the x displacement for the jump number i. We obtain

$$D_{xx} = \tfrac{1}{2} \sum_{i=1}^{n} \frac{\langle x_i^2 \rangle}{2\Delta t} f_x \tag{2.21}$$

$$f_x = 1 + \left(2 \sum_{i=1}^{n-1} \sum_{j=1}^{n-1} \langle x_i x_{i+j} \rangle \Big/ \sum_{i=1}^{n} \langle x_i^2 \rangle \right) \tag{2.22}$$

is the correlation factor. It is a measurable mechanism-dependent parameter.

In a completely random walk the directions of successive displacements are uncorrelated so $\langle x_i x_{i+j} \rangle$ is zero ($j=1, 2, \ldots$) and the correlation factor is just unity. This will be the case for vacancies, interstitials, or divacancies in a pure cubic metal say.

For isotropic crystals we have

$$D = D_{xx} = D_{yy} = D_{zz} . \tag{2.23}$$

If we suppose that there is only one kind of jump e.g. a single vacancy mechanism then

$$\sum_{i=1}^{n} \langle x_i^2 \rangle = n \langle x^2 \rangle = \frac{n}{3} \langle (x^2 + y^2 + z^2) \rangle = \frac{nr^2}{3},$$

where r is the length of one jump. Since $f=1$ we obtain

$$D = \tfrac{1}{6} \Gamma r^2 , \tag{2.24}$$

where $\Gamma = n/\Delta t$ is the mean number of jumps made per unit time for one vacancy (or interstitial etc. as the case may be). For vacancy diffusion in a cubic lattice $\Gamma = z\omega$, where z is the number of nearest neighbours per site and ω is the probability per unit time a vacancy will exchange with a particular neighbour atom.

Hence for vacancies

$$D_v = \tfrac{1}{6} z \omega r^2 . \tag{2.25}$$

For non-cubic lattice with $f=1$ we shall have

$$D_{xx} = \tfrac{1}{2} \sum_{\alpha} \Gamma_\alpha x_\alpha^2 ,$$

where Γ_α is the jump frequency for jumps of type α having a projected displacement x_α along the x-axis.

Much more interesting are the cases where successive jumps are correlated and $f \neq 1$. If we consider just isotropic crystals then we need only one correlation factor f and we find from Equations (2.21–23)

$$D = \tfrac{1}{6} \Gamma r^2 f \tag{2.26}$$

$$f = 1 + \left(2 \sum_{i=1} \sum_{j=1} \langle \mathbf{r}_i \cdot \mathbf{r}_{i+j} \rangle / \sum_{i=1} \langle r_i^2 \rangle \right) . \tag{2.27}$$

When all the jumps have the same length r then $\langle \mathbf{r}_i \cdot \mathbf{r}_{i+j} \rangle$ is independent of i and equal to $r^2 \langle \cos \theta_j \rangle$ and hence

$$f = 1 + 2 \sum_{j} \langle \cos \theta_j \rangle , \tag{2.28}$$

where $\langle \cos \theta_j \rangle$ is the average of the cosine of the angle between a jump and the jth following jump. We now summarize some results for f. Detailed derivations can be found in a review by LeClaire [9].

2.2.2.2. *Correlation effect in self-diffusion*

By 'self-diffusion' we mean the case where all the jump frequencies are identical.

This is well approximated by an isotope diffusing in a mixture of isotopes of the same substance e.g. ^{40}K in ^{39}K and ^{41}K.

(i) *Single vacancy mechanism.* If an atom has just exchanged with a vacancy the possible directions of its next jump are not all equally probable. It is most likely to jump back to its original position since the vacancy is available. Hence $\langle\cos\theta_1\rangle \neq 0$. The third jump is correlated with the second and hence with the first, and so on. However, when all jumps are of the same length and the jump vectors are all at least two-fold axes of rotational symmetry it can be shown

$$\langle\cos\theta_j\rangle = \langle\cos\theta_1\rangle^j$$
$$f = (1 + \langle\cos\theta_1\rangle)/(1 - \langle\cos\theta_1\rangle). \tag{2.29}$$

We can do a rough estimate of f. The relative probability that the vacancy will exchange again with the tracer after the first jump is $1/z$, and $\cos\theta_1 = -1$. Hence $\langle\cos\theta_1\rangle$ is $-1/z$ and we obtain

$$f \simeq \frac{z-1}{z+1} \simeq 1 - \frac{2}{z}$$

However there are smaller probabilities that the vacancy will first migrate to one of the other neighbour sites to the tracer and then exchange with it.
The Table II gives exact and approximate values.

TABLE II

	f	1–2/z
diamond	0.5000	0.5000
cubic	0.6531	0.666
b.c.c.	0.7272	0.750
f.c.c.	0.7815	0.833
h.c.p.	$0.7812 = f_x = f_y$	0.833
	$0.7815 = f_z$	

(ii) *Interstitialcy mechanism.* For the NaCl-structure we can distinguish three possible types of jump (see Figure 2).

Name	Frequency	f
direct (collinear)	ω_1	2/3
indirect (non-collinear)	ω_2	32/33
indirect (non-collinear)	ω_3	0.9643

(iii) *Divancy mechanism.* For anion-cation vacancy pairs in NaCl and CsCl f is a function of ω_A/ω_C, the ratio of anion and cation jumps, and hence is a function of T. The contribution of vacancy pairs to the cation self-diffusion coefficient in NaCl is shown in Figure 4.

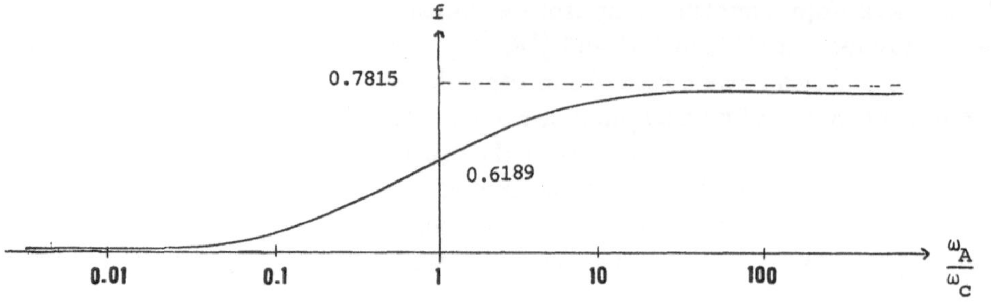

Fig. 4.

For vacancy pairs in cubic metals there will be only one jump frequency and f is a number again e.g. $f=0.475$ for the fcc lattice.

2.2.2.3. *Correlation effect for impurity diffusion*

We suppose that the solute concentration is so low that each atom effectively diffuses through pure solvent i.e. less than 1% solute. f now depends on impurity jump frequencies and solvent jump frequencies. Furthermore the solvent jumps may be affected by the presence of the solute.

It can be shown that for a single vacancy or interstitialcy mechanism

$$f_I = B/(\omega_2 + B) \tag{2.30}$$

provided that the jump direction is at least a two-fold axis of rotational symmetry. Here ω_2 is the impurity jump rate and B is a function of the jump rates of solvent atoms only.

In the five frequency model it is assumed that for the single vacancy mechanism in a f.c.c. lattice only the solvent jumps shown in Figure 5 are affected by the impurity. Measuring positions relative to the impurity:

ω_1 = vacancy-solvent exchange (both n.n.)
ω_2 = vacancy-impurity exchange
ω_3 = first neighbour vacancy – second or more distant neighbour solvent atom exchange.
ω_4 = inverse of ω_3
ω_0 = vacancy-solvent exchange in pure solvent.

At equilibrium we have

$$\frac{\omega_4}{\omega_3} = \exp(g_B\beta),$$

where g_B is the n.n. vacancy-impurity binding energy.

An accurate expression for the five frequency model takes the form

$$B = \omega_1 + \tfrac{7}{2}\omega_3 F,$$

where F is a function of ω_4/ω_0 calculated by Manning.

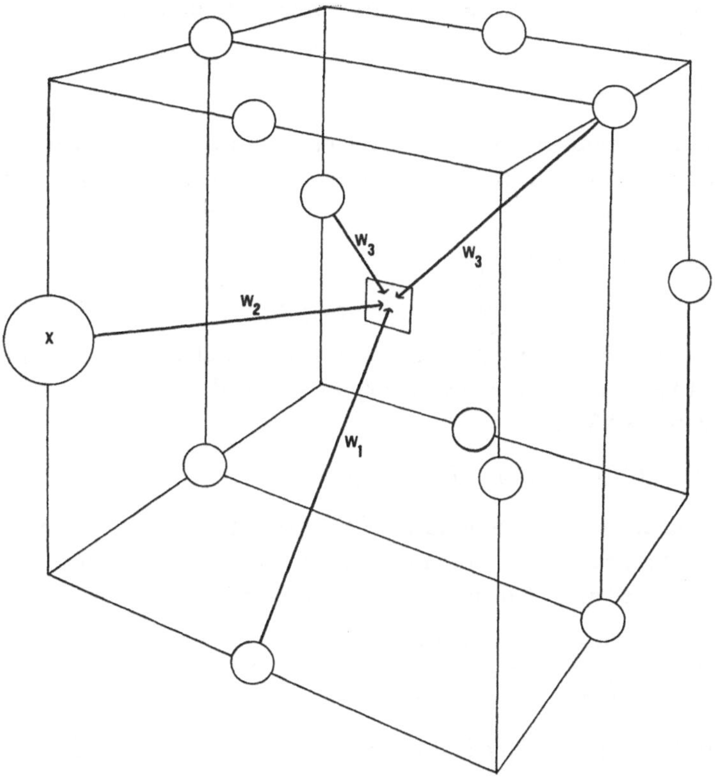

Fig. 5.

Simple limiting cases are

$$\omega_4 \ll \omega_0, \quad F = 1 \qquad \text{`weak binding'}$$
$$\omega_4 = \omega_0, \quad F = 5.15/7$$
$$\omega_4 \gg \omega_0, \quad F = 2/7\,.$$

For ionic crystals the 'tight binding' approximation

$$\omega_3 = 0 \quad f_I = \omega_1/(\omega_1 + \omega_2)$$

is useful because the binding energy is large.

In summary then, measurements of f could be used to distinguish jump mechanisms in self-diffusion and to obtain some information on relative jump frequencies in impurity diffusion. The two older methods of measurement are described by LeClaire [9]. They were the comparison of self-diffusion and ionic conductivity measurements in ionic conductors, and the effect of solute additions on self-diffusion in metals. We consider next the third method, the isotope effect in diffusion.

3. The Isotope Effect and Equilibrium Rate Theory

3.1. Introduction

If we confine ourselves to isotropic self-diffusion or impurity diffusion by a single mechanism characterized by the jump frequency ω_2 of diffusing species then the diffusion coefficient is

$$D = Af\omega_2, \tag{3.1}$$

where A contains geometrical and other factors but does not depend on jump frequencies. We now compare the tracer diffusion coefficients of two isotopes of the diffusing species having different masses m^α and m^β. From Equation (3.1) we have

$$\frac{D^\alpha - D^\beta}{D^\beta} = \frac{f^\alpha\omega_2^\alpha - f^\beta\omega_2^\beta}{f^\beta\omega_2^\beta}. \tag{3.2}$$

For the vacancy mechanism and the interstitialcy mechanism if the jump direction is at least a two-fold rotational symmetry axis then the correlation factors take the form in Equation (2.30):

$$f^\alpha = B^\alpha/(\omega_2^\alpha + B^\alpha), \quad f^\beta = B^\beta/(\omega_2^\beta + B^\beta), \tag{3.3}$$

where B is a function of the jump rates of the host atoms. If we assume that the host atom jumps are independent of the mass of the isotopes α and β then $B^\alpha = B^\beta$ and we can eliminate B^α, B^β and f^β from these equations and obtain the result

$$E \equiv \frac{D^\alpha/D^\beta - 1}{\sqrt{m^\beta/m^\alpha} - 1} = \frac{f^\alpha(\omega_2^\alpha/\omega_2^\beta - 1)}{\sqrt{m^\beta/m^\alpha} - 1} \equiv f^\alpha g. \tag{3.4}$$

The quantity E, 'strength of the isotope effect', is here defined in terms of the measurable tracer diffusion coefficients and the masses of the isotopes. It gives us the product of the correlation factor f^α and the function g defined by

$$g = (\omega_2^\alpha/\omega_2^\beta - 1)/(\sqrt{m^\beta/m^\alpha} - 1). \tag{3.5}$$

For other jump mechanisms e.g. indirect interstitialcy, or where there are several jump frequencies, e.g. anisotropic diffusion, there are more complicated expressions for E [10].

The earliest forms of equilibrium rate theory gave expressions for the jump frequency ω, when only a single atom is permanently displaced, of the form

$$\omega = \nu \exp(-\Delta g/kT), \tag{3.6}$$

where Δg is the free energy of activation and ν an average vibration frequency of the migrating atom. Since Δg will be the same for the two isotopes whereas vibration frequencies vary with mass as $\nu \propto m^{-1/2}$ it follows that $g = 1$. If the migration step involves the simultaneous displacement of n atoms, e.g. $n = 2$ for a direct interstitialcy, $n = 4$ for ring diffusion, then a proposed extension [11] leads to

$$g = \left[\left(\frac{(n-1)m + m^\beta}{(n-1)m + m^\alpha}\right)^{1/2} - 1\right] \Big/ \left[\left(\frac{m^\beta}{m^\alpha}\right)^{1/2} - 1\right], \tag{3.7}$$

where m is the average mass of the host atoms. Notice that for large n we find $g \to 0$.

Roughly speaking g is a measure of the many-body character of the jump. When many atoms are involved, changing the mass of just one of them has little effect so the isotope effect tends to zero. If these simple results for g are accurate then we have a method of determining correlation factors. Alternatively, from measurements of E when the diffusion mechanism and f are well established the value of g may give some insight into the theory of jump frequencies.

3.2. Measurement and results for isotope effect

Since D^α and D^β differ typically by 2% and we shall be looking at deviations of E from unity of only about 20% the need for great precision is obvious. However, simultaneous diffusion of the two isotopes from a thin layer will eliminate errors due to variations in time and temperature of anneal. Applying the thin layer diffusion Equation, (2.20), to each isotope, the ratio of concentrations c^α and c^β at the same position in the crystal is found to be given by

$$\log (c^\alpha/c^\beta) = \text{constant} - (D^\alpha/D^\beta - 1) \log c^\alpha . \tag{3.8}$$

We do not have to know the penetration distance or time! The preferred method of determining the concentrations is the half-life separation method applied to two gamma emitting isotopes. The total gamma radiation is counted as a function of time and fitted by least squares to

$$c(t) = c^\alpha(0) \exp(-\lambda^\alpha t) + c^\beta(0) \exp(-\lambda^\beta t) . \tag{3.9}$$

TABLE III

Isotope effect for f.c.c. and h.c.p. lattices (Self-diffusion)

Substance	Isotopes	$E = gf$	g(single vacancy)	Ref.
Cu (f.c.c.)	64,67	0.684 ± 0.014 (894–1061 °C)	0.87 ± 0.02	[16]
Ag (f.c.c.)	105,110	0.718 ($T \leqslant 750$ °C) 0.718–0.639 ($T > 750$ °C)	$\geqslant 0.918$	[16]
Pd (f.c.c.)	103,112	0.813 ± 0.042 (1450–1500 °C)	1.02 ± 0.04	[17]
γ-Fe (f.c.c.)	55,59	0.57 ± 0.06 (1171–1349 °C) 0.71 ± 0.04 (1452 °C) 0.77 ± 0.05 (1338 °C)	0.74 ± 0.08	[18]
	52,59	0.53 ± 0.01	0.68 ± 0.02	[19]
Zn (h.c.p.)	65,69	(289–418 °C)	$g_A = 0.957 \pm 0.036$ $g_B = 0.926 \pm 0.030$	[20]
	65,69		$g_B = 0.88 \pm 0.03$	[21]
NaCl	22,24 (Na)	0.72 ± 0.07	0.93 ± 0.09	[12]
AgBr	105,111 (Ag)		0.66	[15]

Obviously isotopes of high radiochemical purity must be used and the decay constants λ^α, λ^β accurately known. For accuracy the half-lives should differ by a factor of 3 or more and since the measurements need times of 10–20 half-lives of the fastest decaying species, highly stable counting equipment is required. For discussion of measurement procedures see references [12–14].

The table of results made in 1964 [12] for well established cases of impurity diffusion by the interstitial mechanism ($f=1$) shows little deviation from $g=1$. Exceptions are H^1 and H^2 in α-Fe, γ-Fe, Ni, Cu. This is because the isotopes are light so that quantum effects become important. We will consider only cases where quantum effects can be neglected.

The results for self-diffusion in f.c.c. and h.c.p. lattices in Table III are for systems in which the single vacancy mechanism is thought to be dominant. Assuming this is correct g is in the range 0.7 to 1.0.

The mechanisms for the b.c.c. metals in Table IV are much less certain. If a vacancy mechanism is assumed then g is approximately 0.5 to 0.6.

TABLE IV

Isotope effect for b.c.c. lattices (self-diffusion)

Substance	Isotopes	$E = gf$	g(single vacancy)	Ref.
Na	22,24	0.36 ± 0.02	0.505 ± 0.02	[22]
δ-Fe	52,59	0.34 ± 0.01	0.46 ± 0.01	[19]
α-Fe	52,59	0.43 ± 0.05	0.59 ± 0.07	[19]

There is only one result for an established interstitialcy mechanism. For Ag ions in AgBr for jumps of type 1 and type 2 shown in Figure 1, the results are $g_1 \simeq 1$ and $g_2 \simeq 0.1$ although there are anomalies in the interpretation [15]. Let us next consider the theoretical significance of g in detail.

3.3. Equilibrium Rate Theory

The theory of jump frequencies outlined here follows the discussions of Vineyard [11], LeClaire [23], and Glyde [24]. Imagine for concretness the migration of a particular atom into an adjacent vacancy. Let $\Phi(\underline{x})$ denote the potential energy, where $\underline{x} \equiv (x_1, x_2, \dots x_n)$ denotes the coordinates of the atoms, (x_1, x_2, x_3) for the jumping atom. Φ will have a minimum, A, for the configuration with the atom at its mean position at its initial site and another minimum, B, for the atom at its final site after the jump. In passing from A to B the system passes through a region of increased $\Phi(\underline{x})$. There is one path of the system for which the maximum increase is minimal. The point where Φ is a maximum on this path is called the saddle point P. Through P draw a surface S perpendicular to lines of constant $\Phi(\underline{x})$.if there were only two coordinates the situation would be that sketched in Figure 6 which shows two dimensional contours of constant energy.

The surface S marks a natural boundary between the well centred on A and that

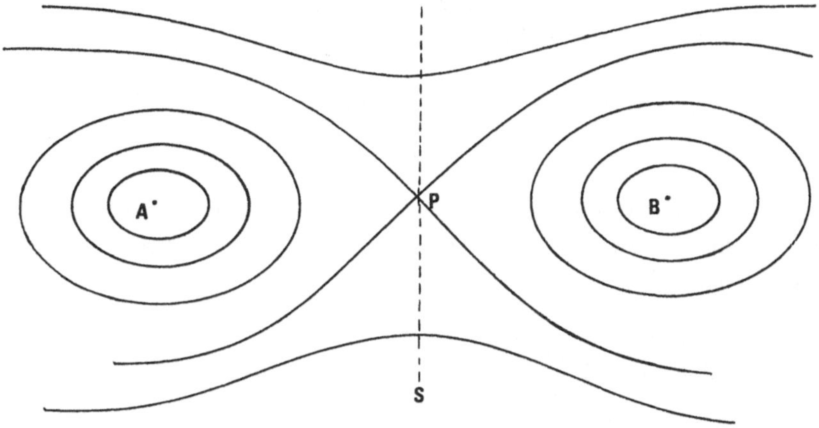

Fig. 6. Surfaces of constant potential energy.

centred on B. The saddle point configuration can be located, given the potential function and a fast computer, by moving the atom to midway between the sites and then adjusting the positions of the surrounding atoms from their mean positions to minimize the potential energy. The positions of about 20 atoms will be adjusted in a typical calculation.

We can assume classical mechanics for the temperatures of interest in Section 3.2. We assume equilibrium statistical mechanics can be used; the validity of this for a transport process is not obvious and is discussed later. The probability of observing the system at position $\underline{x}$ to $\underline{x}+\mathrm{d}\underline{x}$ with velocity in the range $\underline{v}$ to $\underline{v}+\mathrm{d}\underline{v}$ is given by

$$\varrho(\underline{x}, \underline{v})\,\mathrm{d}\underline{x}\,\mathrm{d}\underline{v} = \varrho_0 \exp(-H\beta)\,\mathrm{d}\underline{x}\,\mathrm{d}\underline{v} \tag{3.10}$$

$$H = T + \Phi = \sum_{i=1}^{n} \tfrac{1}{2} m_i v_i^2 + \Phi(\underline{x}). \tag{3.11}$$

The jump frequency ω is then

$$\omega = I/Q, \tag{3.12}$$

where

$$I = \underset{(\underline{v}\cdot \mathrm{d}\underline{S} > 0)}{\iint} \varrho(\underline{x}, \underline{v})\, \underline{v}\cdot \mathrm{d}\underline{S}\,\mathrm{d}\underline{v} \tag{3.13}$$

is the flux of phase points normal to the surface S and in the direction from A to B (the element of surface is $\mathrm{d}\underline{S} = \underline{p}\,\mathrm{d}S$ where $\underline{p}$ is a unit vector normal to the element of area $\mathrm{d}S$ and $\underline{v}$ the velocity of the phase point). Q is a normalization constant giving the number of points originally in A:

$$Q = \iint_A \varrho(\underline{x}, \underline{v})\,\mathrm{d}\underline{x}\,\mathrm{d}\underline{v}. \tag{3.14}$$

To evaluate I make an orthogonal transformation to new coordinates $\underline{y}$ such that the axis y_1 is normal to S at P, i.e. y_1 parallel to $\underline{p}$.

$$(x_i - x_i^P) = \sum_{j=1}^{n} c_{ij} y_j, \quad \text{or} \quad (\underline{x} - \underline{x}^P) = \underline{c}\underline{y}$$
$$\underline{c}^T \underline{c} = 1, \quad C = 1 \tag{3.15}$$

using an obvious matrix notation in which superscript T denotes transpose and C is the determinant of matrix $\underline{c}$ (and a similar convention for all other matrices). The flux is now given by

$$I = \underset{\dot{y}_1 > 0}{\iint} \delta(\underline{y}_1) \, \dot{\underline{y}}_1 \varrho(\underline{y}, \dot{\underline{y}}) \, \mathrm{d}\underline{y} \, \mathrm{d}\dot{\underline{y}}, \tag{3.16}$$

where $\dot{y} \equiv \mathrm{d}y/\mathrm{d}t$. Of course y_1 is normal to S only at P but only small displacements from P are of interest because of the Boltzmann factor so the approximation in taking it as normal everywhere is unimportant. For small displacements about P we can Taylor expand the potential energy about P:

$$\begin{aligned} \Phi - \Phi(P) &= \tfrac{1}{2}(\underline{x} - \underline{x}^P)^T \, \underline{k}^+ \, (\underline{x} - \underline{x}^P) \\ &= \tfrac{1}{2}\underline{y}^T \underline{b}^+ \underline{y}, \end{aligned} \tag{3.17}$$

where $\underline{k}^+$ is the matrix of the force constants and where $\underline{b}^+ = \underline{c}^T \underline{k}^+ \underline{c}$. The + superscript will always denote properties at P. Finally, the kinetic energy is given by

$$T = \tfrac{1}{2}\underline{v}^T \underline{m} \underline{v} = \tfrac{1}{2}\dot{\underline{y}}^T \underline{a}^+ \dot{\underline{y}}, \tag{3.18}$$

where $\underline{a}^+ = \underline{c}^T \underline{m} \underline{c}$.

To do the integrations in (3.17) we use the coordinate transformation (with Jacobian unity) [30]

$$\dot{y}_1 = \dot{\omega}_1; \quad \dot{y}_s = \dot{\omega}_s + A_{1s}^+ \dot{\omega}_1 / A_{11}^+, \quad s = 2, \ldots, n, \tag{3.19}$$

where A^+ is the determinant of $\underline{a}^+$ and A_{rs}^+ is the cofactor of a_{rs}^+. The kinetic energy is then

$$T = \tfrac{1}{2} A^+ \dot{\omega}_1^2 / A_{11}^+ + \tfrac{1}{2} \sum_{2}^{n} \sum a_{rs}^+ \dot{\omega}_r \dot{\omega}_s. \tag{3.20}$$

It is now trivial to do the integration over $\dot{\omega}_1$ and the remaining integrations can be done using the identity

$$\iint_{-\infty}^{\infty} \exp\left(-\tfrac{1}{2}x^T \underline{d} x\right) \mathrm{d}\underline{x} = (2\pi)^{M/2} / \sqrt{D}, \tag{3.21}$$

where $\underline{d}$ is an MxM matrix with determinant D. The integrations over $\underline{y}$ can be done in the same way. Similarly to evaluate Q we Taylor expand the potential energy about A

$$\Phi - \Phi(A) = \tfrac{1}{2}(\underline{x} - \underline{x}^A)^T \underline{k}(\underline{x} - \underline{x}^A) \tag{3.22}$$

and use the identity (3.21) for the integrations. Collecting these results and simplifying by use of the properties of $\underline{c}$, Equation (3.15), we find

$$\omega = \bar{\nu} \exp(-[\Phi(P) - \Phi(A)]\,\beta) \tag{3.23}$$

$$\bar{\nu} = \frac{1}{2\pi}\left(\frac{A_{11}^+}{B_{11}^+}\frac{K}{M}\right)^{1/2}. \tag{3.24}$$

The mass dependence is entirely in the frequency factor $\bar{\nu}$ which can be written in several ways. If we define $\underline{\alpha}$ by

$$\underline{\alpha}^{-1} \equiv \underline{a}^+ = \underline{c}^T \underline{m} \underline{c}$$

then using (3.15) one finds $\underline{\alpha} = \underline{c}^T \underline{m}^{-1} \underline{c}$ and so

$$\alpha_{11} = A_{11}^+/A^+ = A_{11}^+/M = \sum_{i=1}^{n} c_{i1}^2/m_i .$$

Hence one finds Glyde's result [24]

$$\bar{\nu} = \frac{1}{2\pi}\left(\sum_{i=1}^{n} c_{i1}^2/m_i\right)^{1/2}\left(\frac{K}{B_{11}^+}\right)^{1/2}. \tag{3.25}$$

Alternatively, the results can be expressed in terms of normal modes. Given a Hamiltonian such as

$$H = \tfrac{1}{2}\dot{\underline{y}}^T \underline{a}\dot{\underline{y}} + \tfrac{1}{2}\underline{y}^T \underline{b}\underline{y}$$

which is not diagonal in $\dot{\underline{y}}$ or $\underline{y}$ we can always diagonalize H by a linear transformation to normal coordinates $\underline{Q}$ whose frequencies $\nu_1, \nu_2, \ldots$ satisfy [30]

$$\det(\underline{a}2\pi\nu - \underline{b}) = 0, \quad \prod_{i=1}^{n} (2\pi\nu_i)^2 = B/A .$$

Applied to Equation (3.24) this product nule gives

$$\bar{\nu} = \prod_{i=1}^{n} \nu_i \Big/ \prod_{i=2}^{n} \nu_i' , \tag{3.26}$$

where the ν are the n frequencies for motion about the initial configuration A and the ν' are the $(n-1)$ frequencies for motions about P constrained to the plane S. This expression, due to Vineyard [11] hides the mass dependence. A more convenient result is to use the product rule for unconstrained motion about P, and about A:

$$\prod_{i=1}^{n} (2\pi\nu_i')^2 = B^+/A^+ = K^+/M$$

$$\prod_{i=1}^{n} (2\pi\nu_a)^2 = K/M .$$

Applied to Equation (3.26) this gives [23]

$$\bar{\nu} = \nu_1' (K/K^+)^{1/2} \tag{3.27}$$

ν_1' is the frequency of the unstable mode which leads to decomposition of the saddle-point configuration. It contains all the mass dependence.

3.4. Calculations based on equilibrium rate theory

The isotope effect has been calculated from the mass dependence of the decomposition mode assuming that the coupling between the migrating isotope and the lattice is short ranged [25–27]. One first does a relaxation calculation to find the saddle-point configuration and then writes the equations of motion for the isotope and the atoms coupled to it. For example for Na^+ in NaCl coupled only to the two nearest anions the displacements will be as shown in Figure 7.

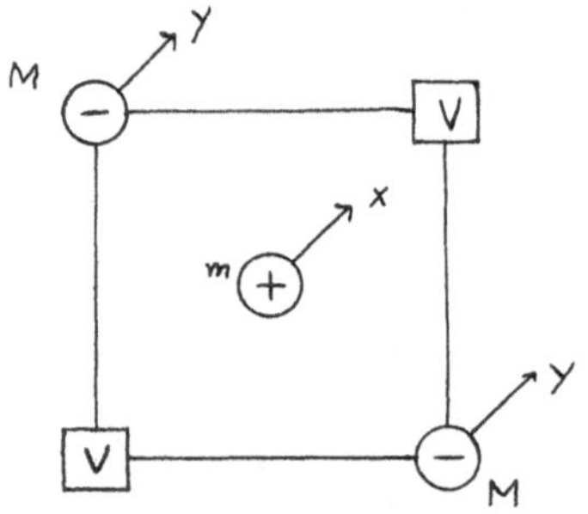

Fig. 7.

and the equations of motion have the form

$$m\ddot{x} = ax + cy$$
$$2M\ddot{y} = cx - by.$$

where a, b, c are appropriate force constants. We can write down and solve the secular equation to find the vibration $i\nu$ (imaginary since we are at the saddle point and the mode is unstable). The result is

$$g \simeq -\frac{m}{\nu^2}\frac{\partial \nu^2}{\partial m} = \tfrac{1}{2}(1 + [1 + 8mMc^2/(2Ma + mb)^2]^{-1/2}).$$

When there is no coupling of the jumping atom to the lattice ($c=0$) then $g=1$ and $\omega \propto 1/\sqrt{m}$.

When there is coupling of particle and lattice then $g<1$ and it can be shown in general that [23]

$$g \equiv \frac{K\cdot E \text{ in mode } \nu_1' \text{ associated with migrating atom.}}{\text{total } K\cdot E \text{ in mode } \nu_1'}$$

The calculation is of course quite complicated for NaCl because of the polarization of the lattice. Similar calculations for Na [25] and Cu [27] were made assuming a vacancy mechanism. For Cu the dominant coupling is with the eight third nearest neighbours. The results are:

	g(Calc.)	g(Expt.)
Na^+/NaCl	0.998	0.93
Na		0.50
Cu	0.92	0.89

The conclusion was that there is no basic difficulty in reconciling theory and experiment except for sodium which they were unable to interpret on a vacancy model.

However, a dynamical computer simulation has led Huntington *et al.* [28] to different conclusions. They find $g>0.95$ for all physically reasonable situations and potentials, for f.c.c., b.c.c. and h.c.p. lattices, and that g is smaller (0.97) for the b.c.c. lattice than for the f.c.c. lattice (0.975–0.980), other things being equal. The potential functions employed were a Born-Mayer potential, and a more realistic potential devised to fit the elastic constants of Na. The former can actually give $g=0.65$ for b.c.c. lattice but it was felt that this reflected the inadequacy of the potential function. They finally concluded that the equilibrium rate theory may be inadequate.

Glyde's equation, (3.25), can also be used for the isotope effect. If it were necessary to specify only motion of the tracer to a new position during the jump, along the x_1 axis say, then one could take $c_{i1}=\delta_{i1}$ and so $\bar{\nu}\propto m_1^{-1/2}$. In fact the $\underline{y}$ coordinates can be taken the same as $\underline{x}(c_{ik}=\delta_{ik})$ so that

$$\bar{\nu} = (1/2\pi m_1^{1/2})\,[K/K_{11}^{+}]^{1/2}\,. \tag{3.28}$$

But in reality there is relaxation about tracer and atom which changes during the jump. Some neighbouring atoms must therefore be specified as moving in particular directions to new positions during the jump i.e. $c_{i1}\neq 0$.

Applied to Ar^{20} equation (3.25) again gives small effects for single vacancies ($g=0.99$) and divacancies ($g=0.88$) but the migration mechanism is still uncertain.

3.5. Concluding Remarks

3.5.1. *Equilibrium Rate Theory.*

In view of the combined uncertainties of calculations (potential functions), of experiments, and of the migration mechanism in the b.c.c. metals it would be premature to say the theory is deficient. However, use of equilibrium theory for a rate process certainly has deficiencies which are probably better tackled by learning how to do a non-equilibrium calculation rather than by patching things up. It is assumed for example that an atom which passes the saddle point will not return and that an equilibrium distribution is maintained at the saddle point. We will illustrate the non-equilibrium program in Section 4; for further discussion of equilibrium theories see References [31–33].

There is an alternative form of equilibrium theory, generally called the dynamical theory, due to Rice [34]. In its original form (outlined in Section 4) it is equivalent in content to Equation (3.28). A modified form in which the criterion for a jump is that the relative displacement of the jumping atom and of the centre of gravity of its neighbours along the jump direction shall exceed a critical value again leads to agreement with experiment for Cu and Ag but not Na and α-Fe [35].

3.5.2. *Uses of Isotope Effect*

The migration mechanism in 'normal' b.c.c. metals is still in doubt [36]. One suggestion is that the failure of isotope effect calculations for Na is because the dominant mechanism is interstitial. Calculations [26] suggest that the most stable interstitial configuration is the crowdion shown in Figure 3a. This would give $g \leqslant \frac{1}{2}$ but $f=0$. It is necessary to suppose that the crowdion can frequently change direction through an intermediate $\langle 110 \rangle$ split interstitial, shown in Figure 3b. In this case $f \simeq 1$ and g is small because more than one atom move at the saddle point.

Since g for a single mechanism is essentially temperature independent any measured temperature dependence can be informative. For example, the high temperature variation for Ag, Table III, suggests a divacancy contribution which is in agreement with other measurements.

For a substitutional impurity f_I can be found from isotope effect measurements if it is assumed that $g_I = g$ (solvent). Combined with measurements of the effect of impurity on the solvent diffusion coefficient as a function of concentration this gives unique values for the frequency ratios $(\omega_4/\omega_0, \omega_2/\omega_1, \omega_3/\omega_1)$ for the five frequency model. This has been done for Zn in Ag and Zn in Cu [37]. The assumption about the g's depends on an empirical correlation between g and the activation volume for diffusion. This kind of information is not only intrinsically interesting but would be invaluable in interpreting results in other kinds of experiments, such as thermotransport.

4. Thermotransport and Non-Equilibrium Rate Theory

4.1. Introduction

A temperature gradient can influence matter transport because the concentration of intrinsic lattice defects will be position-dependent if thermal equilibrium is maintained and because the gradient may bias the jump frequencies. Let us write the jump frequency for an atom jumping from a site at mean temperature T to an adjacent site at temperature $T + \Delta T$ in the form

$$\omega(T, T + \Delta T) = \omega_0(T)\left[1 + \Delta\omega(T)\,\Delta T\right], \tag{4.1}$$

where $\omega_0(T)$ is the isothermal frequency. The equation defines the jump bias parameter $\Delta\omega(T)$.

Consider as an example the flux of vacancies, V, due to a temperature gradient along the X-axis in a pure metal, M, as calculated from the random walk Equation (2.14). If we neglect terms of second order in derivatives the terms in $\langle(\Delta x)^2\rangle$ can be approximated, by using the Einstein equation,

$$\left(n_v \frac{\partial}{\partial x} + \frac{\partial n_v}{\partial x}\right)\left(\frac{\langle(\Delta x)^2\rangle}{2\Delta t}\right) \simeq n_v \frac{\partial D_v}{\partial T}\frac{\partial T}{\partial x} + D_v \frac{\partial n_v}{\partial x}, \tag{4.2}$$

where D_v is the isothermal vacancy diffusion coefficient. To evaluate the term in $\langle \Delta x \rangle$ consider diffusion in a one-dimensional lattice (the final results are the same for the cubic lattices). If the vacancy makes n jumps in time Δt then we can write, in the notation of Section 2,

$$\langle \Delta x \rangle = \sum_{i=1}^{n} \langle x_i \rangle .$$

For a vacancy at temperature T call p_+ the relative probability of a jump to $T+\Delta T$ and p_- the relative probability of a jump to $T-\Delta T$. The mean displacement on the first jump is then

$$\langle x_1 \rangle = r(p_+ - p_-) = rA\left[\omega(T+\Delta T, T) - \omega(T-\Delta T, T)\right],$$

where r is the jump distance and A is fixed by $p_+ + p_- = 1$. Making a Taylor expansion about T one obtains

$$\langle x_1 \rangle = r^2 \left(\frac{\partial \log \omega_0(T)}{\partial T} - \Delta\omega(T)\right)\left(\frac{\partial T}{\partial x}\right) + 0\left(\left(\frac{\partial T}{\partial x}\right)^2\right)$$

and hence

$$\begin{aligned}\frac{\langle \Delta x \rangle}{\Delta t} &= \frac{n\langle x_1 \rangle}{\Delta t} + 0\left(\left(\frac{\partial T}{\partial x}\right)^2\right) \\ &= 2D_v \left(\frac{\partial \log \omega_0(T)}{\partial T} - \Delta\omega(T)\right)\left(\frac{\partial T}{\partial x}\right) + 0\left(\left(\frac{\partial T}{\partial x}\right)^2\right).\end{aligned} \tag{4.3}$$

The mean drift arises from the bias, $\Delta\omega$, *and* simply because ω_0 is temperature dependent. Equations (4.2) and (4.3) substituted into (2.14) give the flux of vacancies, J_v. Since there can be no net transport of lattice sites relative to axes fixed in the lattice $J_M = -J_v$. The flux of metal atoms is found to be

$$J_M = -D_v\left(-\frac{\partial n_v}{\partial x} + \frac{q_m^* n_v}{kT^2}\frac{\partial T}{\partial x}\right), \tag{4.4}$$

where

$$q_M^* = \Delta h_m - 2kT^2\,\Delta\omega \tag{4.5}$$

is called the heat of transport for one jump and we used

$$\Delta h_m = kT^2 \frac{\partial \log \omega_0(T)}{\partial T} = kT^2 \frac{\partial \log D_v}{\partial T}. \tag{4.6}$$

The first equality defines the activation enthalpy which can be determined from isothermal experiments. It is simple to find from the Einstein equation the relation

$$D_M^* = f_M n_v D_v / n_M , \tag{4.7}$$

where D_M^* is the measured tracer self-diffusion coefficient and f_M the correlation factor. Finally the theory of defect concentrations in Section 2 gives, for thermodynamic equilibrium,

$$\mu_v = 0 = g_v + kT \log c_v \tag{4.8}$$

$$\frac{\partial n_v}{\partial x} = \frac{\partial n_v}{\partial T}\frac{\partial T}{\partial x} = \frac{n_v h_v}{kT^2}\left(\frac{\partial T}{\partial x}\right), \tag{4.9}$$

where h_v and g_v, the enthalpy and free energy of vacancy formation are related by

$$h_v \equiv - T^2 \left(\frac{\partial (g_v/T)}{\partial T}\right)_p .$$

Equations (4.4) and (4.7)–(4.9) give

$$J_M = - (D_M^* n_M / f_M k T^2)(q_M^* - h_v)\frac{\partial T}{\partial x} \tag{4.10}$$

h_v can be found from other experiments and so we can hope to determine q_M^* from a thermotransport experiment.

If we consider impurity diffusion the analysis is much more complicated because we have to consider correlation effects. The form of the result will be

$$J_I = - D_I^* \left(\frac{\partial n_I}{\partial x} + \frac{Q_{I\,\mathrm{app}}^* n_I}{kT^2}\frac{\partial T}{\partial x}\right), \tag{4.11}$$

where $Q_{I\,\mathrm{app}}^*$, the apparent heat of transport, is a function of q^* parameters for the various kinds of jump. For the direct interstitial mechanism the apparent heat is just q^* for an intersitial jump. For the five frequency model of impurity diffusion by the vacancy mechanism it is a function [36] of ω_1, ω_3 and heats of transport q_1^*, q_2^*, q_3^*, q_4^* for the jumps $\omega_1 - \omega_4$.

4.2. Measurement Methods and Results

4.2.1. *Marker Experiments*

The principle of the experiment is to place in the crystal a pair of markers which remain fixed relative to the crystal planes in the region of the marker. If there is a flow of matter the fluxes at the two markers will differ because the temperatures are different. The change in separation distance Δl, after time t, of markers at temperatures T_1 and T_2 is given by

$$\Delta l = Kt\,[J_M(T_1) - J_M(T_2)]/n_M , \tag{4.12}$$

where K is a geometric factor allowing for any change in shape of the specimen and must be found from experiment. For the vacancy mechanism $q_M^* - h_v$ can thus be found using Equations (4.10) and (4.12). The markers used are either surface scratches or fine inert wires embedded in the crystal. Unfortunately it is not entirely clear that the boundary conditions are such that the markers remain fixed relative to the local lattice since, for example, scratches can be sinks and sources of vacancies [39]. Thus although most of the available data was obtained by this method some degree of caution is necessary.

4.2.2. *Thin Layer Method*

A thin layer of tracer is deposited on one surface of mating optical flats. The other surface is coated with a non-diffusing inert radioactive marker. The flats are welded together, annealed in the temperature gradient, and the penetration profile determined in the normal way. This profile is not Gaussian because there is a mean drift, $\langle \Delta x \rangle \neq 0$, and the inert tracer provides a necessary origin in the analysis of results. This has been used only for Au^{195} and Sb^{125} in Au with HfO_2 marker. For the gold tracer $Q^*_{I\,\mathrm{app}}$ is approximately $q^*_{Au} - h_v$ for the vacancy mechanism. The result [42] for $q_{Au} - h_v$ unfortunately does not agree with several marker experiments (9.0 kcal mole^{-1} compared with -4.8, -8.6, and 0 kcal mole^{-1}), but is more likely to be correct.

4.2.3. *Soret Effect*

This is the classical experiment in thermotransport. A two component system is annealed in the temperature gradient until a stationary concentration gradient is achieved. For impurity migration by the vacancy mechanism in metals one needs more information about the frequencies of the five frequency model than is at present available to extract q^* parameters with any confidence. We quote one result for an ionic crystal in Table V where the tight binding approximation is probably valid. For light interstitial impurities in metals there is fairly accurate and unambiguous data (no correlation effects), particularly from a variant technique, the thermoosmosis method.

4.2.4. *Thermoelectric Effects in Ionic Crystals*

The theory here is now old established [38]. The method is limited by the lack of reversible electrodes and the need for accurate transport numbers but interpretation is uncomplicated by correlation effects for the pure or lightly doped crystals quoted below.

Detailed reviews of the experimental methods and results are available [38, 36]. Some of these results, with additions, are given in Table V.

4.3. The Heat of Transport for One Jump

There is at present no generally agreed approach to a molecular calculation of q^* although some of the important ingredients of such a calculation have been identified. As an introduction we will first discuss two aspects relevant to all kinds of crystal

TABLE V

Substance	Experiment	Result
	Single vancancy mechanism (self-diffusion)	
Al, Zn, Ag, Au	Marker	$\|q^* - \Delta h\| \lesssim 0.3\Delta h$
Pt, Co	Marker	$q^* - \Delta h > 0.3\Delta h$, T dependent
Cu	Marker	$q^* - \Delta h \simeq \pm 0.2\Delta h$
	Thin layer [42]	$q^* - \Delta h \simeq 0.6\Delta h$
KCl^4, TlCl, AgCl [41]	Thermopower	$q^* - \Delta h \simeq 0.06 - 1.6\Delta h$
	Single vacancy mechanism (impurity)	
Sr^{2+} in NaCl	Soret	$q_2^* - 2q_1^* = -36$ kcal mole^{-1}
		$\Delta h_2 - 2\Delta h_1 = -16$ kcal mole^{-1}
(tight binding model assumed)		
	Interstitial impurities	
		q^*(kcal mole^{-1})/Δh(kcal mole^{-1})
C in γ-Fe		$-2/36$
H in Ni		$-(1.7 - 1.2)/8.4$
H in γ-Fe-Ni		$-(1.5 - 0.2/8.4$
C in α-Fe		$-24/24$
H in α-Fe		$-(8.1 - 5.6)/2.7$
H in α-Zr		6.0/11.4
D in α-Zr		6.5/11.4
For systems with H impurity $dq^*/dT \simeq 10$ cal mole^{-1} deg^{-1}		
	Interstitialcy	
Ag^+ in Ag Cl [41]	Thermopower	Type 1 $-4.14/1.29$
(see Figure 2 for type 1 and type 3 jumps)		Type 2 $-27.6/6.6$

which can be presented within the framework of the dynamical theory of Rice referred to earlier.

The essence of the dynamical theory in its original form is as follows. The jump frequency can be written as a product

$$\omega = \omega_a \omega_b \,. \tag{4.13}$$

ω_a is the frequency with which the migrating atom, number 1, attains the critical displacement q_1 midway between initial and final sites. ω_b is the probability conditional on atom 1 being at q_1 that the bridge atoms (close neighbours to atom 1 at the saddle point) will acquire the critical displacements $(q_2, q_3, \dots q_n)$ or greater displacements in the same directions necessary to allow atom 1 to pass.

4.3.1. ω_b *and the Local Equilibrium Effect*

Originally Rice calculated ω_b using equilibrium distribution function theory and the harmonic approximation. To make the calculation in the temperature gradient Weeks and Shuler [43] find is easier to adopt the ideas of the Lennard-Jones and Devonshire cell theory in which each atom moves within its own cell in the mean field of its neighbours. The probability of finding a set of atoms with atoms displaced from their cell centres by amounts $\mathbf{r}_1, \mathbf{r}_2, \ldots \mathbf{r}_n$ is

$$P^{(n)}(\mathbf{r}_1, \mathbf{r}_2, \ldots \mathbf{r}_n) = \prod_{i=1}^{n} P^{(1)}(\mathbf{r}_i),$$

where $P^{(1)}(\mathbf{r}_i)$ is the probability of observing atom i displaced to $\mathbf{r}_i$. In the crudest approximation, in which the model just reduces to a set of Einstein oscillators, this is given by

$$P^{(1)}(\mathbf{r}_i)\,\mathrm{d}r_i = (a_i^2/\pi)^{3/2} \exp[-a_i^2(x_i^2 + y_i^2 + z_i^2)]\,\mathrm{d}x_i\,\mathrm{d}y_i\,\mathrm{d}z_i$$
$$a_i^2 \equiv k_i/2kT_i, \tag{4.14}$$

where k_i is a force constant and T_i is the mean temperature in cell i. This introduces the local equilibrium assumption: the temperature in the Boltzmann factor of an isothermal calculation is replaced by a suitable local mean temperature calculated from the macroscopic temperature gradient. If the atom jumps toward a higher temperature it will find the bridge atoms in their critical configuration more frequently than in an isothermal system and so one should find $\Delta\omega > 0$ if this is the only effect.

For the cell model one clearly can write

$$\omega_b = \prod_{i=2}^{n} \left(\int_{q_i}^{\infty} \mathrm{d}x_i \int_{-\infty}^{\infty} \mathrm{d}y_i \int_{-\infty}^{\infty} \mathrm{d}z_i P^{(1)}(x_i, y_i, z_i) \right) \tag{4.15}$$

by taking the direction of the critical displacement as the x-axis in each cell. The temperature T_i can be written as

$$T_i = T_1 + \alpha_i r \frac{\mathrm{d}T}{\mathrm{d}x}, \tag{4.16}$$

where r is the jump distance and α_i a suitable geometric factor. Assuming for the moment that ω_a depends only on temperature T_1 one finds by combining equations (4.13)–(4.16) the result

$$q^* - \Delta h_m = -\sum_{i=2}^{n} (k_i q_i^2)\,\alpha_i$$

provided that $\frac{1}{2}k_i\,q_i^2 \gg kT_1$, and in the same approximation for an isothermal system

$$\omega_b \sim \exp\left(-\sum_{i=2}^{n} k_i q_i^2/2kT\right).$$

This result for q^* gives order of magnitude agreement and the right sign for the light interstitial impurities, C and H, in α-Fe, but a detailed conparison has not been made because of the lack of suitable force constants [43].

Closely related but not identical ideas have been discussed by Wirtz, Girifalco, and Huntington. [28]. Clearly one could use the idea of local equilibrium in the equilibrium rate theory of Section 3 but no calculations have been made like this.

Huntington's [28] results suggest that the local equilibrium effect for self-diffusion by the single vacancy mechanism in metals is essentially independent of details of potential and crystal structure and is approximately given by

$$q^* - \Delta h_m \simeq -0.2\,\Delta h_m.$$

If this is correct there must be other contributions for a vacancy mechanism (Table V).

4.3.2. *ω_a and Phonon-Scattering*

In the harmonic approximation the displacement $x_1(t)$ of the migrating atom in the jump direction, is given by

$$x_1(t) = \sum_k \alpha_k \varepsilon_k^{1/2} \cos(\omega_k t + \delta_k),$$

where ε_k, ω_k, and δ_k are the energy, frequency and phase of the kth normal coordinate. The long term average frequency, $L(\varepsilon)$, with which $x_1(t)$ attains some value q_1 when the mode energies are $\varepsilon \equiv (\varepsilon_1, \varepsilon_2, \ldots)$ is given by Kac's formula:

$$L(\varepsilon) = \pi^{-2} \int_0^\infty \int_0^\infty \mathrm{d}y\, \mathrm{d}z \cos(q_1 z) \{\prod_k J_0(\alpha_k \varepsilon_k^{1/2} z) - \prod_k J_0(\alpha_k \varepsilon_k^{1/2} (z^2 + \omega_k^2 y)^{1/2})\}\, y^{-2},$$

where J_0 is a Bessel function of zero order.

If the probability of finding the energy set ε is $\varrho(\varepsilon)$ then the frequency with which the atom attains a displacement q_1 is

$$\omega_a = \int L(\varepsilon)\, \varrho(\varepsilon)\, \mathrm{d}\varepsilon. \tag{4.17}$$

In the isothermal case $\varrho(\varepsilon)$ is given by the Boltzmann distribution which, for classical systems, can be written

$$\varrho(\varepsilon) = \prod_k \beta \exp(-\beta \varepsilon_k) \equiv \prod_k \varrho_k^{(e)}. \tag{4.18}$$

The result from these equations is

$$\omega_a = \nu_a \exp(-\beta U_a)$$
$$\nu_a = (2\pi)^{-1} \sum_k \alpha_k^2 \omega_k^2 / \sum_k \alpha_k^2, \quad U_a = q_1^2 / \sum_k \alpha_k^2.$$

How will this result be modified by a temperature gradient? Compare the frequency of jumps for an atom jumping into a vacancy on its high temperature side with the frequency when the vacancy is on the low temperature side. A phonon from the high temperature side will presumably have a greater effect on the atom in the second case than in the first case since in the latter the vacancy will tend to shield the atom from the phonon by scattering. There is an excess of phonons from the high temperature side so one can naively expect the jump frequency to be biased in favour of jumps towards lower temperatures i.e. $\Delta\omega < 0$. The source of the bias is the asymmetry of the surroundings of the jumping atom. To express the picture quantitatively we need a non-equilibrium distribution function $\varrho(\varepsilon)$.

A crude description of the heat flux i.e. theory of thermal conductivity is as follows. The mean heat flux $\mathbf{Q}$ in a stationary gradient is given by

$$\mathbf{Q} = V^{-1} \sum_k \bar{\varepsilon}_k \mathbf{v}_k, \tag{4.19}$$

where V is the crystal volume, k denotes $(\mathbf{k}, j)$ with $\mathbf{k}$ the wave-number vector and j the polarization index, $\mathbf{v}_k$ is the group velocity, and

$$\bar{\varepsilon}_k = \int \varepsilon_k \varrho(\varepsilon)\, d\varepsilon = \int \varepsilon_k \varrho_k(\varepsilon_k)\, d\varepsilon_k.$$

For a stationary state the distribution function for the simple mode k satisfies the Boltzmann equation. In the relaxation time approximation for a classical system this is by a familiar argument [44]

$$\begin{aligned} -(\varrho_k - \varrho_k^{(e)})/\tau_k &= \mathbf{v}_k \cdot \nabla \varrho_k^{(e)} = \mathbf{v}_k \cdot \nabla T\, \partial \varrho_k^{(e)}/\partial T \\ &= \mathbf{v}_k \cdot \nabla T\, (\beta \varepsilon_k - 1)\, \varrho_k^{(e)}/T, \end{aligned} \tag{4.20}$$

where τ_k is the relaxation time. For an isotropic system these equations give the well-known result for the lattice thermal conductivity κ

$$\begin{aligned} \mathbf{Q} &= -\kappa \nabla T \\ \kappa &= k(3V)^{-1} \sum_k v_k^2 \tau_k. \end{aligned}$$

At high temperatures the lattice conductivity varies at T^{-1} and if we take τ_k as independent of k then $\tau \sim T^{-1}$.

Schottky [45] proposed that one might replace $\varrho_k^{(e)}$ in (4.18) of the Rice theory by ϱ_k from (4.20). Applied to a one dimensional chain where $\omega_a \equiv \omega$ since there are no bridge atoms this gives

$$q^* - \Delta h_m = 2 \Delta h_m \Lambda / r \quad \text{for} \quad \Delta h_m \beta \gg 1$$

where

$$\Lambda = -\sum_k \alpha_k^2 v_k \tau_k / \sum_k \alpha_k^2$$

is essentially a mean free path and r is the jump distance. The amplitude factors α_k were found analytically for a chain of identical coupled harmonic oscillators the presence of the vacancy being allowed for by a different coupling constant between the neighbours of the vacancy. Applied to copper at the melting point with suitably adjusted force constants this gives $(q^* - \Delta h_m)/\Delta h_m$ in the range 2.3–2.8 dependent on the approximation for τ_k, compared with ~ 0.6 experimentally. Huntington [46] suggests that a three dimensional calculation would give a much smaller value ($\ngtr 0.1$), but a proper calculation has never been done. The value will be reduced slightly by the local equilibrium effect in ω_b. The most discouraging thing is that

$q^* - \Delta h_m$ will vary as T^{-1} whereas such temperature dependences as have been found are in the other direction.

This calculation has not been applied to impurity diffusion but a crude calculation suggests large effects [47]. The mean force on the impurity at saddle point is written as

$$\bar{\mathbf{F}} = V^{-1} \sum_k \bar{\mathbf{p}}_k v_k A_k = - q_p^* \nabla T / T , \tag{4.21}$$

where A_k is the scattering cross section and $\bar{\mathbf{p}}_k$ the momentum contribution from mode k. Using the Boltzmann Equation (4.20) and the Rayleigh scattering cross-section predicts appreciable positive contributions q_p^* to q^*. Unfortunately this kind of calculation also makes the approximation of replacing real momentum by crystal momentum which is difficult to assess (see later).

The effects so far discussed appear to be the only ones relevant to ionic conducting crystals but there are no theoretical calculations for them. For metals several electronic effects have been studied theoretically. The most obvious candidate, electron scattering by the jumping atom, appears to be almost negligible for self-diffusion in single band metals e.g. Cu but may be the cause of the larger positive values of q^* in multi-band metals e.g. Pt, Co [46]. For substitutional impurities in single band metals the electron scattering effect may theoretically be large [47]. For various views for metals we refer to the papers of a recent conference [3].

Our discussion of the local equilibrium and phonon scattering effects proceeded by a rather informal grafting of non-equilibrium ideas on to an isothermal equilibrium rate theory which itself is not entirely satisfactory. Rather than pursuing the description of the complex electronic effects just mentioned let us set up a non-equilibrium calculation neglecting electronic effects and see what differences and advantages the program has from that so far considered both isothermally and non-isothermally.

4.4. An Example of Non-Equilibrium Theory

4.4.1. *Derivation of a Kinetic Equation*

One needs a kinetic equation for $f_1(\underline{R}, \underline{P}, t)$, the probability of finding the atom of interest with coordinates $\underline{R}$, $\underline{P}$ at time t, derived from the Liouville equation for $\varrho(\{\underline{N}\}, \underline{R}, \underline{P}, t)$, the system distribution function. $\{\underline{N}\}$ denotes the coordinates of the N other atoms.

If the atom of interest is a heavy atom of mass M in a lattice of light atoms of mass m then there is an obvious perturbation parameter, $\lambda = (m/M)^{1/2}$, and it is well-known that the kinetic equation is the generalized Fokker-Planck equation

$$\left(\frac{\partial}{\partial t} + \frac{\underline{P}}{M}\cdot\nabla_R + \langle \underline{F} \rangle \cdot \frac{\partial}{\partial \underline{P}}\right) f_1(\underline{R}, \underline{P}, t)$$
$$= \frac{\partial}{\partial \underline{P}}\cdot\left\{\underline{\xi}\cdot\left(kT\frac{\partial}{\partial \underline{P}} + \frac{\underline{P}}{M}\right) + \underline{\eta}\cdot\nabla \log T\right\} f_1(\underline{R}, \underline{P}, t) \tag{4.22}$$

$\underline{F}$ is the intermolecular force on the neavy atom and $\langle \underline{F} \rangle$, defined by

$$\langle \underline{F}(\underline{R}) \rangle = \int \mathrm{d}\{\underline{N}\}\, \underline{F} \varrho^{(e)}(\{\underline{N}\}, \underline{R}, \underline{P}) / f_1^{(e)}(\underline{R}, \underline{P}), \tag{4.23}$$

is the mean force at equilibrium (indicated by superscripts (e)). The friction tensors which characterize the coupling of the particle to the lattice are defined by

$$\underline{\xi} = \beta \int_0^\infty \mathrm{d}s \langle \underline{\mathscr{F}}(0)\, \underline{\mathscr{F}}(-s) \rangle \tag{4.24}$$

$$\underline{\eta} = \beta \int_0^\infty \mathrm{d}s \langle \underline{\mathscr{F}}(0)\, \underline{J}_H(-s) \rangle - \beta \langle \underline{F} \underline{\mathscr{A}} \rangle, \tag{4.25}$$

where

$$\underline{\mathscr{F}} = \underline{F} - \langle \underline{F} \rangle, \underline{\mathscr{A}} = \underline{A} - \langle \underline{A} \rangle$$

$$\underline{A} = \sum_{i=0}^{M} \underline{R}_i (e_i - \bar{H}_i)$$

e_i is the total energy (kinetic and potential) of atom i, each atom in a pairwise interaction taking half the energy, and $\bar{H}_i$ the partial molecular enthalpy. Atom 0 is the heavy atom ($\underline{R}_0 \equiv \underline{R}$). $\underline{J}_H$ is the molecular expression for the heat flux. The time evolutions are to be calculated with the heavy atom static and the averages $\langle\ \rangle$ are defined as in (4.23), ξ and η are R dependent in general.

A very crude derivation [48] is to imagine a system maintained in a stationary state by external resevoirs. It can be characterized by the local equilibrium distribution function

$$f^{(0)} = C \exp \left(- \sum_{i=0}^{N} (kT(R_i))^{-1} [e_i(\underline{R}_i) - \mu_i(\underline{R}_i)] \right), \tag{4.26}$$

where C is a normalization constant, μ_i is the chemical potential of atom i and $T(\underline{R}_i)$ the temperature at the position of atom i. The system is then isolated and approaches equilibrium. By the use of the Zwanzig identity an equation for f_1 can be found from the solution of the Liouville equation with initial condition at $t=0$ of $f(0)=f^{(0)}$. This equation is the generalized Fokker-Planck equation if only terms of lowest order in λ^2 and ∇T are retained and 'long times', of the order of a few vibration periods, are retained. The difficulty is not in extracting the lowest order terms but justifying the neglect of the higher order ones. More elaborate discussions [49, 50] appear to justify the result; from the solid state viewpoint the interest lies less in the derivation than the use of the result. Before passing to this we note that 'weak coupling' Brownian motion equations have also been used in this problem. The atom of interest is then a harmonic oscillator weakly coupled to the lattice and λ now measures the strength of the coupling. A full discussion of the relation to the example studied here has been given recently [50].

4.4.2. *Calculation of Jump Rate*

We can define the potential of the mean force $W(R)$ on the heavy atom at R at equlibrium such that [51]

$$\langle \underline{F}(\underline{R}) \rangle = -\nabla_{\underline{R}} W(\underline{R})$$

The kinetic equation describes the state of the atom vibrating about lattice site A (the origin) and about to jump to a vacant lattice site at B. Let both sites be on the x-axis. The potential of the mean force as we pass from A to B will look as shown in Figure 8.

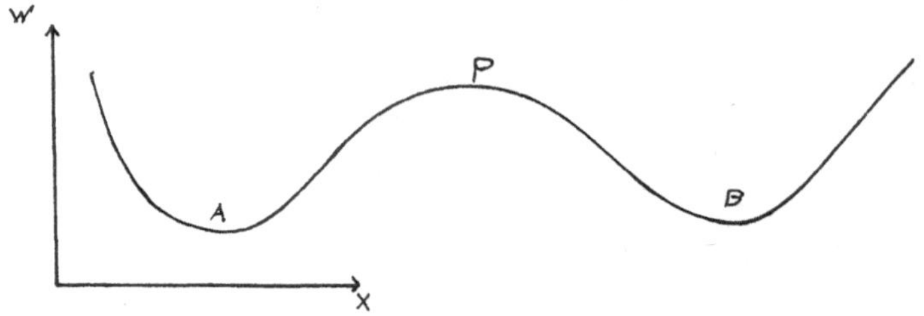

Fig. 8.

If the atom could not leave the well at A then in the limit of long times a stationary distribution satisfying (4.22) would be valid. If the atom can excape to B and the barrier is large compared with kT then the stationary solution, $f_1^{(s)}$ say, still obtains to a high degree of accuracy near A but will overestimate the ensemble population of atoms near the saddle point P. This suggests a solution of the form

$$f_1(\underline{R}, \underline{P}, t) = f_1^{(s)}(\underline{R}, \underline{P})\, G(\underline{R}, \underline{P}, t)$$

with boundary conditions

$$\begin{aligned} G(\underline{R}, \underline{P}, t) &\simeq 1, \quad \underline{R} \sim \underline{R}_A \\ G(\underline{R}, \underline{P}, t) &\simeq 0, \quad x \gg x_p \end{aligned}$$

for small t.

The expression for the jump frequency is

$$\omega = I/Q$$

$$I = \int\int \mathrm{d}\underline{P}\, \mathrm{d}\underline{R}\, \delta(x - x_p) \frac{P_x}{M} f_1(\underline{R}, \underline{P}, t)$$

$$Q = \int\int_A \mathrm{d}\underline{P}\, \mathrm{d}\underline{R}\, f_1(\underline{R}, \underline{P}, t)$$

where $\underline{R} \equiv (x, y, z)$, $\underline{P} \equiv (P_x, P_y, P_z)$.

Note that in defining ω in terms of f_1 there is no chance to say anything about the bridge atoms, dynamical coupling with them enters through the friction constants only.

Contrast this with the arbitrary choices available in equilibrium theory e.g. Section 3.6.1 compared with 3.3.

Consider for simplicity a one-dimensional system. A stationary solution is

$$f_1(x, P) = C_1 \exp[-\beta(P^2/2M + W(x) + \theta(x)\nabla T/T)],$$

where

$$\theta(x) = \int_0^x \eta(x')\,dx'$$

and C_1 is a normalization constant. If as a starting approximation we take $G=1$ everywhere we find $\omega=0$. This can be remedied in the manner of equilibrium rate theory by integrating only over positive values of P_x. If we then use the harmonic approximation to calculate W for the evaluation of Q then W will have the form

$$W(x) = W(0) + \tfrac{1}{2}k_A x^2,$$

where k_A is a function of the harmonic force constants of the crystal. The final result is then found to be

$$\omega_0(T) = \frac{1}{2\pi}\left(\frac{k_A}{M}\right)^{1/2} \exp[-\beta(W(x_p) - W(0))]$$

$$q^* - \Delta h_m = (\theta(x_P) - \langle\theta\rangle)\, x_P^{-1}$$

$$\langle\theta\rangle \equiv \int_{-\infty}^{\infty} \exp(-k_A x^2\beta/2)\,\theta(x)\,dx \Big/ \int_{-\infty}^{\infty} \exp(-k_A x^2\beta/2)\,dx.$$

It is as simple to do this approximation in three dimensions. The isothermal frequency ω_0 is just the same as that for equilibrium rate theory when we neglect coupling between particle and lattice i.e. Equations (3.23) and (3.28). More interesting is that we have for the first time a formula for q^* derived without premature specialization to harmonic approximation, relaxation time approximation, etc. and comprising two distinct contributions. The first, arising from $\langle F\rangle$ in Equation (4.25), contains no dynamical effects but contains the 'local equilibrium' effects discussed crudely in Section 4.2. If in the *isothermal* kinetic equation we had replaced $\langle F\rangle$ by F averaged over the local equilibrium distribution $f^{(0)}$, Equation (4.26), and linearized in ∇T this would be the sole contributor to q^*; it describes the distortion of the mean potential surface W. The other contributor to q^* comes from the time correlation integral in η and contains all phonon scattering effects. For example, it is straightforward to recover the results of Gerl's calculation, Equation (4.21) *et seq.*, if similar approximations are made. But we should be able to do better than this by extending the methods whereby the evaluation of the time correlation formula for thermal conductivity is reduced to the solution of a Boltzmann integral equation [52]. The difficulty of the calculation comes from the need to do lattice dynamics with the heavy particle static at an arbitrary point on its jump trajectory. So far only a one dimensional nearest-neighbour lattice has been studied and the results are almost certainly atypical [50].

Only an approximate solution in one dimension is known for G [48]. It is necessary to assume (a) a high barrier so that $\partial G/\partial t$ can be neglected, (b) ξ is independent of x near P, (c) η can be approximated near P by

$$\eta(x) = \eta(x_P) + (x - x_P)\,\eta'$$

(d) harmonic approximation for W near P

$$W(x) = W(x_P) - \tfrac{1}{2}k_P(x - x_P)^2 .$$

With this solution we do not need to restrict integration to positive P_x; there really is a net flux from A to B. The result is that the previous rate, $\omega^{(s)}$ say, is replaced by

$$\omega(T, T + \Delta T) = \omega^{(s)}(T, T + \Delta T)\,(1 - \xi/Ma)^{1/2},$$

where a is the positive root of

$$a(Ma - \xi) = k_P + \eta' \nabla \log T .$$

In the isothermal case the correction factor reduces to 1 for $\xi \to 0$, whereas for large friction, $\xi^2 \gg 4Mk_P$, the frequency is

$$\omega = \omega^{(s)}(Mk_P)^{1/2}/\xi \ll \omega^{(s)} .$$

For a one dimensional harmonic nearest-neighbour chain and neglecting the effect of the vacancy the evaluation of ξ is straightforward. The result is $\xi = 2(mk)^{1/2}$ where k is the force constant. Consequently ω is independent of M and the isotope effect is zero ($g = 0$). Equilibrium theory makes the same prediction (Equation (3.25)). Rockmore and Turner [53] have studied this large friction limit in detail. The simpler Smoluchowski equation for

$$\eta(\underline{R}, t) = \int f_1(\underline{R}, \underline{P}, t)\,\mathrm{d}\underline{P}$$

can then be used to calculate the jump rate. A crude extension to three dimensions for a simple cubic lattice leads to the same result for the isotope effect and the non-equilibrium solution can be found even when the barrier is not large compared with kT. They quote results for Cd in Ag and Cu for which the isotope effect is zero though the mass ratio is rather too close to unity for the theory to be strictly applicable.

In non-equilibrium theories the coupling on which the isotope effect depends requires accurate evaluation of the time correlation function $\xi(R)$. The difficulty is the same as that for η and very little is known for stable three-dimensional lattices. For the models mentioned above ξ turns out position independent but this is probably not true in general. There are certainly no calculations accurate enough for the difference in frequency factors $\bar{\nu}$ of equilibrium and non-equilibrium theories to be distinguished by comparison with experiment. When the additional difficulties of obtaining solutions of the kinetic equation are remembered it appears that progress by this route to the isotope effect although very desirable will remain slow. Useful results for q^* are more likely to be obtained particularly as alternative discussions are rather primitive.

5. Defect Interactions

So far we have considered the measurement and theoretical treatment of detailed information about jump mechanisms, mainly in metals. When one makes experiments on ionic conducting crystals one soon realizes that defect interactions are relatively important because the interactions are now long-range (coulombic). The description of a doped ionic crystal generally made was outlined in Section 2 and will not be repeated. The point defects (cation vacancies, divalent impurity cations) are classified into nearest-neighbour impurity-vacancy pairs (clusters) and unpaired defects. The concentrations of unpaired and pair defects were related by the law of mass action, Equation (2.11), with suitable activity coefficients inserted ad-hoc. These concentrations can be measured experimentally.

Recently several papers have claimed to detect small systematic discrepancies between this model for doped crystals [54, 55] and also a similar one for pure crystals [56] and very accurate conductivity and diffusion measurements. Possible explanations are the trapping of vacancies by larger defect clusters, migration of other change carriers e.g. trivacancies, dependence of activation energies on temperature etc. An alternative explanation which has been difficult to rule out is that the rather crude description of defect interaction is not good enough. The correct description of simultaneous short range clustering and long range (collective) effects in coulomb systems is subtle, not peculiar to defects in simple ionic crystals [57] and worth trying to understand. However, an attempt [58] to derive the mass action equations for cluster concentrations from the theory of the distribution functions for point defects was not very successful. One needs not only the pair correlation functions but higher order correlation functions as well. Furthermore when these correlation functions are expressed as cluster expansions in concentrations of point defects following the standard methods of Mayer cluster theory [59] the convergence of the expansions is unacceptably poor over much of the concentration range. In this final lecture we outline the elements of a new treatment [60] which allows one to put definite limits on the range of the naive theory of Section 2 and to go a little further. The idea is simply to derive a general law of mass action which contains exact formal cluster expansions for the activity coefficients. The expansions are in concentrations of the entities appearing in the law of mass action, not point defects, and they converge much better than the conventional cluster expansions used earlier.

To derive a law of mass action between cluster concentrations we need a general definition of a cluster. We therefore define for any pair of point defects a and b a characteristic distance q_{ab}. Then a and b are said to be in the same physical cluster if and only if $R_{ab} \leqslant q_{ab}$. For the doped ionic crystal a convenient choice, adopted in the example calculation later, is to take q as the nearest neighbour distance for oppositely charged defects and zero for defects of like charge.

Every configuration of defects can be uniquely partitioned into a collection of physical clusters. Each kind of physical cluster can be treated as a new species as follows. The canonical ensemble partition function, Equation (2.22), can be used to construct the grand partition function Ξ. The result is

$$\Xi/\Xi_0 = \sum_{\underline{N} \geqslant \underline{0}} \frac{\underline{z}^{\underline{N}}}{\underline{N}!} \int \mathrm{d}\{\underline{N}\}\, F_{\underline{N}}(\{\underline{N}\}), \tag{5.1}$$

where, for a crystal with σ kinds of point defect,

$$\underline{z}^{\underline{N}} = \prod_{i=1}^{\sigma} z_i^{N_i}, \quad \underline{N}! = \prod_{i=1}^{\sigma} N_i!$$

$$\sum_{\underline{N} \geqslant \underline{0}} = \sum_{N_1=0} \sum_{N_2=0} \cdots \sum_{N_\sigma=0}$$

$$F_{\underline{N}}(\{\underline{N}\}) = \exp(-\beta A_{\mathrm{int}}(\{\underline{N}\}))$$

$$z_i = \exp[(\mu_i - \mu_{i0} - A_i)\beta]$$

for an interstitial μ_i = Gibbs chemical potential, of interstitial atom, $\mu_{i0} = 0$,

for a vacancy $\mu_i = 0$

for a substitional impurity μ_i = Gibbs chemical potential of impurity atom

(for a vacancy and a substitional impurity:) μ_{i0} = chemical potential of host atom replaced by point defect.

Ξ_0 is the partition function of the perfect crystal. The other symbols are as in Section 2. Now this form of partition function is the same as for a gas or electrolyte solution [59] and so all the techniques of Mayer cluster theory can be applied to get the earlier results [58].

However one can prove the following basic result:

$$\Xi/\Xi_0 = \sum \frac{\underline{y}^{\underline{n}}}{\underline{n}!} \int \mathrm{d}\{\underline{n}\}\, \underline{\lambda}(\{\underline{n}\})\, F_{\underline{n}}(\{\underline{n}\}) \tag{5.2}$$

$\underline{n}$ denotes a set of n_1 physical clusters of kind 1, n_2 of kind 2 etc. The activity y_i of cluster species i is just the product of z functions one for each of the point defects in it. λ_i is defined by

$$\lambda_i(\{i\}) = \exp[-\beta A_{\mathrm{int}}(\{\underline{s}_i\})]/\underline{s}_i!$$

where $\underline{s}_i$ denotes the set of point defects comprising cluster i.
Finally

$$F_{\underline{n}}(\{\underline{n}\}) = \exp(-\beta[A_{\mathrm{int}}(\{\underline{n}\}) + U(\{\underline{n}\})]), \tag{5.3}$$

where now $A_{\mathrm{int}}(\{\underline{n}\})$ refers only to interactions between defects in different physical clusters of the set $\underline{n}$. U is defined to be positive infinite for any configuration in which two point defects in different physical clusters are the distance q or less apart.

Comparing Equations (5.1) and (5.2) we see that we can apply Mayer (mathematical) cluster theory to (5.2) just as we did for (5.1) provided we allow for the differences that now our species have internal energy (the λ-functions) and unpleasant configuration restrictions (the U energy). Formally these complications cause little trouble. Application of the classical cluster method of McMillan and Mayer leads to the generalized law of mass action:

$$\frac{a_i(\{i\})}{\prod\limits_{j \subseteq \underline{n}} a_j(\{j\})} = \frac{\lambda_i(\{i\})}{\prod\limits_{j \subseteq \underline{n}} \lambda_j(\{j\})}, \tag{5.4}$$

where i is a physical cluster which can be formed by bringing together a set of $\underline{n}$ physical clusters each comprising only one point defect. Each activity a_i is the product of an activity coefficient γ_i and a concentration ϱ_i both dependent of the configuration $\{i\}$ of the cluster. By familiar methods γ can be expressed as a cluster expansion in the concentrations of physical clusters. For systems with coulomb interactions the integrals of the expansion diverge. One therefore resums to a modified cluster expansion due to Mayer [59]. All these steps are very similar to those of electrolyte theory and the earlier theory of defect interactions.

To illustrate the results we consider the application to the law of mass action for the equilibrium between paired and unpaired defects in a doped crystal. The general law of mass action (5.4) then reduces to the Equation (2.11) except that we now have exact formal cluster expansion for the activity coefficients of each of the three species appearing there. The leading terms are as follows:

5.1. Cycle Contributions

Define $q(\{ij\})$ to be $-\beta$ times the total coulombic interaction energy between physical clusters i and j. The cycle diagram contribution to $\log\gamma$ is

$$-\log\gamma_\alpha^{(c)}(\{i_\alpha\}) = \tfrac{1}{2}\sum_\beta \int \mathrm{d}\{j_\beta\}\, q(\{i_\alpha j_\beta\})\, \varrho(\{j_\beta\})\, m(\{j_\beta i_\alpha\}),$$

where α, β label the different species of physical cluster and m is a function called the chain sum. We can represent the integral in a conventional way by a diagram

$-\log\gamma_\alpha^{(c)} = q$  m

The definition of the chain sum m is then equivalent to

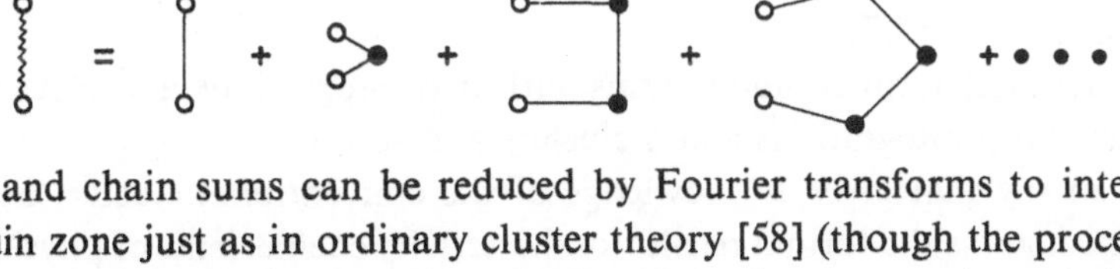

The cycle and chain sums can be reduced by Fourier transforms to integrals over the Brillouin zone just as in ordinary cluster theory [58] (though the process is more complicated – for details see elsewhere [60]). When there are only unpaired defects this reduces in the limit $\kappa a \to 0$, where a is the cation-cation lattice spacing, to the familiar Debye-Hückel limiting law

$$-\log\gamma^{(c)} = \kappa b/2,$$

where

$$b = e^2/DkT, \quad \kappa^2 = 4\pi b \sum_i c_i z_i^2,$$

D is the dielectric constant and c_i is the site fraction of defect with effective change $z_i e$. However at finite concentrations the collective effects described by this law are modified by the finite space occupied by the sites. When there are only unpaired defects the numerical results for $\kappa a < 1.3$ can be represented by

$$-\log\gamma^{(c)} = \tfrac{1}{2}\kappa b\left[1 - 0.570\kappa a + 0.21(\kappa a)^2\right].$$

Notice that this is different from the Debye-Hückel activity coefficient used in the simple mass action theory, Equation (2.11).

$$-\log\gamma_{DH} = \tfrac{1}{2}\kappa b/(1 + \kappa a). \tag{5.5}$$

When clusters of pairs, triplets, etc. of defects are included it is difficult to get a useful analytic representation but it is straightforward to calculate $\gamma^{(c)}$ numerically.

In like manner the chain sum is just the Debye-Hückel limiting value for unpaired defects,

$$m(R) = -(b/R)\exp(-\kappa R)$$

for $\kappa a \to 0$.

5.2. Modified virial expansion contributions

The rest of $\log\gamma$ is an expansion in concentrations but the coefficients are cluster integrals which are themselves concentration dependent. The leading term contains interactions between pairs of clusters:

$$\begin{aligned}-\log\gamma_\alpha^{(2)}(\{i_\alpha\}) = \sum_\beta \int d\{j_\beta\}\, \varrho(\{j_\beta\})\, [\exp(m(\{i_\alpha j_\beta\}) - \\ - \beta A^{(s)}(\{i_\alpha j_\beta\}) - \beta U(\{i_\alpha j_\beta\})) - 1 - \\ - m_{ij}^2(\{i_\alpha j_\beta\})].\end{aligned}$$

$A^{(s)}(\{ij\})$ is A_{int} for the pair of clusters i and j minus the coulombic part of the interaction, so it is zero except at very small separations, and U is the cluster restriction energy introduced in (5.3).

The next (triplet) term contains terms with two integrations and three interacting clusters, the next 3 integrations and 4 clusters and so on.

It is fair to say that given a knowledge of the conventional cluster theory results one can almost guess the formal results for the activity coeffiecients of physical clusters. The new results are more difficult to use because (1) cycle and chain sums have to be worked out numerically, (2) the cluster restrictions, represented by U, in the virial integrals make the calculation very tedious for triplet and higher integrals, (3) we have a set of highly non-linear integral equations to solve for the concentrations of clusters. The scheme will be useful only if a small number of kinds of cluster dominate.

A useful approximation for a doped crystal at not too high a concentration or low a temperature has been to assume that only unpaired defects and nearest-neighbour impurity-vacancy pairs are present, and the naive law of mass action, Equation (2.11), holds. To test this we first assume that these are the only clusters in the present formalism and use the new law of mass action and activity coefficients. We define the 'pair' and 'triplet' approximations as retaining in $\log \gamma$ the cycle and pair terms, and the cycle pair and triplet terms, respectively. Table VI shows some results for the fraction p of impurity ions which are in nearest-neighbour complexes for AgCl doped with a total concentration c of $CdCl_2$. The superscripts L, p, t refer to Equation (2.11), and the pair and triplet approximations respectively.

TABLE VI

κa	$p^{(L)}$	$p^{(p)}$	$p^{(t)}$	$10^3 c^{(t)}$
250°C				
0.15	0.202	0.205	0.203	2.00
0.20	0.262	0.269	0.261	2.87
0.30	0.347	0.370	0.345	4.77
0.40	0.402	–	0.389	6.94
0.60	0.474	0.544	0.437	11.31
50°C				
0.10	0.689		0.604	1.36
0.15	0.746		0.609	2.08
0.30	0.766		0.522	3.40

One sees that the three approximations agree well up to $c \simeq 2 \times 10^{-3}$ at 250°C. At higher concentrations triplet and pair approximations diverge but considerable cancellations occur which makes agreement between the conventional equation and the triplet approximation (which is more accurate than the pair of course) persist to higher concentrations. The range over which the conventional method is justified appears remarkable when it is remembered that the new method includes short range interactions between unpaired and paired defects and the effects of complexes and finite lattice spacing on the Debye-Hückel shielding. These are not negligible at high concentrations [60] and do not appear in the conventional formalism. Furthermore, Fuoss and Onsanger [61] have expressed doubts about the fundamental significance of the factor $(1+\kappa a)^{-1}$ appearing in the Debye-Hückel expression (5.5) used in the usual formalism! The conclusion so far is that the conventional formalism will be accurate over a good concentration range although this is fortuitous and that the deviations for the alkali halides noted earlier probably have some other cause. Of course at low enough temperatures 'measurable' deviations begin at lower concentrations e.g. the 15% deviation of $p^{(L)}$ from $p^{(t)}$ at $c \simeq 1.4 \times 10^{-3}$ and 50 °C in Table VI. But larger clusters form in appreciable quantities at only slightly higher c,e.g., a rough calculation gives 0.00, 0.01 and 0.03 for the fraction of impurity in physical clusters of three defects at 50 °C for the concentrations in the table. Such precise calculations as were made for the triplet approximation would be very time consuming.

Probably a more difficult problem than formation of larger clusters at low temperatures, for which there is experimental evidence, is the problem of the state in crystals like AgCl and AgBr at higher temperatures e.g. $T \geqslant 250$ °C where impurity solubilities greater than 1% are often found but where, so far as I am aware, discrete clusters larger than pairs have not been identified in transport measurements [62]. Whether the new formalism can be extended to this range remains to be seen.

References

1. Howard, R. E. and Lidiard, A. B.: *Rept. Progr. Phys.* **27**, 161 (1964).
2. Kröger, F. A.: *The Chemistry of Imperfect Crystals*, North-Holland, Amsterdam, 1964.
3. Lodding, A. and Lagerwall, T. (eds.): *Atomic Transport in Solids and Liquids*, Verlag der Zeitschift für Naturforschung, Tübingen, 1971.
4. Allnatt, A. R. and Jacobs, P. W. M.: *Proc. Roy. Soc.* **A 260**, 350 (1961).
5. Christy, R. W.: *Amer. J. Phys.* **40**, 40 (1972).
6. LeClaire, A. D.: *Phil. Mag.* **3**, 921 (1958).
7. Allnatt, A. R. and Rice, S. A.: *J. Chem. Phys.* **33**, 543 (1960).
8. Manning, J. R.: *Phys. Rev.* **124**, 470 (1961); **139A**, 126 (1965); *Acta Met.* **15**, 817 (1967).
9. LeClaire, A. D.: *An Advanced Treatise on Physical Chemistry*, Vol. 10 (ed. by W. Jost), Academic Press, New York, 1970.
10. LeClaire, A. D.: *Treatise on Physical Chemistry*, Vol. 10, Ch. 5, Academic Press, London, 1970.
11. Vineyard, G. H.: *J. Phys. Chem. Solids* **3**, 121 (1957).
12. Barr, L. W. and LeClaire, A. D.: *Proc. Br. Ceram. Soc.* **1**, 109 (1964).
13. Rothman, S. J. and Peterson, N. L: *Phys. Rev.* **154**, 552 (1967).
14. Peterson, N. L. and Rothman, S. J.: *Phys. Rev.* **2**, 1540 (1970).
15. LeClaire, A. D.: *Atomic Transport in Solids and Liquids* (ed. by A. Lodding and T. Lagerwall), Verlag der Zeitschrift für Naturforschung, Tübingen, 1972, 265.
16. Rothman, S. J. and Peterson, N. L.: *Atomic Transport in Solids and Liquids* (ed. by A. Lodding and T. Lagerwall), Verlag der Zeitschrift für Naturforschung, Tübingen, 1972, 245.
17. Peterson, N. L.: *Phys. Rev.* **136**, A568 (1964).
18. Heumann, Th. and Imm, R.: *J. Phys. Chem. Solids* **29**, 1613 (1968).
19. Walter, C. M. and Peterson, N. L.: *Phys. Rev.* **178**, 922 (1969).
20. Peterson, N. L. and Rothman, S. J.: *Phys. Rev.* **103**, 645 (1967).
21. Batra, A.: *Phys. Rev.* **159**, 487 (1967).
22. Mundy, J. M., Barr, L. W., and Smith, F. A.: *Phil. Mag.* **44**, 783 (1966).
23. LeClaire, A. D.: *Phil Mag.* **14**, 1271 (1966).
24. Glyde, H. R.: *Phys. Rev.* **180**, 722 (1969).
25. Brown, R. C., Worster, J., March, N. H., Perrin, R. C., and Bullough, R.: *Proceedings of the Thomas Grahame Memorial Symposium*, University of Strathclyde, 1969, Gordon and Breach, New York, 1970.
26. Brown, R. C., Worster, J., March, N. H., Perrin, R. C., and Bullough, R.: *Phil. Mag.* **23**, 555 (1971).
27. Brown, R. C., Worster, J., March, N. H., Perrin, R. C., and Bullough, R.: *Atomic Transport in Solids and Liquids* (ed. by A. Lodding, and T. Lagerwall), Verlag der Zeitschrift für Naturforschung, Tübingen, 1972, 209.
28. Huntington, H. B., Feit, M. D., and Lortz, D.: *Crystal Lattice Defects*, **1**, 193 (1970).
29. Burton, J. J.: *Phys. Rev.* **B 2**, 5010 (1970).

30. Slater, N. B.: *The Theory of Unimolecular Reactions*, Cornell University Press, Ithaca, 1959.
31. Weiner, J. H.: *Phys. Rev.* **169**, 570 (1968).
32. Rice, S. A. and Frisch, H.: *J. Chem. Phys.* **32**, 1026 (1960).
33. Franklin, W.: *J. Phys. Chem. Solids* **28**, 829 (1967).
34. Rice, S. A.: *Phys. Rev.* **112**, 804 (1958).
35. Achar, B. M.: *Phys. Rev. B* **2**, 3848 (1970).
36. Peterson, N. L.: *Solid State Phys.* **22**, 409 (1968).
37. Peterson, N. L. and Rothman, S. J.: *Atomic Transport in Solids and Liquids* (ed. by A. Lodding and T. Lagerwall), Verlag der Zeitschrift für Naturforschung, Tübingen, 1972, 248.
38. Allnatt, A. R. and Chadwick, A. V.: *Chem. Rev.* **67**, 681 (1967).
39. Biersack, J. and Diez, W.: *Phys. Stat. Sol.* **27**, 139 (1968).
40. Jacobs, P. W. M. and Kright, P. C.: *Trans. Faraday Soc.* **66**, 1227 (1970).
41. Corrish, J. and Jacobs, P. W. M. (to be published).
42. Crolet, J. L. and Lazarus, D.: *Solid State Comm.* **9**, 347 (1971).
43. Weeks, J. D. and Shuler, K. E.: *J. Chem. Phys.* **56**, 1883 (1972).
44. Klemens, P. G.: *Solid State Phys.* **7**, 1 (1958).
45. Schottky, G.: *Phys. Stat. Sol.* **8**, 357 (1965).
46. Huntington, H. B.: *J. Phys. Chem. Solids* **29**, 1641 (1968).
47. Gerl, M.: *Z. Naturforsch.* **26a**, 1 (1971).
48. Allnatt, A. R.: *Atomic Transport in Solids and Liquids* (ed. by A. Lodding and T. Lagerwall), Verlag der Zeitschrift für Naturforschung, 1972, 36.
49. Nicolis, G.: *J. Chem. Phys.* **43**, 1110 (1965).
50. Allnatt, A. R. and Rowley, L. A.: *Molecular Physics* **24**, 993 (1972); **24**, 1073 (1972); **25**, 361 (1973).
51. Hill, T. L.: *Statistical Mechanics*, McGraw Hill New York, 1958.
52. Schieve, W. C. and Leaf, B.: *Physica* **30**, 1208 (1964).
53. Rockmore, D. M. and Turner, R. E.: *Physica* **29**, 873 (1963).
54. Allnatt, A. R., Pantelis, P., and Sime, S. J.: *J. Phys. C.* **4**, 1778 (1971).
55. Fuller, R. G., Marquardt, C. L., Reilly, M. H., and Wells J. C.: *Phys. Rev.* **176**, 1036 (1968).
56. Jacobs, P. W. M. and Pantelis, P.: *Phys. Rev. B* **4**, 3757 (1971).
57. Cohen, M. H. and Thompson, J. C.: *Advan. Phys.* **17**, 857 (1968).
58. Allnatt, A. R.: *Advan. Chem. Phys.* **11**, 1 (1967).
59. Friedman, H. L.: *Ionic Solution Theory Based on Cluster Expansion Methods*, Interscience, New York, 1962.
60. Allnatt, A. R. and Loftus, E.: (to be published).
61. Fuoss, R. M. and Onsager, L.: *Proc. Natl. Acad. Sci. (U.S.)* **47**, 818 (1961).
62. Lidiard, A. B.: *Handbuch der Physik*. **20**, 246 (1957).

THE FERMI SOLID

R. A. GUYER

Dept. of Physics and Astronomy, University of Massachusetts, Amherst, Mass. 01002, U.S.A.

1. The Fermi Solid; Introduction

These lectures will attempt to describe some of the physics of the solid systems that are called the fermi solids. The fermi solids are *dielectric* solids which have properties which depend upon the fermi nature of the particles from which they are constructed. As such the fermi solids are a subclass of the quantum solids [1]. Consider the sequence of dielectric solid systems; the neutron solid, solid ^{3}He, solid ^{4}He, solid H_2, solid HD, solid D_2, solid neon, solid argon, ... In this sequence the noble gas solids Ne, A, Kr, Xe, Rd are essentially non-quantum solids that form a useful classical limit against which to test ones understanding of the quantum solids [2]. The solid hydrogens, H_2 HD and D_2 are weakly quantum mechanical. The neutron solid and the helium solids, ^{3}He and ^{4}He, are strongly quantum mechanical. Solid ^{4}He is a bose solid; it has been speculated that this *solid* might exhibit superfluidity [3]. Solid ^{3}He and the neutron solid are the fermi solids. ^{3}He is a laboratory solid about which a great deal is known. The neutron solid is believed to be present in some of the pulsars.

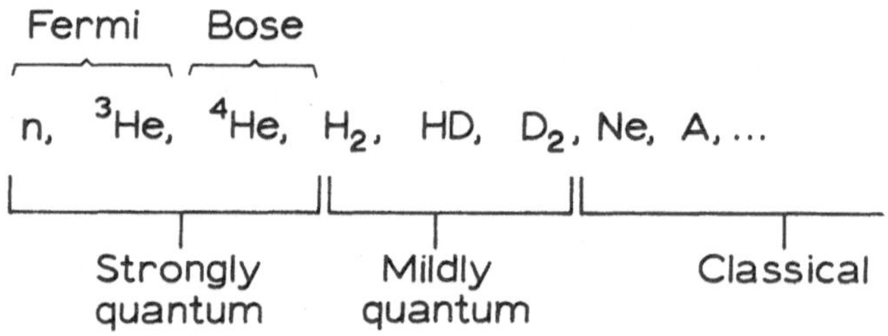

Fig. 1. Dielectric solids. The most quantum of these systems are the neutron solid and the helium solids. The hydrogen solids are mildly quantum; the rare gas solids are essentially classical.

The most obvious physical feature of the sequence of systems we are discussing is seen in the evolution of the phase diagram as one goes from the classical to the quantum limit. See Figure 2.

At $P=0$ the neon system undergoes a phase transition from liquid to solid at $T\approx 25$ K. At a negative pressure, $P_m \approx -1000$ atm, neon undergoes a liquid-solid transition at zero temperature. The hydrogen systems undergo a liquid-solid phase transition at $T=0$ K at $P<0$. (Recently Ginsboro and Sobyanin have speculated that liquid parahydrogen under negative pressure, will undergo a normal to superfluid transition [3].) None of the strongly quantum systems has a liquid to solid phase transition at $P=0$. For solid ^{3}He and ^{4}He external pressures in excess of 25–30 atm are required to produce a solid. The attendant low temperature and low pressure liquids are exhaus-

Abecassis de Laredo and Jurisic (eds.), Selected Topics in Phys. Astrophys. and Biophys. 44–86.

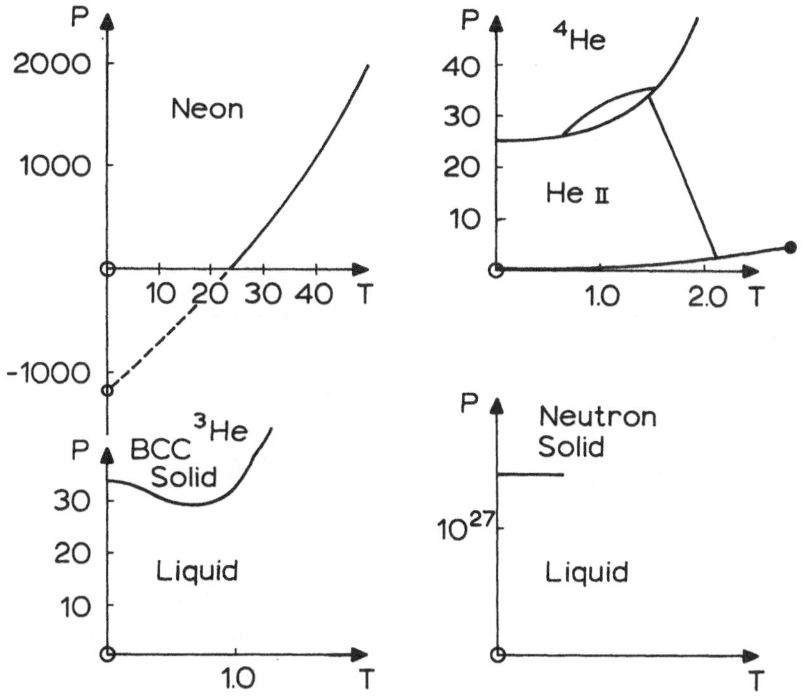

Fig. 2. Phase diagrams. Solid neon is a classical solid that forms at $P=0$ at $T \approx 25$ K. A zero temperature melting pressure can be extrapolated; it is about minus 1000 atm. The helium solids and the neutron solid are quantum solids. These solids are formed at $T=0$ under a positive external pressure.

tively studied. For neutron matter the 'astronomical' pressure of 10^{27} atm is required to produce a solid.

The rare gas systems, the hydrogens, the heliums, and neutrons form dielectric solids. The basic building blocks of these solids are electrically neutral inert objects-atoms of ^{3}He, ^{4}He, Ne, H_2 ... or neutrons. These systems are described by the Hamiltonian

$$\mathscr{H} = \sum_i \frac{p_i^2}{2m} + \frac{1}{2} \sum_{ij}' v(r_{ij}), \tag{1}$$

where $v(r_{ij})$ is the Lennard-Jones interaction between a pair of helium atoms, neon atoms, neutrons, etc. This interaction for the rare gas systems

$$v(r) = 4\varepsilon\left[(\sigma/r)^{12} - (\sigma/r)^6\right] \tag{2}$$

is shown in Figure 3. It is characterized by 2 numbers, the strength ε and the length σ. The attractive part of $v(r)$, proportional to r^{-6}, is due to the induced dipole-induced dipole interaction; the repulsive part of $v(r)$, proportional to r^{-12}, is due to the Pauli principle attempting to prevent overlap of the electron orbitals of adjacent-atoms. The values of ε and σ for the systems we are discussing are listed in Table I.

At zero pressure the Hamiltonian describing these systems is characterized by 4

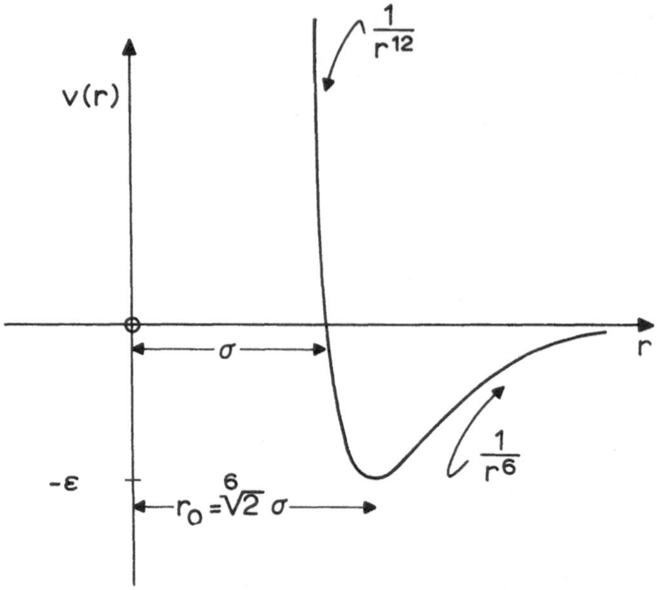

Fig. 3. The Lennard-Jones potential. At large values of r this interaction is attractive, r^{-6}; at small values of r, $r < \sigma$, it is strongly repulsive. The potential minimum occurs at $r_0 = \sqrt[6]{2}\sigma$.

parameters, h, m, σ and ε. From these numbers we have two energies; ε, the strength of $v(r)$, and $h^2/m\sigma^2$, a quantum mechanical energy. There is a dimensionless parameter that measures the ratio of the 'quantum energy' to the strength of the interaction. This parameter is called the de Boer parameter; it is denoted by Λ,

$$\Lambda^2 = h^2/m\sigma^2\varepsilon. \tag{3}$$

The quantum mechanical law of corresponding states is the statement that for systems described by a single Hamiltonian like Equation (1) certain scaled functions, a volume, an energy, etc., should be a universal function of Λ [4]. For example

$$V^* = V/\sigma^3 = f_1(\Lambda), \tag{4}$$
$$E^* = E/\varepsilon = f_2(\Lambda), \tag{5}$$
$$\Theta_D^* = \tfrac{9}{8}k_B\Theta_D/\varepsilon = f_3(\Lambda), \tag{6}$$

TABLE I

	m (amu)	ε (Kelvin)	σ (Ångström)
^{3}He, ^{4}He	3, 4	10	2.56
H_2, HD, D_2	2, 3, 4	37	2.93
Ne	20	36	2.75
A	40	119	3.40
Kr	84	159	3.60
Xe	131	228	4.10
n	1	~10 MeV	~1 Fermi

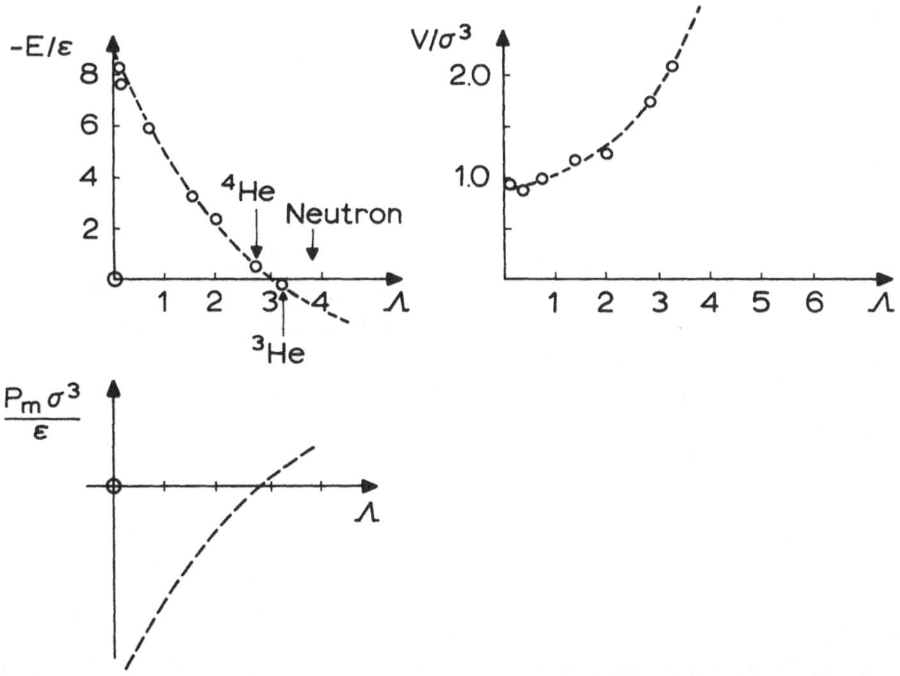

Fig. 4. The law of corresponding states. The energy, volume per particle, etc. when scaled by an appropriate factor fall on a universal curve as a function of Λ, the de Boer parameter. If the melting pressure, P_m, is also a function of Λ, then, it also scales. From the melting pressure scaling it is possible to estimate the pressure of the liquid-solid phase transition in neutron matter.

where V is the volume per particle, E is the energy per particle, $\frac{9}{8}k_B\Theta_D$ is an approximate zero point energy per particle, etc. The idea then is that if one knows the functions $f_n(\Lambda)$ it is possible to infer the properties of a system (V, E, $\frac{9}{8}k_B\Theta_D$, ...) from a knowledge of its microscopic parameters (m, σ, ε). One might know $f_n(\Lambda)$ from a theory of systems described by the Hamiltonian in Equations (1) and (2) or from data on a few systems described by this Hamiltonian. In Figure 4 we show V^* and E^* as a function of Λ for the noble gas systems and the hydrogen systems.

In this figure we also show P_m^*, the scaled melting pressure, $P_m^*=P_m\sigma^3/v$, where P_m is the melting pressure at $T=0$ K. Our purpose in discussing the law of corresponding states is that we would like to use it to provide a set of parameters (σ and ε) for neutron matter and to elucidate the qualitative features of solid neutron matter. If we find that solid neutron matter is a quantum crystal of about the same degree of quantumness as solid helium then as we go on to describe the properties of the labaratory quantum crystals we can expect that qualitatively and quantitatively (but scaled) similar properties are present in the astrophysical quantum crystals.

In suggesting the application of the law of corresponding states above to solid neutron matter – or for that matter to solid helium – we apply the law to systems that are not at $P=0$. In the case of solid helium external pressures of about 30 atm are involved; in the case of solid neutron matter external pressures (the weight of the rest of the neutron star) of order of magnitude 10^{27} atm are involved. These finite pressure systems are correctly described by a Hamiltonian like Equation (1) but also containing external forces

$$\mathscr{H}(P) = \mathscr{H} + \sum_i \mathbf{F}_i \cdot \mathbf{x}_i \tag{7}$$

$$F_i = P\Delta^2 \sum_{j(\neq i)} e^{\alpha}_{ij}, \tag{8}$$

P is the external pressure, Δ is the interparticle spacing and e^{α}_{ij} is the α-component of a unit vector pointing from particle i to the *near neighbor* particle at j. In the bulk of a crystal $F^{\alpha}_i=0$. Since $\sum_{j(\neq i)} e^{\alpha}_i=0$. The forces in Equation (7) are on the surface. In the presence of external forces the properties of a system depend upon the magnitude of these forces (the external pressure). Thus there is a second length introduced into the characterization of the system. This length is

$$l = (V/N)^{1/3}, \quad \text{where} \quad V = \partial \langle E \rangle / \partial P. \tag{9}$$

There is no longer a *unique* quantum energy like that used to define Λ^2 Equation (3). Thus there are difficulties in principle with the law of corresponding states at $P>0$. Let us put them aside and regard the law as empirical – we use data on known systems – chosen in a suitably consistent way to construct empirical universal curves to permit us to look at systems at a unique $P \neq 0$. For example we take the noble gas crystals and the hydrogens at P_m and construct the curves shown in Figure 4.

We use the curves in Figure 4 and the data on nuclear matter, energy per particle ≈ -16 MeV, volume per particle $\approx 6\,fm^3$, to find an ε and σ to describe this system by an effective Lennard-Jones 6-12 potential. The results are shown in Figure 4 and correspond to $\varepsilon \approx 50$ MeV and $\sigma \approx 1.7\,fm$. This procedure originally applied by Anderson and Palmer [5], refined by Clark and Woo [6], would be highly suspect were it not for the detailed calculations of Canuto and Chitre [7] that have verified its conclusions. We take over the values of ε and σ obtained by this procedure and note

$$\Lambda_n^2 \approx 4.0. \tag{10}$$

Thus we argue that the degree of quantumness of solid neutron matter is much the same as that of solid helium.

The fermi solids, ^{3}He and solid neutron matter, are quantum solids. A quantum solid (crystal) is a crystal in which the zero point motion of a particle is a large fraction of the near neighbor distance. The quantum crystals are an interesting and unusual class of system because of their large zero point motion [1]. Some appreciation for the reason certain crystals are quantum crystals and the physics that is a consequence of their being quantum crystals can be gleened from an attempt to construct a crystal. Consider an aggregate of neon atoms. To build a crystal from this aggregate of atoms we must lay them down in space in a regular array so that their energy is a minimum. Since the potential of interaction has a minimum at $r_0 = \sqrt[6]{2}\,\sigma$ we can begin to build a crystal by placing an atom at $\mathbf{x}=0$ and surrounding it with z ($z=\#$ of near neighbors) atoms at distance $r_0=\sqrt[6]{2}\,\sigma$ and so on. Then, the energy of the atom at $\mathbf{x}=0$ (the same as the energy of any other atom) is given by

$$E = \mathcal{U}_0 + \varepsilon_\mathrm{d}, \tag{11}$$

where $\mathcal{U}_0$ is the depth of the potential well the atom sits in and ε_d, the dynamic energy, is the energy of motion of the atom about its lattice site at $\mathbf{x}=0$. We estimate these two energies. For the potential energy we take

$$\mathcal{U}_0 = -\,z\varepsilon \tag{12}$$

since the atom at $\mathbf{x}=0$ sits at the bottom of the potential well of z near neighbors. For the dynamic energy ε_d we take the energy of the oscillation of the atom about the point $\mathbf{x}=0$. This energy is the energy eigenvalue of

$$\left(-\frac{\hbar^2}{2m}\nabla^2 + \tfrac{1}{2}kr^2\right)\phi_\mathrm{d}(r) = \varepsilon_\mathrm{d}\phi_\mathrm{d}(r) \tag{13}$$

the Schroedinger equation that describes the atom. Here $k=zv''(r_0)$. For the Lennard-Jones interaction given by Equation (2) we have $v''(r_0)=56\varepsilon/\sigma^2$. The solution to Equation (13) leads to Einstein oscillator wavefunctions whose size is governed by $\alpha^2=\sqrt{mk}/\hbar$ and to the energy eigenvalue

$$\varepsilon_\mathrm{d} = \tfrac{3}{2}\hbar\sqrt{\frac{k}{m}} = \frac{3}{4\pi}\sqrt{\frac{56z\varepsilon h^2}{m\sigma^2}}\,. \tag{14}$$

Now as a minimum condition that our recipe has produced a solid we require that the atom we are looking at (that at $\mathbf{x}=0$) be bound in the potential well of its neighbors. That is, the potential well in which it sits, $\mathcal{U}_0$, must be deeper than the positive energy of its motion in the bottom of the well. We require

$$\frac{\varepsilon_\mathrm{d}}{|\mathcal{U}_0|} < 1\,. \tag{15}$$

We may use Equations (12) and (14) to write

$$\frac{\varepsilon_\mathrm{d}}{|\mathcal{U}_0|} = \mathrm{const}\,\Lambda\,, \tag{16}$$

where $\mathrm{const}=3\sqrt{56z}/4\pi z\approx 1$. Thus the solid is stable for $\Lambda<1$; it is unstable for $\Lambda>1$. We know that for neon $\Lambda^2\approx 1/3$, thus a neon crystal can be built according to our recipe. A neutron crystal or a helium crystal cannot because for these systems $\Lambda>1$. How is a helium crystal built? We will use the experimental results on solid helium to tell us the answer. At zero pressure ($P=0$) a stable helium crystal cannot be built. At a pressure of about 30 atm a crystal exists [8]. For this crystal it is observed that the interparticle spacing is 3.75 Å. This spacing is far greater than the optimium distance $r_0=\sqrt[6]{2}\,\sigma$ that will minimize the potential energy. The reason for this is that if

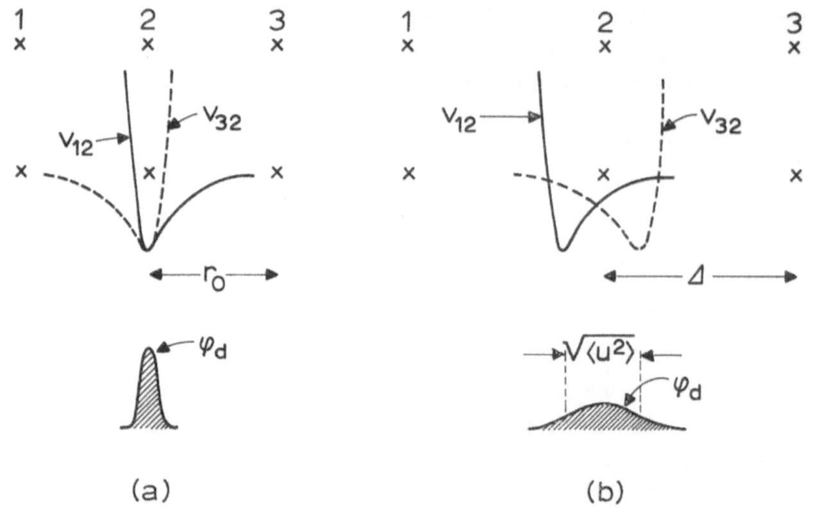

Fig. 5. Crystal models. The potential energy is a minimum for the left hand arrangement, (a), particle 2 is at distance r_0 from 1 and 3. Due to the compactness of ϕ_d the zero point energy is large and the particle not bound by its neighbors. In (b) the experimental arrangement for a ^{3}He crystal at 30 atm is shown. Particle 2 sits away from the potential minimum in order to have more room and reduce its zero point energy.

built like neon – the atoms in solid helium find themselves confined to the small region of space between $r_0 = \sqrt[6]{2}\,\sigma$ and σ; i.e. to a volume of space approximately $\sigma^3(\sqrt[6]{2}-1)^3$. A particle of small mass has too much zero point energy when so confined – it is not bound. If the neighbors are moved further apart – to $\Delta > r_0$ – although some expense in potential energy is involved the reduction in energy of localization is substantial and a stable system can be formed. See Figure 5.

It is apparent from the discussion above and Figure 5 that the zero point motion of a particle in solid helium is a large fraction of the near neighbor distance. As a measure of the zero point motion we take $\sqrt{\langle u^2 \rangle}$ where u is the displacement of an atom from its lattice site and $\langle \ldots \rangle$ is the average over the single particle wave-function for the atom. For solid helium at low pressure we have $\sqrt{\langle u^2 \rangle}/\Delta \approx 1/3$; for solid neon $\sqrt{\langle u^2 \rangle}/\Delta \approx 1/10$.

It is the large zero point motion of the particles in a quantum crystal that lead to the interesting properties of these systems. In these lectures we will consider three consequences of this zero point motion:

(1) The particles in the crystal meet one another in the space between lattice sites (often) at distances $\lesssim \sigma$ – thus the short range motions of neighboring particles is correlated. (Short range correlation energy is important in atomic and molecular physics; see the lectures by Pincus, this volume, 138.

(2) The displacements of particles away from their lattice sites is large. A conventional theory of phonons does not work. A sophisticated alternate theory of the phonons must be developed.

(3) There are excitations present in the systems that are due to the zero point motion that are interesting and quite unique.
We will review the progress made on (1) and (2) and then concentrate on the physics associated with the systems of excitations in (3).

2. Short Range Correlations

Because of the large zero point motion of the particles in a quantum crystal pairs of particles encounter one another at distances on the order of σ midway between neighboring lattice sites. As a consequence it is necessary to account for the detailed relative motion of pairs of particles that interact very strongly. This point is made quite clear if we attempt to make an improvement on the $P=0$ calculation of the energy of a solid that we made in the introduction. Let us at least take over the experimental near neighbor distance and crystallographic structure and attempt a Hartree calculation of the energy. To do this we choose as a total wavefunction a product of single particle wavefunctions, e.g. $\phi_{\mathbf{R}_i}(\mathbf{x}_i)$, each of which localize a particle on a lattice site,

$$\psi_{\mathrm{H}}(1 \ldots N) = \prod_{i=1}^{N} \phi_{\mathbf{R}_i}(\mathbf{x}_i). \tag{17}$$

The Hartree energy is

$$E_{\mathrm{H}} = \frac{\langle \psi_{\mathrm{H}} | \mathscr{H} | \psi_{\mathrm{H}} \rangle}{\langle \psi_{\mathrm{H}} | \psi_{\mathrm{H}} \rangle}, \tag{18}$$

where the optimum choice of $\phi_i(i)$ is found by varying the Hartree energy, E_{H}, with respect to $\phi_i(i)$. In carrying out a Hartree calculation involving particles that interact with a hard core, it is necessary to restrict the motion of the particle localized near lattice site $\mathbf{R}_i$ such that its wave function does not overlap the wavefunction of a second particle. Should such an overlap occur, the Hartree contribution to the potential energy, given by

$$\int \mathrm{d}\mathbf{x}_1\, \mathrm{d}\mathbf{x}_2 \phi_1(1)^2\, \phi_2(2)^2\, v(\mathbf{r}_{12})$$

will be singular. Recall $v(r) \to r^{-12}$ as $r \to 0$. [This hard core energy can be contrasted with the Coulomb case; the coulomb repulsion has a mild 'hard core' so that the Hartree potential energy is finite, e.g.

$$\int \mathrm{d}\mathbf{x}_1\, \mathrm{d}\mathbf{x}_2 \varrho(1)\, \varrho(2) \frac{1}{r_{12}} \to \int \mathrm{d}\mathbf{R} \int r^2\, \mathrm{d}r\, \mathrm{d}\Omega \varrho(1)\, \varrho(2) \cdot \frac{1}{r}.$$

In writing the second line we use the center of mass and relative coordinate to show that the phase space for relative motion $4\pi r^2\, \mathrm{d}r$ cancels out the repulsion for potential less singular than r^{-3}.] To avoid the singular hard core energy we erect rigid barriers midway between adjacent lattice sites as shown in Figure 6. Subject to this constraint (the presence of barriers) we vary ϕ in Equation (18) to minimize E_{H}. A procedure

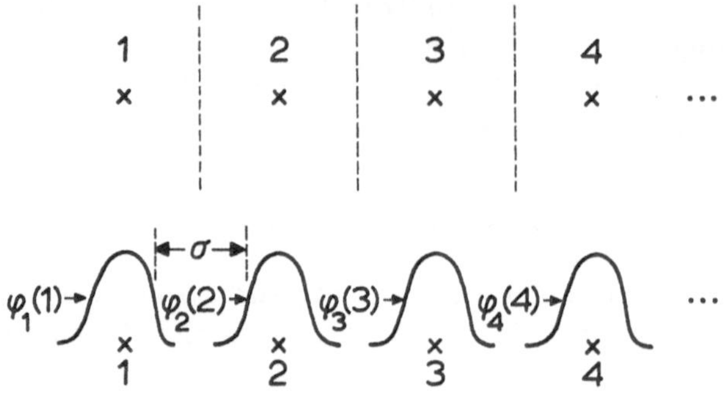

Fig. 6. Hartree calculation. A Hartree calculation of the energy uses a product of single particle states. Because of the hard core repulsion at $r<\sigma$ the particles try to stay near the center of their respective cells. The Hartree energy is much larger than the experimental energy.

much the same as that we use here was employed by Nosanow and Shaw [9] in their Hartree calculations. For ^{4}He Nosanow and Shaw found an energy per particle of order +20 K. This result does not compare favorably with the experimental energy of about 0 K. This is particularly apparent when one realizes that the components of this energy, a potential energy and a kinetic energy, are about the same size as the error in the Hartree energy. The problem with this calculation is made clear upon looking at the single particle wavefunction. (See Figure 6). The excess positive energy in the Hartree calculation is due to the localization of the single particle wavefunctions near the center of the cell. This localization is forced on the wavefunction even though we have made the near neighbors distance Δ. The particles are kept in their respective cells by the barriers we have erected. But they still attempt to stay near the center of their cells to avoid the repulsive energy that would occur when the particles on opposite sides of a barrier approach the barrier at the same time. So we see that both of our attempts to treat the particles as independent have been unsatisfactory.

Thus we introduce short range correlations into the wavefunction. For a pair of particles we replace $\phi_1(1)\,\phi_2(2)$ by $\phi_1(1)\,\phi_2(2)\,f(r_{12})$, where the function $f(r_{12})$ correlates the relative motion of particles 1 and 2. The function $f(r_{12})\to 0$ as $r_{12}\to 0$; $f(r_{12})\to 1$ or is a very mild function of r_{12} as $r_{12}\to +\infty$. See Figure 7. The correlation function serves two purposes; (1) since $f(r)\to 0$ as $r\to 0$ (and $f(r)\,v(r)\to 0$ as $r\to 0$) it is no longer necessary to erect the rigid barriers midway between lattice sites; (2) a particle may wander quite far from its lattice site ($f(r_{12})$ will correlate the motion of the neighboring particles such that this is possible) and reduce it's energy of localization.

The first successful calculation of the ground state properties of the quantum crystals that employed short range correlated wavefunctions was that of Nosanow [10]. This pioneering work opened a flood gate of activity on this problem. A physical discussion of the short range correlation problem has been given by Guyer [1, 11]; a more rigorous mathematical discussion and comparison of various 'theories' of short range correlation has been given by Brandow [12]. Here we will reproduce a discussion similar

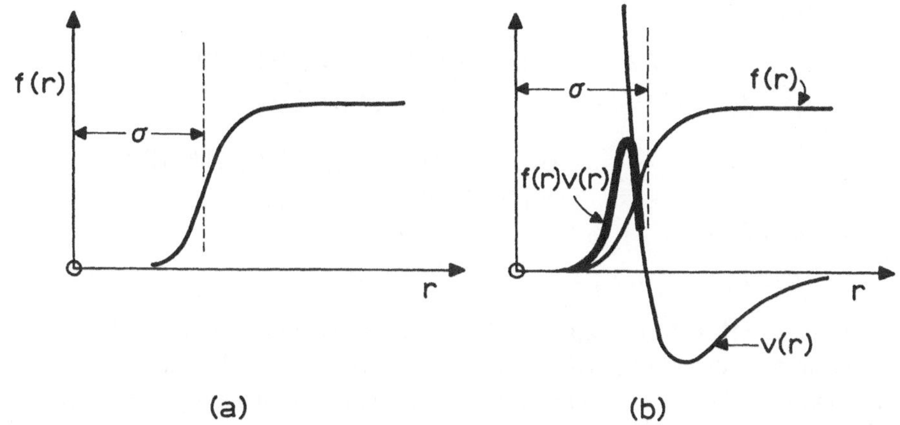

Fig. 7. Short range correlation function. The function $f(r) \to 0$ for $r < \sigma$ sufficiently rapidly that the product $f(r)v(r) \to 0$ as $r \to 0$. Thus a calculation of the average potential energy is finite; e.g. $\int \mathrm{d}r\, f(r)v(r)$ is finite.

that given previously that attempts to point out the key physical elements that are desirable in a theory of short range correlations. Our approach will use a cluster expansion, but this is by no means the only way to achieve the results we will obtain. Since, we are motivated by physical considerations it is apparent that any formalism seriously applied can achieve the same result. It is our view that the present approach permits one to introduce the essential physics most simply.

The expectation value of the energy of a system described by $\mathscr{H}$ may be written in the form

$$E = \frac{\langle \psi | \mathscr{H} | \psi \rangle}{\langle \psi | \psi \rangle} = \lim_{\gamma \to 0} \frac{\mathrm{d}}{\mathrm{d}\gamma} \log \langle \psi | e^{\gamma \mathscr{H}} | \psi \rangle . \tag{19}$$

In the spirit of the cluster expansion of Van Kampen [10, 13] we write

$$\langle \psi | e^{\gamma \mathscr{H}} | \psi \rangle = \prod_i \langle \psi(i) | e^{\gamma h(i)} | \psi(i) \rangle \prod_{i<j} \times \\ \times \frac{\langle \psi(ij) | e^{\gamma h(ij)} | \psi(ij) \rangle}{\langle \psi(i) | e^{\gamma h(i)} | \psi(i) \rangle \langle \psi(j) | e^{\gamma h(i)} | \psi(j) \rangle} + \cdots , \tag{20}$$

where $h(i)$, $h(ij)$, ... are a sequence of Hamiltonians, a single particle Hamiltonian, a pair Hamiltonian, etc. and $\psi(i)$, $\psi(ij)$, ... are a sequence of wavefunctions, a single particle wavefunction, a pair wavefunction, etc. The idea contained in the factorization of $\langle \psi | e^{\gamma \mathscr{H}} | \psi \rangle$ in Equation (19) is that the system can be approximated sequentially as N single particles, $N(N-1)/2!$ pairs of particles, $N(N-1)(N-2)/3!$ triplets of particles etc. i.e. a sequence of n-particle clusters. For each cluster there is a corresponding Hamiltonian and wavefunction yet to be specified. The denominator in the pair term in Equation (20) divides out the single particle approximation to the energy (of the particles in that term) that was made in the single particle term. Putting Equation (20) into

Equation (19) yields

$$E = \sum_i \frac{\langle\psi(i)|\, h(i)\, |\psi(i)\rangle}{\langle\psi(i)\,|\,\psi(i)\rangle} + \sum_{i<j} \left(\frac{\langle\psi(ij)|\, h(ij)\, |\psi(ij)\rangle}{\langle\psi(ij)\,|\,\psi(ij)\rangle} - \right.$$

$$\left. - \frac{\langle\psi(i)|\, h(i)\, |\psi(i)\rangle}{\langle\psi(i)\,|\,\psi(i)\rangle} - \frac{\langle\psi(j)|\, h(j)|\psi(j)\rangle}{\langle\psi(j)\,|\,\psi(j)\rangle} \right) + \cdots. \tag{21}$$

We also write this as

$$E = E_1 + E_2 + E_3 + \cdots.$$

The usefulness of this cluster expansion for the energy depends upon (1) the correlation energy of larger clusters of particles being well approximated by that of smaller clusters and (2) making a reasonable choice of the sequence of Hamiltonians and wavefunctions to be employed in carrying out the expansion. To get an idea how we might wish to proceed in constructing a sequence of Hamiltonians we write $\mathscr{H}$ in the form

$$\mathscr{H} = \sum_i T(i) + \sum_{ij}{}' \omega(ij) + \tfrac{1}{2} \sum_{ij}{}' \tilde{v}(ij), \tag{22}$$

where the $\omega(ij)$ are interactions between i and j of a convenient analytic form which will permit us to shape the single particle wavefunction to our convenience and $\tilde{v} = v - 2w$. For example in dealing with a problem for which localization is important we might choose $\omega(ij)$ such that

$$\sum_{j \neq (i)} w(ij) = \mathscr{U}_i(i) \approx \mathscr{U}_0 + \tfrac{1}{2} k (\mathbf{x}_i - \mathbf{R}_i)^2. \tag{23}$$

Let us make this choice and also choose the sequence of Hamiltonians to be

$$\begin{aligned} h(i) &= T(i) + \mathscr{U}_i(i) \\ h(ij) &= h(i) + h(j) + \tilde{v}(ij) \\ &\vdots \end{aligned}$$

and the sequence of wavefunctions to be the eigenfunctions of the corresponding Hamiltonians i.e.

$$h(i)\,\psi(i) = \varepsilon\psi(i) \tag{24-1}$$

$$h(ij)\,\psi(ij) = \varepsilon_{ij}\psi(ij) \tag{24-2}$$

$$\vdots$$

With this choice the energy E given by Equation (21) is

$$E = N\varepsilon + \sum_{i<j} (\varepsilon_{ij} - 2\varepsilon) + \cdots \tag{25}$$

To make this description complete we need to specify the single particle potential in Equation (23). To motivate a reasonable choice of $\mathscr{U}_i(i)$ we consider the solution to the pair problem i.e. Equation (24-2). We have

$$[h(i) + h(j) + \tilde{v}(ij)]\, \psi(ij) = \varepsilon(ij)\, \psi(ij). \tag{26}$$

We write the pair wavefunction in the form $\phi_1(1)\,\phi_2(2)\,g_{12}(r_{12})$ and calculate the energy

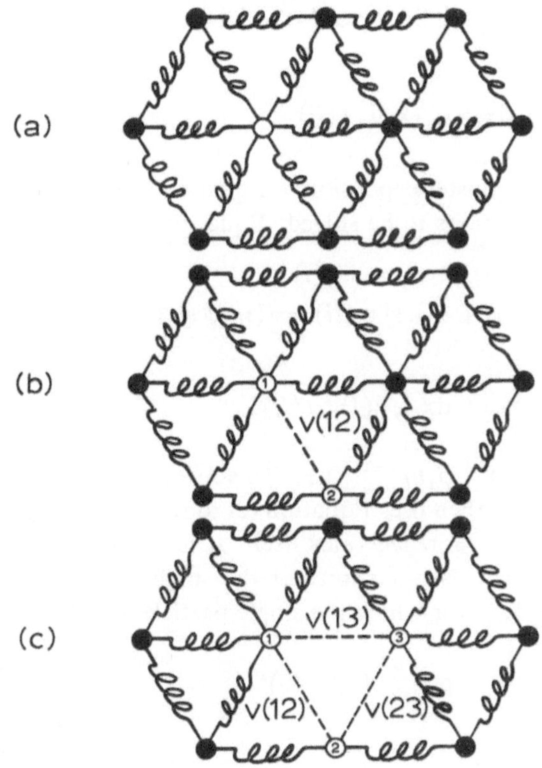

Fig. 8. Sequence of approximations. The idea of the cluster expansion is to approximate the system of particles as an aggregate of single particles, (a), as an aggregate of pairs, (b), as an aggregate of triplets. In each case the particles in a cluster see one another through the Lennard-Jones 6-12 potential – they see the rest of the particles through a convenient effective interaction.

shift in Equation (26) due to $\tilde{v}(12)$. (Note $g_{12}(12)$ and $\varepsilon(12)$ depend upon $\mathbf{R}_1$ and $\mathbf{R}_2$.)

$$\Delta\varepsilon(12) = \varepsilon(12) - 2\varepsilon = \frac{\langle\phi_1(1)\,\phi_2(2)|\,v(12) - 2w(12)\,|\psi(12)\rangle}{\langle\phi_1(1)\,\phi_2(2)\,|\,\psi(12)\rangle}. \tag{27}$$

The important term, – the first term – can be written in the form

$$\frac{\langle\phi_1(1)\,\phi_2(2)|\,v(r)\,g_{12}(r)\,|\phi_1(1)\,\phi_2(2)\rangle}{\langle\phi_1(1)\,\phi_2(2)\,|\,\psi(12)\rangle}. \tag{28}$$

In this form we can regard $t = vg$ as an effective interaction between the uncorrelated single particles states (we have used $\langle\phi\,|\,\psi\rangle = 1$). As with $g_{ij}(ij)$ and $\varepsilon(ij)$ above, this interaction depends upon $\mathbf{R}_1$ and $\mathbf{R}_2$. Now we take $t\,(ij)$ so defined to make a plausible construction of the single particle potential. We write

$$\mathscr{U}_{\mathbf{R}_i}(\mathbf{x}_i) = \sum_{j \neq (i)} \int \mathrm{d}\mathbf{x}_j \, |\phi_{Rj}(\mathbf{x}_j)|^2 \, t(ij). \tag{29}$$

and

$$w(ij) = \int \mathrm{d}\mathbf{x}_j \, |\phi_{\mathbf{R}_i}(\mathbf{x}_j)|^2 \, t(ij). \tag{30}$$

Thus if we truncate the cluster expansion in Equation (21) after the 2-particle term we have a system of 3 equations to be solved; Equations (24-1), (24-2) and (29)

$$[T(i) + \mathscr{U}_i(i)] \, \phi_i(i) = \varepsilon\phi_i(i) \tag{a}$$

$$[h(i) + h(j) + \tilde{v}(ij)] \, \psi(ij) = \varepsilon(ij) \, \psi(ij) \tag{b}$$

and

$$\mathscr{U}_i(i) = \sum_{(j \neq i)} \int \mathrm{d}\mathbf{x}_i \, |\phi_j(j)|^2 \, t(ij), \tag{c}$$

where $t(ij) = v(ij) \, \phi_i(i)^{-1} \, \phi_j(j)^{-1} \, \psi(ij)$. To solve these equations one starts with $\mathscr{U}$ to generate ϕ and ψ and then uses Equation (c) to reconstruct $\mathscr{U}$ [14]. The equations are solved when a given $\mathscr{U}$ regenerates itself. It is necessary in implementing these equations to make a convenient choice of the single particle potential. We choose $\mathscr{U}_i(i) = \mathscr{U}_0 + \frac{1}{2}\gamma(\mathbf{X}_i - \mathbf{R}_i)^2$ for which the single particle wavefunction is

$$\phi_{R_i}(\mathbf{x}_i) = A_0 \exp - \frac{\alpha^2}{2} (\mathbf{x}_i - \mathbf{R}_i)^2$$

and

$$\hbar^2\alpha^2/m = \hbar\sqrt{\gamma/m}.$$

It is important to recognize that there are approximations of two kinds in dealing with Equations (a)–(c); a set of approximations that generate these equations and a set of approximations that make the solution of these equations relatively simple.

Calculations carried out using the formalism we have described here have been quite successful in obtaining reasonable value for the energy, pressure and compressibility of the quantum crystals. The literature in which this success is demonstrated can be reached through References [1] and [12].

3. Phonons

We stated above that the conventional picture of phonons did not work. From Figure 5 we see that the spring constant is of the wrong sign [15] and the anharmonic terms were relatively large invalidating the usual Taylor series expansion in $\mathbf{u}_R = \mathbf{x}_R - - \mathbf{R}$. [16] Our task in this section is to indicate the outlines of an adequate theory of the phonons in a quantum crystal. A point of philosophy is in order before we embark on this task. The system we are talking about, solid ^{3}He or solid ^{4}He, is described exactly by the Hamiltonian in Equation (1). The properties of the normal liquid, of the superfluid at $T < T_\lambda$, of the magnetically ordered solid at $T < 1$ mK, etc. are all in this Hamiltonian. To get a particular piece of physics out of Equation (1), e.g. superfluidity, a magnetically ordered solid, etc. requires having a good idea about the essential motions of the particles permitted by Equation (1) that lead to that physics. Then, in place of Equation (1) one takes a model Hamiltonian that contains a simple representation of these essential motions. Working with the model Hamiltonian one can make a reasonably compact description of a particular aspect of the system. One can always return to the full Hamiltonian, Equation (1), to provide parameters for the elaboration of the models or, as a guide for how to handle the models. Beginning with our discussion here of the phonons we will deal with model Hamiltonians. We do this to stress the importance of abstracting the essential motions from the full Hamiltonian and as a pedagogic device to permit us to exhibit some of the techniques that are useful in dealing with model Hamiltonians. As a consequence of this approach in some respects our ability to exhaustively treat a problem will be limited. We will attempt to point to the source of a complete discussion when this occurs.

There are two important physical ideas that will lead to our picture of a phonon [17, 18]. These are:

(1) Each particle in a lattice sits at a lattice site in a selfconsistent field due to its neighbors. Motions of the neighbors of a particle will induce motions of the particle itself in response to the self-consistent field.

(2) The motions of a particle in the vicinity of its lattice site are manifested in transitions of the particle among the Hartree states available to it.

From the discussion above about short range correlations we know that the particles in the solid can be regarded in leading approximation as Hartree particles interacting through the effective interaction

$$v_e(ij) = t(ij) = v(ij)\, g_{ij}(ij). \tag{30}$$

A particle at lattice site i sees its neighbors through the Hartree field they contribute at $\mathbf{R}_i$, i.e.

$$\mathscr{U}_{\mathbf{R}_i}(\mathbf{x}_i) = \sum_{j \neq (i)} \int \mathrm{d}\mathbf{x}_j \varrho_j(j)\, v_e(\mathbf{r}_{ij}), \tag{31}$$

where $\varrho_j(j) = |\psi_j(j)|^2$ is the single particle density at $\mathbf{R}_j$. The particle at $\mathbf{R}_i$ is described by the Hartree states $\phi^\alpha_{R_i}(\mathbf{x}_i)$ which are the solutions of

$$\left[-\frac{\hbar^2}{2m}\nabla_i^2 + \mathscr{U}_i(i)\right]\phi^\alpha_{R_i}(\mathbf{x}_i) = \varepsilon_\alpha \phi^\alpha_{R_i}(\mathbf{x}_i). \tag{32}$$

The ground state of the system corresponds to each particle occupying its ground state wavefunction $\phi^0_i(i)$. The phonons correspond to collective excitations of the particles among the higher excited states in the Hartree field. To describe the excitations above the ground state we need (to good approximation) only two of the Hartree states [19]. To see this recall that we may choose $\mathscr{U}_i(i) = \mathscr{U}_0 + \gamma(\mathbf{x}_i - \mathbf{R}_i)^2$ for which

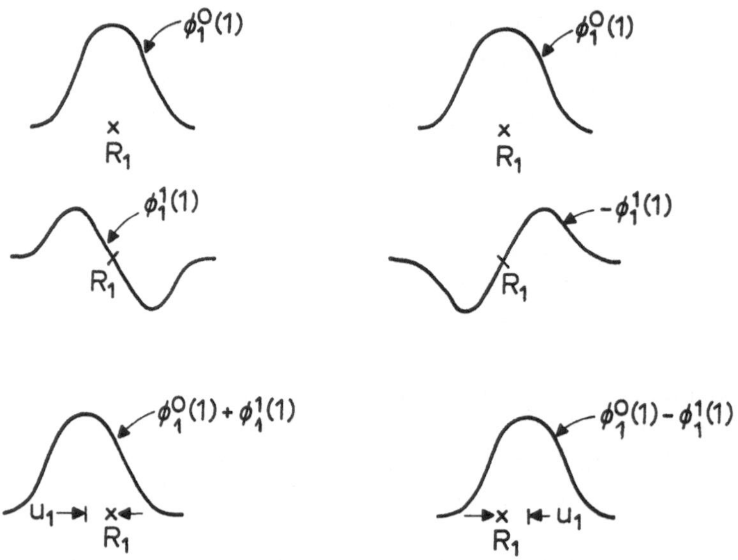

Fig. 9. Displacement. The two low lying states of lattice site R_1 are the Hartree states $\phi_1{}^0(1)$ and $\phi_1{}^1(1)$. A displacement of the particle at this lattice site requires the superposition of ϕ^0 and ϕ^1. The superposition in (a) yields a displacement to the left – that in (b) yields a displacement to the right.

the first few eigenstates are

$$\phi_{R_i}^0(\mathbf{x}_i) = A_0 \exp - \alpha^2 (\mathbf{x}_i - \mathbf{R}_i)^2/2, \quad \varepsilon_0 = \frac{\hbar^2\alpha^2}{2M};$$

$$\phi_{R_i}^1(\mathbf{x}_i) = A_1 \alpha x_i \exp - \alpha^2 (\mathbf{x}_i - \mathbf{R}_i)^2/2, \quad \varepsilon_1 = \frac{3}{2}\frac{\hbar^2\alpha^2}{2M}.$$

The phonons represent displacement motions of the particle about their respective lattice sites. If we construct a wavefunction for a particle of the form

$$\psi_i(i) = c_0\phi_i^0(i) + c_1\phi_i^1(i)$$

we see that the particle is displaced from its equilibrium site by

$$\langle x_i\rangle = \langle\psi_i(i)|\, x_i\, |\psi_i(i)\rangle = 2c_0c_1 \langle\phi_1^1(i)|\, x_i\, |\,\phi_1^0(i)\rangle.$$

In Figure 9 we illustrate this point. More complex motions than the simple displacements that can be represented using ϕ^0 and ϕ^1 are possible using the higher excited states of the Hartree Hamiltonian.

For a description of the phonons we make use of ϕ^0 and ϕ^1. The phonons are coherent transitions $0 \to 1$ which are coupled among the lattice sites by the Hartree field.

For the purpose of discussion the phonons we take a model Hamiltonian of the form

$$\mathscr{H}_P = \sum_R \varepsilon_0(R)\, n(R)_0 + \varepsilon_1(R)\, n(R)_1 - \sum_{RR'} |m(RR')|\, (b_{R1}^+ b_{R0} + b_{R0}^+ b_{R1})\,(b_{R'1}^+ b_{R'0} + b_{R'0}^+ b_{R'1}), \tag{33}$$

where b_{R1}^+ creates a particle in the state 1 at lattice site R, i.e. in ϕ_R^1, b_{R0}^+ creates a particle in the ground state at lattice site R, ε_α is the energy of a particle in the state α

$$\varepsilon_\alpha = \kappa_{\alpha\alpha} + \sum_{j\neq(1)} \int\int \mathrm{d}\mathbf{x}_1\, \mathrm{d}\mathbf{x}_j\, |\phi_j^0(j)|^2\, |\phi_1^\alpha(1)|^2\, v_e(1j), \tag{34}$$

$\kappa_{\alpha\alpha}$ is the α-α matrix element of the kinetic energy, and

$$M(RR') = \int \mathrm{d}\mathbf{x}_R \int \mathrm{d}\mathbf{x}'_{R'}\, \phi_R^0(\mathbf{x}_R)\, \phi_R^1(\mathbf{x}_R)\, v_\varrho(\mathbf{x}_R - \mathbf{x}_{R'})\, \phi_{R'}^0(\mathbf{x}_{R'})\, \phi_{R'}^1(\mathbf{x}_{R'}). \tag{35}$$

The operators that multiply $M(RR')$ permit the processes shown in Figure 10.

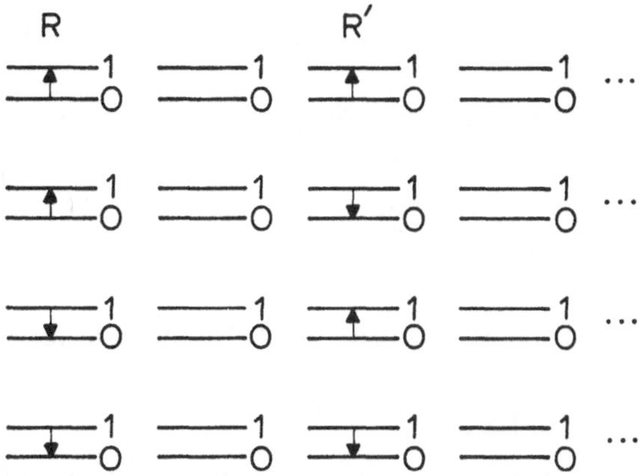

Fig. 10. Lattice excitation. There are 4 possible excitation creation and destruction processes. These include two double excitation processes (a) and (d), and two excitation transfer processes, (b) and (c).

The first two terms in Equation (33) simply measure the energy of the particle at each lattice site according to how it occupies the two states available to it. The last term in Equation (33) represents the 4 possible processes shown in Figure 10 by which an excitation $(b_{R1}^+ b_{R0})$ or a de-excitation $(b_{R0}^+ b_{R1})$ at a given lattice site can induce an excitation of de-excitation at another lattice site.

To find the phonons we treat the Hamiltonian in Equation (33) as follows:

(a) Since, there is one particle at each lattice site we use $n(R)_0 + n(R)_1 = 1$ to write the first two terms as $(\varepsilon_1 - \varepsilon_0)\, n(R)_1$.

(b) The important events, excitation and de-excitation, are represented by $c_R^+ = b_{R1}^+ b_{R0}^+$ and $c_R = b_{R0}^+ b_{R1}$. We also can write $n(R)_1 = c_R^+ c_R$ and note that we have $[c_R, c_{R'}^+] = \delta_{R'R}$. In terms of the excitation annihilation and creation operators we have

$$\mathscr{H}_P = \sum_R (\varepsilon_1 - \varepsilon_0)\, c_R^+ c_R - \sum_{RR'} |M(RR')|\, (c_R^+ + c_R)\, (c_{R'}^+ + c_{R'}). \tag{36}$$

(c) We use the Bloch transformation

$$c_R^+ = \frac{1}{\sqrt{N}} \sum_k \beta_k^+ e^{ikR}, \quad c_R = \frac{1}{\sqrt{N}} \sum_k \beta_k e^{-ikR} \tag{37}$$

to write Equation (36) in the form

$$\mathscr{H}_P = \sum_k (\varepsilon_1 - \varepsilon_0)\, \beta_k^+ \beta_k - \sum_k |M(k)|\, [\beta_k^+ \beta_{-k} + \beta_k^+ \beta_k + \beta_k \beta_k^+ + \beta_k \beta_{-k}],$$

where (38)

$$M(k) = \sum_{R' \neq (R)} |M(RR')|\, e^{ik(R-R')}.$$

Equation (38) is the phonon Hamiltonian in wave vector representation. Let us treat this Hamiltonian in approximation and exactly.

Case I. We drop the double excitation and de-excitation terms, thus

$$\mathscr{H}_P^1(1) = \sum_k [(\varepsilon_1 - \varepsilon_0) - 2|M(k)|]\, \beta_k^+ \beta_k - \sum_k |M(k)|. \tag{39}$$

We have used $[\beta_k, \beta_{k'}^+] = \delta_{kk'}$ to reverse the third $M(k)$ term. We identify the phonon-frequencies with the $\varepsilon_1 - \varepsilon_0 - 2|M(k)|$;

$$\omega(k) = \frac{\varepsilon_1 - \varepsilon_0 - 2|M(k)|}{\hbar}. \tag{40}$$

Now for a one dimensional system with $M(RR')$ non-zero for near neighbors only we have

$$M(k) = 2M(\Delta) \cos k\Delta;$$

$M(\Delta)$ is the amplitude of the near neighbor coupling. We write the phonon dispersion relation in the form

$$\hbar\omega(k) = \varepsilon_1 - \varepsilon_0 - 4|M(\Delta)| + 8|M(\Delta)| \sin^2 \frac{k\Delta}{2}.$$

We note that at $k=0$; $\omega(0) = \varepsilon_1 - \varepsilon_0 - 4|M(\Delta)|$; as $k \to 0$, $\hbar\omega(k) \to \hbar\omega(0) + k^2$. Thus the phonon dispersion relation does not have $\omega(0) \to 0$ and it does not have $\omega(k)$ proportional to k for small k. We can make the dispersion relation touch 0 at $k=0$ by the choice $\varepsilon_1 - \varepsilon_0 = 4|M(\Delta)|$; but we have no freedom to get rid of the parabolic phonon dispersion curve, $\omega(k)$ proportional to k^2 for small k. To get phonons that behave correctly we need to include the double excitation terms.

Case II. We write the full Hamiltonian in the form

$$\mathscr{H}_P = \sum_k \left\{ \left[\frac{(\varepsilon_1 - \varepsilon_0)}{2} - |M(k)| \right] (\beta_k^+ \beta_k + \beta_{-k}^+ \beta_{-k}) - - |M(k)|\, (\beta_k^+ \beta_{-k}^+ + \beta_k \beta_{-k}) \right\} \tag{41}$$

$$\mathscr{H}_p = \sum_k A_k (\beta_k^+ \beta_k + \beta_{-k}^+ \beta_{-k}) + B_k (\beta_k^+ \beta_{-k}^+ + \beta_k \beta_{-k}).$$

We make a transformation of this Hamiltonian as described in Kittel [20]; i.e. we write

$$\alpha_k^+ = u_k\beta_k^+ - v_k\beta_{-k} \tag{42-a}$$

$$\alpha_k = u_k\beta_k - v_k\beta_{-k}^+ \tag{42-b}$$

and attempt to find u_k and v_k (for each k) such that (1) $[\alpha_k, \alpha_k^+]=1$ and (2) $\mathscr{H}_P$ has the form

$$\mathscr{H}_P = \sum \varepsilon(k)\alpha_k^+\alpha_k + \text{const.} \tag{43}$$

The commutation relations imply $u_k^2 - v_k^2 = 1$. The requirement that $\mathscr{H}_P$ has the form given by Equation (43) means that $[\alpha_k^+, \mathscr{H}_P] = -\varepsilon(k)\,\alpha_k^+$. This latter requirement leads to a pair of relation between u_k and v_k; i.e.

$$A_k u_k + B_k v_k = +\,\varepsilon(k)\,u_k \tag{44-a}$$

$$B_k u_k + A_k v_k = -\,\varepsilon(k)\,v_k\,. \tag{44-b}$$

The solution to this pair of equations yields

$$\varepsilon(k) = \sqrt{A_k^2 - B_k^2} \tag{44}$$

or

$$\mathscr{H}_P = \sum_k \sqrt{A_k^2 - B_k^2}\;\alpha_k^+\alpha_k + \text{const.} \tag{46}$$

Now from Equation (46) and Equation (41) using the one dimensional near neighbor model employed above we have ($\Delta\varepsilon = \varepsilon_1 - \varepsilon_0$)

$$\hbar\omega(k) = \tfrac{1}{2}\Delta\varepsilon\sqrt{1 - \frac{8\,|M(\Delta)|}{\Delta\varepsilon} + \frac{16\,|M(\Delta)|}{\Delta\varepsilon}\sin^2\frac{k\Delta}{z}}\,. \tag{47}$$

If we choose $\Delta\varepsilon = 8\,|M(\Delta)|$ we have $\omega(0)=0$ and

$$\hbar\omega(k) = \tfrac{1}{2}8\,|M(\Delta)|\sqrt{2}\sin\frac{k\Delta}{z}\,. \tag{48}$$

Now for a linear chain with near neighbor harmonic interactions of strength γ the dispersion relation is

$$\hbar\omega(k) = 2\sqrt{\frac{\gamma}{m}}\sin\frac{k\Delta}{z} \tag{49}$$

Comparing Equations (48) and (49) permits us to define an effective spring constant $\hbar\sqrt{\gamma_e/m} = 8\,|M(\Delta)|^2$ or

$$\gamma_e = \frac{m\Delta\varepsilon\,|M(\Delta)|}{\hbar^2}\,. \tag{50}$$

Thus we have found phonons in the system to have an effective spring constant $\gamma_e = m\Delta\varepsilon\,|M(\Delta)|/\hbar^2$. What is this spring constant? We can find out be returning to the equations for the ground state and first excited state wavefunctions, for ε_1 and ε_0 and to Equation (35) for $M(\Delta)$. In doing this we use

$$\phi_R^1(x) = A_1 \alpha (x - R) \exp - \frac{\alpha^2}{2}(x - R)^2 = - \frac{1}{\alpha}\frac{A_1}{A_0}\frac{\mathrm{d}}{\mathrm{d}x}\phi_R^0(x)$$

and

$$\phi_R^0(x)\phi_R^1(x) = - \frac{1}{2\alpha}\frac{A_1}{A_0}\frac{\mathrm{d}}{\mathrm{d}x}|\phi_R^0(x)|^2 .$$

Thus we have

$$M(RR') = \frac{A_1^2}{\alpha^2 A_0^2}\int \mathrm{d}x \int \mathrm{d}x' v_e(x - x')\frac{\mathrm{d}}{\mathrm{d}x}|\phi_R^0(x)|^2 \frac{\mathrm{d}}{\mathrm{d}x'}|\phi_{R'}^0(x')|^2 . \tag{51}$$

We integrate by parts on x and x' and we use $\mathrm{d}/\mathrm{d}x' v(x-x') = -\mathrm{d}/\mathrm{d}x v(x-x')$ to obtain

$$M(RR') = - \frac{A_1^2}{\alpha^2 A_0^2}\int \mathrm{d}x \int \mathrm{d}x' \, |\phi_R^0(x)|^2 |\phi_{R'}^0(x')|^2 \frac{\mathrm{d}^2}{\mathrm{d}x^2} v_e(x - x') . \tag{52}$$

Note the instrinsic minus sign; we have already exhibited this explicitly in writing the model Hamiltonian, Equation (35). The normalization constants are related by $A_1^2/A_0^2 = 1$, $\Delta\varepsilon = \hbar^2\alpha^2/m$ so that Equation (50) becomes [19]

$$\gamma_e = \int \mathrm{d}x \int \mathrm{d}x' \, |\phi_R^0(x)|^2 |\phi_{R'}^0(x)|^2 v_e''(x - x') . \tag{53}$$

Now for extremely well localized particles we have $|\phi_R^0(x)|^2 = \delta(x - R)$; in this case Equation (53) reduces to the usual result:

$$\gamma_e^L = v_e''(R - R') \tag{54}$$

in which the second derivative of the interaction is evaluated at the equilibrium sites. When the single particle density is not strongly localized the appropriate spring constant is the second derivative of the interaction averaged over the relative motion of the pair of particles. Using the center of mass and relative coordinate $(\mathbf{R} = \mathbf{x} + \mathbf{x}')/2$ and $\mathbf{r} = \mathbf{x} - \mathbf{x}'$) we may write

$$\gamma_e = \frac{\int \mathrm{d}r \, e^{(-\alpha^2/2)(r - \Delta)^2} v''(r)}{\int \mathrm{d}r \, e^{(-\alpha^2/2)(r - \Delta)^2}} \tag{55}$$

which explicitly exhibits this fact.

The model we have employed yields phonons with a spring constant given by Equation (55). The generalization of the result of this model to three dimensions and to arbitrary neighbor interactions is simple but tedious. The results for these more general cases are straightforward. The application of the generalization to detailed numerical calculations for solid helium yields very satisfactory results. First, the spring constant defined in Equation (55) is positive for solid helium, whereas, the spring constant γ_e^L is negative. Secondly the phonon dispersion curves so generated are in excellent agreement with experiment [21]. The most comprehensive and rigorous discussion of phonons in solid helium is that of Horner [22]; Horner discusses the self-consistent harmonic phonons, anharmonic phonons and the relation of the treatment of the phonons to short range correlations. A recent application by Horner [23] of the very powerful formalism he has developed has succeeded in finding evidence for a small anomalous dispersion in solid ^{3}He. This anomalous dispersion provides an explanation of a low temperature specific heat anomaly [24].*

4. Excitations

We now go beyond the relatively small displacement motions manifested in the phonons and consider the excitations associated with large displacement motions that are present in a quantum crystal. As noted earlier these large displacement motions are a consequence of the large zero point motions. To discuss them we take a model Hamiltonian that describes the rudimentary motions of the particles that are important. These are:

(1) the large displacement motions that permit a particle to move from one lattice site to another;

(2) the hard core repulsion that tends to prevent the double occupation of a lattice site.

We use the Hubbard Hamiltonian

$$M = \sum_{RR'\sigma}^{(nn)} t(RR')\, b^+_{R\sigma} b_{R'\sigma} + U \sum_{R\sigma\sigma'}^{nn} b^+_{R\sigma} b^+_{R\sigma'} b_{R\sigma'} n_R\,, \tag{56}$$

where the operator $b^+_{R\sigma}$; creates a particle in the ground state at lattice site R of spin σ; i.e. in $\phi^0_R(x)$. [This operator would be $b^+_{R0\sigma}$; in our discussion of the phonons. Now, because we are ignoring the small displacements, we no longer need the states created by $b^+_{R1\sigma}$, so we can drop the state label without ambiguity.]

The matrix element for the large displacement motion is

$$t(RR') = K(RR') + V(RR'),$$

where

$$K(RR') = \int \mathrm{d}\mathbf{x}\, \phi^0_R(\mathbf{x})\, KE\, \phi^0_{R'}(\mathbf{x}) \tag{57}$$

is the off-diagonal kinetic energy $(R \neq R')$ and

$$V(RR') = \sum_{R''} \int \mathrm{d}\mathbf{x} \int \mathrm{d}\mathbf{x}'\, \phi^0_R(\mathbf{x})\, \phi^0_{R'}(\mathbf{x})\, v_e(\mathbf{x}-\mathbf{x}')\, |\phi^0_{R''}(\mathbf{x}')|^2 \tag{58}$$

is an off-diagonal matrix element of the potential energy. The hard core energy U is

* **Note added in proof.** But see, Castles and Adams, *Phys. Rev. Letters* **30**, 1125 (1973).

given by

$$U = \int \mathrm{d}\mathbf{x} \int \mathrm{d}\mathbf{x}' \, |\phi_R^0(\mathbf{x})|^2 \, |\phi_R^0(\mathbf{x}')|^2 \, t_{RR}^{00}(\mathbf{x} - \mathbf{x}') \,. \tag{59}$$

[Here we denote the effective interaction that leads to U by a notation that indicates the Hartree states that are involved in its calculation. In a similar notation Equation (29) would be written with $t_{RR'}^{00}(\mathbf{x}-\mathbf{x}')$ in place of $t(ij)$.] The first term in Equation

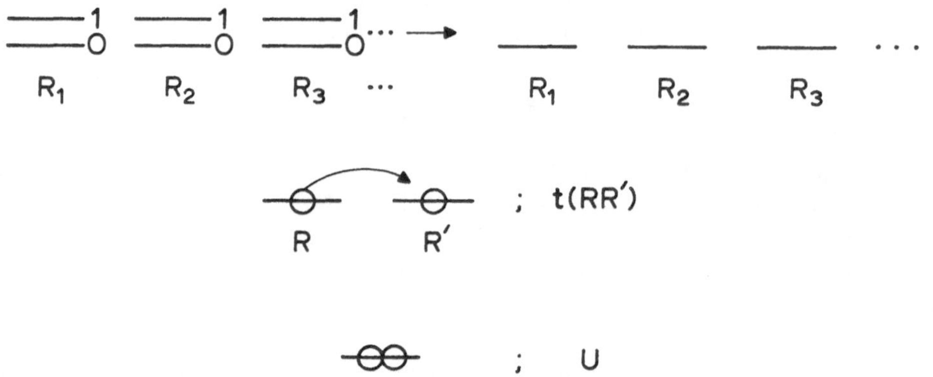

Fig. 11. Model Hamiltonian. To describe the extreme displacement motions we abandon the two state per site model (used for the phonons) and go to one state per site. Particles are transferred from site to site by $t\,(RR')$; a doubly occupied site costs energy U.

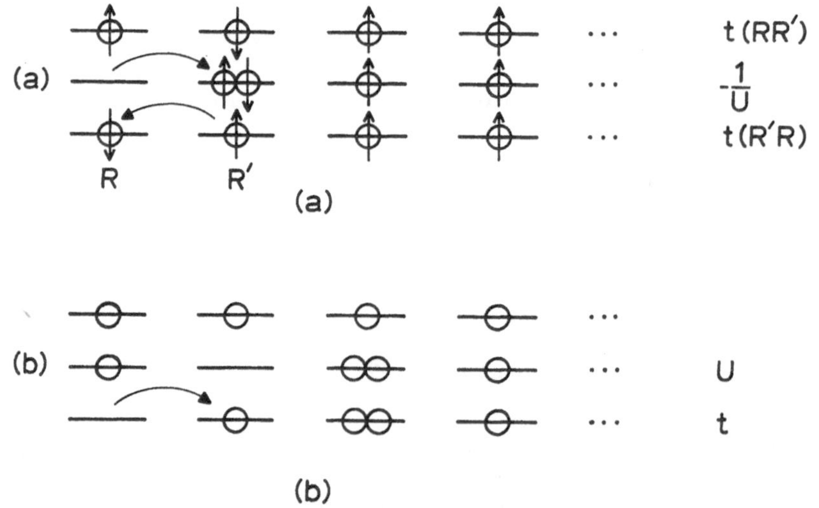

Fig. 12. Excitations. The exchange process is shown in (a). In step 1 the spin up particle at R is transferred to R' ($t(RR')$ is the matrix element) – the intermediate state is lattice site R' doubly occupied (the energy denominator $(E - \mathscr{H}_0)^{-1} = -U^{-1}$) then the spin down particle returns to R. The vacancy creation and motion process is shown in (b). By convention the doubly occupied site is moved to the surface – the creation of this site costs energy U. The hole moves counter to the particle with matrix element $t(RR')$.

(56), involving $t(RR')$, transfers a particle from R' to R with the spin orientation of the particle in tact. The second term in Equation (56), involving U, works to prevent double occupation of a lattice site. See Figure 11.

The large displacement motion processes that we want to discuss are shown in Figure 12. There are two possible motions of the particles described by Equation (56) that are of interest to us, the exchange motions and the motions of vacancies.

The exchange motions correspond to the process shown in Figure 12a and are described by the Hamiltonian

$$\mathscr{H}_X = \sum_{RR'\sigma\sigma'} t(RR') \frac{1}{-U} t(R'R)\, b^+_{R\sigma} b_{R'\sigma} b^+_{R'\sigma'} b_{R\sigma'}. \tag{60}$$

The amplitude for this process, $-t^2/U$, involves $t(RR')$ twice (to transfer the particles from R to R' and back) and $-U$, the energy denominator for the intermediate state. This Hamiltonian, Equation (60), can be put into pseudo spin form upon using the substitutions $b^+_{R\uparrow}\, b_{R\downarrow} = \sigma^+_R$, $b_{R\downarrow}\, b_{R\uparrow} = \sigma^-_R$ and $n(R)_\uparrow - n(R)_\downarrow = \sigma^z_R$. Then the motions corresponding to the exchange process are described by the pseudo-spin Hamiltonian [28]

$$\mathscr{H}_X = -2 \sum_{R<R'} J(RR')\left(\tfrac{1}{2} + \sigma_{-R}\cdot\sigma_{-R'}\right), \tag{61}$$

where $J(RR') = -t(RR')^2/U$. We will return shortly to discuss the physics in this Hamiltonian.

The vacancy creation process (in the model) and the motion of a vacancy through the system are shown in Figure 12b. The Hamiltonian, Equation (56) will describe the motion of the hole, a vacancy, provided that we leave the 'lump' (the doubly occupied site) behind and stationary [29, 30]. This 'lump' is the models representation of the extra structure required in the creation of a vacancy. We put it on the surface of the solid.

The order of magnitude of the basic energies that enter the Hamiltonians, Equations (56) and (61) are

$$t \approx 0.1 \text{ K}$$
$$U \approx 10 \text{ K}$$
$$J = \frac{t^2}{U} \approx 10^{-3} \text{ K}.$$

The energy t^2/U characterizes the exchange process; the energy U is associated with the vacancy creation process; the energy t is associated with the process of vacancy motion. On the phase diagram in Figure 13 we show the various regions of interest and the relation of these regions to the basic parameters. There are five distinct regions listed in Table II. The nature of each region and of the transition between these regions is also listed. The physics of the 3 solid regions, regions I, II and III, will be discussed in detail below. This physics is determined by the excitations present

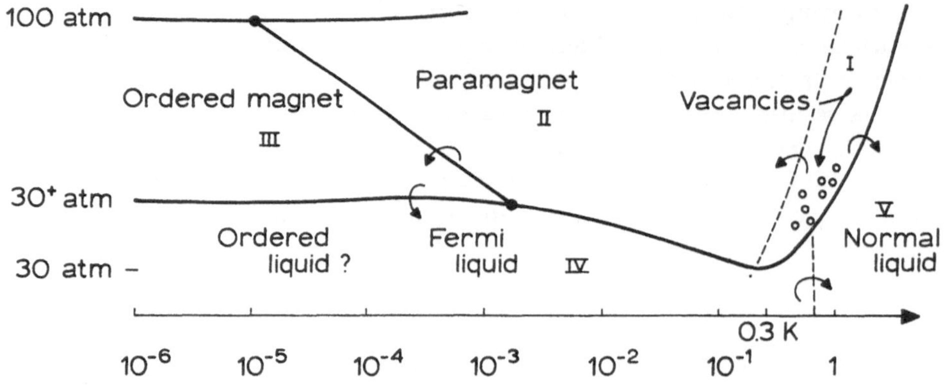

Fig. 13. The phase diagram. The phase diagram of ^3He is very complicated. Depending upon k_BT and the basic energies, t, U and t^2/U, a variety of outcomes are possible. The various regions are listed in Table II along with brief remarks about each.

and by the relationship of k_BT to the characteristic energy for the excitations. We will discuss this physics in the context of a variety of experiments. In particular we will consider thermodynamic evidence for the presence of the exchange process and as an extension of that discussion we will describe Pomeranchuk cooling (cooling by adiabatic compression). We will also sketch the variety of NMR experiments, T_1, T_2 and D, that provide the bulk of our quantitative information about the exchange process and vacancies; i.e. about t, U and J.

4.1. Thermodynamics

For the purpose of doing calculations that examine the thermodynamic consequences of the exchange process we use the pseudo-spin Hamiltonian in Equation (61). In terms of Z the partition function, quantities of interest are

$$
\begin{aligned}
Z &= \mathrm{Tr}_S\, e^{-\beta\mathscr{H}_x} \\
E &= \frac{-\partial}{\partial\beta}\log Z \\
S &= \frac{\partial}{\partial T}(k_BT\log Z) \\
P &= k_BT\frac{\partial}{\partial V}\log Z\,.
\end{aligned}
\tag{62}
$$

We will consider the thermodynamics due to $\mathscr{H}_x$ in region II, $J \ll k_BT < 1/2T_m$. Thus we may use a high temperature expansion of the partition function [31]; e.g.

$$\mathrm{Tr}_S\, e^{-\beta\mathscr{H}_x} = \mathrm{Tr}_S\left(1 - \beta\mathscr{H}_x + \frac{\beta^2}{2}\mathscr{H}_x^2 + \cdots\right).$$

TABLE II

Region	
I	Vacancy dominated solid (paramagnetic solid)
II	Paramagnetic solid
III	Anti-ferromagnetic solid
IV	Fermi liquid
V	Normal liquid
Transition	
I $\to$II	$T\to 0$ disappearance of vacancies, $n_v \sim \exp -\beta U$.
II $\to$III	$\beta t^2/U \leqslant 1$ (magnetic ordering)
III$\to$IV	$t/U \gtrsim 1/10$ (Mott transition induced by lowering P).
IV$\to$V	$\lambda_T > \Delta$ (Fermi gas becomes degenerate)
I $\to$V	(Mott transition induced by $k_B T$.)

Carrying out this procedure we find

$$\frac{E}{N} \propto - J(\beta J) \tag{63}$$

$$\frac{S}{Nk_B} \propto \log 2 - 0(\beta J)^2 \tag{64}$$

$$P \propto \gamma_J \frac{J(\beta J)}{V}. \tag{65}$$

We may understand these results quite simply.

For the energy: as $T \to +\infty$ the average energy per particle is zero. A typical particle has neighbors with spin parallel to itself (energy $+J$) or anti-parallel to itself (energy $-J$) with equal probability; its energy is zero. As $T \to 0$ K a typical particle has all of its neighbors aligned anti-parallel (energy $-zJ$); its energy is $-zJ$. At $\beta J \ll 1$ the situation departs mildly from the case of equally probable parallel and anti-parallel neighbors. The departure is of order $\beta J \ll 1$; thus $E/N - J(n_\uparrow - n_\downarrow) \propto -J(\beta J)$.

For the entropy: At high temperatures each particles has too possible spin orientations ($S \propto \log 2$); at low temperatures each particle is locked into a spin pattern that exists throughout the whole crystal. $S \propto \log 1 = 0$. The degradation of the complete spin disorder that is present at high temperature goes as $(\beta J)^2$; thus $S/Nk_B \approx \log 2 - - 0(\beta J)^2$.

For the pressure: The pressure depends upon the volume derivative of the partition function. The partition function is volume dependent because J, the basic parameter in $\mathscr{H}_x$, is volume dependent. (See later for details). Recall $P = k_B T \partial \log Z/\partial V = - \partial E/\partial V$. Thus the leading temperature-dependence in P is the same as that in E. The parameter γ_J in the pressure equation is

$$\gamma_J = \frac{\partial \log |J|}{\partial \log V}. \tag{66}$$

For solid helium γ_J is positive and of order 20 [32]. This means that J increases as the volume increases. The sign of the pressure is determined by the sign of γ_J. Note J itself appears in Equation (65) squared. At a fixed temperature the energy may be decreased (made more negative) by an increase in the magnitude of J. Since, $\gamma_J > 0$, the system will tend to push outward to increase V and correspondingly J and achieve a reduction in energy. If we had $\gamma_J < 0$ the spin pressure would be negative; the system would be able to lower its energy by increasing its density.

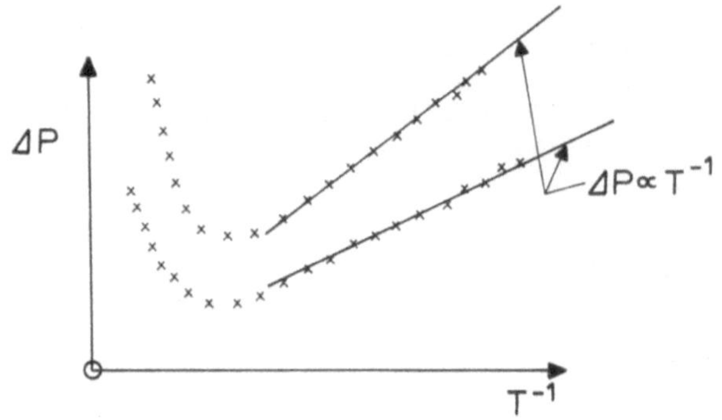

Fig. 14. Excess pressure. Adams and co-workers have measured the excess pressure $P(T) - P(T_F)$, (T_F is a fiducial temperature) and find it to vary as T^{-1} as $T \to 0$. Measurements at several pressures (molar volumes) permit one to extract $|J|$ and γ_J.

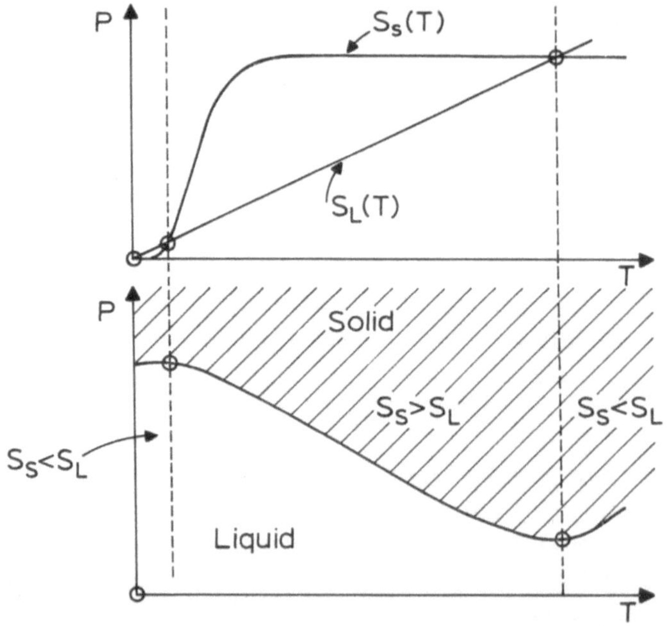

Fig. 15. The ^{3}He melting curve. The melting curve of helium is determined by the relative entropy of the liquid and solid. At $T > 0.3$ K the entropy of the liquid is greater than the solid; $(dP/dT)_m > 0$. In the temperature interval 10^{-3} K $\lesssim T < 0.3$ K the solid entropy is greater than the liquid entropy; $(dP/dT)_m < 0$. At very low temperatures $(dP/dT)_m > 0$ once again.

It is important to remember that the energy, entropy and pressure we are discussing are due to the exchange motion of the particles only. They are the spin energy, spin entropy and spin pressure. The total energy, entropy and pressure have other contributions due for example to the phonons, etc. Recall at $T=0$ K a pressure of 30 atm is required to form a crystal. At $T\approx T_m$ a phonon pressure, $\propto T^4$, is exerted. As $T\to 0$ this pressure decreases as T^4; and the spin pressure becomes more important. All the while an ambient pressure of 30 atm is required to keep the system together.

Adams and co-workers have done a series of experiments using a very sensitive capacitive strain gauge to measure the spin pressure [32, 33]. These experiments are done at several molar volumes as a function of T. Then, using

$$P = 3\gamma_J J\,(\beta J) \tag{67}$$

they are analyzed to determine $J(V)$. Adams and co-workers find $P(T)$ to obey Equation (67) down to quite low temperature. They find the magnitude of J to be in the range 10^{-3} K to 10^{-5} K with $\gamma_J\approx +20$. Thus the thermodynamic of solid ^{3}He in region II is quite closely that of a Heisenberg magnet [34, 35].

From a knowledge of the behavior of the entropy of solid ^{3}He and liquid ^{3}He at low temperatures it is possible to understand the unusual shape of the low temperature liquid-solid melting curve (see Figure 13). Recall the melting curve has $\mathrm{d}P/\mathrm{d}T<0$ at temperatures below 0.3 K. It is this feature that is unusual. The melting curve is determined from the Clausius-Clapeyron equation [36]

$$\left.\frac{\mathrm{d}P}{\mathrm{d}T}\right|_m = \frac{S_L - S_S}{V_L - V_S}, \tag{68}$$

where S_L and S_S are the entropy of liquid and solid respectively and V_L and V_S are the volumes of liquid and solid respectively. Now from experiment we have $V_L > V_S$ and $V_L - V_S$ approximately constant. How do the entropies of the liquid and solid behave? For solid ^{3}He the entropy is $\log 2$ at $\beta J \ll 1$; as $k_B T$ approaches 0 it approaches 0 qualitatively as shown in Figure 15.

For liquid ^{3}He the entropy is given by

$$\frac{S(T)}{Nk_B} = \gamma\,\frac{T}{T_F}, \tag{69}$$

where γ is a constant of order 1. For the ^{3}He Fermi liquid the Fermi temperature $T_F\approx 1$ K so that $S_L(T)/Nk_B$ is of order 1 at $T\approx 1$ K. Below $T\approx 1$ K the entropy of the liquid is less than the solid – the liquid is spin ordered, whereas the solid is spin disordered. The reason for the relative amount of spin order in liquid and solid is simply that in the liquid the single particle wavefunctions extend over the size of its container whereas in the solid the single particle wavefunctions are localized. Each particle in the liquid is aware of each other perticle – consistent with this the Pauli principle forces pairs of particles with the same wavefunction to choose an antiparallel spin alignment. In the solid the particles 'condense' onto lattice sites – there is almost no overlap of single particle wavefunctions from one lattice site to the next – the Pauli principle need not be invoked. Each particle has two degrees of spin freedom available to it. See Figure 16.

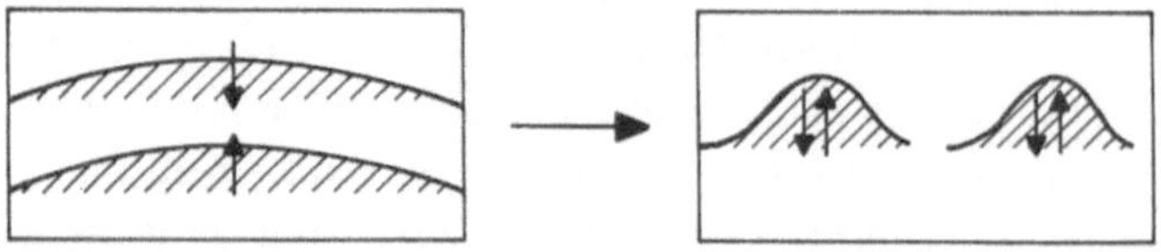

Fig. 16. Liquid and solid entropy. The solid entropy is greater than the liquid entropy because the solid wavefunctions are localized – each particle in the solid has 2 spin degrees of freedom available to it, whereas, the liquid wavefunctions extend over the system – the particles in the liquid must obey the Pauli principle.

Now from Figure 15 we see that at $T > 0.3$ K $S_L > S_S$ and $dP/dT > 0$. The high temperature minimum in the melting curve, at 0.3 K, occurs where $S_L(T) = S_S(T)$. Below 0.3 K, $S_S > S_L$, $dP/dT < 0$ down to $T \approx 0.001$ K. At $T < 10^{-3}$ K, once again one suspects that the solid entropy (that of anti-ferromagnetic spin waves?) becomes less than the liquid entropy.

It is possible to use the unusual shape of the low temperature melting curves of liquid ^3He to build a refrigerator. The basic idea, due to Pomeranchuk, has been demonstrated by Anufriyev [38] and used very successfully in a number of recent experiments by Richardson and co-workers [39] at Cornell and Wheatley and co-workers [40] at La Jolla. We will discuss Pomeranchuk cooling or adiabatic compression in a way that will emphasize its similarity to adiabatic demagnetization. Let us begin with a review of this latter process.

4.2. Adiabatic Demagnetization

Adiabatic demagnetization involves a 4 step process that we consider schematically in Figure 17a and in terms of an entropy diagram in Figure 17b. The steps carried out by an experimentalist are: (1) establish the system (an electron or nuclear paramagnet plus phonon and other excitations; e.g. a CMN salt pill) at T_1 in zero external field; (2) turn the external field to H_0; (3) isolate the system so that from this point onward it is at constant entropy; (4) return the magnetic field to zero. The progress of the system through these steps may be followed on the entropy diagram. By step (3) the system is established at point 3 on Figure 17b in isolation. As the field is returned to zero – the system evolves at constant entropy and ends up at point 4 on the figure. The temperature drops from T_1 to T_2, the temperature appropriate to point 4. The actual implementation of this recipe requires some care: see e.g. Bierlein and Bertman [41]. In Figure 17c we attempt to provide a physical explanation for how this cooling process works. We represent the phonons and other non-magnetic excitations in the system by an oscillator (a mass at the end of a spring) – the paramagnetic spins are represented by a 2 level spin system. At step (1), the energy in the oscillator is $k_B T_1$; the population of the 2 spin states is determined by T_1 and the internal magnetic field.

Step (2) drives the upper spin energy level to an energy $\mu H \gg k_B T$ so that it empties and the population of the lower energy level increases. Since, the system is still attached to the reservoir, the energy lost by raising this energy level must leave the system; it goes into the reservoir. At this point the spins are effectively at $T=0$ K. Now the system is isolated, step (3), and then the field is turned back to zero. As the field returns to zero – the upper spin energy level – inaccessible when the field was at $H=H_0$

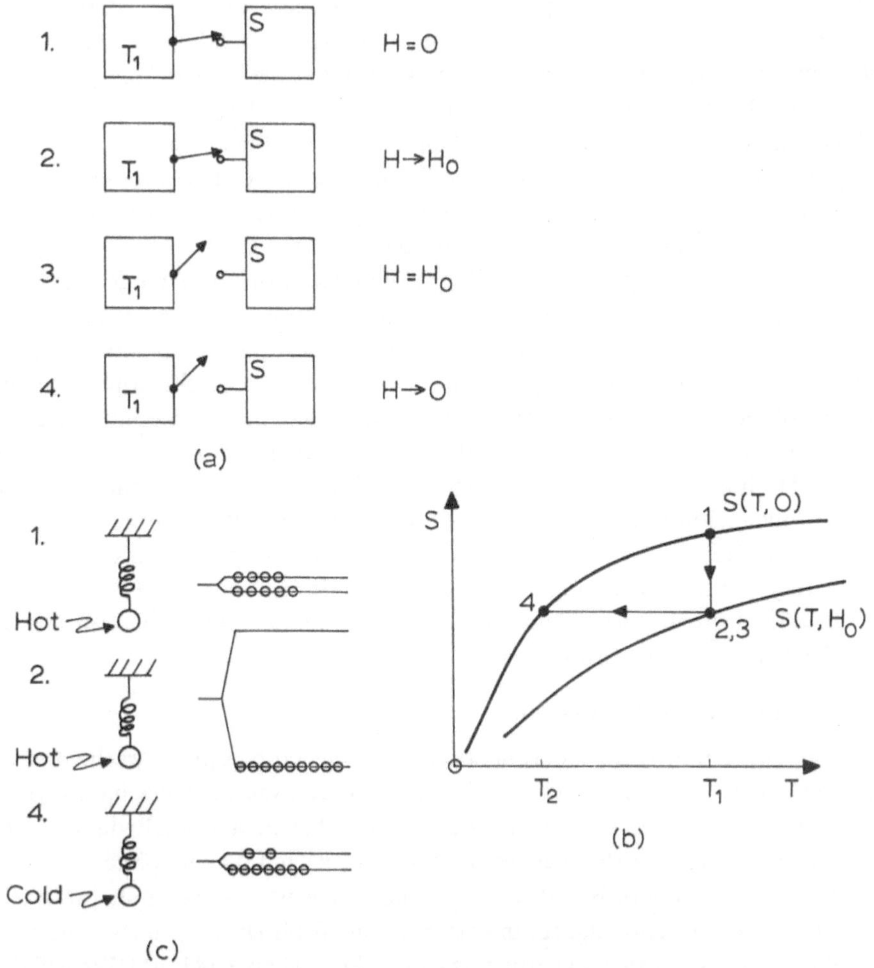

Fig. 17. Adiabatic demagnetization. The steps of an adiabatic demagnetization are: (1) come to equilibrium with a reservoir at T_1 and $H=0$, (2) while in contact with the reservoir turn up H to H_0; (3) break contact with reservoir (from this point onward the process is adiabatic), (4) turn H down from H_0 to 0. This sequence of steps can be followed on an entropy diagram as shown in (b). As the process takes place the energy in the excitations in the system and in the spins change as shown in (c) Step (2), in which $H \to H_0$ raises the upper energy level to such an energy that it must empty. This is the moment when the energy leaves the System. The system is then isolated – when the upper energy level is returned as in (4) the reservoir is no longer available to supply the energy to fill it. The excitations expend energy to fill the upper level and they cool.

– becomes accessible and the oscillator (excitations in the system) expends some of its energy to raise spins from the lower energy level to the upper energy level. The oscillator cools. The final temperature is less than T_1. The basic idea: remove an energy – isolate the system – re-introduce the energy level.

4.3. Adiabatic compression

The entropy diagram in terms of which we discussed the melting curve of ^3He looks qualitatively similar to the entropy diagram we used above to discuss adiabatic demagnetization. We see that pressure – its increase – plays a role analogous to the reduction of H in the latter process. An increase in pressure takes the system from a low entropy state to a high entropy state. We may cool by using the last 3 of the 4 steps above. Establish the system in the low entropy liquid phase at T_1. Isolate the system – increase the pressure. Because the system is isolated during compression it moves at constant entropy and finally ends up at T_2 where it is completely solid. The microscopic mechanism of cooling – analogue of the re-introduction of the upper spin energy level that was removed by the magnetic field – is the 'condensation' of extended fermi liquid wave-functions into localized single particle wavefunctions. This happens as pieces of solid form. The spin degree of freedom is available in the pieces of solid to take up the energy in the excitations in the system. It was not available in the liquid. The implementation of the recipe requires a highly sophisticated technology. Recently Richardson and co-workers at Cornell have used this kind of refrigerator to study the melting curve of ^3He itself; i.e. the refrigerator itself. They have found a number of usual features of the melting curve that they are able to identify with changes in the entropy of both solid ^3He and liquid ^3He at temperatures below about 3 mK [39, 42]. The study of the melting curve solid ^3He and liquid ^3He in the range $T<3$ mK will no doubt be one of the important undertakings of the major low temperature laboratories in the next few years. See Figure 19.

4.4. Nuclear magnetic resonance

Nuclear magnetic resonance experiments can provide a direct probe of the motions of the particles in a system. Thus it is possible that the analysis of NMR data will permit the experimental determination of t, U and J. Further in the NMR data one might discover new phenomena that suggest that an unexpected or new piece of physics is operating. A large body of NMR data on solid ^3He exists; it has done both of these jobs. Here we will describe the rudiments of a pulsed NMR experiment and indicate the kind of information such experiments can make available [43, 44]. We will discuss experiments on solid ^3He in some detail.

An atom of ^3He has an unpaired nuclear spin and thus a nuclear magnetic moment. A piece of solid ^3He is an aggregate of approximately localized nuclear magnetic moments that can be manipulated by magnetic fields. To do a pulsed NMR experiment on a piece of solid ^3He one carries out the following sequence of steps: (We speak here of T_1-measurements; for an exhaustive review of these and other kinds of measurements T_2, D_z and D_E measurements on solid ^3He see Guyer *et al.* [43].)

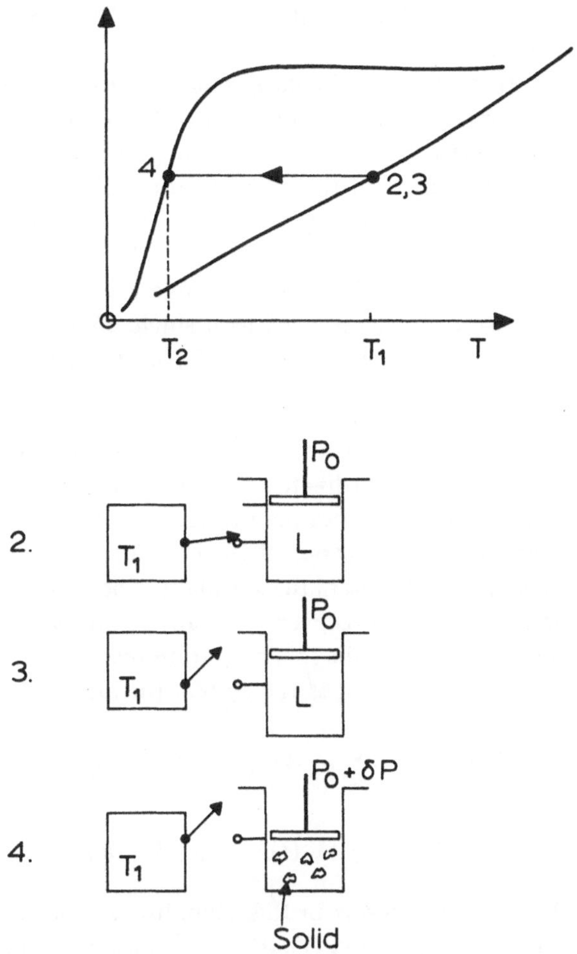

Fig. 18. Pomeranchuk cooling. The process involves starting the liquid in equilibrium at T_1 – isolate it – and compress at constant entropy. The system cools because energy in the excitations goes into the spin degree of freedom that becomes available upon converting liquid to solid.

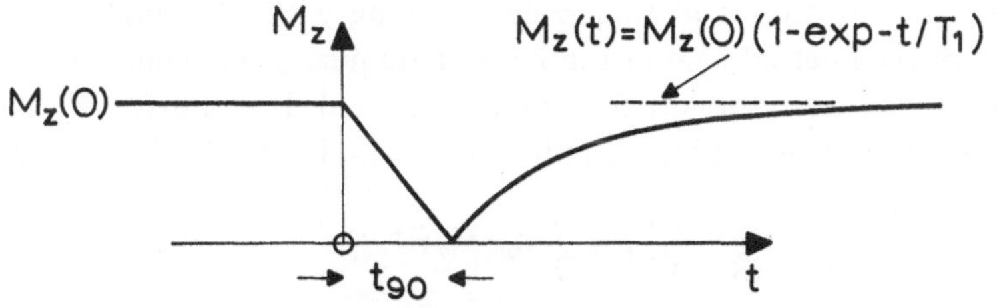

Fig. 19. Definition of T_1. The time T_1 is defined in terms of the asymptotic approach of $M_z(t)$ to $M_z(0)$; $M_z(t) = M_z(0)\,(1 - \exp - t/T_1)$.

(1) Place the sample, at equilibrium at temperature T, in a d.c. magnetic field of strength H_0, $\mathbf{H}=(0, 0, H_0)$.

(2) At time $t=0$, turn on a weak-circularly polarized r.f. field in the x-y plane of strength H_1 at the Larmor frequency $\omega_0=\gamma H_0$; $\mathbf{H}_1=H_1(\cos\omega_0 t, \sin\omega_0 t, 0)$. Turn this field off after the time interval $t_{90}=\pi/2\omega_1$, where $\omega_1=\gamma H_1$.

(3) Observe the time evolution of the magnetization $M_z=\sum_R \mu_R^z$, as time evolves beyond t_{90}. Measure T_1, the rate of recovery of $M_Z(t)$ back to its value at $t\leqslant 0$. See Figure 19 for a precise definition of T_1.

How does the magnetization evolve in time as one goes through the sequence of steps (1)→(3). The magnetization, regarded as a single classical magnetic moment, precesses about the d.c. magnetic field $\mathbf{H}_0$ at the Larmor frequency

$$\omega_0 = \gamma H_0\,, \tag{68}$$

where $\gamma\approx 2\times 10^4$ rad g^{-1}. At time $t=0$ an r.f. magnetic field in the plane perpendicular to $\mathbf{H}_0$ is turned on. Viewed in a coordinate system rotating with M_z at frequency ω_0 the magnetization appears static as does the r.f. magnetic field. In the rotating coordinate system, the magnetization precesses about H_1 – after time interval $t_{90}=\pi/2\omega_1$ (or $\omega_1 t_{90}=\pi/2$) the magnetization lies in the x-y plane. Then the r.f. field is turned off and one watches the magnetization recover to M_z (0) as time evolves.

What is the mechanism by which the magnetization recovers to its $t\leqslant 0$ valve? Let us look at a Hamiltonian that describes the system (including the magnetic fields). We have

$$\mathscr{H} = \mathscr{H}_\mathrm{L} - \sum_R \boldsymbol{\mu}_R\cdot\mathbf{H}_0 - \sum_R \boldsymbol{\mu}_R\cdot\mathbf{H}_1(t) + \sum_{RR'} \boldsymbol{\mu}_R\cdot\boldsymbol{\mu}_{R'} C(RR') \tag{69}$$

or

$$\mathscr{H} = \mathscr{H}_\mathrm{L} - \mathbf{M}_z\cdot\mathbf{H}_0 - \mathbf{M}_z\cdot\mathbf{H}_1(t) + \sum_R \boldsymbol{\mu}_R\cdot\mathbf{H}_\mathrm{d}(R), \tag{70}$$

where $C(RR')\propto|\mathbf{R}-\mathbf{R}'|^3$. The terms in this Hamiltonian are: $\mathscr{H}_\mathrm{L}$ the particle or lattice Hamiltonian given by Equation (1), $\mathscr{H}_Z=-\sum_R \boldsymbol{\mu}_R\cdot\mathbf{H}_0$ is the Zeeman Hamiltonian that describes the precession of M_z in the presence of $\mathbf{H}_0$, the third term describes the coupling of the magnetization to the r.f. field, the last term describes the internal coupling between the magnetic moments of the solid sample – it is the nuclear dipolar field – its strength is about 1 G or 10^{-7} K. The recovery of the magnetization back to the $t\leqslant 0$ value is due to the internal magnetic field on each particle – the dipolar field. The r.f. field was particularly effective in driving M_z to zero because it was tuned by the experimenter to precess at frequency ω_0 along with M_z. Similarly, the Fourier component of the internal field at the Larmor frequency will couple strongly to the magnetization and undo the job done by the r.f. field. It will drive M_z back toward its t$\leqslant$0 value. What does the internal dipolar field look like? We have (schematically)

$$\mathscr{H}_\mathrm{d} \approx \sum_{RR'} \boldsymbol{\mu}_R\cdot\frac{\boldsymbol{\mu}_{R'}(t)}{|\mathbf{R}-\mathbf{R}'|^3} = \sum_R \boldsymbol{\mu}_R\cdot\mathbf{H}_\mathrm{d}(\mathbf{R}, t), \tag{71}$$

where we have written $\boldsymbol{\mu}_{R'}(t)$ to indicate that because of excitations of the kind we are considering the magnetic moment of the particle at lattice site R' can fluctuate in time

– the local internal field at R, $H_d(R, t)$, fluctuates in time. The ω_0 Fourier component of the local field fluctuations at R causes the recovery of the magnetization; the recovery of the magnetization is due to

$$\mathbf{H}_d(\mathbf{R}, \omega_0) = \int dt e^{-i\omega_0 t} \mathbf{H}_d(\mathbf{R}, t). \tag{72}$$

This Fourier component is due to the motions of the particles (carrying their magnetic moments) that are the neighbors of **R**. Thus in observing the recovery of M_z back to $M_z(0)$ one is seeing evidence for the motions of these particles. These motions are due to the excitations in the system. See Figure 20.

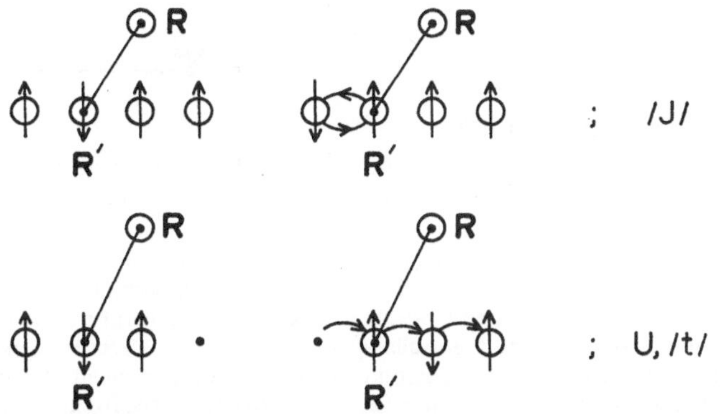

Fig. 20. Relaxation processes. Relaxation of the magnetization is due to the ω_0 Fourier component of the local dipolar field – the time evolution of the local dipolar field is due to the exchange process and to the presence of vacancies in the system.

We will not work through the details of a calculation of T_1 for solid ^{3}He. They can be found in the paper by Guyer *et al.* [43] or in the literature that paper uses. Instead we will exhibit the result of a T_1 experiment, make an explanation of the qualitative behavior of the experiment and indicate how the exchange process and vacancy motion are quantified by an analysis of the data. In Figure 21 we show the results of a T_1 experiment on solid ^{3}He at $P \approx 100$ atm; we plot T_1 vs T^{-1}.
We have divided the temperature range into regions; we will discuss region I and II; region III requires a relatively exotic explanation that we will not be able to devote space to. In region I the relaxation of the magnetization is due to particle motions that are possible because of the presence of vacancies in the system. A particle, e.g. at R' in Figure 20, will move due to the presence of vacancies at a rate

$$\omega_v = \frac{|t|}{\hbar} n_v, \tag{73}$$

where n_v, the vacancy concentration, is given by

$$\frac{n_v}{N} = \exp - \beta U. \tag{74}$$

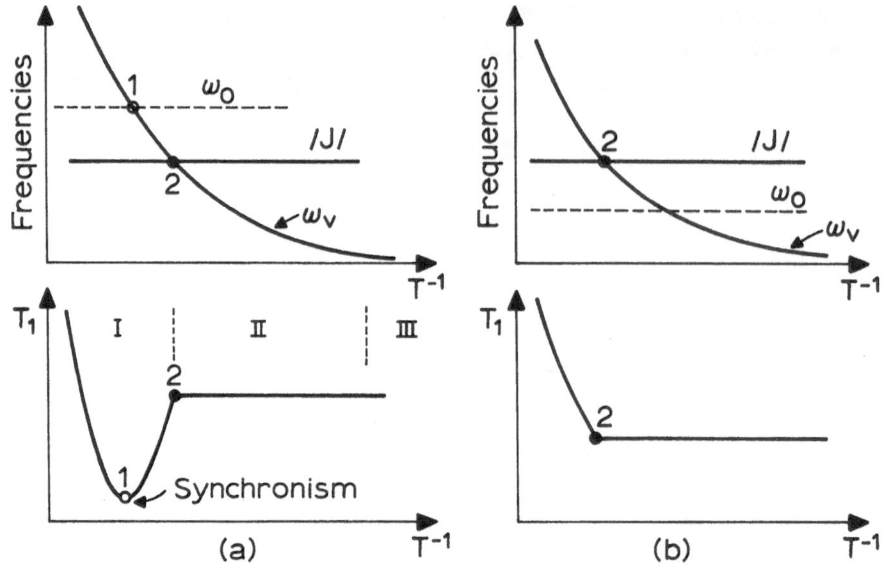

Fig. 21. T_1 as a function of T^{-1}. The value of T_1 depends upon the characteristic frequency for motion among the particles and on the Larmor frequency, ω_0. At temperatures above 2 the important particle motions are due to the presence of vacancies – below the temperature corresponding to 2 the important particle motions are due to the exchange process. T_1 is temperature dependent down to 2 and temperature independent thereafter. The outcome of a T_1 measurement also depends upon the comparison of ω_0 with the frequence of motion. See (a) and (b). In (a) the frequency ω_0 equals ω_v at 1; the minimum relaxation time occurs at the temperature corresponding to 1.

In Equation (73) n_v is the probability that a neighbor of the particles at R' is a vacancy; $|t|$ (see Equation (56)) is the rate at which the particle tunnels into the vacant neighboring lattice site. In terms of the characteristic frequency of motion of the particles that contribute to the fluctuations in the local dipolar field the relaxation time is given by

$$\frac{1}{T_1} \sim \frac{\omega_d^2}{\omega_v} = \frac{(\hbar\omega_d)^2}{|t|} \exp \beta U . \tag{75}$$

Here ω_d, the dipolar frequency, is related to the average strength of the dipolar coupling by

$$\hbar\omega_d \approx \frac{|\mu_N|^2}{\Delta^3}, \tag{76}$$

where Δ is the near neighbor distance. Since $\hbar\omega_d \approx 1\ \mathrm{G} \approx 10^{-7}$ K; we have $\omega_d \approx 10^4$ s^{-1}. From Equation (75) we see that the temperature dependence of T_1^{-1} is the temperature dependence of the vacancy excitation. The magnitude of T_1^{-1} depends upon $|t|$. On Figure 21 we show the rate ω_v vs T^{-1}. In region II the relaxation of the magnetization is due to the exchange process. This process takes place at the rate $|J|$. In analogy with Equation (75) we may write

$$\frac{1}{T_1} \sim \frac{(\hbar\omega_d)^2}{|J|}\,. \tag{77}$$

On Figure 21 we show the rate $|J|$ vs T^{-1}. The outcome of the experiment shown on the T_1 vs T^{-1} plot is understood in terms of the rates indicated on Figure 21. In region I the rate of particle motion due to vacancies is greater than that due to the exchange process – thus the relaxation time in region I is temperature dependent with the temperature dependence of the vacancy concentration. At point 1 the rate of vacancy motion is synchronous with the Larmor frequency of the magneitc moments $\omega_0 \approx \omega_v$. The energy transfer process occurs most rapidly and T_1 is a minimum. At temperatures above and below T_{min} the value of T_1 is greater because the spins and particle motions are out of synchronism. At 2 the rate of motion of particles due to the exchange process (it is constant at $|J|$ independent of T) becomes greater than the rate of motion due to the vacancies. The number of vacancies has become too few. Then T_1 becomes temperature independent at the value given by Equation (77). From the analysis of data in this region $|J|$ may be found.

This discussion has been overly simplified to indicate the essential result – a detailed discussion is found in Guyer *et al.* [43].

The results of the analysis of a large body of NMR data – similar to that we have discussed here – yields values of t, U and J that we show on the phase diagram in Figure 22.

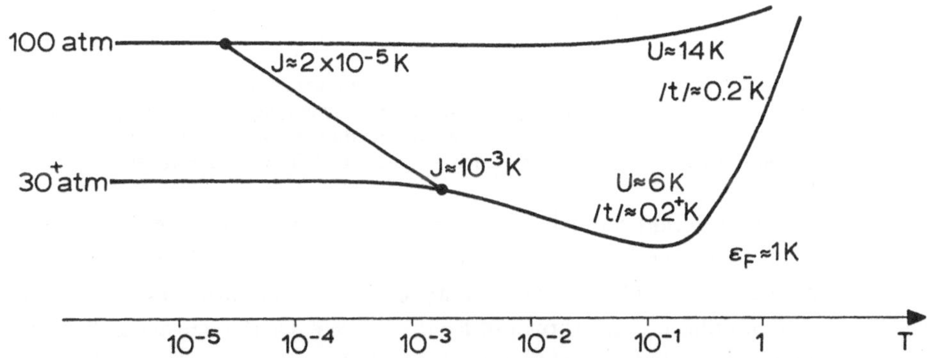

Fig. 22. The Parameters t, U and J. On a phase diagram we show the values of t, U and J determined by NMR experiments.

There is quite good agreement between experimental determination of $|J|$ by thermodynamic methods and by NMR methods. There is relatively little good thermodynamic data on the vacancies. What data there is is consistent with the NMR determination of U and $|t|$.

4.5. Theory of t and J

Above we have described a variety of experiments in which one sees evidence for the exchange process and the presence of vacancies. Here we will briefly sketch the elements of a theory of the basic parameters that characterize these excitations – i.e. the parameters t and J that are appropriate to the Hamiltonians, Equation (56) and Equation (61). We will not discuss the theory of U; see for example References [29] and [30]. The theory of J is a theory of the exchange interaction. A rather extensive literature is addressed to this question in electron systems and in solid ^{3}He [45]. We will not attempt to review this literature in detail. We will take from it the important ideas that must be discussed [46]. Let us consider the exchance process and how it might be described in terms of a configuration space in which we can represent the state of an N-particle system. We will begin with a simple example – a one dimensional two particle system involving the potential shown in Figure 23.

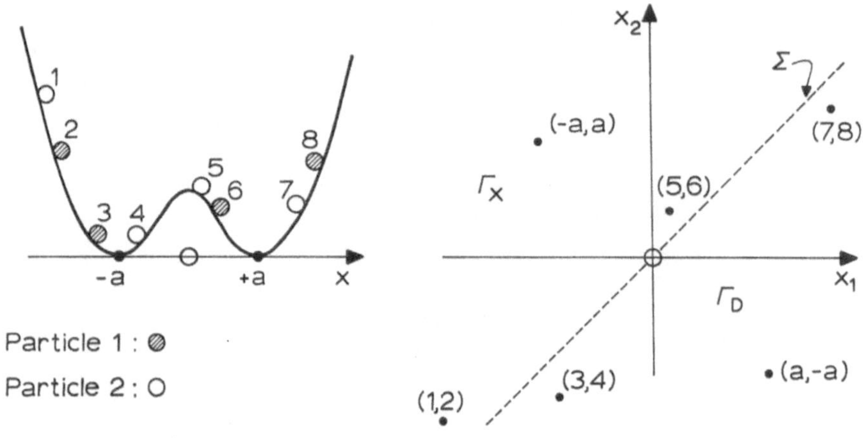

Fig. 23. Configuration space. For a pair of particles in a one-dimensional double well the possible relative orientations of the particles can be displayed on a simple two dimensional configuration space. The pairs of particle (1,2), (3,4), (5,6), (7,8) are shown on the diagram in their appropriate location. When the particles pass Σ – they exchange order – 1 is to the left of 2 → 1 is to the right of 2.

We assume there is a mild repulsive short range potential between the pair of particles. The ground state wave functions – in leading approximation they are degenerate – are $\phi_{-a}(1)\,\phi_{+a}(2)$ and $\phi_{+a}(1)\,\phi_{-a}(2)$. We may represent configurations of the system in the 2 dimensional phase space shown in Figure 22. We use the density in this phase space to represent the probability of finding the particles in a particular configuration. Because of the repulsive short range potential interaction between the wavefunctions for the pair is large in the regions of configuration space near $(a, -a)$ and $(-a, a)$. Below the dotted line, $x_1 > x_2$, particle 1 is to the right of particle 2. We call this region of the configuration space the *direct* region; denote it by Γ_D.Above the dotted line, $x_2 > x_1$, particle 1 is to the left of particle 2. We call this region of configuration space the *exchange* region; denote it by Γ_x. Several examples of direct and exchange configurations are shown in Figure 23. The pair of particles in the potential $U(x)$ can be said to exchange place when they cross the dotted line which divides the direct space from the exchange space. We can show that the rate at which the pair of particles crosses the surface between Γ_D and Γ_x is given by $J/\hbar$ where

$$J = \frac{\hbar^2}{m} \int_{\Sigma} d\mathbf{S} \cdot \Phi_x \nabla \Phi_D \tag{78}$$

and ϕ_D is the wavefunction for the pair of particles in the direct space ($\phi_D \approx \phi_{-a}(1) \times \times \phi_{-a}(2)$), Φ_x is the wavefunction for the pair of particles in the exchange space ($\Phi_x \approx \phi_{-a}(1)\ \phi_{+a}(2)$), Σ denotes the surface that divides the two spaces (in this case the dotted line). The result in Equation (78) is suggestive of the formula for the probability current – although it is not derived from that formula. For the more complex problem of the exchange of a pair of particles that are part of an N-particle system a straight forward generalization of Equation (78) is valid. Of course to achieve this generalization one must get involved in a discussion of the $3N$-dimensional configuration required to describe the system, Γ_D, Γ_x and Σ must be carefully defined for a multiple dimensional geometry etc. All of this can be done in exact analogy to our discussion above. One finds

$$J = \frac{\hbar^2}{m} \int d\mathbf{\Sigma} \cdot \psi_{12}(12; 3 \cdot N)\, \nabla \psi_{12}(21; 3 \cdot N), \tag{79}$$

where $\psi_{12}(12; 3 \ldots N)$ is the many body wavefunction for the particle arrangement 1 near $\mathbf{R}_1$, 2 near $\mathbf{R}_2, \ldots$ and $\psi_{12}(21; 3 \ldots N)$ is the many body wavefunction for the particle arrangement 2 near $\mathbf{R}_1$, 1 near $\mathbf{R}_2, \ldots$.

To see what is in Equation (79) consider the pair approximation to the many body wavefunction for which we have

$$\psi_{12}(12) = \phi_1(1)\, \phi_2(2)\, g(r_{12}),$$
$$\psi_{12}(21) = \phi_1(2)\, \phi_2(1)\, g(r_{12}),$$

where $\phi_1(1) = A_0 \exp -(\alpha^2/2)\ (\mathbf{x}_1 - \mathbf{R}_1)^2$. Upon substituting into Equation (79) and carrying out the manipulations we obtain

$$J = -\frac{\hbar^2 \alpha^2 \Delta}{m} \int ds \psi_{12}(12)\, \psi_{21}(21). \tag{80}$$

This is essentially the overlap integral between the wavefunctions representing the two configurations. Let us look at it in some detail. In the integrand we have

$$I(R) \equiv \phi_1(1)\, \phi_2(2)\, g(r_{12})\, \phi_2(1)\, \phi_1(2)\, g(r_{12}) = \\ = \phi_1(1)\, \phi_2(1)\, \phi_2(2)\, \phi_1(2)\, g(r_{12})^2 .$$

For this integrand to be large we need $\phi_1(1)$ to overlap $\phi_2(1)$, $\phi_2(2)$ to overlap $\phi_1(2)$ and $|\mathbf{x}_1 - \mathbf{x}_2|$ to be of order σ or greater (due to the factor $g(r_{12})^2$). In Figure 24 we show that these three conditions are filled at a point along the line bisecting a pair of lattice sites; i.e. at $|\mathbf{x}_1| = |\mathbf{x}_2| = l = \alpha\sqrt{\Delta^2 + \sigma^2}/4$. Thus the order of magnitude of the

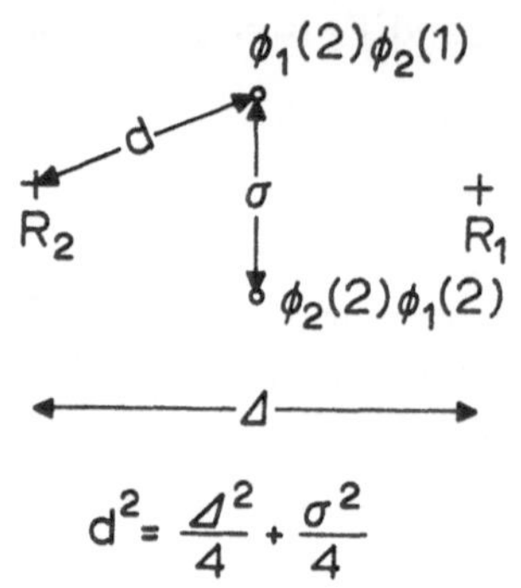

Fig. 24. Exchange process. The overlap integral is a maximum along the line that bisects R_1 and R_2 when the relative separation of the pair of particles is about σ.

surface integral is

$$I(12) = \phi_1(l)\,\phi_2(l)\,\phi_2(l)\,\phi_1(l) = \left(\exp -\frac{\alpha^2}{2}\frac{\Delta^2+\sigma^2}{4}\right)^4 =$$

$$= \exp -\frac{\alpha^2}{2}(\Delta^2+\sigma^2).$$

The order of magnitude of J is given by

$$J \approx -\frac{\hbar^2\alpha^2}{m}\exp -\frac{\alpha^2}{z}(\Delta^2+\sigma^2). \tag{81}$$

Numerical calculations of J using formulae like Equation (80) or the exact expression, Equation (79) with many body wavefunctions, lead to theoretical values of J that are of the same order of magnitude as experiment. There continue to be a number of details that must be taken into account to bring about excellent agreement. Although these certainly need to be attended to there is little doubt that the essential physics has been identified and incorporated into the above description of the calculation.

Formulae and results very similar to those described here for J also apply to the calculation of $|t|$ and to the comparison of those calculations with experiment.

Thus the many body wavefunctions we worked with in the early part of these lectures can be successfully used to calculate the parameters that characterize the excitations that we are describing by the model Hamiltonians. The model Hamiltonians can be provided with parameters from experiment or theory.

4.6. Vacancy Waves

The second of the excitations we described above was the vacancy excitation [29, 30]. As was the case with the exchange process, these excitations appear in the thermodynamics, in NMR experiments (see above) etc. They are characterized by the energy required to create a static vacancy, U, and the energy associated with their motion. It is conventional to place the 'lump' associated with the model vacancy on the surface

Then the hole that is left behind can be at any lattice site in the bulk of the crystal. We take the vacancy to be at R and designate that state by

$$\psi_R = b_R |0\rangle . \tag{82}$$

This state is N fold degenerate, i.e. ψ_{R_1} has the same energy as ψ_{R_2}, for $|t|=0$. Since, $|t|\neq 0$ the N-fold degeneracy of the ground state is split by vacancy tunneling into a spectrum of vacancy waves. We construct a vacancy wave as a superposition of the ψ_R

$$\psi_k = \frac{1}{\sqrt{N}} \sum_R e^{i\mathbf{k}\cdot\mathbf{R}} \psi_R = \frac{1}{\sqrt{N}} \sum_R e^{i\mathbf{k}\cdot\mathbf{R}} b_R |0\rangle .$$

The energy of the state ψ_k is calculated from the model Hamiltonian, Equation (56), and is

$$\varepsilon(k) = U + 2|t| \cos k\Delta$$

for a 1-dimensional system with a near neighbor t. In Figure 25 we show this dispersion relation.

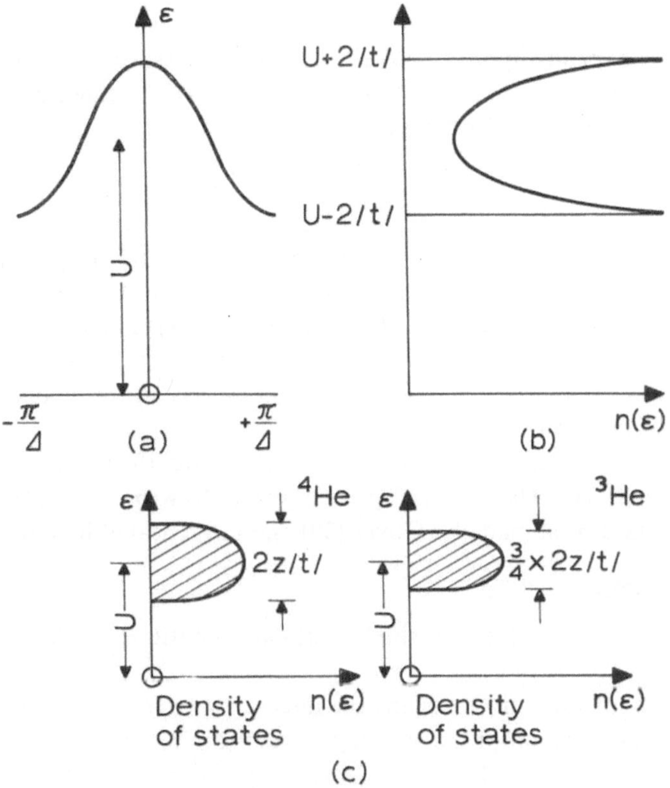

Fig. 25. Vacancy waves. The dispersion relation for vacancy waves in one dimension is shown in (a). It spans the interval $U \pm 2|t|$. The corresponding density of states is shown in (b). For solid ^{3}He and ^{4}He (in 3-dimensions) the density of states behaves as shown in (c).

It is an inverted single particle dispersion relation – the maximum energy occurs at $k=0$, the minimum energy at the Brillouin zone boundary. A simple argument will explain this feature of the spectrum: the creation of a vacancy wave corresponds to removal of a density fluctuation. A short wavelength density fluctuation is more energetic than a long wavelength density fluctuation. Thus the creation of a short wavelength vacancy corresponds to removal of more energy than the creation of a long wavelength vacancy. The qualitative behavior of the vacancy wave dispersion relation in a 3-dimensional system is the same as shown above. The details differ according to the crystallographic structure. In Figure 25 we also show schematically the essential features of the 3-dimensional dispersion relation. The spectrum of vacancy wave energies spans the interval $U-z|t|$ to $U+z|t|$, where z is the number of near neighbors. An experiment that measures the thermal activation measures $U-z|t|$.

The thermodynamics due to the vacancy waves is calculated according to the prescription.

$$E(T, V) = \sum_k \varepsilon(k)\, e^{-\beta\varepsilon(k)}$$

$$P(T, V) = -\frac{\partial}{\partial V} E(T, V).$$

Since $|t|$ and U depends upon V, there is an excess pressure due to the presence of vacancies. Since the vacancies are thermally activated their concentration decreases with decreasing temperature according to

$$x_V \sim \exp - \beta(U - z|t|).$$

The thermodynamic measurements on solid helium at temperatures near the melting temperature find evidence for an anomalous contribution to the specific heat and to the pressure [47]. These experiments see the presence of vacancies against a background of phonons that are not well understood. Thus the thermodynamic data on these excitations is of qualitative use only. The bulk of the useful data on these excitations is the NMR data discussed earlier – even so that data is scarce compared to similar data on the exchange process.

Particular interest in the vacancy waves is derived from the comparison of vacancy waves in ^{3}He and ^{4}He. The essential points, which draw on the work of Brinkman and Rice [48], have been made by Guyer [30] and are repeated here for completness.

4.6.1. *Vacancy Waves in* ^{4}He

(a) the vacancy wave in ^{4}He is a single particle excitation: a relation of the form $\omega=\omega(k)$ exists.

(b) the density of state in ^{4}He spans the interval $U\pm z|t|$ as shown in Figure 25.

(c) the diffusion constant of the vacancy waves is limited by vacancy-vacancy scattering;

$$D_v(4) \simeq \bar{v}\lambda_{vv},$$

where $\bar{v}=\Delta|t|/\hbar$.

4.6.2. *Vacancy Waves in* ^{3}He

(a) the vacancy waves in ^{3}He are many body excitations; a relation of the form $\omega=\omega(k)$ does not exist.

(b) the density of states spans an interval approximately $U\pm\frac{3}{4}z|t|$ as shown in Figure 25.

(c) the diffusion constant of the vacancy waves is limited by spin fluctuation scattering:

$$D_V(3)\simeq\bar{v}\Delta$$

where $\bar{v}=\Delta|t|/\hbar$.

The particularly dramatic difference between the vacancy wave as a single particle excitation in ^{4}He and as a many body excitation in ^{3}He appears in the diffusion constant or mobility. A direct measure of this difference is possible in a light scattering experiment. We consider light scattering from the hydrodynamic modes of an aggregate of vacancy waves [49].

Let us begin with a brief description of a light scattering experiment. See Figure 26. Light characterized by $(\mathbf{k}_0, \omega_0)$ is incident on the sample from the left. A detector is placed at angle θ where it measures the amount of light scattered at frequency ω at wave vector $\mathbf{k}$. (Since the light energy is large compared to the intrinsic energies of motion in the system we have (to good approximation) $|\mathbf{k}|=|k_0|=k_0$ and $\mathbf{K}=\mathbf{k}-\mathbf{k}_0$

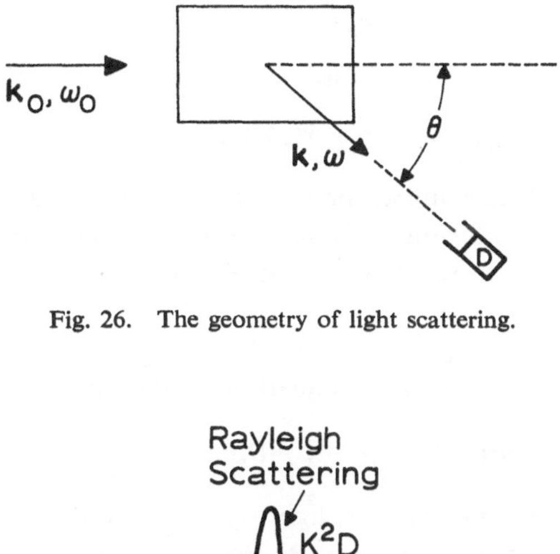

Fig. 26. The geometry of light scattering.

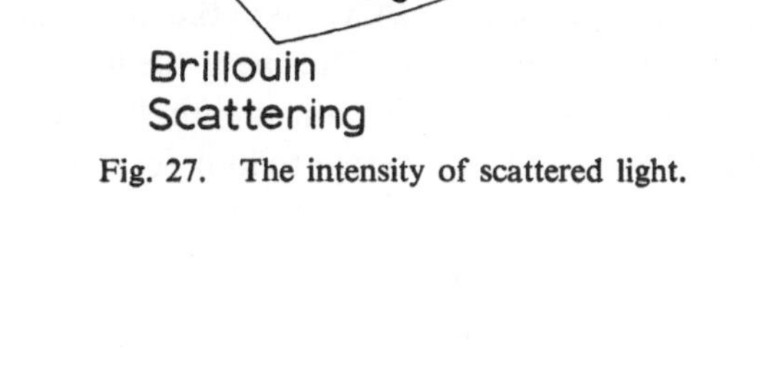

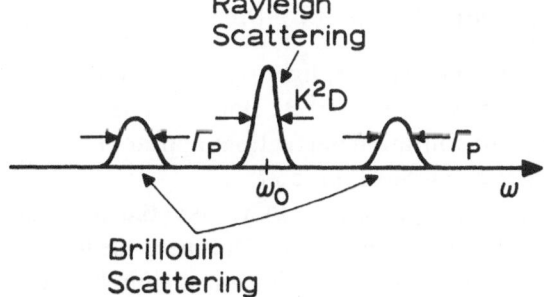

Fig. 27. The intensity of scattered light.

$\sin\theta/2$; **K** is called the momentum transfer.) A typical profile of light intensity scattered at ω is shown in Figure 27.

This scattered intensity involves frequency shifts of zero $\omega=\omega_0$ and of values small compared to ω_0. Light is scattered from the sample because of the density fluctuations in it; the frequency profile of the scattered light provides information about the time evolution of the density fluctuations. For a system containing vacancies there are 3 sources of density fluctuations, pressure, temperature and vacancy concentration. Pressure fluctuations are accompanied by density fluctuations through $\partial\varrho/\partial P$; temperature fluctuations are accompanied by density fluctuations through $\partial\varrho/\partial T$; a fluctuation in the concentrations of vacancies corresponds to a density fluctuation, $N(\mathbf{x})=N(1-x_v(\mathbf{x}))$. Density fluctuations due to pressure fluctuations have amplitude [50]

$$(\delta\varrho^2)_P = \varrho_0^2 \frac{k_B T}{K_T v}$$

and decay away by the propogation of a mildly damped sound wave, i.e.

$$\delta\varrho_P(\mathbf{x}, t) = \partial\varrho\,(\mathrm{o}\cdot\mathrm{o})\exp(\mathbf{K}\cdot\mathbf{x} - \omega(k)\,t)\,e^{-\Gamma(k)t}.$$

Light scattered from them is Doppler shifted by $\omega(k)$ and has width $\Gamma(k)$ centered at $\omega(k)$. See the Brillouin components in Figure 27. Density fluctuations due to temperature fluctuations have amplitude

$$(\delta\varrho^2)_S = \varrho_0^2 \frac{k_B T}{K_T N}\cdot\left(1 - \frac{c_v}{c_p}\right)$$

and decay away according to the diffusion equation for temeprature:

$$\delta\varrho_T(\mathbf{x}, t) \sim \delta\varrho\,(\mathrm{o}, \mathrm{o})\exp - x^2/4D_T t.$$

Light scattered from them is not Doppler shifted but centered at $\omega=\omega_0$ with a width $\Gamma\sim K^2 D_T$. The thermal diffusion constant is proportional to the thermal conductivity. Density fluctuations due to vacancy concentration fluctuations have amplitude

$$(\delta\varrho^2)_v \sim \varrho_0^2 x_v$$

and decay according to the diffusion equation for concentration:

$$\delta\varrho_v(\mathbf{x}, t) = \delta\varrho\,(\mathrm{o}, \mathrm{o})\exp - x^2/4D_v t.$$

As with the temperature fluctuations, light scattered from these fluctuations appears as an unshifted line with width $\Gamma_v\sim K^2 D_v$. Both scattering from the thermal fluctuations and the vacancy concentration fluctuations appear in the Rayleigh component of the light scattering spectrum. See Figure 27.

At temperatures near T_m(e.g. 1.8 K in BCC ^{4}He) the amplitude of light scattering from $(\delta\varrho^2)_v$ is greater than that from $(\delta\varrho^2)_s$. Thus near the melting curve the Rayleigh component of the scattered light profile is almost entirely due to fluctuations in the concentration of vacancy waves. An experiment to look for light scattering is presently underway at the University of Massachusetts. That experiment will attempt to look at ^{3}He and ^{4}He at the optimium temperature and pressure that will produce large amplitude scattering.

5. Conclusion

These lectures have attempted to describe physical phenomena in the fermi solids (particularly solid ^{3}He). These phenomena have been described at the microscopic level (of many body wave-functions) and in terms of models. An attempt was made in the modeling of phenomena to retain only their essential features. Quantification of the models was accomplished using the many body wavefunctions.

References

1. Guyer, R. A.: *Solid State Physics,* Vol. 23, Seitz, Turnbull and Ehrenreich, Academic Press, N.Y., N.Y., 1969, 402. This paper is a reasonably complete survey of the literature up to 1970; it emphasizes the short range correlation problem. N. R. Werthamer, *Am. J. Phys.* **37**, (1969). 763 This paper is a comprehensive survey of the phonon problem (see later).
2. Pollack, G. L.: *Rev. Modern Phys.* **36**, 748 (1964). This paper contains extensive data on the rare gas solids; it is the source of the parameters for these systems listed in Table I. There is no equivalent review for the hydrogen systems – details about these must be found from the literature. One might begin with Raich and Etters, *J. Low Temp. Phys.* **6**, 229 (1972).
3. Leggett, A. J.: *Phys. Rev. Letters,* **25**, 1543 (1970); and T. L. Ginsberg and A. A. Sobyanin, *JETP Letters* **15**, 242 (1972).
4. Hillier, J. H., Serajul Islam, M., and Walkley John: *J. Chem. Phys.* **43**, 3705 (1965).
5. Anderson, P. W. and Palmer, R. G.: *Nature (Physical Science)* **231**, 145 (1971).
6. Clark, J. W. and Chao, N. C.: (preprint).
7. Canuto, V. and Chitre, K.: *Phys. Rev. Letters* **30**, 999 (1973).
8. Wilks, J.: *Liquid and Solid Helium,* Oxford University Press, Oxford, England 1967. This book serves as an excellent source for data on the quantum liquids and solids.
9. Nosanow, L. H. and Shaw, G. L.: *Phys. Rev.* **128**, 546 (1972).
10. Nosanow, L. H.: *Phys. Rev.* **146**, 120 (1966).
11. Guyer, R. A. and Zane, L. I.: *Phys. Rev.* **188**, 445 (1969). See also the thesis by Sarkissian, Duke Univ., 1969.
12. Brandow, B. H.: *Phys. Rev. A4*, 422 (1971); *Annals of Physics* **74**, 112 (1972).
13. Van Kampen, N. G.: *Physica* **27**, 783 (1961).
14. This system of equations was first derived by Iwamoto, F. and Namaizawa, H.: *Prog. Theor. Phys. Suppl.* **37**, 234 (1966). The fact that these equations should be taken seriously was demonstrated by Sarkissian (see Reference [11] and Reference [1]).
15. de Wette, F. W. and Nijboer, B. R. A.: *Phys. Letters* **18**, 19 (1965). In this paper it is shown that, at physical volumes, the conventional phonon frequencies in part of the Brillouin zone are imaginary.
16. The conventional treatment of phonons is illustrated in Carruthers, P. A.: *Rev. Mod. Phys.* **33**, 92 (1961).
17. Brenig, W.: *Z. Physik* **171**, 60 (1963).
18. Fredkin, D. R. and Werthamer, N. R.: *Phys. Rev.* **138**, A1527 (1965). Further extensions of this work is reviewed in Reference [1] (Werthamer, N. R.).
19. Nosanow, L. H. and Werthamer, N. R.: *Phys. Rev. Letters* **15**, 618 (1965).
20. Kittel, C.: *Quantum Theory of Solids,* Wiley, New York, 1963, p. 25.
21. Osgood, E. B., Minkiewicz, V. J., Kitchens, T. A., and Shirane, G.: *Phys. Rev. A***5**, 1537 (1972).
22. Horner, H.: *Z. Physik* **242**, 432 (1971).
23. Horner, H.: *Phys. Rev. Letters* **25** 147 (1970).
24. Guyer, R. A.: *Phys. Rev. Letters* **24**, 810 (1970); the data is referenced and described in this letter.

25. This Hamiltonian was first introduced into the quantum crystal problem by Gersch and co-workers. See for example Gersch, H. A. and Fernandez, J. F.: *Phys. Rev.* **149**, 514 (1966).
26. Guyer, R. A. and Zane, L. I.: *Phys. Rev. Letters* **24**, 660 (1970).
27. The 'electron problem' for which the Hubbard Hamiltonian was invented is described by Wilson, J. A.: *Adv. Phys.* **21**, 143 (1972) and many other places. The theoretical literature can be found from this reference.
28. Anderson, P. W.: *Concepts in Solids*, W. A. Benjamin, New York, 1963.
29. Hetherington, J. H.: *Phys. Rev.* **176**, 231 (1968).
30. Guyer, R. A.: *J. Low Temperature Physics* **8**, 427 (1972).
31. Baker, G., Gilbert, H. E., Eve, J., and Rushbrooke, G. S.: *Phys. Rev.* **164**, 800 (1967).
32. Panczyk, M. F., Scribner, R. A., Straty, G. C., and Adams, E. D.: *Phys. Rev. Letters* **19**, 1102 (1967).
33. Panczyk, M. F. and Adams, E. D.: *Phys. Rev.* **187**, 321 (1969). A comprehensive review of thermodynamic measurements on the quantum crystals appears in *Reviews of Modern Physics* **44**, 668 (1972).
34. Kirk, W. P. and Adams, E. D.: *Phys. Rev. Letters* **27**, 392 (1971). The data presented in this paper suggests that under extreme conditions (large values of the external field) departures from a simple near neighbor magnet occur. See Reference [35].
35. Zane, L. I.: *Phys. Rev. Letters* **28**, 421 (1972) also a more extensive article to be published in *J. Low Temperature Physics*.
36. Callen, H. B.: *Thermodynamics*, Wiley, New York, 1960.
37. Thompson, J. R. and Meyer, H.: *Cryogenics* **7**, 296 (1967).
38. Anufriyer, Y. D.: *JETP Letters* **1**, 155 (1965).
39. Sites, J. R. Osheroff, D. O., Richardson, R. C., and Lee, D. M.: *Phys. Rev. Letters* **23**, 836 (1969). Johnson, R. T., Rosenbaum, R., Symko, O. G., and Wheatley, J. C.: *Phys. Rev. Letters* **22**, 449 (1969). A measure of the progress made in such experiments can be gleened from Osheroff, D. O, Richardson, R. C., and Lee, D. M.: *Phys. Rev. Letters* **28**, 885 (1972).
40. Johnson, R. T., Lounasmaa, O. V., Rosenbaum, R., Symko, O. G., and Wheatley, J. C.: *J. Low Temp. Phys.* **2**, 203 (1970).
41. Bierlein, R. and Bertman, B.: *Am. J. Phys.* **37**, 101 (1969). K. Mendelsohn, *The Quest for Absolute Zero*, World Univ. Library, London, 1966.
42. See the third item in Reference [39] and a paper to be published by the same authors in *Phys. Rev. Letters*.
43. Guyer, R. A., Richardson, R. C., and Zane, L. J.: *Rev. Mod. Phys.* **43**, 532 (1971). This paper is an extensive review of NMR data.
44. Abragam, A.: *The Principals of Nuclear Magnetism*, Oxford Univ. Press, Oxford, England, 1961.
45. McMahan, A. K.: *J. Low Temp. Phys.* **8**, 115 and 159 (1972).
46. The ideas here are a generalization of those of Herring, C. [*Magnetism* Vol. IV, (ed. by G. T. Rado and H. Suhl), Academic Press, New York, 1966] due to McMahan, Reference [45].
47. Adams, E. D., Straty, G. C., and Wall, E. L.: *Phys. Rev. Letters* **15**, 549 (1965).
48. Brinkman, Wm. and Rice, M.: *Phys. Rev.* *B***1**, 1324 (1970).
49. We consider scattering from the hydrodynamic mode because it is not apparent that the vacancy creation process – required for a neutron experiment – is possible.
50. Frenkel, J.: *Kinetic Theory of Liquids*, Dover Publications Inc., New York, 1955.

RADIATIVE AND NONRADIATIVE PROCESSES IN MOLECULES

R. LEFEBVRE

Laboratoire de Photophysique Moleculaire 91 Orsay, France*

1. Introduction

The study of radiative processes (absorption and emission of radiation) in atoms and molecules is the purpose of spectroscopy. Interesting situations arise when nonradiative processes are in competition with the radiative ones. This subject has been quite intensively studied in the past five years, with particular emphasis on polyatomic molecules [1, 2]. The aim of these lectures is not to give a complete account of these developments, but to select a few representative situations, the theory of which can be treated completely. This will serve to illustrate some of the techniques which are of use in this field. It can be observed that in fact these techniques are not limited to spectroscopic problems. They can also serve as the basis for treating certain types of chemical reactions [3, 4].

1.2. Experimental background

We enumerate now a number of situations involving radiative and nonradiative processes, with emphasis on their similarities.

1.2.1. *Radiative Decay*

This is the simplest situation one can think of where a process is taking place; that is to say we observe a change in the properties of the system as time goes on. Consider an atom (or a molecule) with the kets $|0\rangle$ and $|r\rangle$ representing the ground state and the first excited state. If we assume that at the initial time (say $t=0$) the system is in the excited state with no photons present in the electromagnetic field (this being represented by $|r; \text{vac.}\rangle$, there will be a time evolution toward the continua of states of representation $|0; \mathbf{k}, \mathbf{e}\rangle$, where $\mathbf{k}$ is the wave-vector of a photon and $\mathbf{e}$ its polarization (two polarizations exist for every choice of $\mathbf{k}$). These functions form continua because $\mathbf{k}$ can have both an arbitrary direction and an arbitrary modulus. The most significant states are those for which $\hbar c|\mathbf{k}| = E_r - E_0 = \hbar\omega_0$. We picture this as shown in Figure 1, and we shall say that the discrete level is *facing* the continua. The law for this process is of course that the probability for finding the atom at time t in the excited state is given by an exponential:

$$|\langle r; \text{vac} \mid \Psi(t)\rangle|^2 = \exp[-At] \tag{1.1}$$

with A (the Einstein coefficient) equal to $\omega_0^3|\mathbf{M}_{0r}|^2/3\pi\varepsilon_0 c^3\hbar$, where $\mathbf{M}_{0r}$ is the transition

* Laboratoire du C.N.R.S.

Abecassis de Laredo and Jurisic (eds.), Selected Topics in Phys. Astrophys. and Biophys. 87–137. All Rights Reserved.

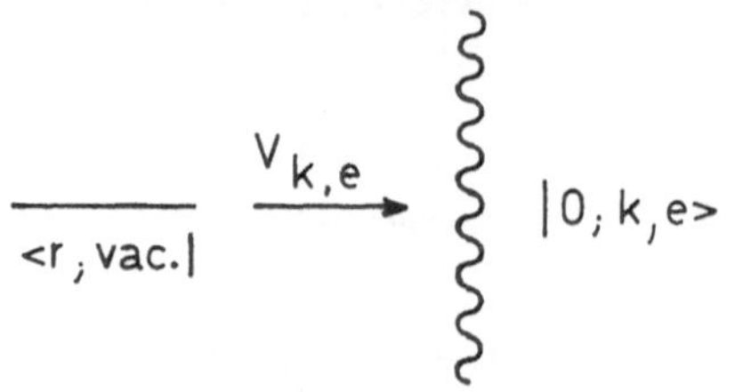

Fig. 1. An example of a discrete state facing continua: spontaneous emission. $V_{k,e}$ is the matrix element coupling the discrete state to a one-photon state.

moment. A complementary experiment consists in measuring the line width (natural line width here), instead of a change in time of the probability. The line width is $\hbar A$.

1.2.2. *Autoionization*

We take another example of a level facing a continuum, but a nonradiative process is now involved. Consider an atom such as He, and let us attempt to build the various states using the Hartree-Fock scheme (Figure 2). The first ionization potential is obtained when one electron is removed, the other being left in a $1S$ orbital. This occurs at about 54 eV above the ground state. But we can also build excited states by promoting two electrons. For instance at about -27 eV (i.e. above the first ionization energy) there is the $(2S)(2P)$ state. Because of the failure of the Hartree-Fock method to give exact wave-functions, there is a coupling between the discrete state and the continuum facing it. Thus the atom, if placed initially in the discrete state, can decay toward the continuum. This will lead to the ejection of an electron (Auger effect). Or instead we can observe an effect of this coupling on the absorption line shape. Note that the radiative decay mentioned above is obviously in competition with this process.

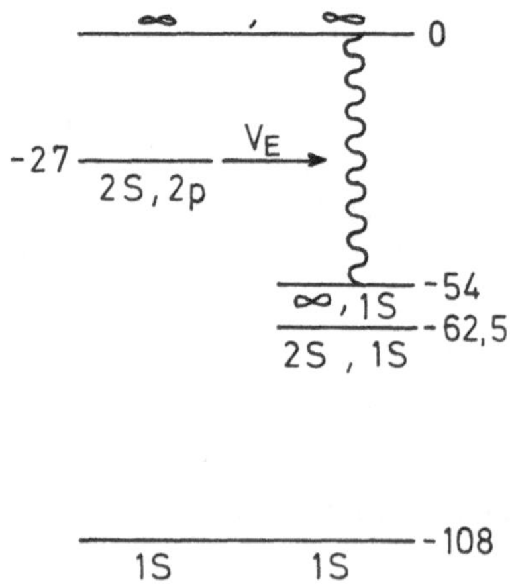

Fig. 2. He orbital scheme (energies in eV). V_E is the matrix element coupling the discrete state to the continuum.

1.2.3. *Predissociation*

A quite similar situation occurs in diatomic molecules when a discrete vibrational state is above the dissociation limit of a different electronic state. A recently examined case [5] is the molecule H_2 where we have the disposition of energy curves given in the Figure 3. We have just indicated two vibrational states belonging respectively to the $B''^1\Sigma_u^+$ state ($v=1$) and to the $D^1\Pi_u$ state ($v'=3$). They are both above the dissociation limit of the $B'^1\Sigma_u^+$ state, so that they are facing a continuum of states. Electrostatic and spin-orbit interactions are here responsible for the coupling. This results in very typical line shapes, first investigated by Fano [6] in the case of autoionization.

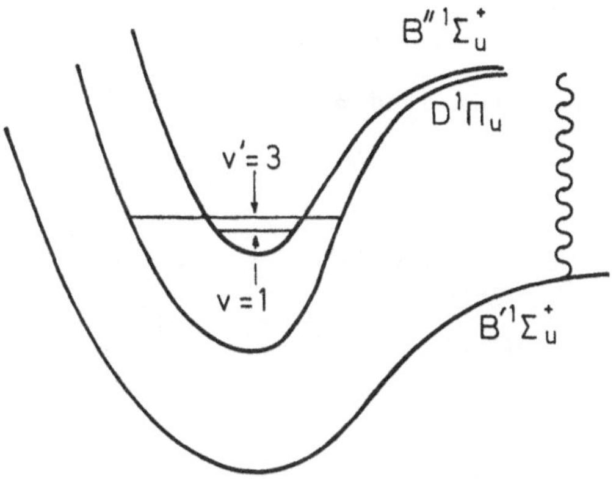

Fig. 3. Some of the potential curves of the H_2 molecule.

1.2.4. *Photodissociation*

We turn now to polyatomic molecules. A photochemical decomposition (or photodissociation) is quite comparable to predissociation. Consider a molecule which, after being excited to a bound level, will dissociate into two fragments A and B [3]. This can again be represented as due to discrete states facing continua. If X is the interfragment coordinate, there exists the potential energy for the relative motion shown in Figure 4. This potential energy may be deep enough to accomodate bound states (such as $\chi_m(X)$ on the figure). It will also possess continuum states (such as $\chi_\varepsilon(X)$). We may be able to build functions of nearly the same energy such as $\xi^i\chi_m$ and $\xi^j\chi_\varepsilon$, where ξ^i and ξ^j are two descriptions of the internal states of the fragments (this requires the difference ΔE shown on the figure to be compensated by the fragments). The elaboration of this model forms the basis for one of the theories of unimolecular decomposition [4].

1.2.5. *Internal Conversion and Intersystem Crossing*

We envisage now a somewhat different case since the processes involve quasi-continuous sets of levels instead of continuous ones. Figures 5 gives the socalled

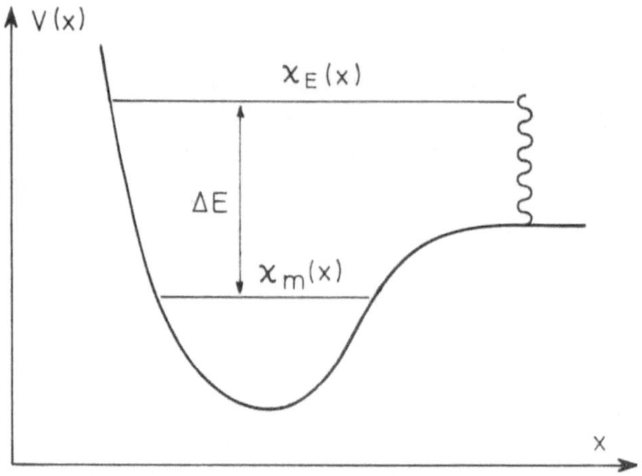

Fig. 4. Potential curve and energy levels for interfragment motion.

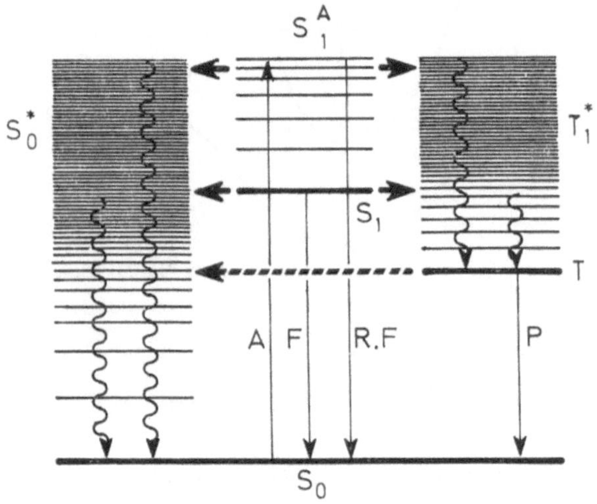

Fig. 5. The Jablonski diagram. Simple arrows: radiative processes (absorption and emission). Horizontal arrows: intramolecular processes (internal conversion and intrasystem crossing). Curved arrows: intermolecular processes (vibrational relaxation). The stars indicate states with an excess vibrational energy.

Jablonski diagram of a normal molecule (singlet ground state). A polyatomic molecule of large size (say 12 or more atoms) has a large number of normal modes. Thus in any of its electronic states (three of them are depicted on the figure, the singlet ground state S_0, the lowest triplet T_1 and the lowest excited singlet S_1) if, starting from the zero-point level we add progressively vibrational energy we get very quickly levels which are extremely close to each other (see densities below). Let us assume the illumination of the molecule has raised it to some vibrationnally excited state S_1^* belonging to the first excited singlet S_1. Two limiting cases can be envisaged.

(a) collision free molecules (the time between two collisions is longer than all the other relaxation times). At least three routes, corresponding to the interaction of the discrete state S_1^* with one continuum and two quasi-continua, are possible. Radiative decay: from S_1^* the molecule goes back to S_0 plus a photon (this is resonance fluorescence, RF). Nonradiative decay: The sets symbolized by S_0^* and T_1^* are so dense that they act effectively as continua, so that the molecule can also decay toward these sets of levels. From S_1^* to S_0^* this is internal conversion (no change in the multiplicity). From S_1^* to T_1^* this is intersystem crossing (a change in the multiplicity). Rate constants can be associated with these two routes, so that the kinetics of the entire process depends on three rate constants. The perturbations responsible for the intramolecular couplings are the nuclear kinetic energy, the spin orbit interaction, and possibly the electronic electrostatic energy.

(b) Molecules in strong interaction with their environment (pressure effects, liquid and solid phases). Vibrational relaxation within a given electronic state is faster than any other process. Thus, after a time of the order of $10^{-11}-10^{-12}$ s, the molecule is in the zero-point level of S_1. Thereafter the situation is very much like the previous one, with the possibilities of either radiative decay (fluorescence) or nonradiative decay toward S_0^* or T_1^*. Once the molecule is either in S_0^* or T_1^* there is again vibrational relaxation. Emission from the zero-point level of T_1 is the usual phosphorescence. In both cases this behaviour is only made possible by the very high densities of vibrational states which can be achieved. For instance [7] in anthracene the intersystem crossing from the $^1B_{2u}$ state (lowest excited singlet) to $^3B_{2u}$ (the lowest triplet) releases a vibrational energy of 12000 cm^{-1}. This leads to an estimate of 10^{10} vibrational levels per cm^{-1} in the triplet. In naphthalene the internal conversion from the second singlet ($^1B_{2u}$) to the first ($^1B_{1u}$) involves a gap of 3400 cm^{-1} only. However this already produces a density of about 10^3 levels per cm^{-1}. It is true that all these levels are not coupled in the same way to the initial discrete state and very interesting work is being done on the evaluation of these intramolecular couplings [8, 9]. In contrast it may also happen that in certain molecules (either because of a small electronic gap [10], or because the molecule is of small size [11]) a given initial level is facing a rather sparse set of levels. A description with only rate constants is not possible. However the techniques to be used are the same in the two situations.

1.2.6. *Photoisomerism*

Finally we say a few words on this process which can also be described with discrete levels facing quasi-continua [12]. However the effect of the nonradiative relaxation does not bring the molecule back to its initial nuclear configuration but to a different one (If the barrier between the two configurations is high enough they can be considered as representing two different chemical species). A well known case is the cis-trans isomerization of stilbene, a compound with two phenyl groups substituting two of the hydrogens of the ethylene molecule. A very schematic diagram indicating the possibility of such an isomerization is given in Figure 6. After excitation to the

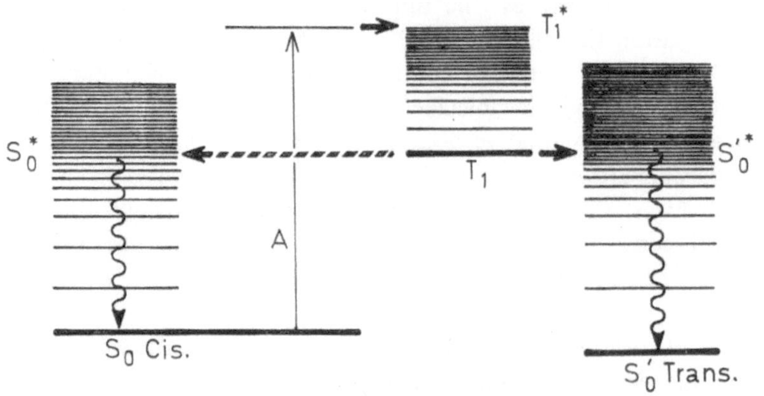

Figure. 6. Schematic energy level diagram to explain photoisomerization.

lowest singlet, the molecule goes to T_1^* which relaxes to T_1. From this point two routes corresponding to intersystem crossing toward vibrationally excited states of the ground singlets of either species (trans and cis) are open. Various other routes (radiative and nonradiative) should of course be added, and competition takes place.

1.3. The Main Characteristics of the Theoretical Models

The theory of radiative and nonradiative processes aims at providing an explicit description of the various phases of the evolution of a molecule interacting with the electromagnetic field. It should indicate, in particular, under which conditions the language which has been used above (the molecule being excited to a certain state of a zeroth order Hamiltonian and then undergoing internal and external processes) is correct. The main characteristics of the models to be discussed in these lectures are as follows:

1. All the processes involve iso-energetic or quasi-iso-energetic transitions. This is equivalent to saying that one uses a time-independent Hamiltonian in the time-dependent Schrödinger equation. Thus if there is a coupling with the electromagnetic field the Hamiltonian of the field is part of the total Hamiltonian which contains also the interaction of the quantum system (atom, molecule) and the field. In case we wanted to account also for the dissipation of the energy of the molecule toward its environment, we would have to introduce the phonon field and write combined wavefunctions for the molecule, plus the photons and the phonons. However this aspect of the problem will not be considered here; we will restrict ourselves to radiative and nonradiative processes in a collision-free molecule.

2. The first step in the description of a process is the enumeration of the sets of levels which are involved, which describe the combined system molecule plus field. These sets may be discrete (sparse or dense) or continuous. One must also decide about the couplings among these levels. The couplings can be treated as mere para-

meters, with the hope that reference to experimental data will allow to determine them if the model is correct. They can also be eventually evaluated from the molecular wave functions.

3. In most of these problems rate constants are only associated with one direction (irreversible decay, and not equilibrium situations). Thus these problems do not require the methods of statistical mechanics. A wave function, to be determined from the Schrödinger equation, will for all times be associated with the system.

4. Many models start by assuming a certain initial form for the wave function (the prepared state). This may be correct since it often happens that the final products of a process do not depend critically on the history of the system. And reference to experimental findings could be used to test this assumption. Another possibility is to describe the entire process, from past to future. Thus for instance this implies starting from the molecule in its ground state, describing the 'collision' of a photon with the molecule and the subsequent evolution (emission of a photon and decay in the other channels).

5. In this presentation, only the vacuum state and the one-photon states of the electromagnetic field are taken into account. Although this is a very particular model for the field, it has the advantage of leading to an explicit treatment of both the excitation and decay phases, without invoking a rather artificial reduction of the wave-packet associated with the composite system molecule plus field in order to leave the excited molecule in the vacuum of photons [13, 14].

2. The Hamiltonian

Since we are going to treat with a time independent Hamiltonian the processes that a molecule can undergo when it interacts with the electromagnetic field, the treatment of the field has to be a quantized one. The complete Hamiltonian can be partitioned as:

$$H = H_{\mathrm{m}} + H_{\mathrm{f}} + V_{\mathrm{i}}, \tag{2.1}$$

where the indices m and f designate the molecule and field respectively. V_{i} is the molecule-field interaction. The molecular Hamiltonian in turn can be divided into:

$$H_{\mathrm{m}} = H_0 + V, \tag{2.2}$$

the basis for this partition being the Born-Oppenheimer method.

2.1. Quantization of the Electromagnetic Field

In order to introduce the quantized version of the electromagnetic field in free space which will be used, we will describe first of all a simple model of a field in a cavity which is given by Scully and Lamb in their quantized treatment of the laser [15].

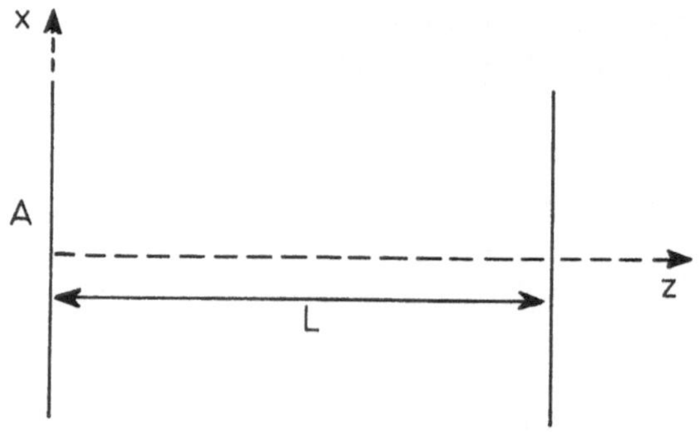

Fig. 7. Cavity with two parallel plates.

Consider (Figure 7) two parallel perfectly conducting plates of area A and at a distance L apart. We look for solutions of the Maxwell equations:

$$\operatorname{div} \mathbf{E} = 0 ; \qquad \operatorname{div} \mathbf{B} = 0 ;$$

$$\operatorname{rot} \mathbf{H} = \frac{\partial \mathbf{D}}{\partial t} ; \qquad \operatorname{rot} \mathbf{E} = -\frac{\partial \mathbf{B}}{\partial t} , \tag{2.3}$$

with $\mathbf{D}=\varepsilon_0\mathbf{E}$ and $\mathbf{H}=\mathbf{B}/\mu_0$. Edge effects are neglected. The boundary conditions are the absence of a tangential component of $\mathbf{E}$ and of a normal component of $\mathbf{B}$ on the plates. We try the following form for $\mathbf{E}$:

$$E_x = q(t) \sin kz ; \qquad E_y = E_z = 0 . \tag{2.4}$$

$E_x$vanishes for $z=0$ and must vanish for $z=L$. Thus $kL=l\pi$, or $k_l=l\pi/L$. We assign also the index l to $q(t)$. Consideration of the equation $\Delta\mathbf{E}-c^{-2}\partial^2\mathbf{E}/\partial t^2=0$ leads, for $q(t)$ to the equation: $\ddot{q}_l+\omega_l^2 q_l=0$, with $\omega_l=ck_l$. This is the equation of a harmonic oscillator. We know that the amplitude is arbitrary (it is fixed by the initial conditions). Thus we may multiply our solution by an arbitrary factor. We may also, because of the linearity of the Maxwell equations, combine together solutions corresponding to the various modes. We write therefore:

$$E_x = \sum_l [2\omega_l^2 m_l/(LA\varepsilon_0)]^{1/2} q_l(t) \sin k_l z . \tag{2.5}$$

We have taken advantage of the arbitrary factor to introduce a 'mass' m_l in the coefficients. The associated magnetic field is obtained from:

$$-\frac{\partial H_y}{\partial z} = \frac{\partial D_x}{\partial t} . \tag{2.6}$$

This gives:

$$H_y = \sum_l [2\omega_l^2 m_l/(LA\varepsilon_0)]^{1/2} \dot{q}_l(t) k_l^{-1} \varepsilon_0 \cos k_l z . \tag{2.7}$$

The total energy can be expressed as:

$$H_f = \tfrac{1}{2} \int\limits_{\text{cavity}} d\tau \left[\varepsilon_0 E_x^2 + \mu_0 H_y^2\right]. \tag{2.8}$$

The integration over x and y produces a factor A. The integration over z is facilitated by the use of the relations:

$$\int_0^L \cos k_l z \cos k_{l'} z \, dz = \int_0^L \sin k_l z \sin k_{l'} z \, dz = \frac{L}{2} \delta_{ll'}, \tag{2.9}$$

which ensure that there are no cross-terms. The result is:

$$H_f = \tfrac{1}{2} \sum_l \omega_l^2 m_l q_l^2 + \tfrac{1}{2} \sum_l m_l \dot{q}_l^2. \tag{2.10}$$

Introducting a 'force constant' $K_l = \omega_l^2 m_l$ and a 'momentum' $p_l = m_l \dot{q}_l$ we get:

$$H_f = \sum_l \left[\frac{p_l^2}{2m_l} + \tfrac{1}{2} K_l q_l^2\right]. \tag{2.11}$$

Thus the total energy can be written as the sum of the energies of harmonic oscillators. This is to be quantized by considering that q_l and p_l are no longer numbers, but operators obeying the commutation rules:

$$[q_l, p_{l'}] = i\hbar\delta_{ll'}; [q_l, q_{l'}] = [p_l, p_{l'}] = 0. \tag{2.12}$$

Let us define two operators a_l and a_l^+ by the relations:

$$a_l = [2\hbar m_l \omega_l]^{-1/2} [m_l \omega_l q_l + i p_l], \tag{2.13a}$$

$$a_l^+ = [2\hbar m_l \omega_l]^{-1/2} [m_l \omega_l q_l - i p_l]. \tag{2.13b}$$

A straightforward calculation using the commutation rules of the q_l's and p_l's shows that:

$$[a_l, a_{l'}^+] = \delta_{ll'}; [a_l, a_{l'}] = [a_l^+, a_{l'}^+] = 0. \tag{2.14}$$

Inverting the two relations defining a_l and a_l^+ we get:

$$q_l = [\hbar/2m_l\omega_l]^{1/2} (a_l^+ + a_l), \tag{2.15a}$$

$$p_l = i[\hbar m_l \omega_l/2]^{1/2} (a_l^+ - a_l). \tag{2.15b}$$

Introducing these operators into the Hamiltonian H_f, with attention paid to the order of the factors leads to:

$$H_f = \tfrac{1}{2} \sum_l \hbar\omega_l (a_l^+ a_l + a_l a_l^+) = \sum_l \hbar\omega_l (a_l^+ a_l + \tfrac{1}{2}), \tag{2.16}$$

where the second form is obtained on using $a_l a_l^+ = 1 + a_l a_l^+$. Since this is a sum of Hamiltonians, we can consider each of them in turn. The total wave function will be a product of wave functions, one for each Hamiltonian. To determine the eigenvalues and eigenfunctions, we take one term and drop the index l:

$$h = \hbar(a^+a + \tfrac{1}{2}). \tag{2.17a}$$

Instead of working with h we form the operator:

$$N = h/\hbar\omega - 1/2 = a^+a. \tag{2.17b}$$

Consider the two commutators $[a, N]$ and $[a^+, N]$. We have:

$$[a, N] = aa^+a - a^+aa = [a, a^+] = a,$$

or

$$a^+aa = aa^+a - a \tag{2.18a}$$

$$[a^+, N] = a^+a^+a - a^+aa^+ = a^+[a^+, a] = -a^+,$$

or

$$a^+aa^+ = a^+a^+a + a^+. \tag{2.18b}$$

Suppose we have an eigenfunction $|\alpha\rangle$ of N, of eigenvalue α:

$$N|\alpha\rangle = \alpha|\alpha\rangle \quad \text{or} \quad a^+a|\alpha\rangle = \alpha|\alpha\rangle. \tag{2.19}$$

We form:

$$Na|\alpha\rangle = a^+aa|\alpha\rangle = (aa^+a - a)|\alpha\rangle = (\alpha - 1)a|\alpha\rangle, \tag{2.20a}$$

$$Na^+|\alpha\rangle = a^+aa^+|\alpha\rangle = (a^+a^+a + a^+)|\alpha\rangle = (\alpha + 1)a^+|\alpha\rangle; \tag{2.20b}$$

thus $a|\alpha\rangle$ and $a^+|\alpha\rangle$ are also eigenfunctions with eigenvalues $\alpha \pm 1$ We now form:

$$\langle\alpha| N|\alpha\rangle = \langle\alpha| a^+a|\alpha\rangle = \alpha\langle\alpha | \alpha\rangle. \tag{2.21}$$

Since $\langle\alpha| a^+a|\alpha\rangle$ and $\langle\alpha | \alpha\rangle$ are both norms of eigenfunctions ($\langle\alpha| a^+$ being the *bra* corresponding to the ket $a|\alpha\rangle$), α must be a positive or null number. Thus α can only be a positive or null *integer* number in order avoiding the construction of eigenfunctions of negative eigenvalues by the successive applications of the operator a. The notation α will now be changed into n. We can write:

$$a|n\rangle = c_n|n-1\rangle \quad \text{and} \quad a^+|n\rangle = c'_n|n+1\rangle$$

Also

$$\langle n| a^+a|n\rangle = n\langle n | n\rangle = |c_n|^2\langle n-1 | n-1\rangle$$

and

$$\langle n| aa^+|n\rangle = |c'_n|^2\langle n+1 | n+1\rangle = (n+1)\langle n | n\rangle.$$

We can choose the phases so that c_n and c'_n are real numbers. Therefore $c_n = \sqrt{n}$ and $c'_n = \sqrt{n+1}$, and we have:

$$a|n\rangle = \sqrt{n}\,|n-1\rangle \quad \text{and} \quad a^+|n\rangle = \sqrt{n+1}\,|n+1\rangle. \tag{2.22}$$

If we go back to the Hamiltonian h we obtain the eigenvalues:

$$\langle n| h|n\rangle = \hbar\omega\langle n| a^+a + 1/2|n\rangle = \hbar\omega(n + 1/2), \tag{2.23}$$

which are the usual eigenvalues of the quantum harmonic oscillator. The eigenfunctions of H_f can be written:

$$|n_1\rangle|n_2\rangle\cdots|n_l\rangle\cdots \quad \text{or simply} \quad |n_1, n_2, \ldots, n_l, \ldots\rangle$$

the corresponding energy being $\sum_l \hbar\omega_l(n_l+1/2)$. Since the number of modes of the field is infinite, the energy contains an infinite term which does not play a role in our problems. Application of a_l or a_l^+ on such a function gives:

$$a_l|n_1, n_2, \ldots, n_l, \ldots\rangle = \sqrt{n_l}\,|n_1, n_2, \ldots, n_l - 1, \ldots\rangle \tag{2.24a}$$

and

$$a_l^+|n_1, n_2, \ldots, n_l, \ldots\rangle = \sqrt{n_l + 1}\,|n_1, n_2, \ldots, n_l + 1, \ldots\rangle \tag{2.24b}$$

The resulting functions give an energy either decreased or increased by $\hbar\omega_l$. This the reason for calling a_l and a_l^+ *annihilation* and *creation* operators since they change the energy by $\hbar\omega_l$, the energy of a photon.

The electric field is now an operator:

$$E_x = \sum_l \varepsilon_l(a_l + a_l^+)\sin k_l z\,, \tag{2.25}$$

with $\varepsilon_l = [\hbar\omega_l/LA\varepsilon_0]^{1/2}$. It can have matrix elements only between states differing by one photon.

We now introduce the formulas and the notations which will be used for treating a problem in free space. The derivation is quite parallel to that given above [16]. One may consider either a box with perfectly conducting walls, or boxes which fill all space and require periodicity of the fields. In both cases the dimensions of the boxes will be made infinite at the end of the calculations. In the latter presentation which is the one we choose, the periodicity of the fields is obtained when the wave-vector $\mathbf{k}$ of a plane wave is of the form:

$$\mathbf{k}_l = (2\pi/L)(l_1\mathbf{U} + l_2\mathbf{V} + l_3\mathbf{W})\,, \tag{2.26}$$

where l_1, l_2 and l_3 are three integers and $\mathbf{U}$, $\mathbf{V}$ and $\mathbf{W}$ are unit vectors along Cartesian axes. Thus l is used to designate this triplet of numbers. In addition for every choice of $\mathbf{k}_l$, two polarizations are possible. Two unit vectors $\mathbf{e}_{l\sigma}(\sigma=1, 2)$, orthogonal to $\mathbf{k}_l$ and mutually orthogonal can be used to specify these polarizations. The most common formulation of the quantization of the field does not consist in associating time-dependent amplitudes to the electric field (as we have done here), but to the potential vector $\mathbf{A}(\mathbf{r}, t)$ which can be written:

$$\mathbf{A}(\mathbf{r}, t) = \sum_{l,\sigma}[\hbar/2\varepsilon_0\omega_l L^3]^{1/2}\,\mathbf{e}_{l\sigma}[a_{l\sigma}(t)\,e^{i\mathbf{k}_l\cdot\mathbf{r}} + a_{l\sigma}^+(t)\,e^{-i\mathbf{k}_l\cdot\mathbf{r}}]\,. \tag{2.27}$$

The electric field $\mathbf{E}(\mathbf{r}, t) = -\partial\mathbf{A}(\mathbf{r}, t)/\partial t$ is therefore:

$$\mathbf{E}(\mathbf{r}, t) = i\sum_{l,\sigma}[\hbar\omega_l/2\varepsilon_0 L^3]^{1/2}\,\mathbf{e}_{l\sigma}[a_{l\sigma}(t)\,e^{i\mathbf{k}_l\cdot\mathbf{r}} - a_{l\sigma}^+(t)\,e^{-i\mathbf{k}_l\cdot\mathbf{r}}]\,, \tag{2.28}$$

since the amplitudes obey $\dot{a}_{l\sigma}(t) = -i\omega_l a_{l\sigma}(t)$ and $\dot{a}_{l\sigma}^+(t) = i\omega_l a_{l\sigma}^+(t)$. The quantization changes $a_{l\sigma}(t)$ and $a_{l\sigma}^{+}(t)$ into $a_{l\sigma}$ and $a_{l\sigma}^{+}$ which annihilate or create a photon of wave-vector $\mathbf{k}_l$, polarization $\mathbf{e}_{l\sigma}$ and energy $\hbar c|\mathbf{k}_l|$. For an atom or a molecule placed at the origin and interacting with a field in the optical range ($\lambda_l = 2\pi/|\mathbf{k}_l| \sim 2000$ Å or more), $\mathbf{k}_l \cdot \mathbf{r} \simeq 0$ over the entire system, so that the electric field operator can be simplified to:

$$\mathbf{E} = i \sum_{l,\sigma} [\hbar\omega_l/2\varepsilon_0 L^3]^{1/2} \mathbf{e}_{l\sigma} (a_{l\sigma} - a_{l\sigma}^{+}). \tag{2.29}$$

The interaction between the field and the system can be written:

$$V_{\mathrm{i}} = -(e/m)\mathbf{A}\cdot\mathbf{p} \quad \text{or} \quad V_{\mathrm{i}} = -e\mathbf{E}\cdot\mathbf{r}, \tag{2.30}$$

(assuming a single optical electron, $\mathbf{r}$ and $\mathbf{p}$ being its position and momentum). The differences introduced by the use of one or the other of the two forms are unessential for the problems treated here.

2.2. The Born-Oppenheimer Method

The Hamiltonian of a molecule is the sum of kinetic energy operators T for the electrons (e) and the nuclei (N), of electrostatic potential energies V and of the spin-orbit interaction:

$$H_{\mathrm{m}} = T_{\mathrm{e}} + T_{\mathrm{N}} + V_{\mathrm{ee}} + V_{\mathrm{eN}} + V_{\mathrm{NN}} + H_{\mathrm{so}}. \tag{2.31}$$

For instance

$$T_{\mathrm{e}} = -\hbar^2/2m \sum_{\mu} \Delta_{\mu}^2 \quad \text{and} \quad V_{\mathrm{eN}} = \sum_{\mu,p} Z_p e^2/r_{\mu p},$$

where the summation μ is over electrons, and the summation p over nuclei of charge $Z_{p}e$. In semi-empirical treatments one may take for the spin-orbit interaction the simple form:

$$H_{\mathrm{so}} = \sum_{\mu,p} \zeta(r_{\mu p}) \mathbf{l}_{\mu p} \cdot \mathbf{s}_{\mu}; \tag{2.32}$$

$\mathbf{l}_{\mu p}$ is the angular momentum of electron μ with respect to nucleus p and $\mathbf{s}_{\mu}$ its spin. In even more simplified treatments:

$$\zeta(r_{\mu p}) = \zeta_p r_{\mu p}^{-3} \tag{2.33}$$

with a different constant ζ_p for different atoms (carbon, nitrogen, etc. ...).

Following Born [17] we first of all look for solutions of the equations:

$$H_{\mathrm{el}} \varphi_m(q, Q) = E_m(Q) \varphi_m(q, Q), \tag{2.34}$$

where

$$H_{el} = H_{\mathrm{m}} - T_{\mathrm{N}}, \tag{2.35}$$

q and Q being used to designate collectively the variables associated with the electrons and the nuclei. Since (2.34) amounts to solving for the motion of the electrons while Q is kept fixed, the electronic energies $E_m(Q)$ are dependent upon Q. We may then solve the equations:

$$[T_N + E_m(Q) + \langle m| T_N |m\rangle_q]\, \chi_{mi}(Q) = W_{mi}\chi_{mi}(Q); \tag{2.36}$$

$\langle m| T_N |m\rangle_q$ is the so-called diagonal correction to the potential energy $E_m(Q)$. An explicit form for it is:

$$\langle m| T_N |m\rangle_q = \int \varphi_m^*(q, Q)\, \{-\hbar^2/2 \sum_p M_p^{-1}\partial^2/\partial Q_p\}\, \varphi_m(q, Q)\, \mathrm{d}q\,. \tag{2.37}$$

Note that the functions $\chi_{mi}(Q)$ are characterized by two indices since for every value of m one gets a series of solutions with $i=0, 1, 2, \ldots$. The next step consists in forming the product wave-functions $\varphi_m(q, Q)\, \chi\alpha_i(Q)$. A test for the quality of these functions is the examination of the matrix of the molecular Hamiltonian. Two types of matrix elements can be constructed:

(a) Matrix elements with the same electronic state on both sides of H_m. On account of the equations obeyed by the electronic and nuclear factors there is found that:

$$\langle \varphi_m(q, Q)\, \chi_{mi}(Q)| H_m |\varphi_m(q, Q)\, \chi_{mj}(Q)\rangle_{q,Q} = W_{mi}\delta_{ij}\,. \tag{2.38a}$$

Thus there are no off-diagonal matrix elements connecting product wave-functions (also called vibronic functions from *vibr*ational electr*onic*) belonging to a given electronic state.

(b) Matrix elements with different electronic states. These are:

$$\begin{aligned}&\langle \varphi_m(q, Q)\, \chi_{mi}(Q)| H_m |\varphi_n(q, Q)\, \chi_{nj}(Q)\rangle_{q,Q} = \\ &= \langle \chi_{mi}(Q)\, \langle m| T_N |n\rangle_q\, \chi_{nj}(Q)\rangle_Q - \\ &- \hbar^2 \sum_p M_p^{-1} \langle \chi_{mi}(Q)\, \langle \varphi_m(q, Q)| \partial/\partial Q_p |\varphi_n(q, Q)\rangle_q\, \partial/\partial Q_p \chi_{nj}(Q)\rangle_Q.\end{aligned} \tag{2.38b}$$

The quantity $\langle m| T_N |n\rangle_q$ is obtained by an obvious generalization of formula (2.37). We conclude that there are off-diagonal matrix elements connecting vibronic states belonging to two different electronic states.

We limit the following discussion to two electronic states since this is by far the

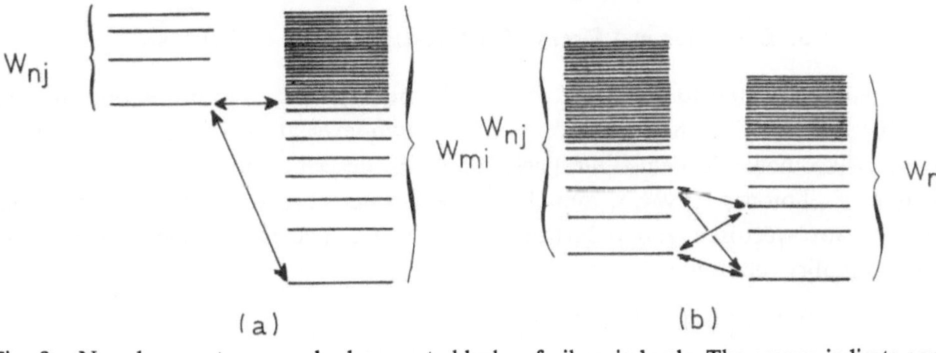

Fig. 8. Non degenerate or nearly degenerate blocks of vibronic levels. The arrows indicate some of the interblock couplings.

most important case. If a diagram is built of the two sequences W_{m0}, W_{m1}, $W_{m2}\ldots$ and W_{n0}, W_{n1}, $W_{n2},\ldots$ there are the two possibilities shown in Figure 8.

In case (a) which corresponds to the most common molecules, the onset of a block is far above the onset of another block (well separated electronic states). Thus if m is the ground electronic state, the lower vibrational levels of this state are very little perturbed by the interactions with the other block, this being due to two reasons: large energy differences between interacting states and weak matrix elements, this latter circumstance depending on the fact that the critical quantities in these matrix elements are proportional to the inverse of the electronic gap since they can be written:

$$\langle\varphi_m(q, Q)|\,\partial/\partial Q_p\,|\varphi_n(q, Q)\rangle_q = \frac{\langle\varphi_m(q, Q)|\,\partial H_{el}/\partial Q_p\,|\varphi_n(q, Q)\rangle_q}{E_n(Q)-E_m(Q)}. \tag{2.39}$$

However when we reach levels which are facing each other, the interactions, although weak, may be very significant for degenerate or quasi-degenerate states, particularly in the case of the high densities of vibrational states mentioned in Section 1.

In case (b) the electronic gap is small or zero (accidental or essential degeneracy). Two subcases can be distinguished: (b′) symmetry or spin (or both) make the matrix elements very small. Only interactions between degenerate or nearly degenerate vibronic states have to be considered; (b″) the matrix elements are large. The product wavefunctions are of little use and a different approach is needed (vibronic formulation of the Jahn-Teller effect, etc....). This case has received only little attention in the theory of non-radiative processes. We will, in the following, consider that our systems always belong to the classes (a) and (b′).

We end these remarks about the Born-Oppenheimer method by emphasizing that the spin-orbit coupling has been incorporated in the electronic Hamiltonian. Thus in $\partial H_{el}/\partial Q_p$ we have the term $\partial H_{so}/\partial Q_p$. The zeroth order functions we consider are not characterized by a given spin multiplicity. However there are cases where it is more convenient to define the zeroth order functions by leaving the spin-orbit interaction outside the electronic Hamiltonian so that it will act as a perturbation[18].

3. Excitation and Decay of a Two-Level Atom or Molecule

As a preliminary step to the study of the behaviour of a system presenting internal couplings we describe the time-evolution of a two-level atom or molecule coupled to the electromagnetic field. Our basis functions describe the excited atom in the vacuum of photons: $|\mathrm{r};\text{vac.}\rangle$, and the ground state atom with one photon in the field, of wave-vector $\mathbf{k}$ and polarization $\mathbf{e}$:$|0;\mathbf{k},\mathbf{e}\rangle$. The latter functions obey the orthonormality relations:

$$\langle 0;\mathbf{k},\mathbf{e}\,|\,0;\mathbf{k}',\mathbf{e}'\rangle = \delta_{\mathbf{ee}'}\delta(\mathbf{k}-\mathbf{k}'). \tag{3.1}$$

We have adopted here a simpler notation $\mathbf{k}$ and $\mathbf{e}$ instead of $\mathbf{k}_l$ and $\mathbf{e}_{l\sigma}$. In order to work in E-normalization there is to affect these functions by a factor. Since:

$$\delta(\mathbf{k}-\mathbf{k}') = |\mathbf{k}|^{-2}\,\delta(|\mathbf{k}|-|\mathbf{k}'|)\,\delta(\Omega-\Omega'), \tag{3.2a}$$

where Ω stands for (θ, φ) giving the direction of $\mathbf{k}$, and:

$$\delta(|\mathbf{k}|-|\mathbf{k}'|) = \hbar c\delta(E-E'), \tag{3.2b}$$

where $E=\hbar c|\mathbf{k}|$, the functions:

$$|\varphi_E^i\rangle = |\mathbf{k}|/\sqrt{\hbar c}\,|0;\mathbf{k}, e\rangle \tag{3.3}$$

obey the E-orthonormality relations:

$$\langle\varphi_{E'}^j \mid \varphi_E^i\rangle = \delta_{ij}\delta(E-E'). \tag{3.4}$$

In $|\varphi_E^i\rangle$ E is the energy label, while i collects the other informations on the photon (direction of wave-vector and polarization). An alternative notation for $|r; \text{vac.}\rangle$ will be $|\varphi_r\rangle$. There are off-diagonal matrix elements of the total Hamiltonian given by

$$V_E^i = \langle\varphi_r|\,H\,|\varphi_E^i\rangle. \tag{3.5}$$

Only the atom-field interaction $-\mathbf{E}\cdot\boldsymbol{\mu}$ contributes ($\boldsymbol{\mu}$ being the electric dipole operator). The states $\langle\varphi_E^i|$ and $|\varphi_{E'}^j\rangle$ for different photons are not coupled. We say that the continua are *prediagonalized.*

3.1. Lippmann-Schwinger solutions of the wave-equation

Our first step is to find the stationary wave-functions in the case of a discrete state facing the continua $|\varphi_E^i\rangle$, $|\varphi_{E'}^j\rangle$, etc.... The stationary wave functions are solutions of:

$$(E-H)\,|\psi_E\rangle = 0. \tag{3.6}$$

Following Lippmann and Schwinger [19] one may form formal solutions of the wave equation. The resolvent (or Green function) method will then be used to give an explicit form to these solutions.

3.1.1. *The Lippmann-Schwinger Equation*

The wave equation can also be written:

$$(E-H)\,|\psi_E\rangle = \lim_{\varepsilon\to 0} + i\varepsilon\left[|\varphi_E^i\rangle - |\psi_E\rangle\right], \tag{3.7}$$

since the right hand side goes to zero as the positive quantity ε goes to zero. This suggests [20] manipulating:

$$(E-H)\,|\psi_E\rangle = +\,i\varepsilon\left[|\varphi_E^i\rangle - |\psi_E\rangle\right], \tag{3.8}$$

and requiring ε to go to zero at the end of the calculations. This relation can also be written:

$$(E+i\varepsilon-H)\,|\psi_E\rangle = (E+i\varepsilon-H)\,|\varphi_E^i\rangle + (H-E)\,|\varphi_E^i\rangle. \tag{3.9}$$

Since

$$(H - E)\,|\varphi_E^i\rangle = (H_0 + V - E)\,|\varphi_E^i\rangle = V\,|\varphi_E^i\rangle\,, \tag{3.10}$$

where H_0 is the Hamiltonian defining the zeroth order levels and V the perturbation (in the present case the atom-field coupling), we obtain by inverting this relation:

$$|\psi_E^{+1}\rangle = |\varphi_E^i\rangle + \frac{1}{E + i\varepsilon - H}\,V\,|\varphi_E^i\rangle\,. \tag{3.11}$$

These functions are the so-called Lippmann-Schwinger (L.S.) solutions of the wave equation. The label i indicates that the function has been obtained from $|\varphi_E^i\rangle$. The label $+$ is related to the choice $+i\varepsilon$ in (8) and has a meaning which will be discovered after we explicit the solutions.

Instead of the explicit form (11), the Lippmann-Schwinger can also be given in the implicit form:

$$|\psi_E^{+i}\rangle = |\varphi_E^i\rangle + \lim_{\varepsilon\to 0}\frac{1}{E + i\varepsilon - H_0}\,V\,|\psi_E^{+i}\rangle\,. \tag{3.12}$$

For this to be correct we must have:

$$\begin{aligned}\frac{1}{E + i\varepsilon - H}\,V\,|\varphi_E^i\rangle &= \frac{1}{E + i\varepsilon - H_0}\,V\,|\psi_E^{+i}\rangle = \\ &= \frac{1}{E + i\varepsilon - H_0}\,V\,|\varphi_E^i\rangle + \frac{1}{E + i\varepsilon - H_0}\,V\,\frac{1}{E + i\varepsilon - H}\,V\,|\varphi_E^i\rangle\,,\end{aligned} \tag{3.13}$$

the latter expression resulting from the replacement of $|\psi_E^{+i}\rangle$ by its explicit form. Thus if we can proove that:

$$\frac{1}{E + i\varepsilon - H} = \frac{1}{E + i\varepsilon - H_0} + \frac{1}{E + i\varepsilon - H_0}\,V\,\frac{1}{E + i\varepsilon - H}, \tag{3.14}$$

the above relation will result. We write z instead of $E + i\varepsilon$ and define the two *resolvents*:

$$G(z) = \frac{1}{z - H} \text{ which fulfills } (z - H)\,G(z) = G(z)\,(z - H) = \mathbf{1}$$

and:

$$G_0(z) = \frac{1}{z - H_0} \text{ which fulfills } (z - H_0)\,G_0(z) = G_0(z)\,(z - H_0) = \mathbf{1}\,.$$

The operator identity (14) amounts to:

$$G(z) = G_0(z) + G_0(z)\,V\,G(z) \tag{3.15a}$$

to be abbreviated as:

$$G = G_0 + G_0 V G\,. \tag{3.15b}$$

This is easily deduced from:

$$G_0(z - H_0 - V)\, G = G_0(z - H_0)\, G - G_0 V G = G_0(z - H_0)\, G_0 = G_0 . \tag{3.16}$$

An akin relation is:

$$G = G_0 + G V G_0 , \tag{3.17}$$

which can be obtained from:

$$G(z - H)\, G = G(z - H + V)\, G_0 = G(z - H)\, G + G V G_0 . \tag{3.18}$$

Iterating any one of the two relations (15) or (17) leads to the Born expansion of the resolvent:

$$G = G_0 + G_0 V G_0 + G_0 V G_0 V G_0 + \cdots \tag{3.19}$$

Combining (15) and (17) gives us:

$$G = G_0 + G_0 V G_0 + G_0 V G V G_0 . \tag{3.20}$$

3.1.2. *The Amplitudes of the Stationary Functions in the Basis of Zeroth Order Functions.*

We rewrite the L.S. equation as:

$$|\psi_E^{+i}\rangle = |\varphi_E^i\rangle + \lim_{\varepsilon \to 0} G(z)\, V\, |\varphi_E^i\rangle . \tag{3.21}$$

We will note $G^+(E)$ the operator $\lim \varepsilon \to 0\ G(z)$. The amplitudes of $|\psi_E^{+i}\rangle$ in the original basis are:

$$\langle \varphi_r\, |\psi_E^{+i}\rangle = \langle \varphi_r|\, G^+(E)\, |\varphi_r\rangle\, V_E^i \tag{3.22a}$$

$$\langle \varphi_{E'}^j\, |\psi_E^{+i}\rangle = \delta_{ij}\, \delta(E - E') + \langle \varphi_{E'}^j|\, G^+(E)\, |\varphi_r\rangle\, V_E^i . \tag{3.22b}$$

The task of finding the stationary wave functions has therefore been changed into that of calculating the matrix elements of $G^+(E)$.

3.1.3. *Determination of the Matrix Elements of the Resolvent*

The starting point is $(z - H)\, G(z) = \mathbf{1}$ with z above the real axis. A resolution of the identity in our problem is:

$$\mathbf{1} = |\varphi_r\rangle \langle \varphi_r| + \sum_j \int \mathrm{d}E''\, |\varphi_{E''}^j\rangle \langle \varphi_{E''}^j| . \tag{3.23}$$

This form for the identity is introduced between $(z - H)$ and $G(z)$ and two kinds of matrix elements are taken. First on 'multiplying' on the left by $\langle \varphi_{E'}^i|$ and on the right by $|\varphi_r\rangle$, this is:

$$\langle \varphi_{E'}^i|\, (z - H)\, \{|\varphi_r\rangle \langle \varphi_r| + \sum_j \int \mathrm{d}E''\, |\varphi_{E''}^j\rangle \langle \varphi_{E''}^j|\}\, G(z)\, |\varphi_r\rangle = \langle \varphi_{E'}^i\, |\varphi_r\rangle \tag{3.24}$$

which is easily reduced, on using:

$$\langle\varphi^i_{E'}\,|\varphi^j_{E''}\rangle = \delta_{ij}\,\delta(E'-E''),\ \langle\varphi^i_{E'}|\,H\,|\varphi^j_{E''}\rangle = E'\,\delta_{ij}\,\delta(E'-E''), \tag{3.25}$$

to:

$$-\,V^{i\,*}_{E'}\langle\varphi_r|\,G(z)\,|\varphi_r\rangle + (z-E')\,\langle\varphi^i_{E'}|\,G(z)\,|\varphi_r\rangle = 0\,. \tag{3.26}$$

Next, on multiplying on the left by $\langle\varphi_r|$ and on the right by $|\varphi_r\rangle$, there is obtained:

$$\langle\varphi_r|\,(z-H)\,\{|\varphi_r\rangle\,\langle\varphi_r| + \sum_i \int \mathrm{d}E'\,|\varphi^i_{E'}\rangle\,\langle\varphi^i_{E'}|\}\,G(z)\,|\varphi_r\rangle = \langle\varphi_r\,|\varphi_r\rangle\,, \tag{3.27}$$

which can be changed into:

$$(z-E_r)\,\langle\varphi_r|\,G(z)\,|\varphi_r\rangle - \sum_i \int \mathrm{d}E' V^i_{E'}\,\langle\varphi^i_{E'}|\,G(z)\,|\varphi_r\rangle = 1\,, \tag{3.28}$$

with $E_r = \langle\varphi_r|\,H\,|\varphi_r\rangle$. The introduction of:

$$\langle\varphi^i_{E'}|\,G(z)\,|\varphi_r\rangle = \frac{V^{i\,*}_{E'}}{z-E'}\,\langle\varphi_r|\,G(z)\,|\varphi_r\rangle$$

into (30) leaves $\langle\varphi_r|\,G(Z)\,|\varphi_r\rangle$ as the only unknown. It is given by:

$$\langle\varphi_r|\,G(z)\,|\varphi_r\rangle = \frac{1}{z-E_r-\int \mathrm{d}E'\,(\sum_i |V^i_{E'}|^2/(z-E')}\,, \tag{3.29}$$

and the other matrix element $\langle\varphi^i_{E'},\,|G(z)\,|\varphi_r\rangle$ can also be determined once this one has been worked out.

3.1.4. *Properties of the Matrix Element* $\langle\varphi_r|\,G(z)\,|\varphi_r\rangle$

Since in the definition of the stationary functions there is implied that ε should go to zero and since $z = E + i\varepsilon$ we must investigate the properties of $\langle\varphi_r|\,G(z)\,|\varphi_r\rangle$ as z goes to the real axis. Let us consider the integral:

$$\int \mathrm{d}E'\,\frac{\sum_i |V^i_{E'}|^2}{z-E'}\,. \tag{3.30}$$

We are going to make use of the identity:

$$\lim_{\varepsilon\to 0}\frac{1}{E\pm i\varepsilon-E'} = \frac{P}{E-E'} \mp i\pi\,\delta(E-E')\,, \tag{3.31}$$

where $P/(E-E')$ is the principal part distribution. It has the property that for any function $f(E')$:

$$\int \mathrm{d}E'\,\frac{P}{E-E'}\,f(E') = \mathscr{P}\int \mathrm{d}E'\,\frac{f(E')}{E-E'} = \lim_{\eta\to 0}\left\{\int^{E-\eta} + \int_{E+\eta}\right\}, \tag{3.32}$$

if $E' = E$ is between the two limits of integration. Thus:

$$\lim_{\varepsilon \to 0} \int \mathrm{d}E' \frac{\sum_i |V^i_{E'}|^2}{z - E'} = F(E) - i\Gamma(E), \tag{3.33}$$

with the definitions:

$$F(E) = \mathscr{P} \int \mathrm{d}E' \frac{\sum_i |V^i_{E'}|^2}{E - E'} \tag{3.34a}$$

and:

$$\Gamma(E) = \pi \sum_i |V^i_E|^2 . \tag{3.34b}$$

We obtain finally:

$$\langle \varphi_r | G^+(E) | \varphi_r \rangle = \frac{1}{E - E_r - F(E) + i\Gamma(E)} . \tag{3.35}$$

We may remark at this point that if we had started with z below the real axis in the calculation of the matrix elements of $G(z)$ we would have ended on the real axis with:

$$\langle \varphi_r | G^-(E) | \varphi_r \rangle = \frac{1}{E - E_r - F(E) - i\Gamma(E)} = \langle \varphi_r | G^+(E) | \varphi_r \rangle^* \neq \langle \varphi_r | G^+(E) | \varphi_r \rangle . \tag{3.36}$$

This is why the $\pm$ signs are needed to indicate the way the real axis has been approached. Thus $\langle \varphi_r | G(z) | \varphi_r \rangle$ is a two-valued function with a cut along the real axis. If the continua start at some energy E_m (in the present case $E_m = 0$), the cut extends from E_m to infinity since for E below E_m the delta function $\delta(E - E')$ cannot operate.

$F(E)$ is the level shift (in the present case the radiative shift) and $\Gamma(E)$ the 'damping constant'. Note that they are both functions of E. In practice, as we shall see below, $F(E)$ and $\Gamma(E)$ are often smooth functions of E, and since it is the neighbourhood $E \sim E_r$ which matters in the calculations, one may replace $F(E)$ by $F(E_r)$ and $\Gamma(E)$ by $\Gamma(E_r)$. Thus the atom (or molecule) behaves as if the energy of the excited state was not E_r, but $E_r + F(E_r)$, the energy of the atom dressed with photons. $F(E_r)$ is in fact very small, and we will get rid of it by modifying the definition of the excitation energy. On the other hand, consideration of $\Gamma(E_r)$ is essential for the characterization of the radiative decay.

3.1.5. *Calculation of the Damping Constant*

We recall that V^i_E is a notation for a matrix element which, before the dimensions of the box in which the field is quantized are made infinite, is given by:

$$\frac{|\mathbf{k}_l|}{\sqrt{\hbar c}} \langle r; \text{vac.}| - \mathbf{E} \cdot \boldsymbol{\mu} \, | 0; \mathbf{k}_l, \mathbf{e}_{l\sigma} \rangle , \tag{3.37}$$

with:

$$\mathbf{E} = i \sum_{l', \sigma'} \sqrt{\frac{\hbar\omega_{l'}}{2\varepsilon_0 L^3}}\, \mathbf{e}_{l'\sigma'} (a_{l'\sigma'} - a^{+}_{l'\sigma'}) . \tag{3.38}$$

Of all the terms making up $\mathbf{E}$, only the one containing $a_{l\sigma}$, the operator annihilating the photon $|\mathbf{k}_l, \mathbf{e}_{l\sigma}\rangle$ contributes. Thus this matrix element is:

$$- i \frac{|\mathbf{k}_l|}{\sqrt{\hbar c}} \sqrt{\frac{\hbar\omega_l}{2\varepsilon_0 L^3}}\, \mathbf{e}_{l\sigma} \cdot \mathbf{M}_{r0} = - i \sqrt{\frac{\omega_l^3}{2\varepsilon_0 c^3 L^3}}\, \mathbf{e}_{l\sigma} \cdot \mathbf{M}_{r0} \tag{3.39}$$

$\mathbf{M}_{r0}$ being the transition moment $\langle r| \boldsymbol{\mu} |0\rangle$. We observe that there is indeed a smooth variation of this matrix element as a function of ω_l, and this makes $F(E)$ and $\Gamma(E)$ smooth functions of $E = \hbar\omega_l$. $\Gamma(E_r)$ has to be calculated as the limit when the box becomes infinite of the quantity $\pi \sum_i |V^i_{E_r}|^2$, which is a sum only over the directions of the wave vector and the polarizations, since the modulus of $\mathbf{k}_l$ has to be taken equal to $\hbar c|\mathbf{k}_l| = \hbar\omega_l = E_r - E_0 = \hbar\omega_0$. In a finite box:

$$\mathbf{k}_l = \frac{2\pi}{L} (l_1 \mathbf{U} + l_2 \mathbf{V} + l_3 \mathbf{W}) . \tag{2.26}$$

As $L \to \infty$, l_1/L becomes a continuous variable to be associated with $k_x/2\pi$, etc.... Thus there is a conversion of the sum $L^{-3} \sum_{l,\sigma}$ into the integral $(2\pi)^{-3} \sum_\sigma \iiint \mathrm{d}k_x\, \mathrm{d}k_y\, \mathrm{d}k_z$. In the case of $\Gamma(E_r)$ there is only to integrate over the angles and sum over the polarizations, so that after replacement of ω_l by ω_0 there is obtained:

$$\Gamma(E_r) = \pi \frac{\omega_0^3}{2\varepsilon_0 c^3 (2\pi)^3} \sum_\sigma \int \mathrm{d}\Omega\, |\mathbf{e}_{l\sigma} \cdot \mathbf{M}_{r0}|^2 . \tag{3.40}$$

To proceed to the calculation of $\sum_\sigma \int \mathrm{d}\Omega\, |\mathbf{e}_{l\sigma} \cdot \mathbf{M}_{r0}|^2$ we may assume $\mathbf{M}_{r0}$ to be colinear with the z axis. Let θ and φ be the two angles giving the direction of $\mathbf{k}_l$. A possible choice for the polarization vectors is:

$$\mathbf{e}_{l1} = (\sin\varphi, - \cos\varphi, 0) \text{ and } \mathbf{e}_{l2} = (- \cos\theta \cos\varphi, - \cos\theta \sin\varphi, \sin\theta) . \tag{3.41}$$

Therefore $\sum_\sigma |\mathbf{e}_{l\sigma} \cdot \mathbf{M}_{r0}|^2 = |\mathbf{M}_{r0}|^2 \sin^2\theta$ and:

$$\int \mathrm{d}\Omega \sum_\sigma |\mathbf{e}_{l\sigma} \cdot \mathbf{M}_{r0}|^2 = |\mathbf{M}_{r0}|^2 \int_0^{2\pi} \mathrm{d}\varphi \int_0^{\pi} \mathrm{d}\theta \sin^3\theta = \frac{8\pi}{3} |\mathbf{M}_{r0}|^2 . \tag{3.42}$$

The final formula for the radiative damping constant $\Gamma(E_r)$ is:

$$\Gamma(E_r) = \Gamma_r = \frac{\omega_0^3}{6\varepsilon_0 \pi c^3} |\mathbf{M}_{r0}|^2 . \tag{3.43}$$

Thus there is the approximate relation:

$$\langle\varphi_r|\, G^+(E)\,|\varphi_r\rangle \simeq \frac{1}{E - E_r + i\Gamma_r} \tag{3.44}$$

We may use this expression to calculate $\langle\varphi_r|\, G^+(z)\,|\varphi_r\rangle$ below the real axis (analytical continuation). It is given by:

$$\langle\varphi_r|\, G^+(z)\,|\varphi_r\rangle = \frac{1}{z - E_r + i\Gamma_r}. \tag{3.45}$$

There is a pole of this function at $z = E_r - i\Gamma_r$. It is a general property of these matrix elements that the poles have to be looked for either along the real axis (when there are bound states), or in the lower half-plane.

3.2. Radiative decay of an excited atom or molecule

We apply now this formalism to the study of the radiative decay of an excited atom (or molecule). We assume that in some way the system has been prepared at $t=0$ with the atom in the excited state while the field is in the vacuum state. There is a time evolution since this is not a stationary state. We are going to calculate three quantities referring to different types of experiments:(a) the probability for finding the system at time t in the state $|r;\text{vac.}\rangle$; (b) the differential photon rate; (c) the final spectrum of emitted photons.

3.2.1. *The Decay of the Initial State*

The wave function at time $t=0$ is $|\psi(0)\rangle = |\varphi_r\rangle$. The identity operator in the basis $|\psi_E^{+i}\rangle$ is:

$$\mathbf{1} = \sum_i \int_{E_m}^{\infty} \mathrm{d}E'\, |\psi_{E'}^{+i}\rangle\, \langle\psi_{E'}^{+i}|\,. \tag{3.46}$$

We apply this operator to the initial function and obtain:

$$\begin{aligned} |\psi(0)\rangle = |\varphi_r\rangle &= \sum_i \int_{E_m}^{\infty} \mathrm{d}E'\, |\psi_{E'}^{+i}\rangle\, \langle\psi_{E'}^{+i}\, |\varphi_r\rangle \\ &= \sum_i \int_{E_m}^{\infty} \mathrm{d}E'\, |\psi_{E'}^{+i}\rangle\, V_{E'}^{i*}\, \langle\varphi_r|\, G^+(E')\,|\varphi_r\rangle^*. \end{aligned} \tag{3.47}$$

Since the initial state is now developed as a wave-packet of stationary states, the time evolution is very easily obtained. This is:

$$\begin{aligned} |\psi(t)\rangle &= e^{-(i/\hbar)Ht}\, |\psi(0)\rangle \\ &= \sum_i \int_{E_m}^{\infty} \mathrm{d}E' e^{-(i/\hbar)E't} |\psi_{E'}^{+i}\rangle\, V_{E'}^{i*}\, \langle\varphi_r|\, G^+(E')\,|\varphi_r\rangle^*. \end{aligned} \tag{3.48}$$

The probability amplitude for finding the system at time t in the state $|\varphi_r\rangle$ is:

$$\langle\varphi_r\,|\psi(t)\rangle = \sum_i \int_{E_m}^{\infty} \mathrm{d}E' e^{-(i/\hbar)E't} |V_{E'}^i \langle\varphi_r|\, G^+(E')\,|\varphi_r\rangle|^2, \tag{3.49}$$

since again the amplitude (23a) is involved. We have:

$$\sum_i |V_{E'}^i|^2 |\langle\varphi_r|\, G^+(E')\,|\varphi_r\rangle|^2 = \frac{\Gamma_r}{\pi}\frac{1}{(E'-E_r)^2+\Gamma_r^2}. \tag{3.50}$$

We use the identity:

$$\frac{\Gamma_r}{\pi}\frac{1}{(E'-E_r)^2+\Gamma_r^2} = -\frac{1}{2i\pi}\left\{\frac{1}{E'-E_r+i\Gamma_r}-\frac{1}{E'-E_r-i\Gamma_r}\right\} \tag{3.51}$$

so that $\langle\varphi_r\,|\,\Psi(t)\rangle$ is the combination of two integrals:

$$I = -\frac{1}{2i\pi}\int_{E_m}^{\infty} \mathrm{d}E'\,\frac{e^{-(i/\hbar)E't}}{E'-E_r+i\Gamma_r}; \tag{3.52a}$$

$$II = \frac{1}{2i\pi}\int_{E_m}^{\infty} \mathrm{d}E'\,\frac{e^{-(i/\hbar)E't}}{E'-E_r-i\Gamma_r}. \tag{3.52b}$$

Since the radiative width Γ_r is a very small fraction of the excitation energy E_r, it is possible to extend the lower limit of integration to $-\infty$. These are integrals along the real axis. We complete the contour by a semi-circle in the lower half-plane (Figure 9). Along this semi-circle $\exp[-(i/\hbar)\,z't]$ goes to zero as the radius goes to infinity. Applying Cauchy's theorem to this clockwise closed contour we obtain:

$$I = -2i\pi\left[-\frac{1}{2i\pi}\,e^{-(i/\hbar)(E_r-i\Gamma_r)t}\right] = e^{-(i/\hbar)E_rt}\,e^{-\Gamma_rt/\hbar}, \tag{3.53}$$

and $II=0$ because the pole in the integrand of (52b) is outside the contour. The

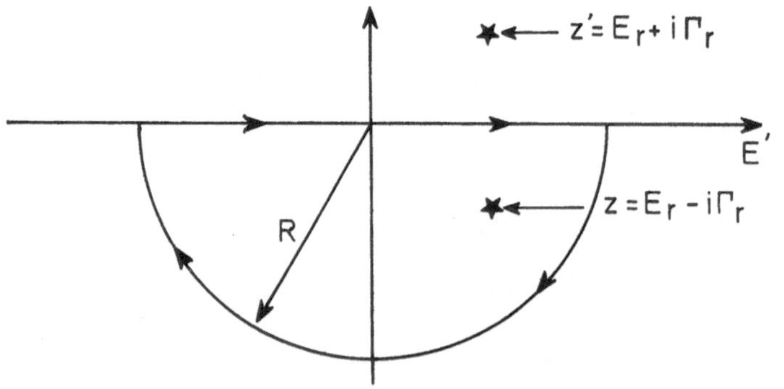

Fig. 9. Contour for the calculation of (49).

probability amplitude for finding the system in the initial state is:

$$\langle \varphi_r \mid \psi(t) \rangle = e^{-(i/\hbar)E_r t} e^{-(\Gamma_r/\hbar)t}, \tag{3.54}$$

and the probability:

$$|\langle \varphi_r \mid \psi(t) \rangle|^2 = e^{-(2\Gamma_r/\hbar)t} = e^{-At} = e^{-t/\tau_r}. \tag{3.55}$$

$$A = \frac{2\pi}{\hbar}\Gamma_r = \frac{\omega_0^3}{3\varepsilon_0 c^3 \pi \hbar} |\mathbf{M}_{r0}|^2$$

is the Einstein coefficient for spontaneous emission, and $\tau_r = A^{-1}$ the radiative lifetime. The system is decaying exponentially with the rate constant A.

3.2.2. *The Differential Photon Rate*

The probability for detecting a photon at time t with a detector which does not differentiate between photons is obtained by summing the probabilities for the continuum states over all possible wave-vectors and polarizations. This is:

$$P(t) = \int_{E_m}^{\infty} dE \sum_i |\langle \varphi_E^i \mid \psi(t) \rangle|^2. \tag{3.56}$$

This quantity does not give the intensity measured at time t, since as $t \to \infty$, $P(t) \to 1$ (the system being certainly no longer in the initial state), while we expect the intensity to go to zero. The intensity is given by the time of change of $P(t)$, that is to say the Differential Photon Rate (DPR) given by:

$$\frac{dP}{dt} = \int_{E_m}^{\infty} dE \frac{d}{dt} \sum_i |\langle \varphi_E^i \mid \psi(t) \rangle|^2. \tag{3.57}$$

We now show that the DPR can be obtained from the probability of finding the system in the discrete state. The time dependent wave function can be written:

$$|\psi(t)\rangle = \tilde{a}(t) e^{-(i/\hbar)E_r t} |\varphi_r\rangle + \int_{E_m}^{\infty} dE \sum_i \tilde{b}_E^i(t) e^{-(i/\hbar)Et} |\varphi_E^i\rangle. \tag{3.58}$$

On introducing (58) into the Schrödinger equation $i\hbar\partial\Psi/\partial t = H\Psi$, there is found that the coefficients obey the coupled equations:

$$i\hbar \frac{d\tilde{a}}{dt} = \int_{E_m}^{\infty} dE \sum_i V_E^i \tilde{b}_E^i e^{-(i/\hbar)(E - E_r)t}, \tag{3.59a}$$

$$i\hbar \frac{d\tilde{b}_E^i}{dt} = V_E^{i*} \tilde{a} e^{-(i/\hbar)(E_r - E)t}. \tag{3.59b}$$

The second of these equations can be integrated formally, the result being *when*

$\tilde{b}_E^i(0)=0$:

$$\tilde{b}_E^i(t) = \frac{V_E^{i*}}{i\hbar}\int_0^t \tilde{a}(t')\,e^{-(i/\hbar)(E_r-E)t'}\,\mathrm{d}t' \tag{3.60}$$

On the other hand:

$$\langle\varphi_E^i \mid \psi(t)\rangle = \tilde{b}_E^i(t)\,e^{-(i/\hbar)Et} \tag{3.61}$$

so that:

$$|\langle\varphi_E^i \mid \psi(t)\rangle|^2 = |\tilde{b}_E^i(t)|^2 \quad \text{and} \quad \frac{d}{\mathrm{d}t}|\langle\varphi_E^i \mid \psi(t)\rangle|^2 = \tilde{b}_E^i\,\frac{\mathrm{d}\tilde{b}_E^{i*}}{\mathrm{d}t} + \text{c.c.} \tag{3.62}$$

Combining (59) and (60) gives:

$$\int_{E_m}^{\infty} \mathrm{d}E\,\tilde{b}_E^i\,\frac{\mathrm{d}\tilde{b}_E^{i*}}{\mathrm{d}t} = \frac{|V_{E_r}^i|^2}{\hbar^2}\,\tilde{a}^*(t)\int_0^{t'}\tilde{a}(t')\,\mathrm{d}t'\int_{E_m}^{\infty}\mathrm{d}E\,e^{(i/\hbar)(E-E_r)(t'-t)}. \tag{3.63}$$

The quantity $|V_E^i|^2$ being a slowly varying function of E is put outside the integral and set equal to $|V_{E_r}^i|^2$. We will admit that E_m can be set equal to $-\infty$ in the integral over E, and use:

$$\int_{-\infty}^{+\infty}\mathrm{d}E\,e^{-(i/\hbar)(E-E_r)(t'-t)} = 2\pi\hbar\delta(t'-t). \tag{3.64}$$

Since:

$$\int_0^t \tilde{a}(t')\,\delta(t'-t)\,\mathrm{d}t' = \tfrac{1}{2}\tilde{a}(t) \tag{3.65}$$

there is obtained:

$$\int_{-\infty}^{+\infty}\mathrm{d}E\,\tilde{b}_E^i\,\frac{\mathrm{d}\tilde{b}_E^{i*}}{\mathrm{d}t} = \frac{\pi|V_{E_r}^i|^2}{\hbar}\,|\tilde{a}(t)|^2, \tag{3.66}$$

and finally:

$$\frac{\mathrm{d}P}{\mathrm{d}t} = \frac{2\pi}{\hbar}\sum_i |V_{E_r}^i|^2\,|\tilde{a}(t)|^2 = \frac{2\Gamma_r}{\hbar}\,|\langle\varphi_r \mid \psi(t)\rangle|^2. \tag{3.67}$$

This shows that the DPR is simply related to the decay of the initial state. It is in fact through the DPR that the decay is generally detected.

3.2.3. *The Spectrum of Emitted Photons*

Another quantity of interest is the energy spectrum of emitted photons. This is:

$$S(E) = \sum_i |\langle\varphi_E^i \mid \psi(+\infty)\rangle|^2 = \sum_i |\tilde{b}_E^i(+\infty)|^2 \tag{3.68}$$

We have:

$$\tilde{b}_E^i(+\infty) = \frac{V_E^{i*}}{i\hbar} \int_0^\infty \tilde{a}(t')\, e^{-(i/\hbar)(E_r - E)t'}\, dt' = \frac{V_E^{i*}}{i\hbar} \int_0^\infty e^{-(i/\hbar)(E_r - i\Gamma_r - E')t'}\, dt'. \tag{3.69}$$

After performing the integration there is obtained:

$$\tilde{b}_E^i(+\infty) = \frac{V_E^{i*}}{E - E_r + i\Gamma_r} \quad \text{and} \quad S(E) = \frac{\Gamma_r}{\pi} \frac{1}{(E - E_r)^2 + \Gamma_r^2}, \tag{3.70}$$

since $\Gamma_r \sim \pi \sum_i |V_E^i|^2$. The spectrum is a Lorentzian centered at $E = E_r$ and of total width $2\Gamma_r = \hbar A$ at half-height. The relation between the lifetime and the width in the energy dispersion is an illustration of the Heisenberg uncertainty relation between time and energy.

3.3. Excitation followed by decay

We consider now a different problem which will serve to indicate when it is legitimate to start from a 'prepared' state with the atom in the excited level. The initial function at time t', say, is taken to be a normalized wave-packet in the continuum $|\varphi_E^i\rangle$, and with energy centered at $\bar{E}$:

$$|\Psi(t')\rangle = \sqrt{\frac{\bar{\Delta}}{\pi}} \int dE' \frac{1}{E' - \bar{E} + i\bar{\Delta}} |\varphi_{E'}^i\rangle. \tag{3.71}$$

Thus the atom is, at that time, with certainty in the ground state, and the photon has a dispersion $2\bar{\Delta}$ in energy, but its direction and polarization are specified. In order to follow the time evolution of this function, we apply the identity operator expressed in the basis of the stationary scattering states $|\Psi_E^{+j}\rangle$. This gives:

$$|\Psi(t')\rangle = \sqrt{\frac{\bar{\Delta}}{\pi}} \int dE' \frac{1}{E' - \bar{E} + i\bar{\Delta}} \int dE'' \sum_j |\psi_{E''}^{+j}\rangle \langle \psi_{E''}^{+j} | \varphi_{E'}^i\rangle. \tag{3.72}$$

The function $|\Psi_E^{+j}\rangle$ can be written:

$$|\psi_E^{+j}\rangle = a^j(E)|\varphi_r\rangle + \sum_i \int dE' b_{E'}^{ji}(E)|\varphi_{E'}^i\rangle. \tag{3.73}$$

Formulae for $a^j(E) = \langle \varphi_r | \Psi_E^{+j}\rangle$ have been given before (Equations 22a and 29). We now need the expression for $b_{E'}^{ji}(E) = \langle \varphi_E^{i'} | \Psi_E^{+j}\rangle$. According to (22b):

$$b_{E'}^{ji}(E) = \delta_{ij}\delta(E - E') + \langle \varphi_{E'}^i | G^+(E) | \varphi_r\rangle V_E^j. \tag{3.74}$$

We have:

$$\langle \varphi_{E'}^i | G(z) | \varphi_r\rangle = \frac{V_{E'}^{i*} \langle \varphi_r | G(z) | \varphi_r\rangle}{z - E'}, \tag{3.75}$$

and:

$$\langle \varphi^i_{E'} | G^+(E) | \varphi_r \rangle = \lim_{\varepsilon \to 0} \langle \varphi^i_{E'} | G(z) | \varphi_r \rangle =$$
$$= \frac{V^{i*}_{E'}}{E - E_r + i\Gamma_r} \left\{ \frac{P}{E - E'} - i\pi\delta(E - E') \right\}. \qquad (3.76)$$

Before going back to our time-dependent problem, we shall pause to show that $|\Psi^{+j}_E\rangle$, which is given by:

$$|\psi^{+j}_E\rangle = \frac{V^j_E}{E - E_r + i\Gamma_r} |\varphi_r\rangle +$$
$$+ \sum_i \int \mathrm{d}E' \left\{ \delta_{ij}\delta(E - E') + \frac{V^{i*}_{E'} V^j_E}{E - E_r + i\Gamma_r} \left[\frac{P}{E - E'} - i\pi\delta(E - E') \right] \right\} |\varphi^i_{E'}\rangle \qquad (3.77)$$

has, when the basis functions describe an ordinary scattering event, an asymptotic behaviour which admits an interesting physical interpretation. In a scattering problem $|\varphi^i_{E'}\rangle$, at large distance between projectile and target (an electron and an atom, two molecular fragments, etc....) has, except for an irrelevant factor, the form $\xi^i \sin(k(E')\, r + \eta)$ where the sine function describes the relative motion (relative coordinate: r, relative kinetic energy: $E' = \hbar^2 k^{1/2}/2\mu$, μ being the reduced mass), and ξ^i the internal motions. $|\varphi^i_E\rangle$ is said to define a channel. Asymptotically $|\varphi_r\rangle$ goes to zero since this is a bound state. Thus:

$$|\psi^{+j}_E\rangle \to \sum_i \int \mathrm{d}E' \left\{ \delta_{ij}\, \delta(E - E') + \frac{V^{i*}_{E'} V^j_E}{E - E_r + i\Gamma_r} \times \right.$$
$$\left. \times \left[\frac{P}{E - E'} - i\pi\delta(E - E') \right] \right\} \xi^i \sin(k(E')\, r + \eta). \qquad (3.78)$$

Integration over E' can immediately be performed, except for:

$$\mathscr{P} \int \frac{V^{i*}_{E'}}{E - E'} \sin(k(E')\, r + \eta)\, \mathrm{d}E' = -\operatorname{Im} \mathscr{P} \int \frac{V^{i*}_{E'}}{E - E'} e^{i(k(E')\, r + \eta)}\, \mathrm{d}E'. \qquad (3.79)$$

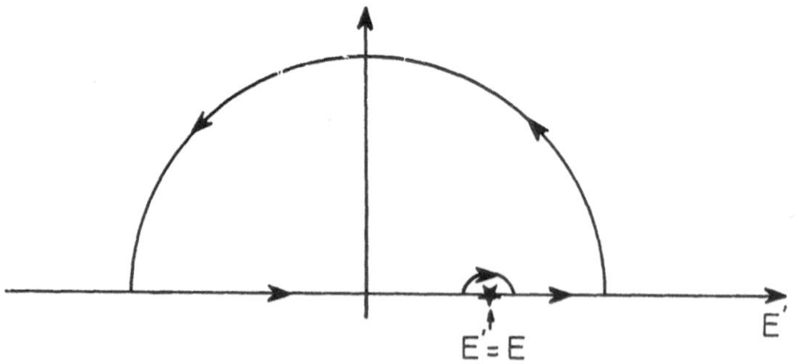

Fig. 10. Contour for the calculation of (79).

With the contour shown in Figure 10, the latter integral is calculated to be $i\pi V_{i*}^{E}$ $\exp[i(k(E)\,r+\eta)]$ (any contribution due to a pole of $V_{E'}^{i*}$ in the upper half-plane vanishes when $r\to\infty$).
Therefore:

$$\mathscr{P}\int \frac{V_{E'}^{i*}}{E-E'}\sin(k(E')\,r+\eta)\,\mathrm{d}E' = -\pi V_E^{i*}\cos(k(E)\,r+\eta). \tag{3.80}$$

When $i\neq j$ in (80) there is found:

$$-\pi\xi^i \frac{V_E^{i*}V_E^j}{E-E_r+i\Gamma_r}\left[\cos(k(E)\,r+\eta)+i\sin(k(E)\,r+\eta)\right] =$$
$$= -2i\pi\,\xi^i \frac{V_E^{i*}V_E^j}{E-E_r+i\Gamma_r}\,\frac{e^{+i(k(E)\,r+\eta)}}{2i}. \tag{3.81}$$

Thus there are only outgoing waves in the channels with $i\neq j$. For $i=j$ there is the term:

$$\xi^i\left\{\frac{e^{+i(k(E)\,r+\eta)}}{2i}\left(1-\frac{2i\pi\,|V_E^j|^2}{E-E_r+i\Gamma_r}\right)-\frac{e^{-i(k(E)\,r+\eta)}}{2i}\right\}. \tag{3.82}$$

The configuration interaction between the discrete state and the continuum states has therefore modified the amplitude of the outgoing wave in the channel i and produced scattered waves in the channels $j\neq i$. The amplitude of the ingoing wave in channel j is that of the unperturbed motion. If in the Lippmann-Schwinger equation we had chosen $-i\varepsilon$ (cf. Equation (8)), we would have obtained an unperturbed outgoing wave in channel j, and ingoing waves in all channels (a very unlikely physical situation).

The amplitudes of the outgoing waves in $|\Psi_E^{+j}\rangle$ give us the jth row of the scattering matrix 'on the energy shell' (conservation of energy is ensured by the form of $|\Psi_E^{+j}\rangle$. We define:

$$\langle\varphi_E^i|\,S\,|\varphi_E^j\rangle = \delta_{ij} - \frac{2i\pi V_E^{i*}V_E^j}{E-E_r+i\Gamma_r}. \tag{3.83}$$

There is in fact a way of calculating the scattering amplitudes which does not require this analysis of the asymptotic behaviour of the scattering states. It can be shown [21] that if a transition operator T is defined by:

$$T = V + VGV \tag{3.84}$$

then S is related to T by:

$$S = \mathbf{1} - 2i\pi T. \tag{3.85}$$

This is easily checked in the present case. V having no matrix elements between continuum states, there is left:

$$\langle\varphi_E^i|\,S\,|\varphi_E^j\rangle = \delta_{ij} - 2i\pi\,\langle\varphi_E^i|\,V\,|\varphi_r\rangle\,\langle\varphi_r|\,G\,|\varphi_r\rangle\,\langle\varphi_r|\,V\,|\varphi_E^j\rangle \tag{3.86}$$

which immediately leads to (83). Although the interaction between an atom or a molecule and the electromagnetic field is not a collision in the ordinary sense, it will be very convenient to use also in this context the language of scattering theory (ingoing and outgoing waves, etc....).

We now apply the evolution operator $e^{-(i/k)Ht}$ to (72) in order to obtain $|\Psi(t)\rangle$ and look for the amplitude $\langle\varphi_r \mid \Psi(t)\rangle$:

$$\langle\varphi_r \, |\psi(t)\rangle = \sqrt{\frac{\bar{\Delta}}{\pi}} \int \mathrm{d}E' \, \frac{1}{E' - \bar{E} + i\,\bar{\Delta}} \int \mathrm{d}E'' \times \\ \times e^{-(i/\hbar)E''(t-t')} \sum_j a^j(E'')\, b_{E'}^{ji*}(E''). \tag{3.87}$$

The integral over E' will first be performed. This is:

$$\int \mathrm{d}E' \, \frac{1}{E' - \bar{E} + i\bar{\Delta}} \left\{ \delta_{ij}\, \delta(E' - E'') + \frac{V_{E'}^i V_{E''}^{j*}}{E'' - E_r - i\Gamma_r} \left[\frac{P}{E'' - E'} \right.\right. \\ \left.\left. + \, i\pi\, \delta(E' - E'') \right] \right\}. \tag{3.88}$$

We ignore as before the energy dependence of the couplings. The contribution of the terms corresponding to scattered waves is zero because:

$$\int \mathrm{d}E' \, \frac{1}{E' - \bar{E} + i\,\bar{\Delta}} \left[\frac{P}{E'' - E} + i\pi\, \delta(E' - E'') \right] = 0. \tag{3.89}$$

This is easily seen from integrating along the contour of Figure 10. Because of the Kronecker symbol δ_{ij} in the other term, we are left with:

$$\langle\varphi_r \, |\psi(t)\rangle = \sqrt{\frac{\bar{\Delta}}{\pi}}\, V_{E_r}^i \int \mathrm{d}E'' \, \frac{e^{-(i/\hbar)E''(t-t')}}{(E'' - \bar{E} + i\bar{\Delta})(E'' - E_r + i\Gamma_r)}, \tag{3.90}$$

where $V_{E''}^i$ has been put equal to V_{Er}^i. The contour we take is that of Figure 9. There are two poles within the contour, one at $z'' = \bar{E} - i\bar{\Delta}$, the other at $z'' = E_r - i\Gamma_r$. The result is:

$$\langle\varphi_r \, |\psi(t)\rangle = (2/i)\sqrt{\pi\bar{\Delta}}\, V_{E_r}^i \, \frac{e^{-(i/\hbar)(\bar{E} - i\bar{\Delta})(t-t')} - e^{-(i/\hbar)(E_r - i\Gamma_r)(t-t')}}{(\bar{E} - i\bar{\Delta} - E_r + i\Gamma_r)} \tag{3.91}$$

We discuss this amplitude in the case where the one-photon wave packet is centered at the excitation energy E_r (this is the most favourable condition for excitation). If $\bar{E} = E_r$, there is the same phase factor in both terms and the probability for finding the atom in the excited state is:

$$P_r(t) = |\langle\varphi_r \, |\psi(t)\rangle|^2 = 4\pi\bar{\Delta}\, |V_{E_r}^i|^2 \left(\frac{e^{-(\Delta/\hbar)(t-t')} - e^{-(\Gamma_r/\hbar)(t-t')}}{\bar{\Delta} - \Gamma_r} \right)^2. \tag{3.92}$$

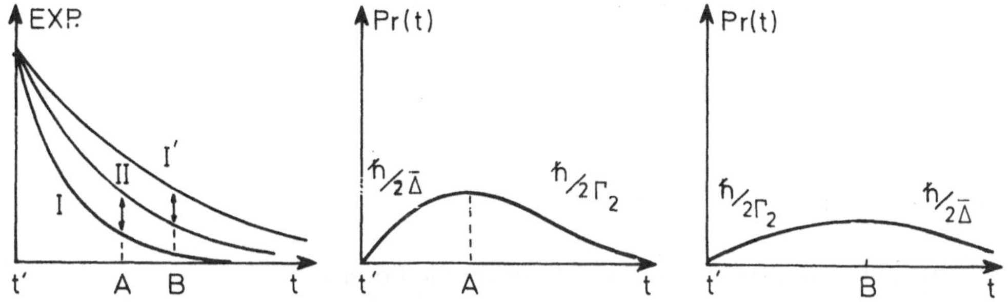

Fig. 11. (a) The exponentials occurring in $P_r(t)$; (b) $P_r(t)$ when $\bar{\Delta} > \Gamma_r$; (c) $P_r(t)$ when $\bar{\Delta} > \Gamma_r$. The arrows in (a) locate the largest differences between the exponentials.

For $t=t'$, $P_r(t')=0$; for $t=+\infty$, $P_r(+\infty)=0$; thus the probability rises from zero to a maximum and decays to zero. In an experiment on a specific atom Γ_r is given, and only $\bar{\Delta}$ can be varied. Consider the two cases $\bar{\Delta} > \Gamma_r$ and $\Delta < \Gamma_r$. The figure 11a gives the function $\exp[-(\bar{\Delta}/\hbar)(t-t')]$ for the two cases (I and I′) together with the function $\exp[-(\Gamma_r/\hbar)(t-t')]$ (II). In the first case $P_r(t)$ will go to a maximum in a time roughly equal to $\hbar/2\bar{\Delta}$ and will decay with the lifetime $\hbar/2\Gamma_r$. In the second case it is the reverse: the rising time is $\hbar/2\Gamma_r$, and the decay time the much longer time $\hbar/2\bar{\Delta}$. Thus we recognize that *the band-width of the exciting pulse to be larger than the width associated with the decay of the system* for the character of the decay to be independent of the mechanism of excitation. This is a simple criterion which will find further applications in our study of systems with internal couplings. The calculation of two auxiliary quantities, the differential photon rate and the spectrum of emitted photons will confirm that when $\bar{\Delta} \gg \Gamma_r$ we recover the results of the previous analysis made with the discrete state as the initial state.

We assume now that we detect with a broad-band detector the photons emitted in a direction (and eventually with a polarization) which is different from that of the incident photon. We are interested by the quantity:

$$\frac{dP^j(t)}{dt} = \int dE \frac{d}{dt} |\langle \varphi_E^j | \psi(t)\rangle|^2 = \int dE \frac{d}{dt} |\tilde{b}_E^j(t)|^2 . \tag{3.93}$$

Our previous Equation (67) is still holding since $\tilde{b}_E^j(t')=0$. Thus:

$$\frac{dP^j(t)}{dt} = \frac{2\Gamma_r^j}{\hbar} |\langle \varphi_r | \psi(t)|^2 ; \tag{3.94}$$

where Γ_r^j is the partial width associated with the decay of the discrete state into the channel j. The partial differential photon rate will rise and decay with $P_r(t)$, and if $\bar{\Delta} \gg \Gamma_r$, the decay will be the same as that calculated in the previous model.

The spectrum of these photons is:

$$S^j(E) = |\langle \varphi_E^j / \psi(+\infty)\rangle|^2 = |\tilde{b}_E^j(+\infty)|^2 . \tag{3.95}$$

Since $j \neq i$ by hypothesis, we have (cf. Equation (62)):

$$\tilde{b}_E^j(t) = \frac{V_{E_r}^{j*}}{i\hbar} \int_{t'}^{t} \tilde{a}(t'')\, e^{-(i/\hbar)(E_r - E)t''}\, dt'', \tag{3.96}$$

with:

$$\tilde{a}(t'') = \langle \varphi_r \, | \psi(t'') \rangle \, e^{(i/\hbar) E_r t''}. \tag{3.97}$$

We use Equation (91) and perform the integration over t'':

$$\tilde{b}_E^j(t) = (2/i)\sqrt{\pi \bar{\Delta}}\, V_E^{j*} V_{E_r}^i \left\{ \frac{e^{-(i/\hbar)E(t-t')}}{(E - E_r + i\Gamma_r)(E - \bar{E} + i\bar{\Delta})} + \right.$$

$$\left. + \frac{e^{-(i/\hbar)(E_r - i\Gamma_r)(t-t')}}{(E_r - i\Gamma_r - E)(E_r - i\Gamma_r - \bar{E} + i\bar{\Delta})} - \frac{e^{-(i/\hbar)(\bar{E} - i\Delta)(t-t')}}{(\bar{E} - i\bar{\Delta} - E)(\bar{E} - i\Delta - E_r + i\Gamma_r)} \right\}. \tag{3.98}$$

Thus after a time $t - t'$ longer than the longer of the two decay times $\hbar/2\Gamma_r$ and $\hbar/2\Delta$ there is left:

$$\tilde{b}_E^j(+\infty) = (2/i)\sqrt{\pi \bar{\Delta}}\, V_E^{j*} V_{E_r}^i \frac{e^{-(i/\hbar)E(t-t')}}{(E - E_r + i\Gamma_r)(E - \bar{E} + i\bar{\Delta})} \tag{3.99}$$

and:

$$S^j(E) = \frac{4\pi \bar{\Delta}\, |V_E^j|^2\, |V_{E_r}^i|^2}{[(E - E_r)^2 + \Gamma_r^2]\,[(E - \bar{E})^2 + \bar{\Delta}^2]}.$$

This is a product of two Lorentzians. If $\bar{\Delta} \gg \Gamma_r$, the Lorentzian of width $2\bar{\Delta}$ may be considered as constant over the range $2\Gamma_r$. Thus the profile is governed by the radiative width, in agreement with the previous treatment.

The 'experiment' which is being described assumes an initial wave-packet of a certain width; photons of various frequencies can be detected. This is not to be confused with another type of measurement where a photon of definite frequency is sent on the atom. In this case conservation of energy requires the energy of an emerging photon to be the same as the energy of the incident photon (this is strictly so only if the recoil energy of the atom is neglected). The quantity of interest is the cross-section $\sigma^j(E)$, a measure of the number of photons of wave-vector and polarization j as a function of the energy of the incident photon. This is:

$$\sigma^j(E) = (2\pi)^4 \left(\frac{\hbar c}{E}\right)^2 |\langle \varphi_E^j | \, T \, | \varphi_E^i \rangle|^2 = (2\pi)^4 \left(\frac{\hbar c}{E}\right)^2 \frac{|V_E^j V_E^i|^2}{(E - E_r)^2 + \Gamma_r^2}. \tag{3.100}$$

This is again a Lorentzian, the maximum in the cross-section occurring when the energy of the incident photon is equal to the excitation energy of the atom.

4. Excitation and Decay of a Multi-Level Molecule

We will now start the study of the more complicated case which is summarized by the diagram of Figure 12. In explicit notation $|\varphi_E^i\rangle$ is again the ground state molecule plus one photon of a given wave-vector and polarization; $|\varphi_r\rangle$ represents the molecule in some excited state $|r\rangle$ and the vacuum of photons; $|\xi_n\rangle$ represents the molecule in the excited state $|n\rangle$ and also the vacuum of photons. The diagram of molecular energy levels is of the type discussed in relation with the application of the Born-Oppenheimer method to molecules. The state $|r\rangle$ belongs to an electronic block, the set $|n\rangle$ to another electronic block (cf. the Figures 8a and 8b). Thus a rather well isolated level of one block (low density of states because of a small vibrational energy) can be facing either a sparse or a dense set of levels of the other block (low or high density of states because of a small or large vibrational energy). There is another circumstance on the diagram which needs some justifications. The states $|\xi_n\rangle$ are supposed not to be coupled to the continua. Since the matrix elements between continuum states and discrete states contain the transition moment (cf. Equation (3.39)), this assumption is equivalent to assuming that the transition moments all vanish. That this may occur frequently in polyatomic molecules can be shown

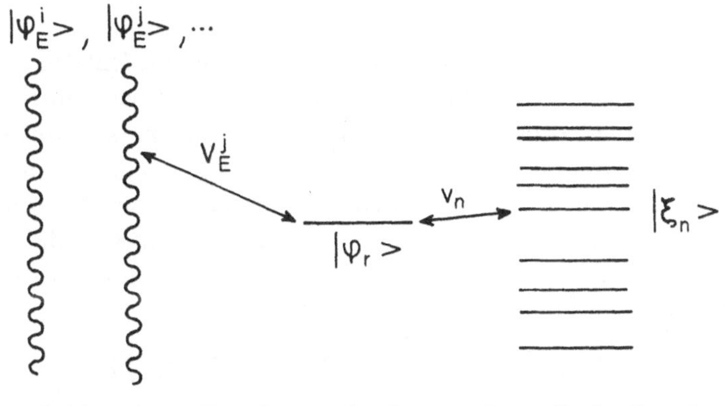

Fig. 12. Energy levels and couplings for a molecule possessing a dipole allowed state coupled to nearby dipole forbidden states. The number of states $|\xi_n\rangle$ effectively coupled to $|\varphi_r\rangle$ can be anything between one and infinity.

as follows. We used previously the notation $\varphi_m(q, Q)\,\chi_{mi}(Q)$ to describe a Born-Oppenheimer state. In the harmonic approximation the potential energy in Equation (2.35) is developed as a Taylor series up to second order in $(Q - Q_0^m)$, where Q_0^m represents the nuclear equilibrium configuration in the mth electronic state. The nuclear factor can then be written as a product of harmonic oscillator wave functions of the $3N$-6 normal modes (N being the number of atoms). The normal modes are the linear combinations of nuclear displacements which transform the Hamiltonian of Equation (2.35) in the harmonic approximation into a sum of independent harmonic oscillator Hamiltonians. Thus $\chi_{mi}(Q)$ can be written:

$$\chi_{mi}(Q) = \prod_{r=1}^{3N-6} \chi_m(v_r, Q_r), \tag{4.1}$$

where v_r stands for the quantum number of the rth harmonic oscillator, and the label m indicates that the function χ_m depends on the electronic state. We assume that in different electronic states the normal modes are the same (this does not amount to assuming the same equilibrium configurations). Thus a transition moment between two vibronic states takes the form:

$$\begin{aligned} \mathbf{M}_{mi,nj} &= \langle \varphi_m(q, Q) \prod_r \chi_m(v_r, Q_r)| \boldsymbol{\mu}_e + \boldsymbol{\mu}_N | \varphi_n(q, Q) \prod_r \chi_n(v'_r, Q_r)\rangle_{q, Q} = \\ &= \langle \prod_r \chi_m(v_r, Q_r) \langle \varphi_m(q, Q)| \boldsymbol{\mu}_e | \varphi_n(q, Q)\rangle_q \prod_r \chi_n(v'_r, Q_r)\rangle_Q . \end{aligned} \tag{4.2}$$

$\boldsymbol{\mu}_e + \boldsymbol{\mu}_N$ is the electric dipole operator of the system of electrons and nuclei, but $\boldsymbol{\mu}_N$ is not contributing because of the orthogonality of the electronic functions. The vector $\mathbf{M}^{el}_{m,n} = \langle \varphi_m(q, Q)| \boldsymbol{\mu}_e | \varphi_n(q, Q)\rangle q$ is the electronic transition moment, which is a function of Q. The Condon approximation amounts to retaining only the zeroth order term in the Taylor expansion near, say, Q^0_m. Under these circumstances:

$$\mathbf{M}_{mi,nj} = \mathbf{M}^{el}_{m,n} |_{Q_m{}^0} \prod_r \langle \chi_m(v_r, Q_r) | \chi_n(v'_r, Q_r)\rangle_{Q_r} \tag{4.3}$$

The transition moment is the electronic transition moment calculated at the equilibrium position of state m multiplied by a product of overlap integrals between vibrational functions. In order to understand the consequences of this factorization we make another assumption which appears to be roughly correct for some classes of molecules [22]: if, when going from state m to state n the electronic surface is displaced along the direction of one normal mode only (this mode will be called the optically active one and denoted Q_a), then the expression (3) for the transition moment reduces to:

$$\mathbf{M}_{mi,nj} = \mathbf{M}^{el}_{m,n} |_{Q_m{}^0} \prod_r \langle \chi_m(v_a, Q_a) | \chi_n(v'_a, Q_a)\rangle_{Q_a} \prod_{r \neq a} \delta_{v_r v_r'} \tag{4.4}$$

The Kronecker symbols arise from the fact that except for Q_a the harmonic oscillator wave functions are the same in both electronic states. Thus we have the selection rule $v'_r = v_r (r \neq a)$, but v_a and v'_a can take any values. For a given v_a and a varying v_a one generates a Franck-Condon progression. We apply now this analysis to the situation pictured in Figure 13. E_0, E_1 and E_2 represent three low-lying electronic states of a molecule. The transition moment between the zero-point level of state 0 and the level r belonging to E_2 is:

$$\mathbf{M}^{el}_{02} |_{Q_0{}^0} \langle \chi_0(0, Q_a) | \chi_2(v_r, Q_a)\rangle_{Q_a} . \tag{4.5a}$$

Between this same state and state n belonging to E_1, it is:

$$\mathbf{M}^{el}_{01} |_{Q_1{}^0} \langle \chi_0(0, Q_a) | \chi_1(v_n, Q_a)\rangle_{Q_a} . \tag{4.5b}$$

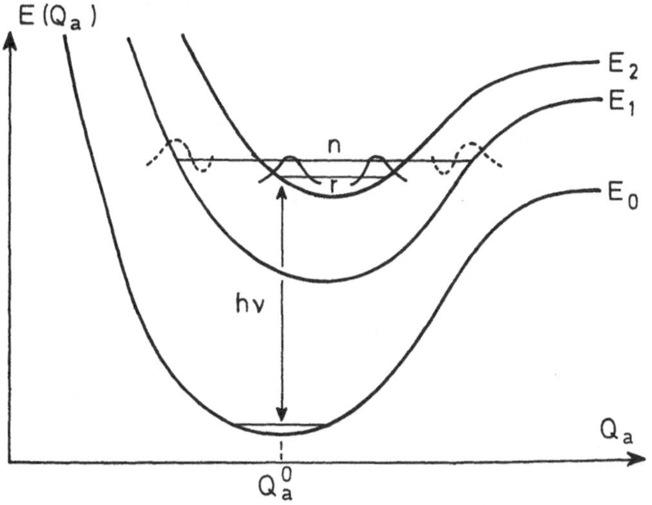

Fig. 13. The transition between states 0 and r is allowed by the Franck-Condon principle. Between 0 and n it is forbidden when the electronic gap between E_2 and E_1 is large.

There are two cases of great practical importance: (a) the electronic gap between E_2 and E_1 is large (E_2 and E_1 being for instance S_2 and S_1, the second and first excited singlets). Both electronic moments may be different from zero, but the overlap integral $\langle\chi_0(0, Q_a) \mid \chi_1(v_n, Q_a)\rangle_{Q_a}$ is very small because the turning points of $\chi_1(v_n, Q_a)$ are out of reach for the ground state vibrational function (cf. Figure 13). We are led in this case to a diagram of molecular energy levels such as shown in Figure 12, with a large number of states $|\xi_n\rangle$. Considerations of more than one optically active mode has no effect on the overall picture. (b) the electronic gap between E_2 and E_1 is small (E_2 and E_1 being for instance S_1 and T_2, the first excited singlet and the second excited triplet). In this case both overlap factors may be important, but the electronic transition moments will be so different that we are again justified in taking the diagram of Figure 12, this time with few (possibly even one only) states $|\xi_n\rangle$.

4.1. Stationary Scattering States for a Molecule with Internal Couplings

We are going now to show that the stationary scattering states corresponding to this level scheme can be calculated exactly (within the approximation which is justified by the smooth variation of V_E^j). We are also going to investigate the case where the set $|\xi_n\rangle$ is so dense (the so-called statistical limit [23]) that, with little error, one may replace this set by a true continuum $|\xi_E\rangle$ (this may of course be the correct model if the state r is above the dissociation limit of the other electronic manifold). Taking any function in one of the continua we define a L.S. stationary scattering state by:

$$|\psi_E^{+i}\rangle = |\varphi_E^i\rangle + \lim_{\varepsilon \to 0} G(z) V |\varphi_E^i\rangle . \quad (3.21)$$

The operator V now contains both the molecule-field interaction and the vibronic coupling. But since $|\varphi_E^i\rangle$ is only coupled to $|\varphi_r\rangle$ we have as before:

$$V\,|\varphi_E^i\rangle = V_E^i\,|\varphi_r\rangle\,. \tag{4.6}$$

We have therefore for the amplitudes:

$$\langle\varphi_r\,|\,\psi_E^{+i}\rangle = \langle\varphi_r|\,G^+(E)\,|\varphi_r\rangle\,V_E^i\,, \tag{4.7a}$$

$$\langle\varphi_{E'}^j\,|\,\psi_E^{+i}\rangle = \delta_{ij}\delta(E-E') + \langle\varphi_{E'}^j|\,G^+(E)\,|\varphi_r\rangle\,V_E^i\,, \tag{4.7b}$$

$$\langle\xi_n\,|\,\psi_E^{+i}\rangle = \langle\xi_n|\,G^+(E)\,|\varphi_r\rangle\,V_E^i\,. \tag{4.7c}$$

The three matrix elements of $G^+(E)$ which are needed can be obtained by a simple extension of the calculations of sections 3.1.3. and 3.1.4. The resolution of the identity is now:

$$\mathbf{1} = |\varphi_r\rangle\,\langle\varphi_r| + \sum_n |\xi_n\rangle\,\langle\xi_n| + \sum_i \int \mathrm{d}E'\,|\varphi_{E'}^i\rangle\,\langle\varphi_{E'}^i|\,. \tag{4.8}$$

Starting from $(z-H)\,G=\mathbf{1}$, we form:

$$\langle\varphi_r|\,(z-H)\,G\,|\varphi_r\rangle = 1\,; \tag{4.9a}$$

$$\langle\varphi_{E'}^i|\,(z-H)\,G\,|\varphi_r\rangle = 0: \tag{4.9b}$$

and

$$\langle\xi_n|\,(z-H)\,|\,G\,|\varphi_r\rangle = 0\,. \tag{4.9c}$$

On introducing the identity operator as expressed in Equation (8) between $(z-H)$ and G in Equations (9), there is obtained:

$$(z-E_r)\,\langle\varphi_r|\,G\,|\varphi_r\rangle - \sum_n v_n\,\langle\xi_n|\,G\,|\varphi_r\rangle -$$

$$-\sum_i \int \mathrm{d}E'\,V_{E'}^i\,\langle\varphi_{E'}|G\,|\,\varphi_r\rangle = 1\,, \tag{4.10a}$$

$$-\,V_{E'}^{i*}\,\langle\varphi_r\,|G|\,\varphi_r\rangle + (z-E')\,\langle\varphi_{E'}^i\,|G|\,\varphi_r\rangle = 0\,, \tag{4.10b}$$

$$-\,v_n^*\,\langle\varphi_r|\,G\,|\varphi_r\rangle + (z-E_n)\,\langle\xi_n|\,G\,|\varphi_r\rangle = 0\,, \tag{4.10c}$$

where v_n and E_n stand for $\langle\varphi_r|\,H\,|\xi_n\rangle$ and $\langle\xi_n|\,H\,|\xi_n\rangle$. Thus:

$$\langle\varphi_{E'}^i|\,G\,|\varphi_r\rangle = \frac{V_{E'}^{i*}}{z-E'}\,\langle\varphi_r|\,G\,|\varphi_r\rangle\,, \tag{4.11a}$$

$$\langle\xi_n|\,G\,|\varphi_r\rangle = \frac{v_n^*}{z-E_n}\,\langle\varphi_r|\,G\,|\varphi_r\rangle\,. \tag{4.11b}$$

With the use of (11a) and (11b), (10a) gives us:

$$\langle\varphi_r|\,G(z)\,|\varphi_r\rangle = \frac{1}{z-E_r-\displaystyle\int \mathrm{d}E'\,\frac{\sum_i |V_{E'}^i|^2}{z-E'} - \sum_n \frac{|v_n|^2}{z-E_n}}\,. \tag{4.12}$$

If z, which is above the real axis, is now allowed to go to the axis, we have, as in the previous problem:

$$\int \mathrm{d}E' \frac{\sum_i |V_{E'}^i|^2}{z - E'} \to F(E) - i\Gamma(E) \tag{4.13}$$

and in addition:

$$\sum_n \frac{|v_n|^2}{z - E_n} \to \sum_n \frac{|v_n|^2}{E - E_n}. \tag{4.14}$$

Thus we obtain an expression for $\langle\varphi_r| G^+(E) |\varphi_r\rangle$. This is:

$$\langle\varphi_r| G^+(E) |\varphi_r\rangle = \frac{1}{E - E_r - F(E) + i\Gamma(E) - \sum_n \frac{|v_n|^2}{E - E_n}}. \tag{4.15}$$

In the calculation of $\langle\varphi_{E'}^i| G^+(E) |\varphi_r\rangle$ and $\langle\xi_n| G^+(E) |\varphi_r\rangle$ from (11a) and (11b), there is to use (15) and to replace $(z-E')^{-1}$ by $P/(E-E') - i\pi\delta(E-E')$. The problem of finding the stationary scattering states has been solved, and any time dependent process can in principle be studied by applying to the initial wave packet the identity operator expressed in this set of states. We are going to test that we have indeed a complete set. We must have obviously:

$$\langle\varphi_r \mid \varphi_r\rangle = \langle\varphi_r| \int \mathrm{d}E \sum_i |\psi_E^{+i}\rangle \langle\psi_E^{+i} \mid \varphi_r\rangle = 1. \tag{4.16a}$$

An explicit form for this is (since $\sum_i |V_E^i|^2 = \pi^{-1}\Gamma(E)$):

$$\int \mathrm{d}E \frac{\pi^{-1}\Gamma(E)}{\left|E - E_r - F(E) + i\Gamma(E) - \sum_n \frac{|v_n|^2}{E - E_n}\right|^2} = 1. \tag{4.16b}$$

This can be changed identically into:

$$-\frac{1}{2i\pi}\left\{\int \mathrm{d}E \frac{1}{E - E_r - F(E) + i\Gamma(E) - \sum_n \frac{|v_n|^2}{E - E_n}} - \right. \tag{4.16c}$$

$$\left. - \int \mathrm{d}E \frac{1}{E - E_r - F(E) - i\Gamma(E) - \sum_n \frac{|v_n|^2}{E - E_n}}\right\} = 1.$$

The first integrand is $\langle\varphi_r| G^+(E) |\varphi_r\rangle$ while the second is $\langle\varphi_r| G^-(E) |\varphi_r\rangle$. We know from the general properties of these matrix elements that $\langle\varphi_r| G^+(z) |\varphi_r\rangle$ can have poles only below the real axis (except if there are bound states, but this is not possible with our level scheme). On the other hand $\langle\varphi_r|\, G^-(z)\,|\varphi_r\rangle$ has poles only above the real axis. We calculate the first integral by choosing the contour of Figure 14a. Since no poles are within the contour, the integral along the real axis plus the integral along the semi-circle gives zero. The latter integral for $|z|\to\infty$ goes to $\int_{s.c.} dz/z = i\pi$ (all other terms being dominated by z in the denominator of the integrand). This leaves $-i\pi$ for our integral. The second integral is calculated with the contour of Figure 14b and $i\pi$ is obtained. Thus the identity (16) has been checked.

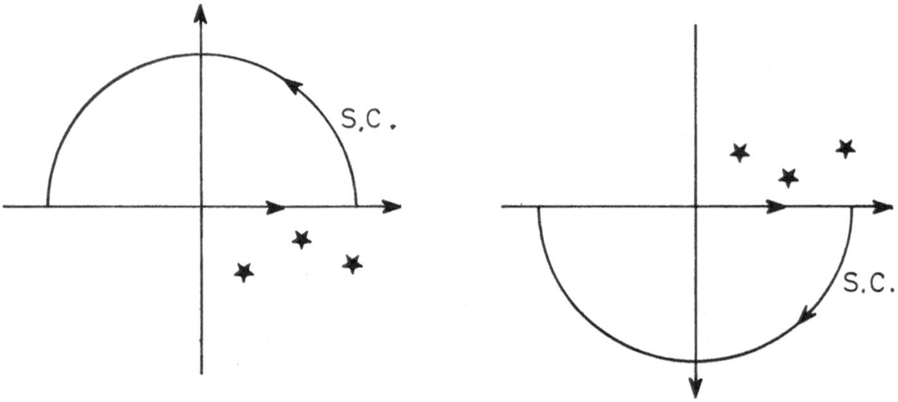

Fig. 14. Contours for the calculation of (4.16). The stars locate the poles of $\langle\varphi_r|\, G^{\pm}(z)\,|\varphi_r\rangle$.

Let us now see what happens if we replace the set of states $|\xi_n\rangle$ by a true continuum $|\xi_E\rangle$. In order to get a manageable problem we imagine that the set is made of equidistant states (separated by δ) with energies $E_n = E_r + n\delta\,(n=0, \pm 1, \pm 2, \ldots)$ and that $v_n = v$. The series in the denominators of $\langle\varphi_r|\, G^{\pm}(z)\,|\varphi_r\rangle$ can then be summed. We have:

$$\sum_n \frac{|v_n|^2}{z - E_n} = |v|^2 \sum_{n=-\infty}^{n=+\infty} \frac{1}{z - E_r - n\delta} = \pi\,|v|^2 \cot\frac{\pi z}{\delta}. \tag{4.17}$$

We must let δ go to zero before having z going to the real axis. From:

$$\cot\frac{\pi z}{\delta} = i\,\frac{e^{i\pi z/\delta} + e^{-i\pi z/\delta}}{e^{i\pi z/\delta} - e^{-i\pi z/\delta}}, \tag{4.18}$$

we observe that $e^{i\pi z/\delta}\to 0$ as $\delta\to 0$ (remembering that z is in the upper half-plane). Thus $\cot\pi z/\delta\to -i$ and calling $|V|^2$ the limit of $|v|^2/\delta$ as v and δ both go to zero (this requires $v/\delta \gg 1$), we have the correspondence:

$$\frac{\pi\,|v|^2}{\delta}\cot\frac{\pi z}{\delta} \to -\,i\pi\,|v|^2 = -\,i\Gamma_{nr}, \tag{4.19}$$

where the nonradiative width is defined as $\Gamma_{nr} = \pi|V|^2$. The quantity $\langle\varphi_r|\, G^+(E)\,|\varphi_r\rangle$

is thus:

$$\langle \varphi_r | G^+(E) | \varphi_r \rangle = \frac{1}{E - E_r - F(E) + i\Gamma(E) + i\Gamma_{nr}}. \tag{4.20}$$

The interaction of $|\varphi_r\rangle$ with the continuum $|\xi_E\rangle$, with a coupling $V = \langle \varphi_r | V | \xi_E \rangle$ has therefore added a width to the radiative width $\Gamma(E) \sim \Gamma(E_r) = \Gamma_r$. However when we go back to relation (16) we observe that now (if we replace $\Gamma(E)$ by Γ_r and ignore the radiative shift):

$$\langle \varphi_r | \int \mathrm{d}E \sum_i |\psi_E^{+i}\rangle \langle \psi_E^{+i} | \varphi_r \rangle =$$

$$= \int \mathrm{d}E \frac{\pi^{-1}\Gamma_r}{|E - E_r + i(\Gamma_r + \Gamma_{nr})|^2} = \frac{\Gamma_r}{\Gamma_r + \Gamma_{nr}} \neq 1. \tag{4.21}$$

The reason for this is that one stationary scattering state per energy is missing. This is obviously the L.S. state:

$$|\psi_E^{+\xi}\rangle = |\xi_E\rangle + \lim_{\varepsilon \to 0} G(z) V |\xi_E\rangle. \tag{4.22}$$

We give below the complete set of amplitudes of the L.S. states to be used in time-dependent problems concerning this case, where one discrete molecular state is coupled intramolecularly to a continuum of states:

$$\langle \varphi_r | \psi_E^{+i} \rangle = \langle \varphi_r | G^+(E) | \varphi_r \rangle V_E^i; \tag{4.23a}$$

$$\langle \xi_{E'} | \psi_E^{+i} \rangle = \langle \xi_{E'} | G^+(E) | \varphi_r \rangle V_E^i; \tag{4.23b}$$

$$\langle \varphi_{E'}^j | \psi_E^{+i} \rangle = \delta_{ji}\delta(E' - E) + \langle \varphi_{E'}^j | G^+(E) | \varphi_r \rangle V_E^i; \tag{4.23c}$$

$$\langle \varphi_r | \psi_E^{+\xi} \rangle = \langle \varphi_r | G^+(E) | \varphi_r \rangle V; \tag{4.24a}$$

$$\langle \xi_{E'} | \psi_E^{+\xi} \rangle = \delta(E - E') + \langle \xi_{E'} | G^+(E) | \varphi_r \rangle V; \tag{4.24b}$$

$$\langle \varphi_{E'}^j | \psi_E^{+\xi} \rangle = \langle \varphi_{E'}^j | G^+(E) | \varphi_r \rangle V. \tag{4.24c}$$

The off diagonal matrix elements of $G^+(E)$ which are needed can be obtained from:

$$\langle \xi_{E'} | G^+(E) | \varphi_r \rangle = V^* \langle \varphi_r | G^+(E) | \varphi_r \rangle \left\{ \frac{P}{E - E'} - i\pi\delta(E - E') \right\}. \tag{4.25a}$$

$$\langle \varphi_{E'}^j | G^+(E) | \varphi_r \rangle = V_{E'}^{j*} \langle \varphi_r | G^+(E) | \varphi_r \rangle \left\{ \frac{P}{E - E'} - i\pi\delta(E - E') \right\}. \tag{4.25b}$$

One can then easily check the following relation:

$$\langle \varphi_r | \int \mathrm{d}E \left[\sum_i |\psi_E^{+i}\rangle \langle \psi_E^{+i}| + |\psi_E^{+\xi}\rangle \langle \psi_E^{+\xi}| \right] |\varphi_r\rangle = 1. \tag{4.26}$$

4.2. Excitation and Decay of a Molecule in the Statistical Limit

We have just seen that in the statistical limit the continuum $|\xi_E\rangle$ which simulates the very dense set of molecular states is to be treated on equal footing with the other continua $|\varphi_E^i\rangle \ldots$. Thus we may suspect that this case requires a very simple extension of the treatment of the two level molecule. Assuming for instance that the system is initially prepared in the state $|\varphi_r\rangle$, we obtain the decay law:

$$\langle\varphi_r|\psi(t)\rangle = \int \mathrm{d}E' e^{-(i/\hbar)E't}\{\sum_i \langle\varphi_r|\psi_{E'}^{+i}\rangle\langle\psi_{E'}^{+i}|\varphi_r\rangle + \langle\varphi_r|\psi_E^{+\xi}\rangle\langle\psi_E^{+\xi}|\varphi_r\rangle\}. \tag{4.27}$$

With the level shift incorporated in E_r and $\Gamma(E)$ replaced by Γ_r this is:

$$\langle\varphi_r|\psi(t)\rangle = \int \mathrm{d}E' e^{-(i/\hbar)E't} \frac{\pi^{-1}(\Gamma_r + \Gamma_{nr})}{(E - E_r)^2 + (\Gamma_r + \Gamma_{nr})^2} = e^{-(i/\hbar)E_r t} e^{-(\Gamma_r + \Gamma_{nr})t/\hbar}. \tag{4.28}$$

Thus:

$$P_r(t) = e^{-2(\Gamma_r + \Gamma_{nr})t/\hbar} = e^{-t/\tau}, \tag{4.29}$$

with:

$$\frac{1}{\tau} = \frac{2\Gamma_r}{\hbar} + \frac{2\Gamma_{nr}}{\hbar} = \frac{1}{\tau_r} + \frac{1}{\tau_{nr}}. \tag{4.30}$$

The radiative and the nonradiative rate constants have to be added. This is due to the independent decay toward the two kinds of channels. The DPR is as before:

$$\frac{\mathrm{d}P}{\mathrm{d}t} = \frac{2\Gamma_r}{\hbar}|\langle\varphi_r|\psi(t)\rangle|^2. \tag{4.31}$$

The spectrum of emitted photons is:

$$S(E) = \sum_i |\langle\varphi_E^i|\psi(t)\rangle|^2 = \frac{\Gamma_r}{\pi}\frac{1}{(E - E_r)^2 + (\Gamma_r + \Gamma_{nr})^2}. \tag{4.32}$$

The width is therefore the sum of the radiative and nonradiative widths. The *quantum yield* for radiative decay is by definition the probability that one photon has been emitted. This is:

$$\mathrm{Y}_{\mathrm{ph.}} = \int \mathrm{d}E\, S(E) = \frac{\Gamma_r}{\Gamma_r + \Gamma_{nr}}. \tag{4.33}$$

The probability for intramolecular conversion with energy E (or possibly fragmentation, if $|\xi_E\rangle$ describes a dissociative state) is:

$$P_{\mathrm{IC}}(E) = \frac{\Gamma_{nr}}{\pi}\frac{1}{(E - E_r)^2 + (\Gamma_r + \Gamma_{nr})^2}, \tag{4.34}$$

and the quantum yield for intramolecular conversion is:

$$\mathrm{Y}_{\mathrm{IC}} = \int \mathrm{d}E\, P_{\mathrm{IC}}(E) = \frac{\Gamma_{nr}}{\Gamma_r + \Gamma_{nr}}. \tag{4.35}$$

Thus if initially the molecule was raised to the excited state by the absorption of a photon, some of the electromagnetic energy is converted into molecular vibrational energy (or translational energy in the case of dissociation). Other processes may follow, either radiative or nonradiative, for example vibrational relaxation, the excess energy being given to the environment. In this case there is finally a conversion of the electromagnetic energy into heat, a very common process of course.

We now assume that initially, at time t', the wave function is a wave packet in the continuum $|\varphi_E^i\rangle$. A straight-forward extension of our previous calculation shows that if the photon wave packet is centered at E_r, the probability for finding the system in state $|\varphi_r\rangle$ is:

$$P_r(t) = 4\pi\bar{\Delta}\,\frac{(e^{-\bar{\Delta}/\hbar\,(t-t')} - e^{-(\Gamma_r+\Gamma_{nr})\,(t-t')^2/\hbar}}{(\bar{\Delta} - \Gamma_r - \Gamma_{nr})^2}. \tag{4.36}$$

Thus if $\bar{\Delta} \gg \Gamma_r + \Gamma_{nr}$ (or in other words if the coherence time $\hbar/2\bar{\Delta}$ of the photon wave packet is much shorter than the shorter of the two molecular decay times $\hbar/2\Gamma_r$ and $\hbar/2\Gamma_{nr}$) it is possible to distinguish the preparation and the decay of the state $|\varphi_r\rangle$, and all the features accompanying the decay of $|\varphi_r\rangle$ are recovered.

4.3. Excitation and Decay Modes of a Molecule with Two Coupled Excited States

We now go to the other limit where only *one* state $|\xi\rangle$ is coupled to $|\varphi_r\rangle$, since this problem can also be solved analytically. There is evidence that this situation is present in certain molecules [26]. For simplicity we assume $|\varphi_r\rangle$ and $|\xi\rangle$ to be degenerate, with energy E_r. Thus the energy level scheme is that of Figure 15. We assume v to be

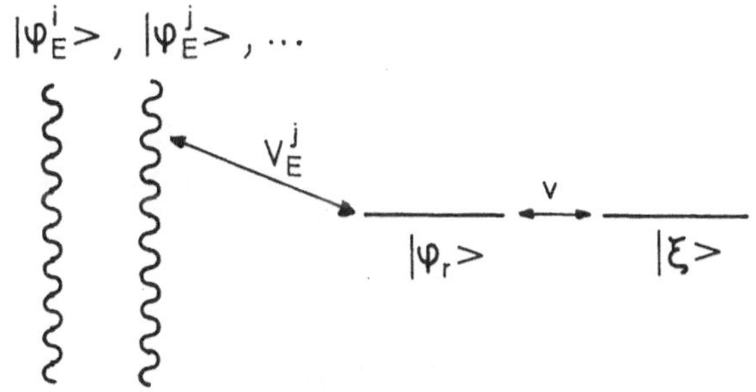

Fig. 15. Energy level scheme for a molecule with two degenerate coupled excited states, one of which is dipole forbidden.

real. Let us start by studying the decay of an arbitrary combination of the two discrete states. Thus:

$$|\psi(0)\rangle = a(0)\,|\xi\rangle + b(0)\,|\varphi_r\rangle. \tag{4.37}$$

At time t the wave function is:

$$|\psi(t)\rangle = a(t)\,|\xi\rangle + b(t)\,|\varphi_r\rangle + \text{continuum components}. \tag{4.38}$$

We are interested for the moment in the calculation of the amplitudes $a(t)=$

$=\langle\xi \mid \Psi(t)\rangle$ and $b(t)=\langle\varphi_r \mid \Psi(t)\rangle$. Applying on the initial wave packet the identity operator in the basis of stationary scattering states and then the evolution operator, there is obtained:

$$|\psi(t)\rangle = a(0)\sum_i \int dE\, e^{-(i/\hbar)Et} |\psi_E^{+i}\rangle \langle\psi_E^{+i}|\xi\rangle + b(0)\sum_i \int dE\, e^{-i/\hbar} |\psi_E^{+i}\rangle \langle\psi_E^{+i}|\varphi_r\rangle. \tag{4.39}$$

Using the definitions of $a(t)$ and $b(t)$, and the formulas (7a, c) for the amplitudes $\langle\varphi_r \mid \Psi_E^{+i}\rangle$ and $\langle\xi \mid \Psi_E^{+i}\rangle$ leads to:

$$a(t) = a(0)\int dE\, e^{-(i/\hbar)Et} \frac{v^2\pi^{-1}\Gamma_r}{|(E-E_r-i\Gamma_r)(E-E_r)-v^2|^2} + b(0)\int dE\, e^{-(i/\hbar)Et} \frac{v\pi^{-1}\Gamma_r(E-E_r)}{|(E-E_r-i\Gamma_r)(E-E_r)-v^2|^2}; \tag{4.40a}$$

$$b(t) = a(0)\int dE\, e^{-(i/\hbar)Et} \frac{v\pi^{-1}\Gamma_r(E-E_r)}{|(E-E_r-i\Gamma_r)(E-E_r)-v^2|^2} + b(0)\int dE\, e^{-(i/\hbar)Et} \frac{\pi^{-1}\Gamma_r(E-E_r)^2}{|(E-E_r-i\Gamma_r)(E-E_r)-v^2|^2}. \tag{4.40b}$$

These integrals can be calculated with a contour made of the real axis and a semi-circle in the lower half-plane. A great simplification results from using the identity:

$$\frac{1}{|(E-E_r+i\Gamma_r)(E-E_r)-v^2|^2} = -\frac{1}{2iv^2\Gamma_r} \times \left\{\frac{E-E_r+i\Gamma_r}{(E-E_r+i\Gamma_r)(E-E_r)-v^2} - \frac{E-E_r-i\Gamma_r}{(E-E_r-i\Gamma_r)(E-E_r)-v^2}\right\} \tag{4.41a}$$

in the first integral, and the identity:

$$\frac{(E-E_r)}{|(E-E_r+i\Gamma_r)(E-E_r)-v^2|^2} = \frac{1}{2i\Gamma_r} \times \left\{\frac{1}{(E-E_r+i\Gamma_r)(E-E_r)-v^2} - \frac{1}{(E-E_r-i\Gamma_r)(E-E_r)-v^2}\right\}, \tag{4.41b}$$

in the last three integrals, because in each case only the first term has poles below the real axis. These poles are obtained by solving:

$$(z-E_r+i\Gamma_r)(E-E_r)-v^2=0 \tag{4.42}$$

and are:

$$z_\pm = E_r - \frac{i\Gamma_r}{2} \pm \frac{\sqrt{\Delta}}{2}, \tag{4.43}$$

with $\Delta = 4v^2 - \Gamma_r^2$ (note that $\sqrt{\Delta}$ can be imaginary). The result is:

$$a(t) = e^{-(i/\hbar)(E_r - i\Gamma_r/2)t} \left\{ a(0) \left[\cos \frac{\sqrt{\Delta} t}{2\hbar} + \right.\right.$$
$$\left.\left. + \frac{\Gamma_r}{\sqrt{\Delta}} \sin \frac{\sqrt{\Delta} t}{2\hbar} \right] - 2ib(0) \frac{v}{\sqrt{\Delta}} \sin \frac{\sqrt{\Delta} t}{2\hbar} \right\}; \quad (4.44a)$$
$$b(t) = e^{-(i/\hbar)(E_r - i(\Gamma_r/2))t} \times$$
$$\left\{ -2ia(0) \frac{v}{\sqrt{\Delta}} \sin \frac{\sqrt{\Delta} t}{2\hbar} + b(0) \left[\cos \frac{\sqrt{\Delta} t}{2\hbar} - \frac{\Gamma_r}{\sqrt{\Delta}} \sin \frac{\sqrt{\Delta} t}{2\hbar} \right]. \quad (4.44b)\right.$$

Several regimes can be distinguished. Consider first the case where $2v \ll \Gamma_r$ (the resonance frequency between $|\varphi_r\rangle$ and $|\xi\rangle$ is much smaller than the radiative width):

(a) Assuming $a(0) = 0$ and $b(0) = 1$ we obtain:

$$b(t) = e^{-(i/\hbar) E_r t} e^{-\Gamma_r t/2\hbar} \left[\cos \frac{\sqrt{\Delta} t}{2\hbar} - \frac{\Gamma_r}{\sqrt{\Delta}} \sin \frac{\sqrt{\Delta} t}{2\hbar} \right]. \quad (4.45)$$

Since $2v \ll \Gamma_r$, $\Delta = i^2 (\Gamma_r - 4v^2)$ and

$$\sqrt{\Delta} \sim i\Gamma_r - 2i \frac{v^2}{\Gamma_r}. \quad (4.46)$$

Thus $\Gamma_r / \sqrt{\Delta} \sim -i$ and:

$$b(t) = e^{-(i/\hbar) E_r t} e^{-\Gamma_r t/\hbar}. \quad (4.47)$$

This is the decay we would have with $|\varphi_r\rangle$ decoupled from $|\xi\rangle$.

(b) Assuming $a(0) = 1$ and $b(0) = 0$ we obtain:

$$a(t) = e^{-(i/\hbar) E_r t} e^{-\Gamma_r t/2\hbar} \left[\cos \frac{\sqrt{\Delta} t}{2\hbar} + \frac{\Gamma_r}{\sqrt{\Delta}} \sin \frac{\sqrt{\Delta} t}{2\hbar} \right] =$$
$$\sim e^{-(i/\hbar) E_r t} e^{-\Gamma_r t/2\hbar} e^{+\Gamma_r t/2\hbar} e^{-v^2 t/\hbar\Gamma_r}. \quad (4.48)$$

The initial state is decaying very slowly, with a lifetime $2\hbar\Gamma_r / v^2$. If $\Gamma_r / v \to \infty$ the initial state becomes a stationary state. This result was noted very easily [24, 25] in the problem of the effect of a static electric field on the metastability of the $2s$ state of hydrogen which is nearly degerate with one of the $2p$ states, and was derived again in some recent work on the effect of pressure on the fluorescence quantum yield of a molecule with two coupled excited states [26].

If on the contrary $2v \gg \Gamma_r$ we have:

$$\sqrt{\Delta} \simeq 2v - \frac{\Gamma_r^2}{4v}. \quad (4.49)$$

(a) If $a(0)=0$ and $b(0)=1$ we obtain (since $\Gamma_r/\sqrt{\Delta}$ is very small):

$$b(t) \simeq e^{-(i/\hbar)E_r t} e^{-\Gamma_r t/2\hbar} \cos \frac{vt}{\hbar}. \tag{4.50}$$

(b) If $a(0)=1$ and $b(0)=0$ we obtain:

$$a(t) \simeq e^{-(i/\hbar)E_r t} e^{-\Gamma_r t/\hbar} \cos \frac{vt}{\hbar}. \tag{4.51}$$

The result is the same in both cases. This is a damped oscillation with a radiative damping which is one half the radiative damping of $|\varphi_r\rangle$. The differential photon rate also shows these damped oscillations. This effect (quantum beats) has been observed [27, 28] when metastable excited hydrogen atoms in the $2s_{1/2}$ state are sent in a region where a static electric field is applied. There is a coupling with the nearby unstable $2p_{1/2}$ state, and if the field is strong enough, $2v \rangle \Gamma_r$. However in this case it is an external perturbation which induces the mixing.

In (50) and (51) there are oscillations at the resonance frequency, which may be very rapid if $2v \gg \Gamma_r$. This suggests looking for another basis which would represent more appropriately the decay modes of the system. We diagonalize the 2×2 matrix of the Hamiltonian built on $|\varphi_r\rangle$ and $|\xi\rangle$. The new functions (molecular wave functions) are:

$$|1\rangle = \frac{1}{\sqrt{2}}\{|\varphi_r\rangle - |\xi\rangle\};\ |2\rangle = \frac{1}{\sqrt{2}}\{|\varphi_r\rangle + |\xi\rangle\}; \tag{4.52}$$

the two energies are $E_1 = E_r - v$ and $E_2 = E_r + v$. We write:

$$|\psi(t)\rangle = C_1(t)|1\rangle + C_2(t)\,12\rangle + \text{continuum components}. \tag{4.53}$$

By identification with the expression (38) for $|\Psi(t)\rangle$:

$$C_1(t) = 2^{-1/2}[-a(t) + b(t)], \tag{4.54a}$$

$$C_2(t) = 2^{-1/2}[a(t) + b(t)], \tag{4.54b}$$

or in explicit form:

$$C_1(t) = e^{-(i/\hbar)(E_r - i\Gamma_r/2)t}\left\{C_1(0)\left[\cos\frac{\sqrt{\Delta}t}{2\hbar} + \frac{2iv}{\sqrt{\Delta}}\sin\frac{\sqrt{\Delta}t}{2\hbar}\right] - \right.$$

$$\left. - C_2(0)\frac{\Gamma_r}{\sqrt{\Delta}}\sin\frac{\sqrt{\Delta}t}{2\hbar}\right\}; \tag{4.55a}$$

$$C_2(t) = e^{-(i/\hbar)(E_r - i\Gamma_r/2)t}\left\{-C_1(0)\frac{\Gamma_r}{\sqrt{\Delta}}\sin\frac{\sqrt{\Delta}t}{2\hbar} + \right.$$

$$\left. + C_2(0)\left[\cos\frac{\sqrt{\Delta}t}{2\hbar} - \frac{2iv}{\sqrt{\Delta}}\sin\frac{\sqrt{\Delta}t}{2\hbar}\right]\right\}. \tag{4.55b}$$

Consider the case where $C_1(0)=1$ and $C_2(0)=0$. We have:

$$C_1(t) = e^{-(i/\hbar)E_r t} e^{-\Gamma_r t/2\hbar} \left[\cos \frac{\sqrt{\Delta} t}{2\hbar} + \frac{2iv}{\sqrt{\Delta}} \sin \frac{\sqrt{\Delta} t}{2\hbar} \right]. \tag{4.56}$$

However $2v/\sqrt{\Delta} \sim 1$ and $\sqrt{\Delta} \sim 2v - \Gamma_r^2/4v$. Thus:

$$C_1(t) = e^{-(i/\hbar)E_r t} e^{(i/\hbar)(v - \Gamma_r^2/8v)t} e^{-\Gamma_r t/2\hbar}. \tag{4.57}$$

The lifetime is twice that of $|\varphi_r\rangle$ decoupled from $|\xi\rangle$. This is a so-called anomalous life-time which is due to the sharing of the transition moment $\langle 0|\boldsymbol{\mu}|r\rangle$ by both states $|1\rangle$ and $|2\rangle$ [11,29].

The DPR in any case is given by:

$$\frac{\mathrm{d}P}{\mathrm{d}t} = \frac{2\Gamma_r}{\hbar} |\langle \varphi_r \, | \psi(t) \rangle|^2, \tag{4.58}$$

and the spectrum of emitted photons can be calculated from the amplitude $\langle \varphi_r \mid \Psi(t)\rangle$. It is not however a Lorentzian. It may show two maxima. Instead of looking at this aspect of the problem, we will show that there are still initial states which lead to a Lorentzian profile.

We will interpret the amplitudes $a(t)$ and $b(t)$ given by Equations (44a) and (44b) in a somewhat different way. Since the poles in the integrands are $z^{\pm}$ (cf. Equations (41a, b)) $a(t)$ and $b(t)$ are linear combinations of:

$$e^{-(i/\hbar)z^+t} \quad \text{and} \quad e^{-(i/\hbar)z^-t}$$

(because of the presence of $\exp[-(i\hbar)Et]$ in the integrands), and they can be written:

$$a(t) = \frac{(\sqrt{\Delta} + i\Gamma_r)\, a(0) + 2vb(0)}{2\sqrt{\Delta}} e^{-(i/\hbar)z^+t} + \\ + \frac{(\sqrt{\Delta} - i\Gamma_r)\, a(0) - 2vb(0)}{2\sqrt{\Delta}} e^{-(i/\hbar)z^-t}, \tag{4.59a}$$

$$b(t) = \frac{2va(0) + (\sqrt{\Delta} - i\Gamma_r)\, b(0)}{2\sqrt{\Delta}} e^{-(i/\hbar)z^+t} + \\ + \frac{-2va(0) + (\sqrt{\Delta} + i\Gamma_r)\, b(0)}{2\sqrt{\Delta}} e^{-(i/\hbar)z^-t}. \tag{4.59b}$$

From $|\Psi(0)\rangle = a(0)\,|\xi\rangle + b(0)\,|\varphi_r\rangle$
and $|\Psi(t)\rangle = a(t)\,|\xi\rangle + b(t)\,|\varphi_r\rangle +$ continuum components we calculate that the probability amplitude describing the decay of the initial state is:

$$\langle \Psi(0) \mid \Psi(t)\rangle = a^*(0)\, a(t) + b^*(0)\, b(t). \tag{4.60}$$

Thus if in (59a) and (59b) we can arrange for one exponential only to be left with an appropriate choice of initial conditions, this will lead to a simple exponential decay. There are two possible choices which define the two decay modes of the system. One is taking:

$$(\sqrt{\Delta} - i\Gamma_r)\, a^+(0) - 2v\, b^+(0) = 0, \tag{4.61a}$$

$$-\,2v\, a^+(0) + (\sqrt{\Delta} + i\Gamma_r)\, b^+(0) = 0. \tag{4.61b}$$

with, when $2v > \Gamma_r$ the normalized solution:

$$a^+(0) = \frac{\sqrt{2}\, v}{\sqrt{\Delta} - i\Gamma_r}\,; \; b^+(0) = \frac{1}{\sqrt{2}}, \tag{4.62}$$

(both coefficients have the same modulus) and when $2v < \Gamma_r$ the normalized solution:

$$a^+(0) = \frac{\sqrt{2}\, v}{\sqrt{\Gamma_r(\Gamma_r - \sqrt{\Delta'})}}\,; \; b^+(0) = -\frac{i}{\sqrt{2}}\sqrt{\frac{\Gamma_r - \sqrt{\Delta'}}{\Gamma_r}}, \tag{4.63}$$

with $\Delta' = -\Delta = \Gamma_r^2 - 4v^2$ (the two coefficients have now different moduli). In any case one calculates that:

$$a(t) = a(0)\, e^{-(i/\hbar)z^+ t}; \; b(t) = b(0)\, e^{-(i/\hbar)z^+ t}, \tag{4.64}$$

and:

$$\langle \Psi(0) \mid \Psi(t) \rangle = [|a^+(0)|^2 + |b^+(0)|^2]\, e^{-(i/\hbar)z^+ t} = e^{-(i/\hbar)z^+ t}. \tag{4.65}$$

The other choice is:

$$(\sqrt{\Delta} + i\Gamma_r)\, a^-(0) + 2v\, b^-(0) = 0, \tag{4.66a}$$

$$2v a^-(0) + (\sqrt{\Delta} - i\Gamma_r)\, b^-(0) = 0. \tag{4.66b}$$

with, when $2v > \Gamma_r$ the normalized solution:

$$a^-(0) = -\frac{\sqrt{2}\, v}{\sqrt{\Delta} + i\Gamma_r}\,; \; b^-(0) = \frac{1}{\sqrt{2}}, \tag{4.67}$$

and when $2v < \Gamma_r$ the normalized solution:

$$a^-(0) = \frac{\sqrt{2}\, v}{\sqrt{\Gamma_r(\Gamma_r + \sqrt{\Delta'})}}\,; \; b^-(0) = -\frac{i}{\sqrt{2}}\sqrt{\frac{\Gamma_r + \sqrt{\Delta'}}{\Gamma_r}}. \tag{4.68}$$

In this case:

$$a(t) = a(0)\, e^{-(i/\hbar)z^- t}; \; b(t) = b(0)\, e^{-(i/\hbar)z^- t} \tag{4.69}$$

and:

$$\langle \Psi(0) \mid \Psi(t) \rangle = e^{-(i/\hbar)z^- t}. \tag{4.70}$$

It is interesting to note that when $2v/\Gamma_r \to \infty$ the two functions $a^+(0)\,|\xi\rangle + b^+(0)\,|\varphi_r\rangle$ and $a^-(0)\,|\xi\rangle + b^-(0)\,|\varphi_r\rangle$ go to $2^{-1/2}[|\xi\rangle + |\varphi_r\rangle]$ and to $2^{-1/2}[|\xi\rangle - |\varphi_r\rangle]$, which are the 'molecular wave functions' (*i.e.* the functions diagonalizing the intramolecular coupling). The condition $2v \gg \Gamma_r$ is of course the condition for having two well isolated molecular states which can separately excited, with subsequent independent decays. On the other hand if $\Gamma_r/2v \to \infty$ the two functions go respectively to $|\xi\rangle$ and $-i|\varphi_r\rangle$, with $|\xi\rangle$ approaching a stationary state while $|\varphi_r\rangle$ decays with its own damping constant.

The Equations (61a, b) and (66a, b) look very much like secular equations, and we shall show that indeed they are of this kind. Let us go back to the equations which define the matrix elements of the resolvent. The four equations obtained from 'multiplying' $(z-H)\,G=\mathbf{1}$ to the left and to the right by either $\langle\varphi_r|$ or $\langle\xi|$ and $|\varphi_r\rangle$ or $|\xi\rangle$ contain the elements $\langle\varphi^i_{E'}|\,G\,|\varphi_r\rangle$ and $\langle\varphi^i_{E'}|\,G\,|\xi\rangle$ but these can be eliminated with the help of:

$$\langle\varphi^i_{E'}|\,G(z)\,|\varphi_r\rangle = \frac{V^{i*}_{E'}}{z-E'}\,\langle\varphi_r|\,G(z)\,|\varphi_r\rangle\,, \tag{4.71a}$$

and

$$\langle\varphi^i_{E'}|\,G(z)\,|\xi\rangle = \frac{V^{i*}_{E}}{z-E'}\,\langle\varphi_r|\,G(z)\,|\xi\rangle\,. \tag{4.71b}$$

The resulting four equations are then (in the limit $\varepsilon\to 0$):

$$\begin{aligned}
&\left(E - E_r - \lim_{\varepsilon\to 0}\int \mathrm{d}E'\,\frac{\sum_i |V^i_{E'}|^2}{z-E'}\right)\langle\varphi_r|\,G^+(E)\,|\varphi_r\rangle - v\langle\xi|\,G^+(E)\,|\varphi_r\rangle = 1\,,\\
&-v\langle\varphi_r|\,G^+(E)\,|\varphi_r\rangle + (E-E_r)\,\langle\xi|\,G^+(E)\,|\varphi_r\rangle = 0\,,\\
&-v\langle\varphi_r|\,G^+(E)\,|\xi\rangle + (E-E_r)\,\langle\xi|\,G^+(E)\,|\xi\rangle = 1\,.\\
&\left(E - E_r - \lim_{\varepsilon\to 0}\int \mathrm{d}E'\,\frac{\sum_i |V^i_{E'}|^2}{z-E'}\right)\langle\varphi_r|\,G^+(E)\,|\xi\rangle - v\langle\xi|\,G^+(E)\,|\xi\rangle = 0\,.
\end{aligned}$$

(4.72a, b, c, d)

We observe that these equations can be obtained from $(z-H)\,G=\mathbf{1}$ by assuming that the identity is: $|\varphi_r\rangle\,\langle\varphi_r| + |\xi\rangle\,\langle\xi|$ and taking for the Hamiltonian, instead of H, an effective Hamiltonian given by:

$$H_{\text{eff.}} = H + \lim_{\varepsilon\to 0}\sum_i \int \mathrm{d}E'\,\frac{H\,|\varphi^i_{E'}\rangle\,\langle\varphi^i_{E'}|\,H}{z-E'}\,. \tag{4.73}$$

What we have done here is a partitioning of the problem [30, 31], by recognizing two classes of states (the discrete and the continuum states) and accounting for the effect of one class on the other (no approximation is involved up to this point). This suggests looking into the properties of the eigenvalues and eigenfunctions of this effective Hamiltonian. We now make the approximation that the level shift $F(E)$ is either negligible or very smooth and incorporated in E_r, and the damping 'constant' $\Gamma(E)$ smooth enough to be replaced by Γ_r.

The functions we are looking for are combinations of $|\varphi_r\rangle$ and $|\xi\rangle$ and have to obey $H_{\text{eff}}.|\Psi\rangle = E\,|\Psi\rangle$. We write $|\Psi\rangle = a\,|\xi\rangle + b\,|\varphi_r\rangle$ and form the secular equations:

$$a\left[\langle\xi|\,H_{\text{eff.}}\,|\xi\rangle - E\right] + b\,\langle\xi|\,H_{\text{eff.}}\,|\varphi_r\rangle = 0\,. \tag{4.74a}$$

$$a\,\langle\varphi_r|\,H_{\text{eff.}}\,|\xi\rangle + b\left[\langle\varphi_r|\,H_{\text{eff.}}\,|\varphi_r\rangle - E\right] = 0\,. \tag{4.76b}$$

Only the element $\langle\varphi_r|\,H_{\text{eff}}.\,|\varphi_r\rangle$ is affected by the term added to H in (73), since there is no coupling between $\langle\xi|$ and the continuum states $|\varphi_E^i\rangle$. This element is:

$$\langle\varphi_r|\,H_{\text{eff.}}\,|\varphi_r\rangle = E_r - i\Gamma_r\,. \tag{4.75}$$

The secular determinant is:

$$\begin{vmatrix} E_r - E & v \\ v & E_r + i\Gamma_r - E \end{vmatrix} = 0\,. \tag{4.76}$$

The roots are *complex* and we therefore change our notation from E to z, with for the two eigenvalues:

$$z^{\pm} = E_r - \frac{i\Gamma_r}{2} \pm \frac{\sqrt{4v^2 - \Gamma_r^2}}{2} \tag{4.77}$$

These are the complex quantities we have met when performing various integrations in the complex plane. When these complex eigenvalues are introduced into the secular Equations (74a, b), the equations so obtained are, except for a factor $-1/2$, the Equations (61) and (66) which have been solved to find out which initial conditions lead to an exponential decay. Thus we have shown that the eigenfunctions of the effective Hamiltonian obtained by partitioning out the continua define the decay modes of the system. This result is of course not limited to two coupled molecular states. The number of decay modes is in any case equal to the number of molecular states introduced in the treatment. However, as we shall see below, it may happen that some of them can hardly be called 'modes' because they are very difficult to excite.

There is an interesting interpretation of the complex energies (77) in terms of ingoing and outgoing waves. We observe on Equations (3. 81) and (3. 82) that the amplitudes of the scattered waves all contain the factor $\langle\varphi_r|\,G^+(E)\,|\varphi_r\rangle$. This quantity becomes infinite when its denominator vanishes. However this condition leads immediately in the present case to the complex energies (77). The meaning of a state with the amplitudes of the outgoing waves infinitely larger than the amplitudes of the ingoing wave is that the system, prepared in some way in the past in a non-stationary state, is undergoing only a decay process [32].

The effective Hamiltonian is not Hermitian (one of its diagonal matrix elements is complex in our example). We cannot expect the usual property of orthogonality to hold for its eigenfunctions $|\pm\rangle$. And in effect, if we calculate $\langle + \mid -\rangle$, in the case $2v > \Gamma_r$ for instance, we obtain:

$$\langle + \mid - \rangle = \frac{1}{2} - \frac{2v^2}{(\sqrt{\Delta} + i\Gamma_r)^2} \neq 0\,. \tag{4.78}$$

It is not possible to write the identity within the basis of discrete states as: $|+\rangle\langle+|+|-\rangle\langle-|$. However such functions possess the property of *biorthogonality* [33]. This means that the identity can be expressed by using the eigenfunctions of both $H_{\text{eff.}}$ and of its Hermitian adjoint $\tilde{H}_{\text{eff.}}$. Calling these eigenfunctions $|\mp\rangle$ and $|\simeq\rangle$ they have the properties $\langle\mp|-\rangle=\langle\simeq|+\rangle=0$. In our case $\tilde{H}_{\text{eff.}}=H^*_{\text{eff}}$. Thus the coefficients in $|\mp\rangle$ and $|\simeq\rangle$ are the complex conjugates of the coefficients given in Equations (62) or (63) and (67) or (68), and the eigenvalues $z^{\pm *}$. For instance, in the case $2v>\Gamma_r$:

$$\langle\mp|-\rangle = -\frac{\sqrt{2}\,v}{\sqrt{\Delta}-i\Gamma_r}\frac{\sqrt{2}\,v}{\sqrt{\Delta}+i\Gamma_r}+\frac{1}{2}=0\,. \tag{4.79}$$

With the help of both sets the identity can be written:

$$\mathbf{1}=\frac{|\mp\rangle\langle+|}{\langle+|\mp\rangle}+\frac{|\simeq\rangle\langle-|}{\langle-|\simeq\rangle}=\frac{|+\rangle\langle\mp|}{\langle\mp|+\rangle}+\frac{|-\rangle\langle\simeq|}{\langle\simeq|-\rangle}. \tag{4.80}$$

It is possible to reformulate entirely our problem with this basis in place of $|\xi\rangle$ and $|\varphi_r\rangle$. We will do this with the purpose of studying the excitation of the molecule. Taking again as the initial wave packet at time t':

$$|\Psi(t')\rangle=\sqrt{\frac{\bar{\Delta}}{\pi}}\int \mathrm{d}E'\,\frac{1}{E'-\bar{E}+i\bar{\Delta}}\,|\varphi^i_{E'}\rangle\,, \tag{3.71}$$

we calculate the coefficients in $|\Psi(t)\rangle$ of the functions $|\pm\rangle$. This is obtained by projecting onto the space of discrete states in the following way:

$$\left\{\frac{|+\rangle\langle\mp|}{\langle\mp|+\rangle}+\frac{|-\rangle\langle\simeq|}{\langle\simeq|-\rangle}\right\}|\Psi(t)\rangle\,. \tag{4.81}$$

After applying the identity in terms of the $|\psi^{+j}_E\rangle$ and the evolution operator on the initial wave packet the coefficients of $|+\rangle$ and $|-\rangle$ are found to be:

$$\frac{\langle\tilde{\pm}|\Psi(t)\rangle}{\langle\tilde{\pm}|\pm\rangle}=\sqrt{\frac{\bar{\Delta}}{\pi}}\int\frac{\mathrm{d}E'}{E'-\bar{E}+i\bar{\Delta}}\int \mathrm{d}E''\,e^{-(i/\hbar)E''(t-t')}\sum_j\times$$
$$\times\frac{\langle\tilde{\pm}|\psi^{+j}_{E''}\rangle}{\langle\tilde{\pm}|\pm\rangle}\langle\psi^{+j}_{E''}|\varphi^i_{E'}\rangle\,. \tag{4.82}$$

We can again perform first the integration over E' and observe that only the unperturbed wave in $|\psi^{+i}_{E''}\rangle$ contributes. There is left:

$$\frac{\langle\tilde{\pm}|\Psi(t)\rangle}{\langle\tilde{\pm}|\pm\rangle}=\sqrt{\frac{\bar{\Delta}}{\pi}}\int \mathrm{d}E''\,e^{-(i/\hbar)E''t}\frac{\langle\tilde{\pm}|\psi^{+i}_{E''}\rangle}{\langle\tilde{\pm}|\pm\rangle}. \tag{4.83}$$

From the definition of $|\psi^{+i}_{E''}\rangle$ there is derived, by interposing the identity as expressed by Equation (80) between G and V:

$$\langle\tilde{\pm}\,|\,\psi_E^{+i}\rangle = \frac{\langle\tilde{\pm}|\,G^+(E)\,|+\rangle}{\langle\mp\,|\,+\rangle}\langle\mp|\,V\,|\varphi_E^i\rangle + \frac{\langle\tilde{\pm}|\,G^+(E)\,|-\rangle}{\langle\simeq\,|\,-\rangle}\langle\simeq|\,V\,|\varphi_E^i\rangle. \tag{4.84}$$

The required matrix elements of $G^+(E)$ are to be obtained from $(z-H)\,G=\mathbf{1}$. After the usual elimination of elements involving continuum states, the equations formed with:

$$\langle +|(z-H)\,G\,|\mp\rangle;\ \langle +|(z-H)\,G\,|\simeq\rangle;$$
$$\langle -|(z-H)\,G\,|\mp\rangle;\ \langle -|(z-H)\,G\,|\simeq\rangle$$

can be put into the form:

$$\langle\mp|\,z-H_{\text{eff.}}\left|\frac{|+\rangle\langle\mp|}{\langle\mp\,|\,+\rangle}+\frac{|-\rangle\langle\simeq|}{\langle\simeq\,|\,-\rangle}\right|G\,|+\rangle = \langle\mp\,|\,+\rangle; \tag{4.85a}$$

$$\langle\mp|\,z-H_{\text{eff.}}\left|\frac{|+\rangle\langle\mp|}{\langle\mp\,|\,+\rangle}+\frac{|-\rangle\langle\simeq|}{\langle\simeq\,|\,-\rangle}\right|G\,|-\rangle = 0; \tag{4.85b}$$

$$\langle\simeq|\,z-H_{\text{eff.}}\left|\frac{|+\rangle\langle\mp|}{\langle\mp\,|\,+\rangle}+\frac{|-\rangle\langle\simeq|}{\langle\simeq\,|\,-\rangle}\right|G\,|+\rangle = 0; \tag{4.85c}$$

$$\langle\simeq|\,z-H_{\text{eff.}}\left|\frac{|+\rangle\langle\mp|}{\langle\mp\,|\,+\rangle}+\frac{|-\rangle\langle\simeq|}{\langle\simeq\,|\,-\rangle}\right|G\,|-\rangle = \langle\simeq\,|\,-\rangle \tag{4.85d}$$

Because of $H_{\text{eff.}}|\pm\rangle = z^{\pm}|\pm\rangle$ and of biorthogonality the equations can immediately be solved. They yield:

$$\langle\mp|\,G(z)\,|+\rangle = \frac{\langle\mp\,|\,+\rangle}{z-z+};\ \langle\mp|\,G(z)\,|-\rangle = 0$$
$$\langle\simeq|\,G(z)\,|+\rangle = 0;\ \langle\simeq|\,G(z)\,|-\rangle = \frac{\langle\simeq\,|\,-\rangle}{z-z-}. \tag{4.86}$$

Thus finally:

$$\langle\mp\,|\,\psi_E^{+i}\rangle = \frac{\langle\mp|\,V\,|\varphi_E^i\rangle}{E-z^+};\ \langle\simeq\,|\,\psi_E^{+i}\rangle = \frac{\langle\simeq|\,V\,|\varphi_E^i\rangle}{E-z^-}. \tag{4.87}$$

We can then perform the integration over E'' (ignoring the energy dependence of the couplings) and there is obtained:

$$\frac{\langle\mp\,|\,\Psi(t)\rangle}{\langle\mp\,|\,+\rangle} = (2/i)\sqrt{\pi\bar{\Delta}}\,\langle\mp|\,V\,|\varphi_{E_r}^i\rangle\,\frac{e^{-(i/\hbar)(\bar{E}-i\bar{\Delta})(t-t')} - e^{-(i/\hbar)z^+(t-t')}}{\bar{E}-i\bar{\Delta}-z^+} \tag{4.88}$$

$$\frac{\langle\simeq\,|\,\Psi(t)\rangle}{\langle\simeq|-\rangle} = (2/i)\sqrt{\pi\bar{\Delta}}\,\langle\simeq|\,V\,|\varphi_{E_r}^i\rangle\,\frac{e^{-(i/\hbar)(\bar{E}-i\Delta)(t-t')} - e^{-(i/\hbar)z^-(t-t')}}{\bar{E}-i\bar{\Delta}-z^-} \tag{4.89}$$

The analogy with the two level molecule is striking, except that now the amplitudes of the *two* (non orthogonal) states $|\pm\rangle$ will grow and decay with time, with expressions similar to (3. 90). Several excitation processes are possible, depending on both the width and the center of the photon wave packet, and the relation between v and Γ_r.

(a) If $2v > \Gamma_r$, the molecular states are well separated with respect to their radiative widths. The two couplings are in this case:

$$\langle \mp | V | \varphi_{E_r}^i \rangle = \langle \simeq | V | \varphi_{E_r}^i \rangle = 2^{-1/2} V_{E_r}^i \tag{4.90}$$

The most favourable choice of parameters to excite, say, state $|+\rangle$ (which goes to the molecular state $|1\rangle$ when $2v/\Gamma_r \to \infty$), is to center the photon wave-packet at the energy of this resonance: $\bar{E} = E_r + \sqrt{\Delta}/2$. If $\bar{\Delta} < 2v$, the state $|-\rangle$ will also be excited, but to a lesser extent. If, on the other hand the wave packet is centered at $\bar{E} = E_r$, then both states are coherently excited. We examine in more detail the case where $\bar{\Delta} > 2v$ (the width of the wave packet is larger than the separation between molecular states, or equivalently the duration of the interaction is shorter than the period of oscillations between the two coupled states). Transforming the amplitudes in order to obtain the development of the wave function in the basis $|\varphi_r\rangle, |\xi\rangle$ we obtain (apart from an uninteresting factor):

$$\begin{aligned} |\Psi(t)\rangle \propto {} & |\varphi_r\rangle \left[e^{-(\bar{\Delta}/\hbar)(t-t')} e^{-(\Gamma_r/2\hbar)(t-t')} \cos \frac{\sqrt{\Delta}}{2\hbar}(t-t') \right] \\ & + i|\xi\rangle \left\{ \frac{\Gamma_r}{2v} \left[e^{-(\bar{\Delta}/\hbar)(t-t')} - e^{-(\Gamma_r/2\hbar)(t-t')} \cos \frac{\sqrt{\Delta}}{2\hbar}(t-t') \right] + \right. \\ & \left. + \frac{\sqrt{\Delta}}{2v} e^{-(\Gamma_r/2\hbar)(t-t')} \sin \frac{\sqrt{\Delta}}{2\hbar}(t-t') \right\} + \cdots \end{aligned} \tag{4.91}$$

Since we assume $2v > \Gamma_r$ and $\bar{\Delta} > 2v$ we have:

$$\hbar/2\bar{\Delta} < \hbar/\sqrt{\bar{\Delta}} < h/2\Gamma_r.$$

Thus the coefficient of $|\varphi\rangle$ will reach its maximum before the oscillations at the resonance frequency $\sqrt{\Delta}/2h$ set in. The coefficient of $|\xi\rangle$ contains the same function of $(t-t')$ affected by the small factor $\Gamma_r/2v$, plus a term which indicates that there is also a damped oscillation, but out of phase with the other one.

(b) If $2v < \Gamma_r$, both resonance energies have the same real part E_r. In this limit:

$$\langle \mp | V | \varphi_{E_r}^i \rangle \to -i \frac{v}{\Gamma_r} V_{E_r}^i, \tag{4.92a}$$

$$\langle \simeq | V | \varphi_{E_r}^i \rangle \to -i V_{E_r}^i. \tag{4.92b}$$

Thus the state $|+\rangle$ (which approaches the dipole forbidden state $|\xi\rangle$) is hardly excited at all, while the state $|-\rangle$ (which approaches $|\varphi_r\rangle$) is excited with an amplitude which is that calculated for the two level case.

5. Appendix: Some References on Radiative and Nonradiative Processes

We give now a short account of the earlier developments of the subject. The proposal that a nonradiative process in a polyatomic molecule is due to the coupling between an isolated zeroth order molecular state and a dense set of other zeroth order states was first made by Robinson and Frosch [34] and Hunt, McCoy and Ross [35], although a similar mechanism had already been used by Kubo [36] and Lax [37] to explain electronic relaxation for impurities in a solid. Lin and Bersohn [38] assigned explicitly the coupling to a break-down of the Born-Oppenheimer approximation, and calculated this coupling for a polyatomic molecule in the harmonic approximation. The Franck-Condon factor occurring in the coupling was correlated with experimental data on aromatic compounds by Siebrand [39]. A formulation of the excitation and of the decay for a discrete set coupled to a set of discrete states was done, following earlier work by Rice [40] and Fano [41], by Bixon and Jortner [42]. A theory of unimolecular decay based on the coupling of discrete states (associated with the activated complex) with continuum states (describing fragmentation) was developed by Mies and Krauss [3]. The mechanism of excitation of a molecule with a dipole allowed state coupled with dipole forbidden states has been studied by Rhodes [43]. These papers have been followed by many others extending the treatment to molecules of small and intermediate size, to photoisomerization and photochemistry (See the reviews [1, 2, 7] for further references).

Alternative approaches to treat time-dependent processes are:

(a) the Fano method [3, 4, 44] which also makes use of the identity expanded in a set of stationary functions, but these are obtained by solving the secular equations expressing the configuration interaction [6], instead of using the Lippmann-Schwinger equation [19];

(b) the Green function or Laplace transform method [45, 46, 47, 48, 49, 13];

(c) the Weisskopf-Wigner method [50, 25, 28].

The elimination carried out in 3.1c is usually performed in operator form [45, 46]. The wave packet formulation of the excitation and decay of a two-level atom is developed by Cohen-Tannoudji [51]. The scattering approach to absorption spectroscopy is reviewed by Shore [52]. Such an approach to treat non-radiative processes is used by Berrondo and Weiss [53].

References

1. Jortner, J., Rice, S. A., and Hochstrasser, R. M.: *Adv. Photochem.* **7**, 149 (1969).
2. Freed, K. F.: *Curr. Top. Chem*, in the press (1972).
3. Mies, F. H. and Krauss, M.: *J. Chem. Phys.* **45**, 4455 (1966).
4. Mies, F. H.: *J. Chem. Phys.* **51**, 787, 798 (1969).
5. Herzberg, G.: *Topics in Modern Physics – A Tribute to E. U. Condon*, Colorado Ass. Un. Press, 1971, p. 191.
6. Fano, U.: *Phys. Rev.* **124**, 1866 (1961).
7. Rice, S. A. and Gelbart, W. M.: *J.I.U.P.A.C.* **27**, 361 (1971).
8. Englman, R. and Jortner, J.: *Mol. Phys.* **18**, 145 (1970).

9. Nitzan, A. and Jortner, J.: *J. Chem. Phys.* **55**, 1355 (1971).
10. Busch, G. E., Rentzepis, P. M., and Jortner, J.: *Chem. Phys. Letters* **11**, 437 (1971).
11. Douglas, A. E.: *J. Chem. Phys.* **45**, 1007 (1966).
12. Gelbart, W. and Rice, S. A.: *J. Chem. Phys.* **50**, 4775 (1969).
13. Freed, K. F.: *J. Chem. Phys.* **52**, 1345, (1970).
14. Beswick, J. A., Lefebvre, R., and Plumejeau, A. M.: *J. Chem. Phys.* **56**, 4011 (1972).
15. Scully, M. O. and Lamb, W. E.: *Phys. Rev.* **159**, 208 (1967).
16. See for instance Louisell, W. H.: *Radiation and Noise in Quantum Electronics*, McGraw-Hill, New York, 1964.
17. Born, M.: *Festschrift Gött, Math. Phys. KI* **1** (1951).
18. Siebrand, W.: *Chem. Phys. Letters* **6**, 192 (1970).
19. Lippmann, B. A. and Schwinger, J.: *Phys. Rev.* **79**, 469 (1950).
20. Goldberger, M. L. and Watson, K. M.: *Collision Theory*, J. Wiley, 1964, Chapter 5.
21. See for instance Levine, R. D.: *Quantum Mechanics of Molecular Rate Processes*, Oxford Univ. Press, 1969, Section 2.5.
22. McCoy, E. F. and Ross, I. G.: *Austr. J. Chem.* **15**, 573 (1962).
23. Jortner, J. and Berry, R. S.: *J. Chem. Phys.* **47**, 2757 (1968).
24. Bethe, H. A.: *Handbuch der Physik* **24/1**, 452 (1933).
25. Lamb, W. E. and Retherford, R. C.: *Phys. Rev.* **79**, 549 (1950).
26. Nitzan, A., Jortner, J., Kommandeur, J., and Drent, E.: *Chem. Phys. Letters* **9**, 273 (1971).
27. Bashkin, S., Bickel, W. S., Fink, D., and Wangness, R. K.: *Phys. Rev. Letters* **15**, 284 (1965).
28. Wangness, R. K.: *Phys. Rev.* **149**, 60 (1966).
29. Bixon, M. and Jortner, J.: *J. Chem. Phys.* **50**, 3284 (1969).
30. Löwdin, P. O.: *J. Chem. Phys.* **19**, 1396 (1951).
31. Fesbach, H.: *Ann. Phys.* **19**, 287 (1962).
32. Siegert, A.: *Phys. Rev.* **56**, 750 (1939).
33. Morse, P. M. and Fesbach, H.: *Methods of Theoretical Physics*, McGraw-Hill, New York, 1953, p. 884.
34. Robinson, G. W. and Frosch, R. P.: *J. Chem. Phys.* **37**, 1962 (1962).
35. Hunt, G. R., McCoy, E. F., and Ross, I. G.: *Austr. J. Chem.* **15**, 591 (1962).
36. Kubo, R.: *Phys. Rev.* **86**, 929 (1952).
37. Lax, M.: *J. Chem. Phys.* **20**, 1752 (1952).
38. Lin, S. H.: *J. Chem. Phys.* **44**, 3759 (1966); Lin, S. H. and Bersohn, R.: *J. Chem. Phys.* **46**, 2732 (1968).
39. Siebrand, W.: *J. Chem. Phys.* **46**, 440 (1967); **47**, 2411 (1967).
40. Rice, O. K.: *J. Chem. Phys.* **1**, 375 (1933).
41. Fano, U.: *Nuovo Cim.* **12**, 156 (1935).
42. Bixon, M. and Jortner, J.: *J. Chem. Phys.* **48**, 715 (1968).
43. Rhodes, W.: *J. Chem. Phys.* **50**, 2885 (1969).
44. Bixon, M., Jortner, J., and Dothan, Y.: *Mol. Phys.* **17**, 109 (1969).
45. Messiah, A.: *Mécanique Quantique*, Dunod, Paris, 1960, Chapters XIX, XXI.
46. Harris, R. A.: *J. Chem. Phys.* **39**, 978, (1963).
47. Mower, L.: *Phys. Rev.* **142**, 799 (1966).
48. Chock, D., Jortner, J., and Rice, S. A.: *J. Chem. Phys.* **49**, 610 (1968).
49. Freed, K. F. and Jortner, J.: *J. Chem. Phys.* **50**, 2916 (1969).
50. Weisskopf, V. F. and Wigner, E.: *Z. Physik* **63**, 54 (1930).
51. Cohen-Tannoudji, C.: *Compléments de Mécanique Quantique*, Univ. of Paris, 1966, unpublished.
52. Shore, B. W.: *Rev. Mod. Phys.* **39**, 439 (1967).
53. Berrondo, M. and Weiss, K.: *J. Chem. Phys.* **52**, 807 (1970).

CHARGE TRANSFER MOLECULAR SOLIDS

P. PINCUS
Dept. of Physics, University of California, Los Angeles, Calif., U.S.A.

1. Introduction

The purposes of these lectures is to discuss a class of organic solids in which arise questions of fundamental solid state physics interest as well as having potential technological applications. Although our focus will be on the solid state physics aspects of the electronic properties of the systems under consideration, the general field of study involves a strong overlap between physics and chemistry.

Solids are often classified as being conductors or insulators (a semi-conductor may be considered to be a small band gap insulator). On a microscopic level, the one electron picture of the distinction between conducting and non-conducting behavior is described as follows:

For example, consider a chain of copper atoms (Figure 1). Copper has an extra $3p$ electron outside its closed shell. We describe this electron in the tight binding model. We then construct a Bloch wavefunction as a linear combination of atomic orbitals $\phi(\mathbf{r}-\mathbf{R}_i)$.

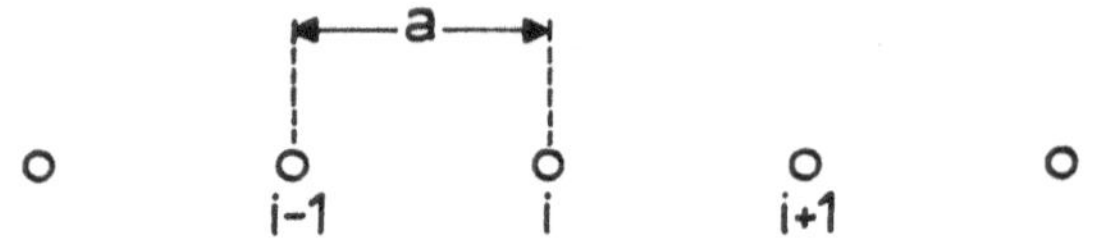

Fig. 1.

$$\psi(k) = \text{const.} \sum_i e^{i\mathbf{k}\cdot\mathbf{R}_i} \phi(\mathbf{r} - \mathbf{R}_i), \tag{1.1}$$

where i labels the position of the atoms in the chain. If we solve the eigenvalue problem with this type of function we will obtain $E = E(k)$ that looks like Figure 2.

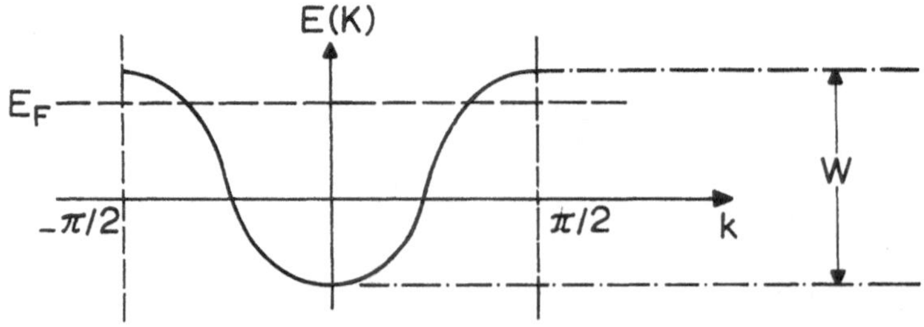

Fig. 2.

Abecassis de Laredo and Jurisic (eds.), Selected Topics in Phys. Astrophys. and Biophys. 138–172. *All Rights Reserved.*

What determines whether or not the system is metallic? Let us fill these energy states with all the available electrons; we thus obtain an energy level defined as the Fermi level E_F. E_F indicates that energy below which all the states are occupied and above which all the states are empty at $T=0$ K. If the only electrons available are one per atom (e.g. copper) we obtain a situation where one half of the states are occupied (because every orbital can hold two electrons). If the Fermi energy lies within the band an electron-hole pair can be excited with arbitrarily small energy. The electrons are then quite free to move and metallic behaviour prevails. If the Fermi levels fall between bands belonging to different atomic orbitals, it requires some minimum band gap energy to create a mobile electron-hole pair and we find insulating behavior. Thus if E_F lies within the band we have a metal and if E_F lies outside the band we have an insulator. However, there exist insulators like e.g. Ni 0 for which a simple one electron band calculation shows to be a metal. Therefore, the independent electron approximation cannot be appropriate for such cases. Indeed, not only is Ni 0 an insulator, but magnetic as well (antiferromagnetic). It is the electronic correlations induced by the Coulomb repulsion between electrons which is at the root of this phenomenon.

What I mean by correlation is that, e.g. because of intra-atomic Coulomb repulsions, it may be energetically too costly for two electrons to occupy the same atomic orbital. Then if, as in Cu, there is one electron per atom, the electrons will tend to become localized and not free to move, because such motion would necessarily force double occupancy of a given site. This localization is essentially the Mott transition and we call such a system, a Mott insulator. This type of correlation and its consequences will be the focus of these lectures.

In such a system, we may then distinguish two competing energies (i) the kinetic energy which tends to spread out the wavefunctions spatially and is measured by the band width W; (ii) a correlation energy E_c which prevents double occupancy of an atomic orbital. Then, for a half-filled band (one electron/site), we expect if $W \gg E_c$ to find metallic behavior and if $W \ll E_c$ a correlated (or Mott) insulator. However, even for a Mott insulator, if the thermal energy $(k_B T \gg E_c)$, we should return to a type of non-degenerate metallic situation. It is then of interest to explore the phase diagram of a correlated solid as a function of temperature. It is for this purpose that the organic systems may shed some light on the problem. It is simply a question of orders of magnitude, as shown in Table I.

Typically for inorganic solids both the band width and intratomic correlation

TABLE I

	Inorganic eV	Organic eV
Band width W	1–10	0.01–0.1
Correlations E_c Energy	1–10	0.1–1

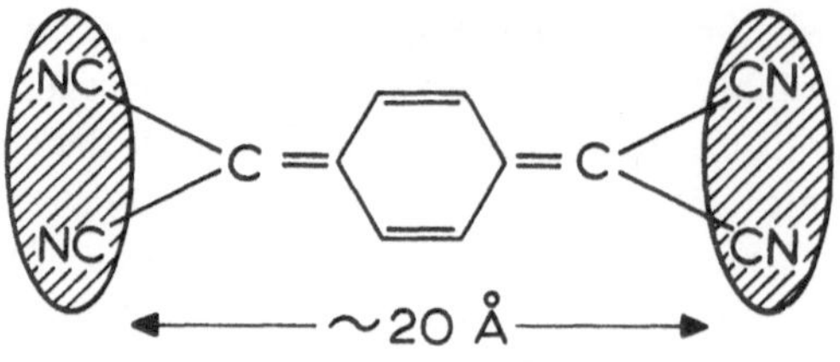

Fig. 3.

energy are in the electron-volt range which corresponds to tens of thousands of degrees Kelvin. Thus, in these simple terms it is rather unlikely for temperature to drive a metal-insulator transition. The situation is somewhat more favorable in some molecular systems. Most molecular solids e.g. anthracene are formed of neutral molecules weakly bound together by Van-der-Waals forces. The charge transfer salts with which we shall be concerned are more ionic in nature. A prototype of such systems is formed by the salts of the molecule tetracyanoquinodimethan (TCNQ).

TCNQ is a rigid planar molecule whose property of most importance to us is its high electrons affinity. For example, in a saline solution, the molecule normally becomes negatively charged with the extra electron mainly localized in the region of the cyano groups. The TCNQ molecular *ion* contains an odd number of electrons and thus one unpaired spin. This is in contrast to a halide ion which closes an atomic shell on becoming ionized. Solid compounds are easily formed of the type X^+TCNQ^- where $X=$(Na, K, Li, NH_4 or any of several organic donors). The compounds typically crystallize in structures with one dimensional face-to-face stacks of TCNQ molecules separated by cation chains

What is meant by the correlation energy E_c in these systems? In first approximation, we mean how much energy it costs to put two electrons on one TCNQ molecule. Because we know the average distance between the two electron sites we can estimate the Coulomb repulsion e^2/r_{TCNQ} where $r_{TCNQ} \simeq 20$ Å. This energy is roughly 1 eV. If we take into account the dielectric screening by the environment this may be reduced by an order of magnitude.

The band width will depend primarily on the molecular states of the electron and on the c-axis lattice constant of the $TCNQ^-$ molecular stacks. The stack separation

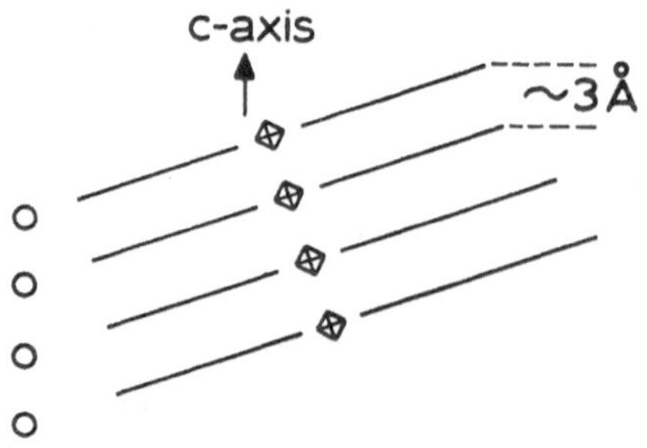

Fig. 4.

is about 3 Å and the electronic states are probably p-like. Estimates of the overlap integrals between such states lead to a band width of the order of $W \simeq 0.01$ eV to 0.1 eV.

We shall see later that the important parameters are W and $E_c/4$. The degree of correlation depends upon the difference $E_c/4 - W$ which may be of the order of few hundred degree Kelvin. The thermal energy kT may then play an important role.

From these estimates we see an advantage of working with molecular solids rather than inorganic compounds. Another advantage is that most inorganic solids that exhibit an insulating-metal transition (e.g. V_2O_3) are transition compounds with orbital degeneracies (d-electrons). Electron correlation problems with orbital degeneracy are very difficult and are just beginning to be studied. The big inconveniences of the organic systems are that (1) we are dealing with systems whose electronic structure we do not know in detail; and (2) the state of the art in the preparation of pure undefected crystals is not as advanced as for the inorganic systems.

We will be often considering a compound NMP^+ $TCNQ^-$ which appears to be a one dimensional metal. By a metal we mean that for this salt the conductivity is about two orders of magnitude worse than copper at room temperature and that the resistance R increases with temperature. At low temperature NMP^+ $TCNQ^-$ is a magnetic insulator. Another type of system that we have not mentioned so far is the Q^+ $TCNQ_2^-$ with a similar structure to the one already mentioned. This substance seems to be a metal at all temperatures (at low temperature the d.c. behavior is like an insulator because of defects and impurities but the a.c. behavior is characteristic of a metal). The alkaline $TCNQ^-$ systems are all insulators at room temperature.

These lectures are mainly devoted to an attempt to describe and classify the various phenomena observed in the charge transfer salts and to relate them to electron correlations.

2. Atomic Hydrogen and Helium [1]

2.1. Atomic Hydrogen: Bohr Atom

We know that a Hydrogen atom consists of a heavy proton and an electron. The potential energy and the electronic kinetic energy are easily written as:

$$V = -e^2/r, \qquad (2.1)$$
$$\text{k.E.} = p^2/2m.$$

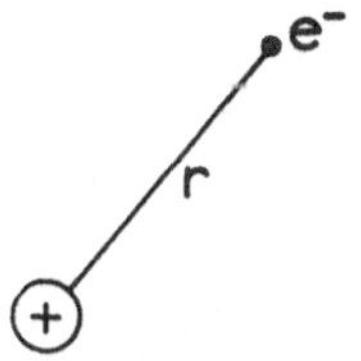

Fig. 5.

The total energy is, of course, the sum

$$E = \frac{p^2}{2m} - \frac{e^2}{r}. \tag{2.2}$$

We shall begin by receiving some of the relevant aspects of the electronic structure of atoms.

Using the Bohr quantization condition

$$\oint \mathbf{p} \cdot d\mathbf{r} = nh = 2\pi pr \quad \text{or} \quad p = \frac{n\hbar}{r}, \tag{2.3}$$

where $n=1, 2, 3, \ldots$

The energy is easily found to be

$$E = -\tfrac{1}{2} \frac{me^4}{n^2\hbar^2}, \tag{2.4}$$

where we define as atomic unit (a.u.) of energy

$$\text{a.u.} \equiv \frac{me^4}{\hbar^2} \simeq 27.2 \text{ eV}$$

The ionization potential of the hydrogen atom is then 1/2 a.u. Another important parameter is the Bohr radius a_0 defined as:

$$a_0 = \frac{\hbar^2}{me^2} \simeq 0.5 \text{ Å}.$$

The ground state wavefunction is easily determined from the Schrödinger equation to be

$$\phi_{1s}(r) \propto e^{-r/a_0}.$$

2.2. Atomic helium

The next degree of complexity is to consider a two electron atoms-helium,

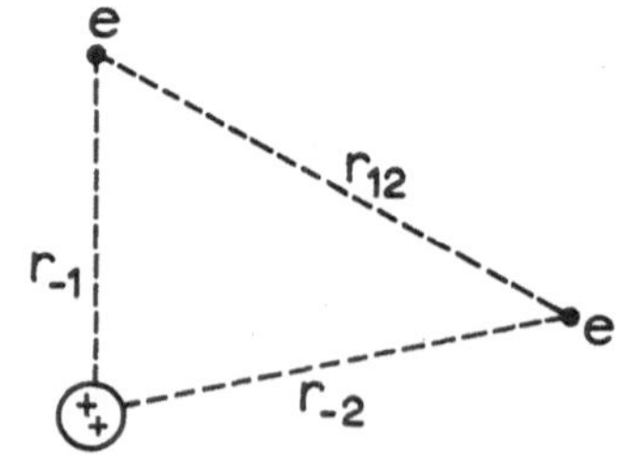

Fig. 6.

The electronic Hamiltonian is

$$H = \underbrace{-\frac{\hbar^2}{2m}\nabla_1^2 - \frac{ze^2}{r_1}}_{H_1} \underbrace{- \frac{\hbar^2}{2m}\nabla_2^2 - \frac{ze^2}{r_2}}_{H_2} + \underbrace{\frac{e^2}{r_{12}}}_{V_I}, \tag{2.5}$$

where $r_{12}=|\mathbf{r}_1-\mathbf{r}_2|$ and $z=2$. If we neglect the interaction term between 1 and 2 we have two independent electrons moving in a Coulomb potential which gives hydrogenic wavefunction but with a different nuclear charge $z=2$. The ground state of the helium atom that results with $V_I=0$ has the two electrons occupying a hydrogenic $1s$ orbital. We may treat V_I by perturbations theory: the perturbation parameter being $1/z$. Let us carry out this program.

Setting $V_I=0$, the orbital ground state wavefunction is

$$\psi(\mathbf{r}_1, \mathbf{r}_2) = \phi_{1s}(\mathbf{r}_1)\,\phi_{1s}(\mathbf{r}_2) \text{ where } \phi_{1s}(r) = ce^{-2r/a_2} \tag{2.6}$$

$\psi(\mathbf{r}_1, \mathbf{r}_2)$ is symmetric since the spins must be in a singlet antisymmetric state. To lowest order in V_I.

$$E = E_0 + E_1$$

where

$$E_0 = -4 \text{ a.u.} \quad \text{and} \quad E_1 = \langle\psi|\, V_I\, |\psi\rangle / \langle\psi\,|\,\psi\rangle . \tag{2.7}$$

The numerator of this expression is

$$\int \phi_{1s}^2(\mathbf{r}_1)\,\phi_{1s}^2(\mathbf{r}_2)\,\frac{e^2}{|\mathbf{r}_1 - \mathbf{r}_2|}\,\mathrm{d}\mathbf{r}_1\,\mathrm{d}\mathbf{r}_2 = \int \frac{\varrho(\mathbf{r}_1)\,\varrho(\mathbf{r}_2)}{|\mathbf{r}_1 - \mathbf{r}_2|}\,\mathrm{d}\mathbf{r}_1\,\mathrm{d}\mathbf{r}_2\,,$$

where the charge density $\varrho(r)$ is given by

$$\varrho(r) = e\phi_{1s}^2(\mathbf{r}).$$

Because of the spherical symmetry involved in the ground state wavefunction it is straightforward carry out the integration using Gauss law. Consider two concentric charged shells in Figure 7.

Because e.g. shell 2 only experiences the potential one to the enclosed charge, we find that the numerator becomes

$$2 \times \int_0^\infty \frac{\varrho(r_2)}{r_2}\,4\pi r_2^2\,\mathrm{d}r_2 \int_0^{r_2} \varrho(r_1)\,4\pi r_1^2\,\mathrm{d}r_1\,,$$

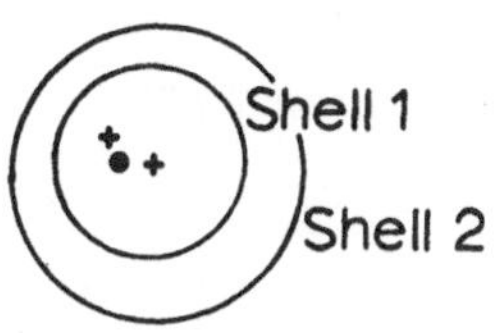

Fig. 7.

where the factor 2 is due to the interchange of shells 1 and 2.

Performing this straightforward integration we find

$$E_1 = \tfrac{5}{8} z = 1.25 \text{ a.u.}$$

so that the total energy, to this order, is

$$E = E_0 + E_1 = -2.75 \text{ a.u.}$$

The difference between the observed ground state energy E_{exp} and the perturbation treatment E_{pert} is approximately 0.15 a.u. that corresponds to roughly 3 eV. From a chemical or solid state physics point of view this discrepancy is too large to neglect. One factor not taken into account very well by perturbation theory is the screening of the nuclear charge as experienced by one electron by the other electron. We can try to correct this situation by replacing the nuclear charge $z=2$ by an effective charge $\tilde{z}$ which is treated as a variational parameter; i.e.

$$\tilde{\phi}_{1s} = c \exp(-\tilde{z}r/a_0).$$

The effective charge $\tilde{z}$ is determined by minimizing the energy,

$$\frac{\mathrm{d}E(\tilde{z})}{\mathrm{d}\tilde{z}} \equiv 0.$$

This is most easily accomplished by rewriting the Hamiltonian in terms of the effective nuclear charge $\tilde{z}$, as

$$H = H_0(\tilde{z}) + \frac{e^2}{r_{12}} - (z - \tilde{z})\frac{e^2}{r_1} - (z - \tilde{z})\frac{e^2}{r_2}, \tag{2.8}$$

where $H_0(\tilde{z})$ is $H_1 + H_2$ but with z replaced by $\tilde{z}$. The term e^2/r_{12} is the electron-electron repulsion term and the other two extra terms restore the initial form of H.

Assuming ψ, is given by hydrogenic wavefunctions for the nuclear charge $\tilde{z}$,

$$\psi = \phi_{1s}(\mathbf{r}_1, \tilde{z})\, \phi_{1s}(\mathbf{r}_2, \tilde{z}), \tag{2.9}$$

we find in terms of atomic units

$$E(\tilde{z}) = -\tilde{z}^2 + \tfrac{5}{8}\tilde{z} - 2(z - \tilde{z})\,\tilde{z}. \tag{2.10}$$

Minimizing $E(\tilde{z})$ with respect to z

$$2\tilde{z}_{\min} - 2z + \tfrac{5}{8} \equiv 0 \quad \text{or} \quad \tilde{z}_{\min} = z - \tfrac{5}{16}$$

Substituting back into the expression for $E(\tilde{z})$,

$$E_{\min} = -(z - \tfrac{5}{16})^2 = -2.85 \text{ a.u.}$$

where we have now set $z=2$.

This result still corresponds to a difference of 0.05 a.u. This is still about 1.3 eV.

which is still not good enough for our purposes. We can make a further improvement by relaxing the assumption that ϕ be of the hydrogenic form.

Let us attempt to find the best independent electron wavefunction to describe the system. This is the Hartree-Fock method.

Assuming a normalized symmetric independent particle wavefunction

$$\psi = \phi(\mathbf{r}_1)\,\phi(\mathbf{r}_2) \tag{2.11}$$

the average energy is

$$E = \int \psi H \psi \, d\mathbf{r}_1 \, d\mathbf{r}_2 ,$$

where H is the hamiltonian

$$H = H_1 + H_2 + e^2/r_{12} .$$

Minimizing the energy with respect to an arbitrary variation of the wavefunctions,

$$\delta\left(\int \psi (H - E)\, \psi \, d\mathbf{r}_1 \, d\mathbf{r}_2\right) \equiv 0$$

we obtain

$$\int \{\delta(\phi_1)\, \phi_2 (H - E)\, \phi_1\phi_2 + \phi_1\delta(\phi_2)\,(H - E)\, \phi_1\phi_2\}\, d\mathbf{r}_1 \, d\mathbf{r}_2 \equiv 0$$

and because ϕ_1, ϕ_2 are considered independent variables we must have independently

$$\int dr_2 \phi_2 (H - E)\, \phi_2 \phi_1 = 0 .$$

And a similar equation with $1 \leftrightarrow 2$.

Substituting H

$$0 = (H_1 - E)\, \phi(\mathbf{r}_1) + \int d\mathbf{r}_2 \phi(\mathbf{r}_2) \frac{e^2}{\mathbf{r}_{12}}\, \phi(\mathbf{r}_2)\, \phi_1(\mathbf{r}_1) \tag{2.12}$$

that is formally equal to the Schrödinger equation for the state $\phi(r_1)$ but with a potential depending on the wavefunction itself.

The integral term can be considered as an effective potential

$$V_{\text{eff}} = \int d\mathbf{r}_2 \, |\phi(\mathbf{r}_2)|^2 \, e^2/\mathbf{r}_{12} . \tag{2.13}$$

This H-F equation has been solved yielding a ground state energy $E_{\text{H-F}} = -2.86$ a.u., only a slight improvement over the $\tilde{z}$ result.

This calculation assumes that a given electron moves in the nuclear potential plus the average potential created by the other electron. The discrepancy between the resulting energy and the exact energy is due to the fact that there is a strong force preventing the two electrons to approach very near to one another. Thus, the repulsive electronic Coulomb energy is reduced by the two electrons avoiding one another.

This is the correlation energy and in the case of helium amounts to about 1 eV. Of course, the correlation effects can be included by writing more complicated wavefunctions taking into account higher hydrogenic configurations. For example, we might try a radical function of the type

$$\psi = c\,|\phi(r_1)\,\psi(r_2) + \psi(r_1)\,\phi(r_2)|\,.$$

Such a wavefunction, which is effectively radial configuration mixing, already reduces the error to about $\frac{1}{2}$ eV.

In general, we define the differences between the Hartree-Fock energy and the exact energy as the correlation energy. In atoms, the correlation energy is typically of the order of 1 eV per electron pair.

3. Hydrogenic Molecules [2]

We now turn to consider the effects of correlations on molecular structures. The elementary two electron problem is then that of molecular hydrogen, H_2. Let us begin first with the hydrogen molecular ion, H_2^+, i.e. one electron and two protons. This is an exactly soluable problem in terms of

$$\begin{array}{ccc} & \leftarrow R \rightarrow & \\ \dot{A} & & \dot{B} \end{array}$$

Fig. 8.

elliptic coordinates, but the method is quite cumbersome and useful only as a check to the more generally applicable approximate schemes (Figure 8). We consider throughout these lectures that R is fixed, in the spirit of the Born-Oppenheimer approximation. As an initial guess, we might try

$$\psi = c\,(\phi_{1s}^A + \phi_{1s}^B) \tag{3.1}$$

i.e. a linear combination of atomic orbitals (LCAO) or tight binding wavefunction. The normalization constant c is given by

$$\begin{aligned} 1 &= \int \psi^2 \mathrm{d}\mathbf{r} = c^2 \int (\phi^A + \phi^B)^2\,\mathrm{d}\mathbf{r} \\ &= c^2 \left[2 + 2 \int \phi^A \phi^B\,\mathrm{d}\mathbf{r}\right] \end{aligned} \tag{3.2}$$

if ϕ^A and ϕ^B are each normalized to unity. We define the overlap integral $S \equiv \int \phi^A \phi^B\,\mathrm{d}\mathbf{r}$, which gives

$$c = \frac{1}{\sqrt{2(1+s)}}.$$

Fig. 9.

This type of wavefunction accounts for about 80% of the binding energy of H_2^+ at the equilibrium distance even if the nuclear charge is optimized, to allow expansion of the wavefunctions. Clearly higher atomic configurations are necessary to give the physically derived distortions (Figure 9).

At large distances the $1s$ type wavefunction become better. We could obviously have begun with another independent LCAO,

$$\psi' = c(\phi^A - \phi^B) \tag{3.3}$$
$$(\langle\psi \mid \psi'\rangle = 0)$$

ψ is called a bonding molecular orbital, ψ' an antibonding orbital. A simple picture convinces us the $E_b < E_{ab}$. Focus on the charge density in the central region for ψ,

Fig. 10.

$\varrho(0) = \psi^2(0)$ is finite (Figure 10). For ψ', $\varrho'(0) = 0$. Thus for the bonding orbital there is more Coulomb attraction to the protons in the intermediate regimes and hence a stronger binding. With this introduction, we may proceed to the two electrons H_2 problem. Kolus and Roothan were able to solve the H-F problem, i.e. no correlation, assuming $\psi = \phi(1)\ \phi(2)$ in terms of the elliptic type wavefunctions. As for He, the correlation energy $\Delta \sim 1$ eV. How about a simpler atomic orbital approach? As with He, we can put two electrons into the H_2^+ bonding molecular orbital

$$\psi_{\text{MO}} = \frac{1}{2(1+s)}\left[\phi^A(1) + \phi^B(1)\right]\left[\phi^A(2) + \phi^B(2)\right]. \tag{3.4}$$

The good feature of this wavefunction is that it allows maximal charge density in the intermediate regions. The bad feature is that it contains 'ionic' configuration $\phi^A(1)\ \phi^A(2)$ which clearly costs 'correlation energy' (Figure 11).

Another approach to the H_2 problem is the Heitler-London or valence bond picture.

$$\psi_{\text{VB}} = C\left[\phi^A(1)\ \phi^B(2) + \phi^A(2)\ \phi^B(1)\right]. \tag{3.5}$$

This is the prototype of a covalent bond which clearly becomes very good at large

$$\overset{-}{\underset{A}{\cdot}} \qquad \overset{+}{\underset{B}{\cdot}} \qquad \text{ionic configuration}$$

Fig. 11.

nuclear separatives. This is a highly correlated wavefunction – no ionic configurations occur. The price one pays for this is that there is no strong piling up of charge density between the protons. An obvious better trial wavefunction, would be a linear combination of the two.

$$\psi_{\mathrm{MO}} = a\psi_{\mathrm{VB}} + b\psi_{1\sim 0}\,. \tag{3.6}$$

Using these wave function, it is a tedious but straight-forward matter to compute the relevant energies. Of course, difficult two center integrals are involved. Here I just quote the results.

TABLE II

$\Delta = E - E_{\mathrm{exact}}$		
Hartree-Fock	1.1 eV	no correlation
M-O	1.3 eV	no correlation
V-B	1.0 eV	highly correlated
ψ'	0.7 eV	

These results all optimize the nuclear charge $\tilde{z}$.

The effects of correlation become rather evident in a very trivial model for a diatomic molecule which is basically the Hubbard hamiltonian. From now on, we shall assume orthonormal atomic orbitals, i.e. $S = \int \phi_A(r)\,\phi_B(r)\,\mathrm{d}\mathbf{r} = 0$ and $\int \phi_A^2\,\mathrm{d}\mathbf{r} = 1$. In addition it is convenient to use 2nd quantized notation. Our model two electron hamiltonian is

$$\mathscr{H} = -\,t \sum_{\sigma} \left(C_{A\sigma}^{+} C_{B\sigma} + C_{B\sigma}^{+} C_{A\sigma}\right) + U\left(n_{A\uparrow} n_{A\downarrow} + n_{B\uparrow} n_{B\downarrow}\right), \tag{3.7}$$

where C_A destroys an electron with spin σ in the atomic orbital centered on atom A and $n_{A\sigma}^{+} = C_{A\sigma}^{+} C_{A\sigma}$ is the number operator. The 'transfer integral' t basically measures the attraction of an electron centered on one nucleus to the other nucleus. The parameter U represents the Coulomb repulsion which operates when two electrons reside on the same atomic orbital relative to that when they reside on different nuclei. The ratio U/t is physically similar to E_c/W of the first lecture. This is a drastic simplification of the real hamiltonian, but is much simpler and clarifies much of the physics.

This hamiltonian is easily diagonalized. The molecule has room for $4 \geqslant N \geqslant 0$ electrons – let's consider each case separately.

(a) $N = 1$

Since there is only one electron, no site is ever doubly occupied, so U plays no role. The bonding orbital is given by

$$\psi_{b\sigma} = \frac{1}{\sqrt{2}} (C_{A\sigma}^{+} + C_{B\sigma}^{+}) \, |0\rangle . \tag{3.8}$$

To check normalization

$$1 = \langle \psi_{b\sigma} \mid \psi_{b\sigma} \rangle = \tfrac{1}{2} \langle 0 | \, (C_{A\sigma} + C_{B\sigma}) \, (C_{A\sigma}^{+} + C_{B\sigma}^{+}) |0\rangle = \tfrac{1}{2} \langle 0 \mid 0 \rangle \times 2 = 1 ,$$

where we have used the anticommutation rules for fermion operators,

$$C_{A\sigma} C_{B\sigma'}^{+} + C_{B\sigma'}^{+} C_{A\sigma} = \delta_{\sigma\sigma' \mathrm{AB}} . \tag{3.9}$$

We now operate with $\mathscr{H}$ on $\psi_{b\sigma}$,

$$\mathscr{H} \psi_{b\sigma} = \frac{-t}{\sqrt{2}} (C_{A\sigma}^{+} + C_{B\sigma}^{+}) \, |0\rangle = - t \psi_{b\sigma} \tag{3.10}$$

$$E_b = - t$$

For the antibonding orbital,

$$\psi_{ab\sigma} = \frac{1}{\sqrt{2}} (C_{A\sigma}^{+} - C_{B\sigma}^{+}) \, |0\rangle . \tag{3.11}$$

$$\mathscr{H} \psi_{ab\sigma} = \frac{-t}{\sqrt{2}} (- C_{A\sigma}^{+} + C_{B\sigma}^{+}) \, |0\rangle = t \psi_{ab\sigma} \tag{3.12}$$

$$E_{ab\sigma} = t$$

$E_{ab} - E_b = 2t$ ———————— $+t$ (antibonding)

———————— $-t$ (bonding)

Fig. 12.

Each orbital is of course double degenerate ($\sigma = \pm\frac{1}{2}$). Thus for this truncated basis and simplified hamiltonian the M-O picture of the molecular ion is exact.

For the moment, let's skip the two electron case and go to the 3 electron or 'H_2^-' situation.

(b) $N=3$

What is a basis set?

$$\begin{aligned} \psi_{A\uparrow} &= |A_\uparrow^+ A_\downarrow^+ B_\uparrow^+ |0\rangle \\ \psi_{B\uparrow} &= |B_\uparrow^+ B_\downarrow^+ A_\downarrow^+ |0\rangle \end{aligned} \tag{3.13}$$

$$\begin{aligned} \mathscr{H} \psi_{A\uparrow} &= - t \, |A_\uparrow^+ B_\downarrow^+ B_\uparrow^+ |0\rangle + U \, |A_\uparrow^+ A_\downarrow^+ B_\uparrow^+ |0\rangle = \\ &= + t \psi_{B\uparrow} + U \psi_{A\uparrow} \\ \mathscr{H} \psi_{B\uparrow} &= - t \, |B_\uparrow^+ A_\downarrow^+ A_\uparrow^+ |0\rangle + U \psi_{B\uparrow} = t \psi_{A\uparrow} + U \psi_{B\uparrow} . \end{aligned} \tag{3.14}$$

Thus $\mathscr{H}$ mixes the states where each orbital is doubly occupied. The eigenstates must therefore be a linear combination of these two functions:

$$\psi_\uparrow = \alpha\psi_{A\uparrow} + \beta\psi_{B\uparrow}, \tag{3.15}$$

where α and β are determined by

$$\begin{aligned}
\mathscr{H}\psi_\uparrow &= E\psi_\uparrow = \alpha\mathscr{H}\psi_{A\uparrow} + \beta\mathscr{H}\psi_{B\uparrow}\\
E(\alpha\psi_{A\sigma} + \beta\psi_{B\sigma}) &= \psi_{A\uparrow}(U\alpha + t\beta) + \psi_{B\uparrow}(U\beta + t\alpha)\\
E\alpha &= U\alpha + t\beta \qquad (3.16)\\
E\beta &= U\beta + t\alpha\\
0 &= \begin{vmatrix} U-E & t \\ t & U-E \end{vmatrix}\\
(U-E)^2 - t^2 &= 0\\
U - E &= \pm t \qquad (3.17)\\
&\boxed{E = U \pm t}
\end{aligned}$$

$$\begin{aligned}
\psi_\uparrow^b &= c(\psi_{A\uparrow} - \psi_{B\uparrow})\\
\psi_\uparrow^{ab} &= c(\psi_{A\uparrow} + \psi_{B\uparrow})
\end{aligned} \tag{3.18}$$

The solutions are again bonding and antibonding orbitals – but we must add U since there is always a doubly occupied site.

———— $U + t$ (antibonding)
———— $U - t$ (bonding)

Fig. 13.

(c) $N=4$ is easy. The only possible wavefunction is

$$\begin{aligned}
\psi &= |A_\uparrow^+ A_\downarrow^+ B_\uparrow^+ B_\downarrow^+|0\rangle.\\
\psi &= 2U\psi \qquad E = 2U
\end{aligned} \tag{3.19}$$

(d) The N=2 or neutral molecule is most interesting. Let's consider a complete basis set.

$$\left.\begin{aligned}
|a\rangle &\equiv |a_\uparrow^+ a_\downarrow^+|0\rangle\\
|b\rangle &\equiv |b_\uparrow^+ b_\downarrow^+|0\rangle\\
|s\rangle &\equiv \frac{1}{\sqrt{2}}(a_\uparrow^+ b_\downarrow^+ + b_\uparrow^+ a_\downarrow^+)|0\rangle
\end{aligned}\right\} \quad \text{singlet } s^2 = 0 \tag{3.20a}$$

$$\left.\begin{aligned}
|1\rangle &\equiv (a_\uparrow^+ b_\uparrow^+)|0\rangle\\
|-1\rangle &\equiv a_\downarrow^+ b_\downarrow^+|0\rangle\\
|0\rangle &\equiv \frac{1}{\sqrt{2}}(a_\uparrow^+ b_\downarrow^+ - b_\uparrow^+ a_\downarrow^+)|0\rangle
\end{aligned}\right\} \quad \text{triplets} \tag{3.20b}$$

Let's check that the states are singlets and triplets. For the singlets $S_z=0$. Consider

$S^+ = S_a^+ + S_b^+$. If $S^+|\alpha\rangle = 0$ and $S_z|\alpha\rangle = 0$, then $|\alpha\rangle$ is a singlet. But $S_a^+ = a_\uparrow^+ a_\downarrow$. For example,

$$\begin{matrix} S_b^+|a\rangle = 0 \\ S_a^+|a\rangle = 0 \end{matrix} \quad \Rightarrow \quad |a\rangle \text{ is singlet}$$

$$S_a^+|s\rangle = \frac{1}{\sqrt{2}}\, b_\uparrow^+ a_\uparrow^+$$

$$\Rightarrow \; {}^+s = 0$$

$$S_b^+|s\rangle = \frac{1}{\sqrt{2}}\, a_\uparrow^+ b_\uparrow^+ = -\frac{1}{\sqrt{2}}\, b_\uparrow^+ a_\uparrow^+$$

This checks.

Similarly one can check that the triplet states are indeed triplets; $S=1$.

Let's first assume a simple M-O description, i.e. for the ground state,

$$\psi_{\text{M-O}} = \tfrac{1}{2}(a_\uparrow^+ + b_\uparrow^+)(a_\downarrow^+ + b_\downarrow^+)|0\rangle \tag{3.21}$$

what is $\langle \psi_{\text{M-O}}|\mathscr{H}|\psi_{\text{M-O}}\rangle$?

$$\mathscr{H} = \langle T + U \rangle \tag{3.22}$$

$$T\psi_{\text{M-O}} = -\tfrac{1}{2}t[|b_\uparrow^+ (a_\downarrow^+ + b_\downarrow^+) + a_\uparrow^+ (a_\downarrow^+ + b_\downarrow^+) + (a_\uparrow^+ + b_\uparrow^+)\, a_\downarrow^+ \ldots] =$$

$$= -t[(a_\uparrow^+ + b_\uparrow^+)(a_\downarrow^+ + b_\downarrow^+)] = -2t\psi_{\text{M-O}} \tag{3.23}$$

$$\langle T \rangle = -2t$$

$$U\psi_{\text{M-O}} = \tfrac{1}{2}U(a_\uparrow^+ a_\downarrow^+ + b_\uparrow^+ b_\downarrow^+) \tag{3.24}$$

$$\langle U \rangle = \frac{U}{4} \times 2 = \frac{U}{2}$$

$$\langle \mathscr{H} \rangle_{\text{M-O}} = \frac{U}{2} - 2t$$

Physically $\frac{1}{4}$ of the time each site is doubly occupied. Thus $U/4$ is the average correlation energy per site, and there are two sites.

Let's estimate V-B energy.

$$\psi_{\text{VB}} = |s\rangle = \frac{1}{\sqrt{2}}(a_\uparrow^+ b_\downarrow^+ + b_\uparrow^+ a_\downarrow^+)|0\rangle \tag{3.25}$$

$$\langle T \rangle = 0$$

$$\langle U \rangle = 0 \tag{3.26}$$

$$\langle \mathscr{H} \rangle_{V\text{-}b} = 0\,.$$

Thus, if $U > 4t$, V-B is better, if $U < 4t$ M-O is better. This is important (note also the factor of $\frac{1}{4}$ that I mentioned earlier). Note that V-B is a localized wavefunction while M-O is a delocalized wavefunction.

However let's do an exact calculation. Since the hamiltonian is spin independent, the singlets and triplets do not mix and may be considered independently. Let's look first at the triplets. Also, since there is no magnetic field, the three triplet components must be degenerate. Thus, we need only look at one component, e.g.

$$|1\rangle = |a_\uparrow^+ b_\uparrow^+|0\rangle$$

$T|1\rangle=0$, $U|1\rangle-0$ thus $|1\rangle$ is an exact eigenstate with zero energy.

$$\boxed{E_{\text{trip}} = 0} \tag{3.27}$$

The singlets are rather more complicated.

$$\begin{aligned} &T\,|a\rangle = -\,t\,(b_\uparrow^+ a_\downarrow^+ + a_\uparrow^+ b_\downarrow^+)\,|0\rangle = -\,\sqrt{2}t\,|s\rangle \\ &U\,|a\rangle + U\,|a\rangle \\ &T\,|b\rangle = -\,\sqrt{2}t\,|s\rangle \\ &U\,|b\rangle = U\,|b\rangle \\ &T\,|s\rangle = -\,\sqrt{2}t\,(|a\rangle + |b\rangle);\ U\,|s\rangle = 0\,. \end{aligned} \tag{3.28}$$

Forming the eigenvalue matrix,

	a	b	s
a	U-E	0	$-\sqrt{2}t$
0 = b	0	U-E	$-\sqrt{2}t$
s	$-\sqrt{2}t$	$-\sqrt{2}t$	$-E$

which gives, the three roots,

$$E = U,\ \tfrac{1}{2}\left[U \pm \sqrt{U^2 + 16t^2}\right]. \tag{3.29}$$

The ground state is clearly

$$E_{\mathrm{g}} = \tfrac{1}{2}\left[U - \sqrt{U^2 + 16t^2}\right]. \tag{3.30}$$

If $t=0$, $E_{\mathrm{g}}=0$ and the singlet and triplet are degenerate. V-B is correct with a localized wavefunction. If $U=0$, $E_{\mathrm{f}}=-2\mathrm{t}$, M-O is correct.

The singlet-triplet splitting is $|E_{\mathrm{g}}|$. We shall mainly be interested in the limit $t \ll U$,

$$E_{\mathrm{g}} \approx -\frac{4t^2}{U}. \tag{3.31}$$

This singlet-triplet splitting is the heart of antiferromagnetism. Consider two spins ($\frac{1}{2}$) with the Heisenberg exchange

$$J\mathbf{S}_1 \cdot \mathbf{S}_2\,. \tag{3.32}$$

The eigenvalues of this operator are easily found by

$$\begin{aligned} &(\mathbf{S}_1 + \mathbf{S}_2)^2 = 2S\,(S+1) + 2\mathbf{S}_1\mathbf{S}_2 \\ &J\mathbf{S}_1\mathbf{S}_2 = \frac{J}{2}\left[(\mathbf{S}_1 + \mathbf{S}_2)^2 - 2S\,(S+1)\right] = \begin{cases} -\,\tfrac{3}{4}J \text{ for the singlet state} \\ \ \ \tfrac{1}{4}J \text{ for the triplet state}. \end{cases} \end{aligned}$$

Thus, there exists a singlet-triplet splitting equal to J which we may identify with $4t^2/U$; i.e. the strongly correlated limit leads to an Heisenberg exchange between localized spins with an exchange integral $J \simeq 4t^2/U$.

4. TCNQ Salts [3, 4]

At this point, I should like to describe some of the typical experimental behavior observed in the charge transfer salts. I shall omit the extensive optical results which are, of course, very important but form a story unto themselves. After finishing this brief survey, I'll attempt to make contact with the Hubbard model.

First of all, the salts may be characterized as: (1) *simple* – i.e.1-1 salts, e.g. Alkali$^+$ $TCNQ^-$, mopholinium$^+$ $TCNQ^-$, NMP^+ $TCNQ^-$

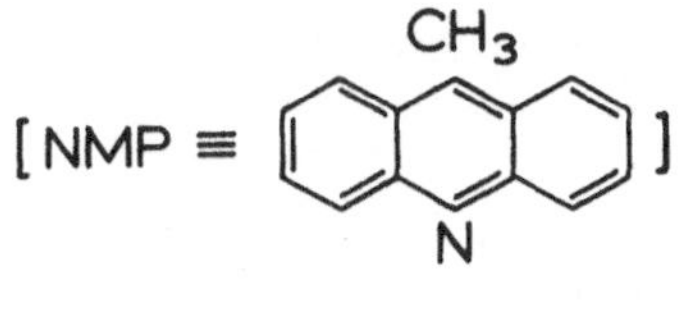

(2) complex – less or more than one electron/TCNQ, e.g.
Q $TCNQ_2$ (Q = quinolinium); $DTC(TCNQ)_2$, Cs_3 $TCNQ_2$...
(3) related systems $TMPD^+$ $TCNQ^-$

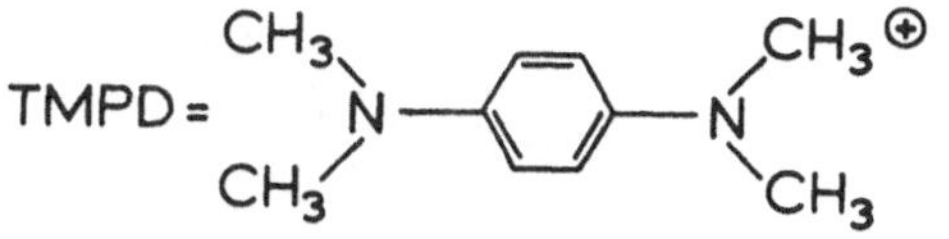

Wurster's Blue perchlorate $TMPD^+$ ClO_4^-

4.1. Simple salts

Alkali's – there are all semiconductors, i.e.

$$\sigma = \sigma_0 e^{-\Delta/kT} \quad \text{with} \quad \frac{\Delta}{k_B} \gtrsim 10^3\,\mathrm{K}\,(\gtrsim \cdot 1\ \mathrm{eV}).$$

They mainly have phase transitives (1st order?) at a few hundred degrees Kelvin but the origin is not well understood. Why are they not metals? Are they Mott insulators? First let's look at the structures. They are dimerized.

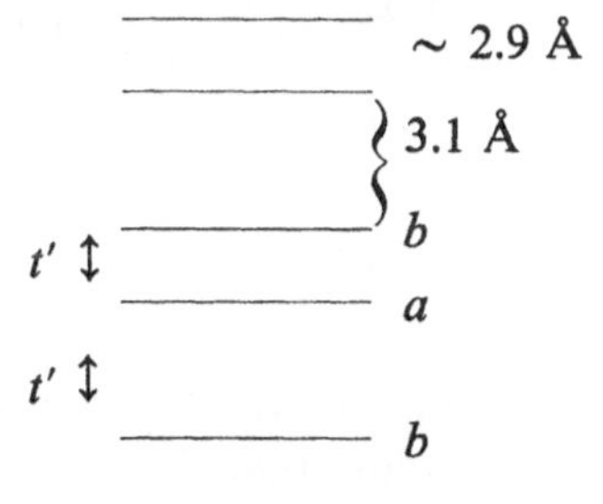

Fig. 14.

But then they should be semiconductors, i.e. 2 TCNQ's per unit all. Suppose we wanted to use band theory. The one electron tight bonding hamiltonian is then

$$\mathscr{H} = -\sum_{i,\sigma} \left[t\left(a_{i\sigma}^{+} b_{i\sigma} + b_{i\sigma}^{+} a_{i\sigma}\right) + t'\left(b_{i\sigma}^{+} a_{i+1\sigma} + a_{i+1\sigma}^{+} b_{i\sigma}\right) \right] \tag{4.1}$$

We Fourier transform the operators:

$$\begin{aligned} a_{j\sigma} &= \frac{1}{\sqrt{N}} \sum_k a_{k\sigma} e^{ik(ja - b/2)} \\ b_{j\sigma} &= \frac{1}{\sqrt{N}} \sum_k b_{k\sigma} e^{ik(ja + b/2)}. \end{aligned} \tag{4.2}$$

Then

$$\begin{aligned} \mathscr{H} &= -t \sum_{k,\sigma} \left(a_{k\sigma}^{+} b_{k\sigma} e^{ikb} + hc\right) - t' \sum_{k,\sigma} a_{k\sigma}^{+} b_{k\sigma} e^{ik(a+b)} + hc \\ &= -\sum_{k,\sigma} a_{k\sigma}^{+} b_{k\sigma} e^{ikb} \left(t + t' e^{-ika}\right) + hc. \end{aligned} \tag{4.3}$$

Writing the equations of motion,

$$\begin{aligned} i\hbar \dot{a}_{k\sigma} &= \varepsilon_k a_{k\sigma} = -b_{k\sigma} e^{ikb} \left(t + t' e^{-ika}\right) \\ i\hbar \dot{b}_{k\sigma} &= \varepsilon_k b_{k\sigma} = -a_{k\sigma} e^{-ikb} \left(t + t' e^{ika}\right); \\ \varepsilon_k^2 &= \left(t + t' e^{-ika}\right)\left(t + t' e^{ika}\right) = t^2 + t'^2 + 2tt' \cos Ka \end{aligned} \tag{4.4}$$

which leads to the spectrum as shown in Figure 15.

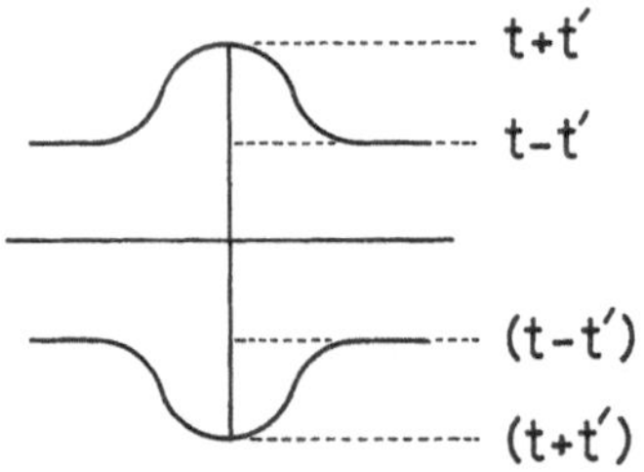

Fig. 15.

$$\varepsilon_k = \pm \sqrt{t^2 + t'^2 + 2tt' \cos Ka}, \tag{4.5}$$

For the case of one electron per atom, the lower band is filled and separated from the upper band by a gap $\Delta = 2(t - t')$. Note that if $t = t'$, $\Delta = 0$ and we have a metal. However, spin resonance doesn't seem to agree with this simple model for two reasons: (1) χ is indeed of the form $\chi = \chi_0 e^{-\Delta'/k_B T}$ but $\Delta' < \Delta$. Second if this were correct there would be no fine structure but *evidence* seems to give *dipolar* splitting of the E.P.R. lines.

$$\sim \frac{\mu^2}{a^3} \sim \frac{10^{-40}}{27 \times 10^{-24}} \approx 10^{-17}\,\text{eg} \approx \mu H_D$$

$$H_D \sim \frac{\mu}{a^3} \sim \frac{10^{-20}}{27 \times 10^{-24}} \sim 10^2\,\text{OE}.$$

[See e.g. Chestnut and Phillips [5] in morpholinium$^+$ TCNQ$^-$]. Thus, it appears that the band model is not appropriate. We are thus led to a Mott localization $U \gg t$ where the electrons are localized on sites.

NMP-TCNQ: This substance is apparently a uniform chain, i.e. equal spacing between TCNQ molecules along a stack. Thus simple band theory would predict 1-d metallic behavior. This has been studied in great detail by the group at Penn [6]. We shall discuss these results later in detail. One question we can answer qualitatively is why is the gap smaller in NMP than in the Alkalis. This is probably due to screening arising from the large electronic polarizibility of the NMP molecules.

4.2. Complex salts

$QTCNQ_2$ also is a uniform chain which seems to be microscopically metallic at all temperatures with a small linear specific heat.

DTC-$TCNQ_2$ is a dimerized insulator with a conductivity activation energy $\sim$0.1 eV and from E.P.R. with a weak exchange interaction of about 10 K.

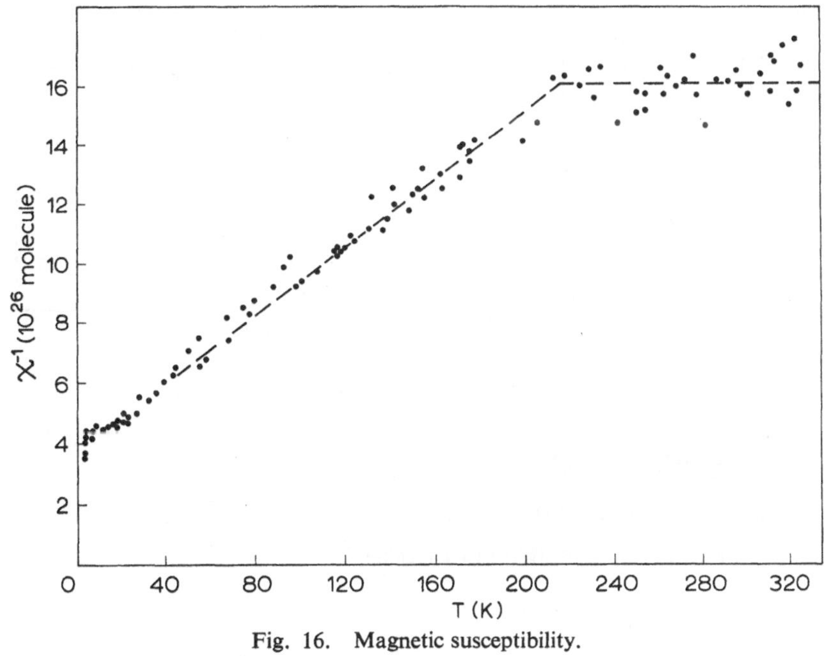

Fig. 16. Magnetic susceptibility.

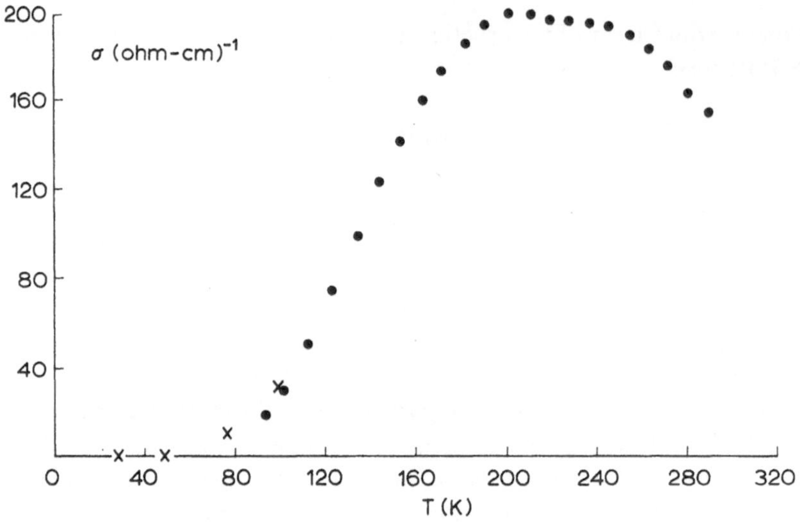

Fig. 17. Conductivity.

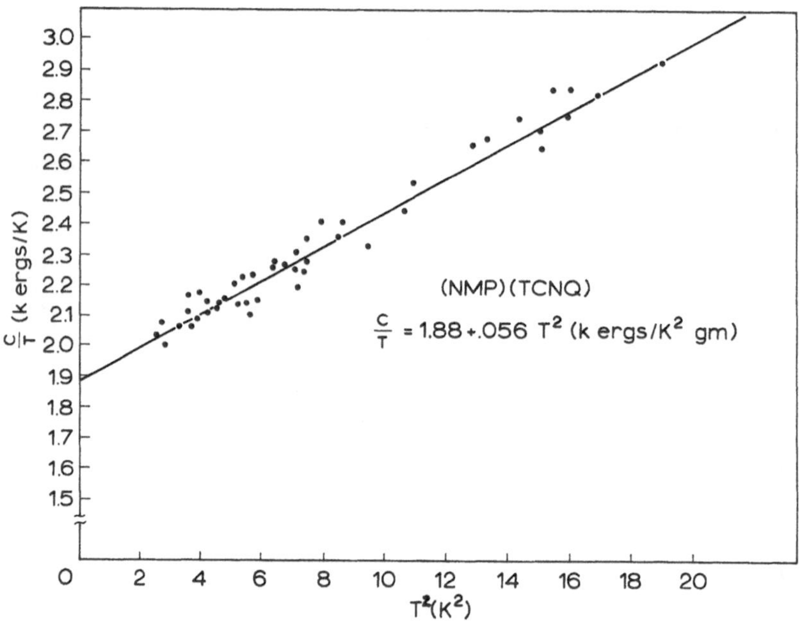

Fig. 18. Specific heat.

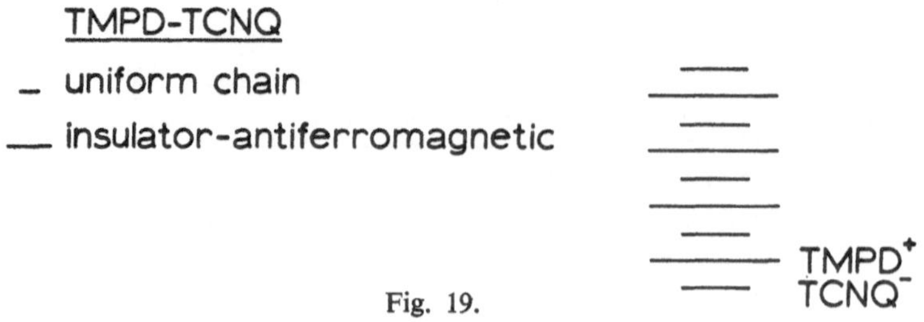

Fig. 19.

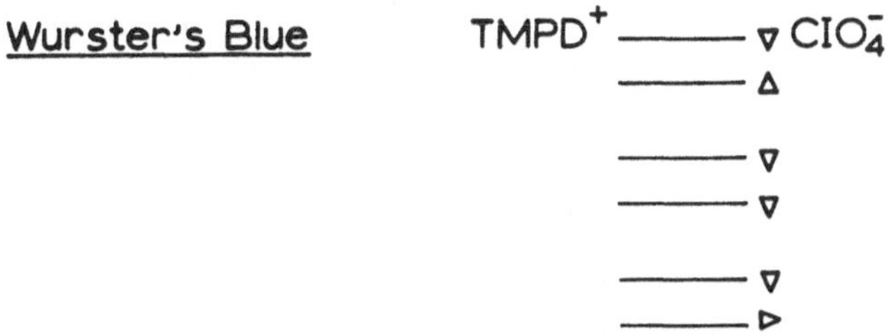

Fig. 20. At low temperatures this is also a dimerized antiferromagnetic insulator.

5. Application of Hubbard Model to Charge Transfer Salts (CTS)

The experimental situations previously described lead us to believe that correlation effects may be dominating the behavior of the charge transfer salts. As we have seen, the Hubbard hamiltonian is a very simple model which includes correlations. Thus we shall try to adapt it as a zeroth order approximation to the CTS systems. This idea was developed by Soos and his co-workers [7] at Princeton.

Our point of view is the following. Let's first consider an isolated neutral TCNQ molecule. Its electronic states may be considered as a set of molecular orbital levels, each doubly occupied up to some chemical potential.

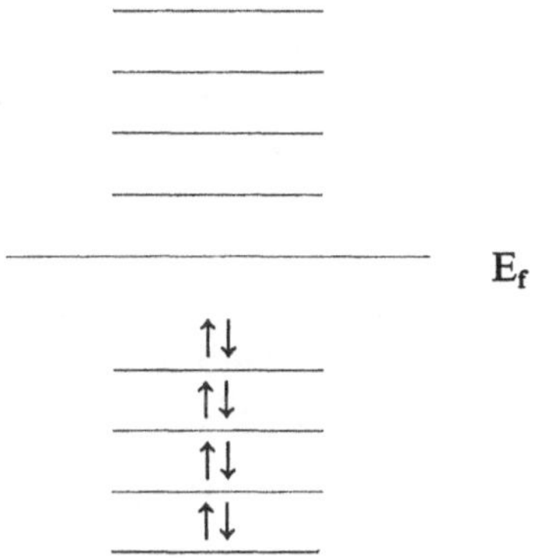

Fig. 21.

If an electron is added to the molecule, it goes into the lowest empty orbital. Indeed intramolecular electron-electron interactions will in general cause rearrangements of the level structure as well as molecular distortions – for the moment we shall neglect such effects. If a second electron is added to the molecule, it should normally enter into the same orbital but at the expense of an additional Coulomb repulsion energy U of interaction with the initial electron. Optical evidence indicates that the next higher unfilled orbital is typically of the order of several electron volts away in energy – again we shall neglect such states. In the salts, we further assume that the cations are completely ionized closed shell configurations, and remain so. This again is suggested by spectroscopic observations on back charge transfer. [Possible important excitonic effects have been suggested by Chaikin, *et al.* [8] when the cations are large polarizible molecules such as NMP.] These considerations lead us to a model for the CTS consisting of passive chains of closed shell cations separated by TCNQ stacks with unpaired electrons on the $TCNQ^-$ ions. The electrons can transfer along the TCNQ chains via the π overlaps. These words describe nothing but a set of one dimensional Hubbard hamiltonians, one for each isolated TCNQ chain;

$$\mathcal{H} = -\sum_{i,\sigma} t_{i,i+1}\left(C^+_{i\sigma}C_{i+1\sigma} + C^+_{i+1,\sigma}C_{i\sigma}\right) + U\sum_i n_{i\uparrow}n_{i\downarrow}, \tag{5.1}$$

where $1 \leqslant i \leqslant N$ indicates the TCNQ site on a chain of N molecules. This would appear to be a reasonable starting point. What predictions as to the physical behavior of a TCNQ salt does one derive from the Hubbard model? Most of our remaining discussions will be based on trying to answer this question.

In contrast to the simple diatomic Hubbard molecule considered earlier, there exist few exact results for the infinite Hubbard chain ($N \to \infty$). If n is the total number of electrons on the chain, then for the half-filled band case ($\varrho \equiv n/N = 1$), Lieb and Wu [9] have calculated the exact ground state energy and have shown the ground state to be a singlet; Takahashi [10] has given the exact zero temperature spin susceptibility; Ovchinnikov [11] has computed some low lying excited states. For arbitrary density (ϱ), Shiba [12] gives the exact ground state energy and zero temperature susceptibility and Coll [13] some low lying excitations. While these studies are very important, the method used is not easily extended to either finite temperatures or dynamic effects. We shall concentrate on various approximation methods and point out the exact results at the appropriate times.

6. Thermodynamics of Dimerized Salts and Speculations of Uniform Chains

Let us begin our discussion of Hubbard model thermodynamics by considering half-filled band dimerized systems, e.g. $alkali^+$ $TCNQ^-$ salts.

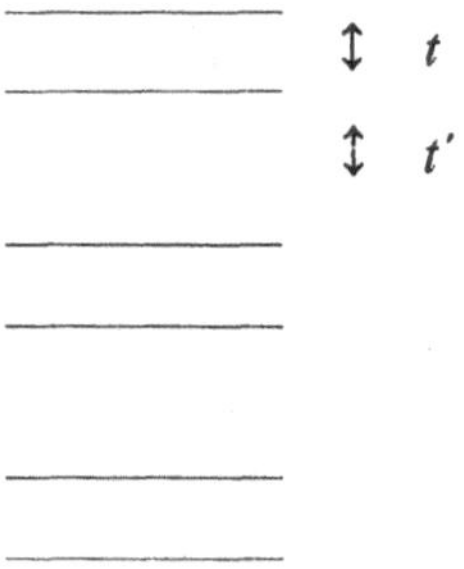

Fig. 22.

One method of attack, is to solve the dimer problem exactly and threat t' as a perturbation. Such a program has been current out to lowest order in t' (Suezaki [14], Bernstein and Pincus [15]). Since a great deal of physics is contained within the dimer problem we shall discuss its thermodynamics in detail. Consider an ensemble of N Hubbard dimers with $2N$ electrons ($\varrho=1$). The interdimer transfer t' is neglected; however, we use the grand cannonical ensemble so that a given dimer contains on the average two electrons, but fluctuations exist with $4 \geqslant n \geqslant 0$ electrons. The grand partition function is then

$$Z = \sum_n \sum_i e^{\beta(\mu n - E_{n,i})} = e^{-\beta\Omega}, \tag{6.1}$$

where μ is the chemical potential which insures the correct average electron density (in this case, $\varrho=1$) and Ω is the Gibbs free energy. The sum $\sum_n$ is over all possible number of electrons i.e. $0 \leqslant n \leqslant 4N$; and the sum $\sum_i$ is over all possible states with n total electrons; $E_{n,i}$ is the corresponding energy. The chemical potential is determined by

$$\varrho = -(2N)^{-1} \frac{\partial\Omega}{\partial\mu}. \tag{6.2}$$

The grand partition function is easily evaluated by noting that

$$e^{\beta(\mu n - E_{n,i})} = e^{\beta \sum_{j=1}^{N} (\mu n_j - E_{j,i})} = \prod_{j=1}^{N} e^{\beta(\mu n_j - E_{j,i})}, \tag{6.3}$$

where j labels a particular dimer $E_{j,i}$ is the energy of the jth dimer when it holds n_j electrons in the ith state. Rearranging Z it is easily shown that

$$Z = \left\{ \sum_{n=0}^{4} \sum_{i(n)} e^{\beta(\mu n - E_i(n))} \right\}^N \equiv z^N, \tag{6.4}$$

where z is effectively a 'single dimer partition function'. Then, in terms of z,

$$\Omega = -Nk_BT \ln z;$$
$$\varrho = \frac{k_BT}{2z} \frac{\partial z}{\partial\mu}. \tag{6.5}$$

Before evaluating, these expressions for the full dimer problem, let us obtain some experience by considering the simpler problem when $t=0$. Then we need only treat individual atoms as our units. Each atom may contain $0 \leqslant n \leqslant 2$ electrons with energies $E_0=0$, $E_1=0$, $E_2=U$. Then, z is simply

$$z = 1 + 2e^{\beta\mu} + e^{\beta(2\mu - U)}. \tag{6.6}$$

The two in the second term arises from the spin degeneracy when the atomic orbital is singly occupied. For the half-filled band case ($\varrho=1$) μ is given by

$$1 = \frac{k_BT}{z} \frac{\partial z}{\partial\mu}$$

or

$$z = 2\left[e^{\beta\mu} + e^{\beta(2\mu - U)}\right] \tag{6.7}$$

which immediately gives $\mu = U/2$. The entropy is

$$S = -\left(\frac{\partial\Omega}{\partial T}\right)_\mu = -Nk_B \ln z + \frac{Nk_BT}{z}\left(\frac{\partial z}{\partial T}\right) \tag{6.8}$$

and the specific heat $C_v = T(dS/dT)$. These expressions are easily evaluated to give

$$C_v = 2Nk_B \frac{(\beta U/4)^2}{ch^2(\beta U/4)}; \tag{6.9}$$

a type of Schottky anomaly at $k_BT \approx U/4$, which essentially involves entropy of delocalization. In order to compute the spin susceptibility, we need only add an external magnetic field H which reproves the spin degeneracy of the singly occupied state by $\pm\mu_BH \equiv \pm h$. The 'atomic partition function' z becomes

$$z = 2(1 + e^{\beta U/2} ch\beta h). \tag{6.10}$$

The total magnetic moment is $M = -(\partial\Omega/\partial H)_\mu$ which leads to $M = +(Nk_BT/z)(\partial z/\partial H)$ and a low field susceptibility per atom

$$\chi = C(T)\,\mu_B^2/kT, \tag{6.11}$$

i.e. a Curie law with a temperature dependent Curie constant $C(T)$ given by

$$C(T) = \frac{e^{\beta U/4}}{2ch^{\beta U/4}} \tag{6.12}$$

$C(T)$ varies from 1 at low temperatures (the value for localized $s = \frac{1}{2}$ electrons) to $\frac{1}{2}$ at high temperatures $k_BT \gg U/4$. In this latter range on the average $\frac{1}{2}$ of the sites are either doubly occupied or vacant – either case being non-magnetic and not contributing to the magnetic moment.

With this example in mind we can easily carry out the same type of calculation for an ensemble of dimers because we've already computed the eigenstates. Let us recall the results

TABLE III

n	E	Degeneracy
0	0	–
1	$-t$ (bonding)	2
	$+t$ (antibonding)	2
2	0	3
	U	1
	$\frac{1}{2}[U \pm \sqrt{U^2 + 16t^2}]$	1
3	$-t+U$	2
	$+t+U$	2
4	$2U$	1

The chemical potential μ remains $U/2$, and the dimer partition function is

$$z = 3 + 8e^{\beta U/2} ch\beta t + 3e^{\beta U} + 2e^{\beta U/2} ch\beta/2\sqrt{U^2 + 16t^2}\,. \tag{6.13}$$

It is a completely straight-forward algebraic exercise to calculate C_v and χ, for arbitrary t and U. However, the physics becomes clearer is we look at some limiting cases.

We have already studied the case $t=0$. What modifications occur if t is finite? With two electrons the energy level picture is as shown in Figure 23.

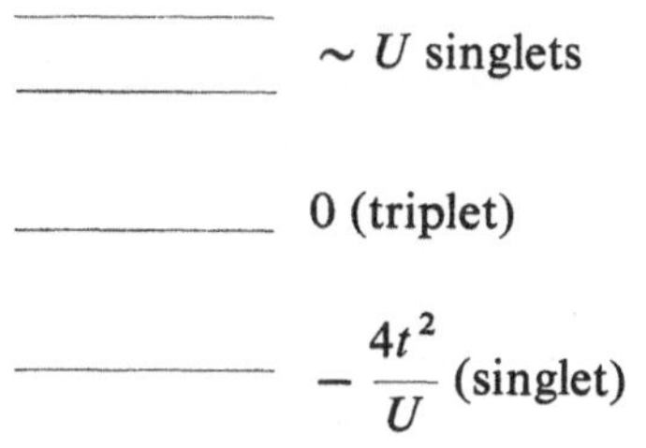

Fig. 23.

The singlet-triplet exchange splitting leads to dramatic modification of C_v and χ at low temperatures; since the ground state is a singlet with a finite singlet-triplet splitting $\chi \to 0$ as $T \to 0$, in contradistinction to the case $t=0$ where the singlet and triplet are degenerate. This leads to Figure 24.

Treating t' as a perturbation leads to only slight modification of these results. One can now rationalize the alkali TCNQ results. At low temperature

$$\chi \propto e^{-\beta \Delta'}.$$

where $\Delta' \sim 4t^2/U$. The activation energy for conductivity is however of order $U/4$ since this requires transfer of an electron from one molecule to another. Also, the

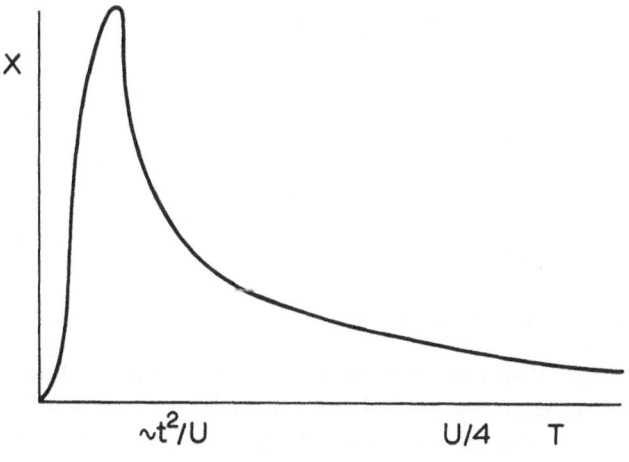

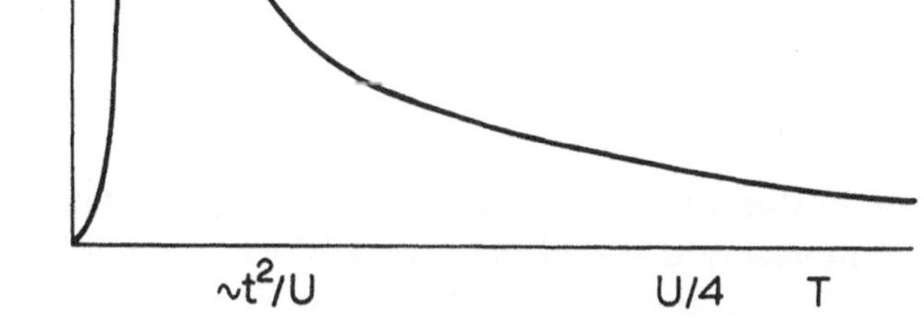

Fig. 24.

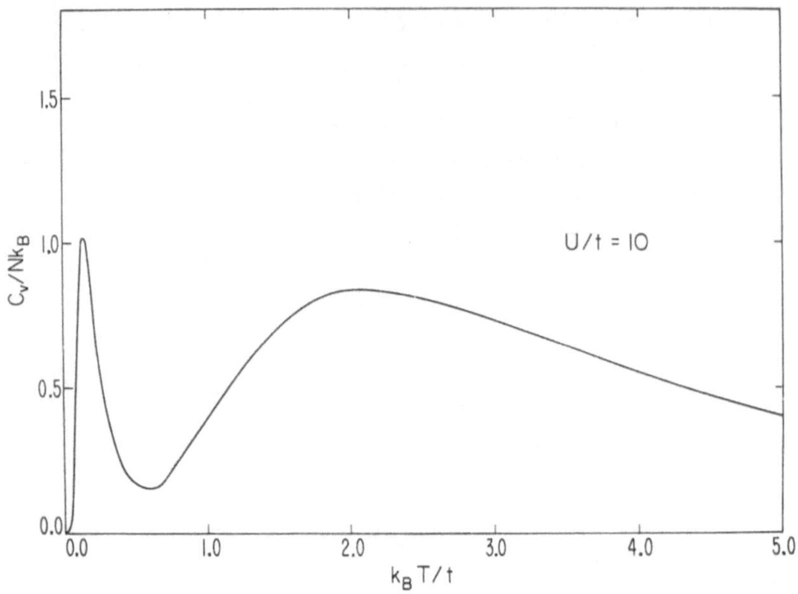

Fig. 25.

dipolar splitting is clear because an excited triplet will be localized on a dimer and with diffuse, only, slowly due to the weaker t'^2/U exchange.

What happens for uniform chains $(t'=t)$ which would be an elementary model of NMP-TCNQ? Our perturbation expansion breaks down. What can we do? In order to obtain a physical feeling for what occurs, let us reconsider the dimer problem in the two limits: (i) non-interacting electrons $U=0$; (ii) nearly atomic limit $U \gg t$. In case (i) the energy level scheme is simply bonding and antibonding orbitals. As t' increases the states spread out into a band of width $\sim t'$. When $t'=t$, the bands

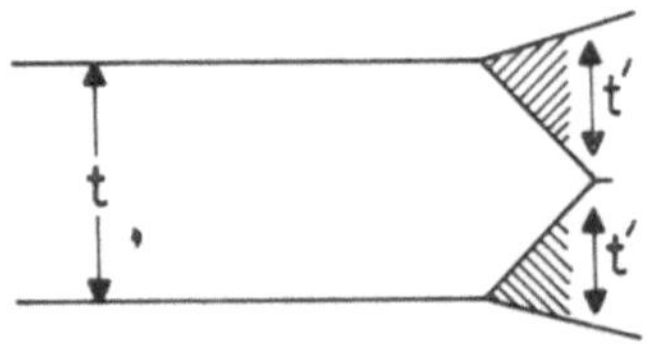

Fig. 26.

overlap as seen previously exactly

$$E_k = \sqrt{t'^2 + t^2 + 2tt' \cos k} \tag{6.14}$$

Thus, at low temperatures there is no gap between excited states and the ground state, i.e. the Fermi energy is within the band and we expect a linear specific heat $C_v \propto k_B T/t$. Similarly while the ground state is a singlet, there is no gap for singlet excitations which leads to a Pauli susceptibility $\chi \sim \mu_B^2/t$. At low temperatures, this behavior is qualitatively different than the semiconducting properties of the dimerized system.

At low temperatures in the atomic limit $U \gg t$, the appropriate energy level scheme is as shown in Figure 27.

Again t' causes a spreading of the levels into bands. At low temperatures only the localized states are occupied and the band widths are $\sim t'^2/U$. As before when $t' \to t$ the singlet and triplet bands overlap leading to a low temperature specific heat $C_v \sim (k_B T)\, U/t^2$ and a finite zero temperature susceptibility $\chi \sim \mu_B^2 U/t^2$. These effects are simply the result of spin fluctuations (or spin waves) on an antiferromagnetic chain. Notice already that we have obtained qualitatively the observed low temperature behavior in NMP-TCNQ. Of course, when $k_B T \sim U/4$ we expect delocalization and

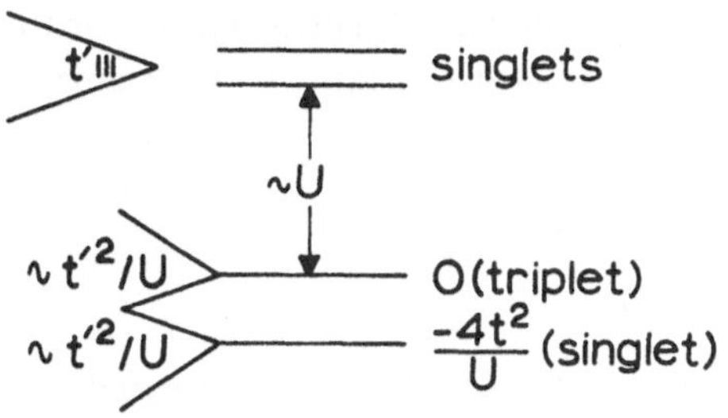

Fig. 27.

corresponding 'metallic' behavior. Indeed, in this limit ($U \gg t$), Ovchinnikov, shows that there exist spin wave excitations

$$\omega_k = \frac{\pi}{2}(4t^2/U)\sin ka \tag{6.15}$$

which agrees with the earlier results of de Cloiseaux and Pearson for the antiferromagnetic Heisenberg chain (making the identification $J = 4t^2/U$). He also showed that the gap for single-particle excitations is $E_g \cong U - 4t$. Takahashi's exact zero temperature susceptibility in this limit is

$$\chi \cong \mu_B^2 U/\pi^2 t^2 \tag{6.16}$$

which also agrees with Griffith's result for the Heisenberg chain. Their results are reasonably consistent with the experimental conductivity activation energy, specific heat, and extrapolated zero temperature susceptibility in NMP-TCNQ if U and t are chosen to be $U \cong 0.14$ eV and $t \cong 0.02$ eV [16].

7. Heisenberg Chain and the *X-Y* Model

Our speculations in the last section for a uniform Hubbard chain in the nearly atomic limit, $t/U \ll 1$, indicated that the low temperature behavior is similar to that of a Heisenberg antiferromagnet. Indeed, in Section 3, we saw that for a diatomic Hubbard molecule, in the same limit, there exists a singlet-triplet splitting which could be expressed in the Heisenberg form. In this section we shall make this more explicit and discuss an approximation to solve for the thermodynamics of a Heisenberg antiferromagnet which is particularly useful for one dimensional systems.

Since we are interested in the atomic limit, we begin with a state in which there is one electron localized at each lattice site, and treat the transfer term in the Hamiltonian by perturbation theory. Because T moves an electron from one site to a neighboring site, the lowest order energy correction involves the transfer of an electron to a nearest neighbor and then a return of an electron to the original site, i.e. second order in t. We then need consider only a pair of neighboring sites (a, b). If the spins of the electrons localized on a and b are parallel to one another, the Pauli principle excludes any transfer process and thus the energy perturbation for this configuration $\Delta E_t=0$. For antiparallel spins there are two possible arrangements

$$\begin{aligned} \phi_1 &= \uparrow \quad \downarrow \\ \phi_2 &= \downarrow \quad \uparrow \\ & \quad a \quad b \end{aligned}$$

The perturbation is $V=-t(a_\sigma^+ b_\sigma + b_\sigma^+ a_\sigma)$ and second order perturbation theory gives

$$\Delta E = -\sum_I \frac{|\langle | \, |I\rangle|^2}{E_I}, \tag{7.1}$$

where here the ground state has zero energy and $|I\rangle$ is the intermediate state. Here we must generalize to degenerate second order perturbation theory. All intermediate states involve a doubly occupied site with energy U. The diagonal terms are then

$$-\frac{1}{U}\sum_I \langle i| \, V \, |I\rangle \, \langle I| \, V \, |i\rangle = -\frac{2t^2}{U}. \tag{7.2}$$

The exchange term is

$$-\frac{1}{U}\sum_I \langle 1| \, V \, |I\rangle \, \langle I| \, V \, |2\rangle = -\frac{2t^2}{U}, \tag{7.3}$$

which leads to the eigenvalue equation

$$\begin{vmatrix} -\dfrac{2t^2}{U} - \lambda & -\dfrac{2t^2}{U} \\ -\dfrac{2t^2}{U} & -\dfrac{2t^2}{U} - \lambda \end{vmatrix} = 0 \tag{7.4}$$

with solutions $\lambda=0$, (triplet) and $\lambda=-4t^2/U$ (singlet). Thus, at least at low temperatures where there are no thermally activated vacancies or doubly occupied sites, we have an effective hamiltonian

$$\mathscr{H} = J \sum_i (\mathbf{S}_i \cdot \mathbf{S}_{i+1}) \quad \text{with} \quad J = 4t^2/U\,, \tag{7.5}$$

which is the Heisenberg antiferromagnet. There thermodynamics for the Heisenberg chain is also extremely difficult and we only have exact results (as previously indicated) at $T=0$ K and for finite systems [17].

One approximate scheme that has proven useful for the one dimensional problem is to rewrite the hamiltonian in the form

$$\mathscr{H} = J \sum_i \left[\tfrac{1}{2}(S_i^+ S_{i+1}^- + S_i^- S_{i+1}^+) + S_i^z S_{i+1}^z\right]; \tag{7.6}$$

then the transverse part is exactly soluable and the longitudinal part (in the representation which diagonalizes the transverse term) is an effective many body interaction. Such a procedure has been discussed by Katsura [18], Katsura and Inawashiro [19], Bulaevskii [20], and Silverstein and Soos [21]. It is certainly not clear why such a procedure should give reasonable results; however, for certain properties, at least, good quantatitive behavior is obtained. Let us sketch the method. We start with the X-Y model (Lieb *et al.* [22]) i.e. the transverse part of (7.6), and transform the spin operators to Fermion's by

$$S_i^+ = b_i^+ \exp\left[i\pi \sum_{j=1}^{c-1} b_i^+ b_i\right];\ S_i^- = \exp\left[-i\pi \sum_{j=1}^{c-1} b_i^+ b_i\right] b_i;$$
$$S_i^z = b_i^+ b_i - 1\,. \tag{7.7}$$

The reason why fermions form a good starting point is that each site (for $S=\frac{1}{2}$ is a two level system and hence $(S_i^+)^2=0$. This transformation preserves the spin commutation rules. The transverse part of the hamiltonian (7.6) $\mathscr{H}_T$ becomes

$$\mathscr{H}_T = J \sum_{i=1}^{N} (b_i^+ b_{i+1}^+ + b_{i+1}^+ b_i)\,. \tag{7.8}$$

This is formally identical to a one-dimensional tight binding metallic chain, so as we've seen previously in Section 4.

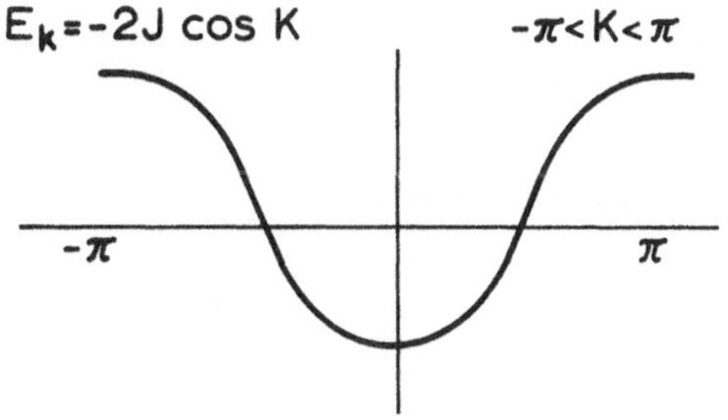

Fig. 28.

What is the ground state? Since there is no rule about conservation of particles the chemical potential, $\mu=0$. Then, the ground state has all negative energy states occupied and all positive energy states empty. The energy necessary to add another excitation is then

$$\hbar\omega_k = 2J \sin k. \tag{7.9}$$

These excitations are the spin waves in this representation. Treating the longitudinal term by H-F theory (Bulaevskii), one finds $E_k=2Jp \sin \mathrm{K}$, $p\approx 1+2/\pi$. On the other hand, des Cloizeaux and Pearson [23] show that the exact $T=0$ K excitations have $p=\pi/2$. It is completely straightforward within this framework to compute the specific heat. The internal energy is given by

$$\begin{aligned} U &= \sum_k n_k E_k \\ &= \sum_k \frac{E_k}{e^{\beta E_k}+1} \\ &= \left(\frac{N}{2\pi}\right)\int_{-\pi}^{\pi} \mathrm{d}k \frac{E_k}{e^{\beta E_k}+1} = \frac{N}{\pi}\int_{-2Jp}^{2Jp} \frac{E}{e^{\beta E}+1}\frac{\mathrm{d}E}{\sqrt{(2Jp)^2-E^2}}. \end{aligned} \tag{7.10}$$

The specific heat at low temperatures is then

$$C_V = \frac{\mathrm{d}U}{\mathrm{d}T} \cong Nk_B^2T/2\pi Jp. \tag{7.11}$$

The spin susceptibility is more interesting. Let's add a field along the z-axis. Then,

$$\mathscr{H} = \sum_k E_k n_k - 2g\mu_B H \sum_k n_k \tag{7.12}$$

which is just a uniform downward shift of the spectrum. From (7.7), the average longitudinal component of spin is $\langle S_z\rangle=2/N\sum_k n_k-1$. For $H=0$, exactly half of the states are occupied leading to $\langle S_z\rangle=0$. With finite H and $T=0$,

$$\sum_k n_k = \frac{N}{\pi}\int_0^{k^*} \mathrm{d}K = \frac{Nk^*}{\pi}, \tag{7.13}$$

where k^* is determined by the condition

$$2J \cos k^* = -2g\mu_B H. \tag{7.14}$$

For small fields, $k^*\simeq\pi/2$ which leads to

$$\langle S_z\rangle = (2/\pi)(g\mu_B H/J) \tag{7.15}$$

and a spin susceptibility

$$\chi \cong (4/\pi)\, \mu_B^2/J \tag{7.16}$$

at $T=0$. This is an overestimate by about a factor of 3 because we've omitted the longitudinal interaction. However, the cited references show that Hartree-Fock or low order perturbation theory treatments of this part of the Heisenberg hamiltonian brings good agreements with Griffith's exact result, (Section 5). Of course, one needs only to replace the n_k's in (7.12) and (7.13) to obtain the temperature dependence of χ.

8. High Temperature Expansions

If we are interested in the strongly correlated limit ($U \gg t$) of the Hubbard model, another useful method to study the high temperature behavior, both static and dynamic, is to carry out a high temperature perturbation expansion in βt and t/U. See Hone and Pincus [24] (half-filled band) and Beni *et al.* [25] (arbitrary density). As a simple example we shall consider here the thermodynamics of a one dimensional half-filled band.

Consider a Hamiltonian

$$\mathscr{H} = \mathscr{H}_0 + \mathscr{H}', \tag{8.1}$$

where $\mathscr{H}_0$ is an exactly soluable unperturbed Hamiltonian and $\mathscr{H}'$ is to be treated as a perturbation. Standard thermodynamics perturbation theory yields, for the Gibbs free energy Ω,

$$e^{-\beta(\Omega-\Omega_0)} = 1 + \sum_{n=1}^{\infty} (-1)^n \int_0^{\beta} \mathrm{d}x_1 \int_0^{x_1} \mathrm{d}x_2 \ldots \int_0^{x_{n-1}} \mathrm{d}x_n \langle \mathscr{H}'(x_1)\, \mathscr{H}'(x_2) \ldots \ldots \mathscr{H}'(x_n)\rangle_0, \tag{8.2}$$

where

$$\mathscr{H}'(x) = e^{x\mathscr{H}_0} \mathscr{H}' e^{-x\mathscr{H}_0} \tag{8.3}$$

and $\langle\ \rangle_0$ denotes an ensemble average with respect to the unperturbed partition function.

We shall now apply (8.2) to the Hubbard model, where $\mathscr{H}'$ is the transfer term. We have already seen (6.10), that the zeroth order ($t=0$) partition function is

$$e^{-\beta\Omega_0} = Z_0 = 2^N (1 + e^{\beta U/2} ch\beta h)^N, \tag{8.4}$$

where N is the number of sites on the chain. The lowest order diagonal term in the expansion (8.2) for the free energy involves the transfer of an electron from one site to a nearest neighbor and then its return, i.e. of order t^2.

$$\Omega - \Omega_0 \cong -(N/\beta) \int_0^{\beta} \mathrm{d}x \int_0^{x} \mathrm{d}y \langle \mathscr{H}'_{i,i+1}(x)\, \mathscr{H}'_{i+1,i}(y)\rangle_0. \tag{8.5}$$

Then, considering all possible occupancies of a pair of sites, (8.5) is easily evaluated to give

$$\Omega \cong \Omega_0 - 8\,(N/U)\,(t/z_0)^2\, e^{\beta U/2}\,[(\beta U/2)\, ch\beta h + sh\,(\beta U/2)]\,, \tag{8.6}$$

where

$$z_0 = 2\,(1 + e^{\beta U/2} ch\beta h)\,. \tag{8.7}$$

For zero field, the resulting specific heat is shown in Figure 29 for various values of t/U. The dominant terms in the spin susceptibility, derived from (8.6) are

$$\chi \approx (N\mu_B^2/2k_BT)\,[e^{(\beta U/4)}/ch\,(\beta U/4)]\,[1 - (2t^2\beta/U)\,th\,(\beta U/4)]\,. \tag{8.8}$$

This result is quite easily understood. The leading ($t=0$) term is simply a Curie law with a temperature dependent magnetic moment arising from the thermally excited vacancies and doubly occupied orbitals, as previously discussed. The transfer contribution may be considered as arising from the first term of an expansion of a Curie-Weiss law

$$\chi = c/(T + \theta) \tag{8.9}$$

with

$$\theta = (2t^2/U)\,th\,(\beta U/4) \tag{8.10}$$

At low temperatures $k_BT \ll U/4$, this simply because proportional to the anitferromagnetic exchange energy as expected from Section 7. For $k_BT \gtrsim U/4$, thermally excited vacancies reduce the average nearest-neighbor exchange; this is the effect of the $th\,(\beta U/4)$ factor in θ.

The type of analysis may be extended to other systems (e.g. $\varrho \neq 1$) and for dynamic responses as well.

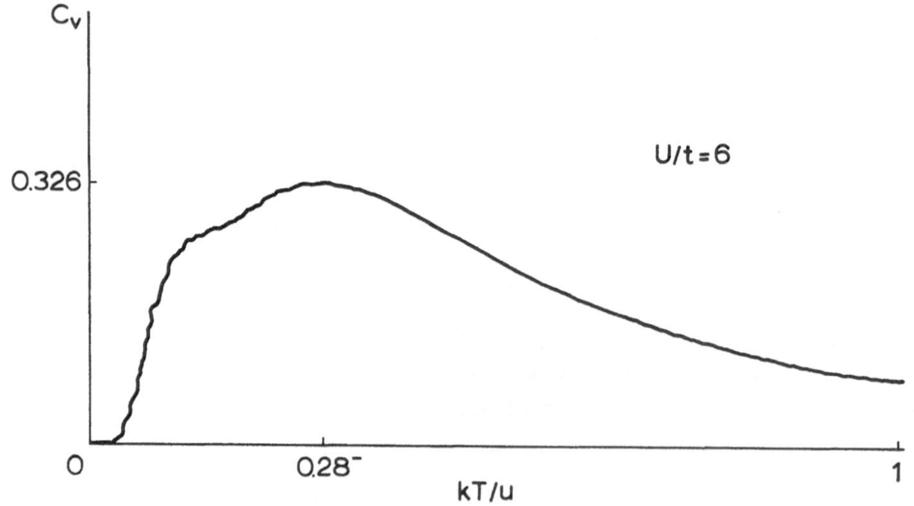

Fig. 29a.

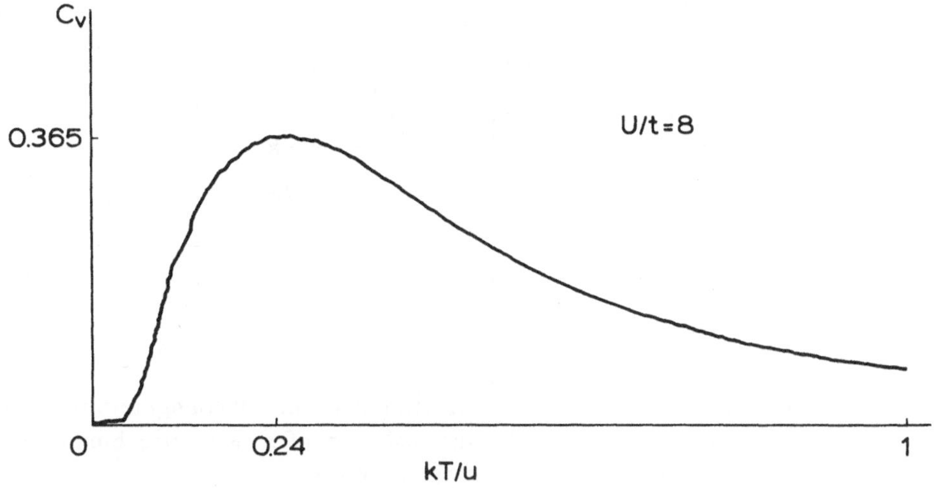

Fig. 29b.

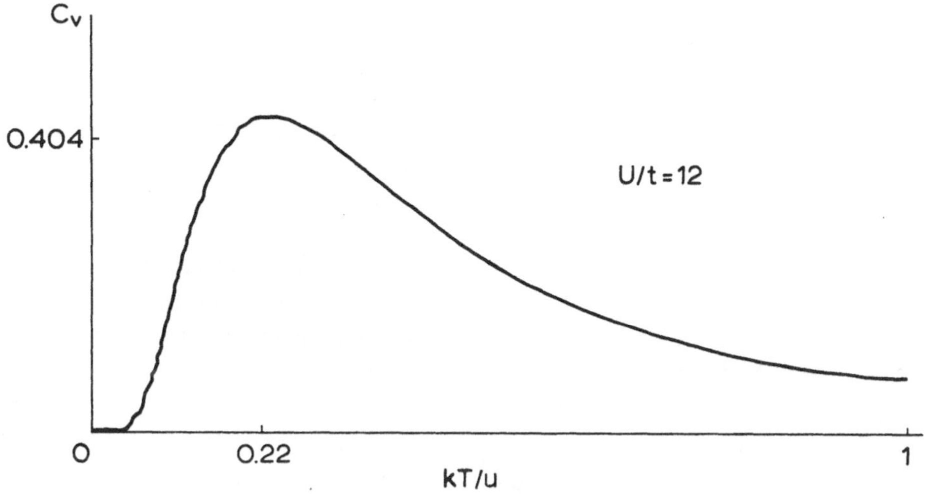

Fig. 29c.

9. Peierls Instability

As we have seen, many of the TCNQ salts crystallize in one dimensional dimerized stacks. It is interesting to pose the question of why dimerized structures seem to be so prevalent relative to uniform chains. One possibility is related to the instability first discussed by Peierls [26]. Peierls considered a uniform chain of atoms, each of which possess one non-degenerate orbital. A tight binding half-filled band is constructed of non-interacting electrons, (Figure 30), with spectrum

$$E(k) = -2t \cos ka \tag{9.1}$$

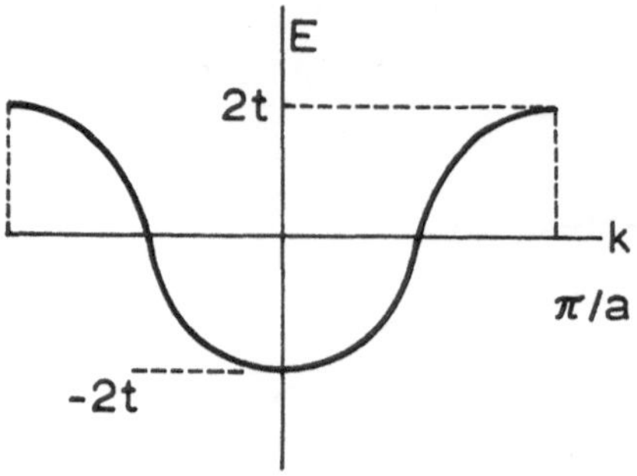

(9.2)

Fig. 30.

Suppose now that the chain distorts to a two atom per unit cell configuration (Figure 31). The band structure is modified by the inclusion of a new zone boundary at π/2a (Figure 32). Note that, at T=0 K, all the occupied

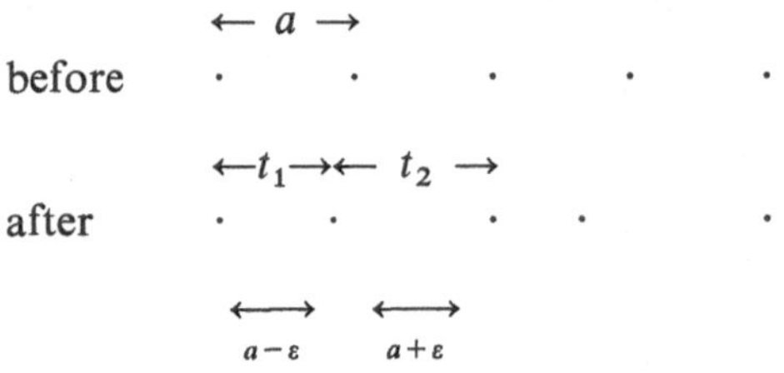

(9.3)

Fig. 31.

states below the Fermi surface have their energies diminished; thus the total electronic energy is reduced by dimerization. This gain is at the expense of an increase in the lattice energy. In one dimension the distorted state generally has lower energy (Adler and Brooks [27], Hallers and G. Vertogen [28].

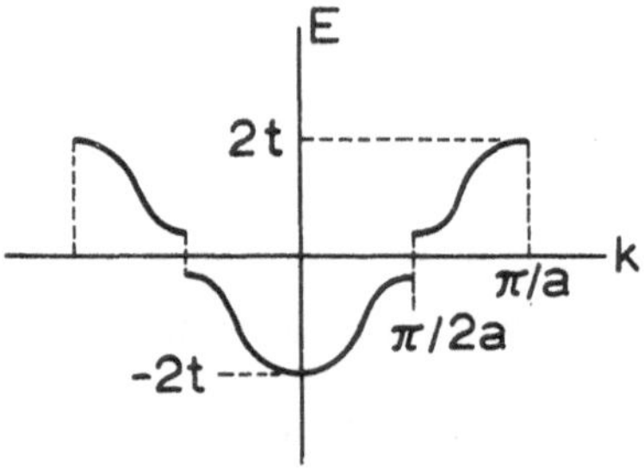

Fig. 32.

This effect may be considered as a collective Jahn-Teller effect where the degeneracy removed is that at the Fermi surface at $ka=\pi/2$.

Let us demonstrate the instability explicitly. Consider the dimerized state where the two different sites in the unit cell are labelled *a* and *b*. Then, the electronic part of the hamiltonian for non-interacting electrons in a tight binding band is given by

$$\mathscr{H} = - t_1 \sum_{i,\sigma} (a^+_{2i-\sigma} b_{2i\sigma} + b^+_{2i\sigma} a_{2i-\sigma}) - t_2 \sum_{i,\sigma} (b^+_{2i\sigma} a_{2i+1,\sigma} + a^+_{2i+1\sigma} b_{2i\sigma}), \tag{9.4}$$

where i labels the unit cell. Fourier transformation of the operators leads to

$$\mathscr{H} = - \sum_{k,\sigma} [t(k)\, a^+_{k\sigma} b_{k\sigma} + t^*(k)\, b^+_{k\sigma} a_{k\sigma}], \tag{9.5}$$

where

$$t(k) = \tfrac{1}{2}(t_1 e^{-ika} + t_2 e^{ikb}). \tag{9.6}$$

This is easily diagonialized in terms of new operators $\alpha_{k\sigma}$ and $\beta_{k\sigma}$ to give

$$\mathscr{H} = - \sum_{k,\sigma} |t(k)|\, (\alpha^+_{k\sigma} \alpha_{k\sigma} - \beta^+_{k\sigma} \beta_{k\sigma}). \tag{9.7}$$

The ground state for the half-filled band is complete occupation of the β states and zero occupancy in the α states, which leads to a ground state energy

$$E_g = - 2 \sum_k |t(k)| = - (N/\pi) \int_{-\pi}^{\pi} |t(k)|\, \mathrm{d}k = - (2N/\pi)(t_1 + t_2)\, E(\lambda^2), \tag{9.8}$$

where $E(\lambda^2)$ is the complete elliptic integral of the second kind and $\lambda^2 = 4t_1 t_2 (t_1 + t_2)^{-2}$.

The distorted chain is derived from the uniform chain by a dimerization with an assumed linear dependence of the transfer integral on the distortion amplitude ε; i.e. $t_1 = t(1+\delta)$; $t_2 = t(1-\delta)$ where $\delta = \alpha\varepsilon$ and α measures the initial spatial variation of the transfer integral. Including the elastic energy of distortion, the total energy of the dimerized chain is

$$U = - (4/\pi)\, NtE(\lambda^2) + NC\varepsilon^2, \tag{9.9}$$

where $\lambda^2 = 1 - \delta^2$ and C is an effective elastic stiffness constant. Minimizing $U(\varepsilon)$ with respect to ε leads, in the weak coupling regime ($\delta \ll 1$) to $\delta \approx \exp(-\pi C/t\alpha^2)$ which explicitly demonstrates the instability.

Thus we have shown the tendency for dimerization in the case of a half-filled non-interacting tight binding chain. What about the Hubbard model for $U \neq 0$? To my knowledge the only considerations of this type have been in the strongly correlated limit $U \gg t$ where the problem reduces to that of a linear Heisenberg antiferromagnet. However, even for that problem there is no exact discussion of the dimerization question. Chestnut [29] has shown that the Ising chain has a similar instability. We have already seen, in Section 7, that another relatively good approximation to the Heisenberg chain is the X-Y model which reduces in one dimension to a system of non-interacting fermions. The previous discussion of the Peierls instability then goes through essentially unchanged (Pincus [30]). This calculation has been extended to finite temperatures leading to (in the adiabatic approximation for the lattice) a second order phase transition. (Beni and Pincus [31].) The longitudinal parts of the interaction has also been included in terms of Hartree-Fock theory (see Section 7) leaving the results essentially unaltered (Beni [32]).

We may only speculate as to whether or not this type of instability is responsible for the frequent dimerization observed in the TCNQ salts. Further experiments should be able to clarify the situation.

Acknowledgement

The author is indebted to Dr M. Puma for taking the initial notes upon which these lectures are based.

References

1. Parr, R.: *Quantum Theory of Molecular Structure,* W. A. Benjamin, Inc. New York, 1964.
2. Parr, R. G.: *Quantum Theory of Molecular Electronic Structure*, W. A. Benjamin, Inc., New York, 1963.
3. Shchegolev, I. F.: Review Article on the 'Highly Conductivity Salts', to appear in *Physica Status Solidi.*
4. Kepler, R. G.: in T. A. Bak (ed.), *Phonons and Phonon Interactions,* W. A. Benjamin, Inc., New York, 1964.
5. Chestnut, D. B. and Phillips, W. D.: *J. Chem. Phys.* **35**, 1002 (1961).
6. Epstein, A. J., Etemad, S., Garito, A. F., and Heeger, A. J.: *Phys. Rev.* **B5**, 952 (1972).
7. Strebel, P. J. and Soos, Z.: *J. Chem. Phys.* **53**, 4077 (1970) and references quoted therein.
8. Chaikin, P. M., Garito, A. F., and Heeger, A. J.: *Phys. Rev.* **B5**, 4966 (1972).
9. Lieb, L. and Wu, F.: *Phys. Rev. Letters* **20**, 1445 (1968).
10. Takahashi, M.: *Prog. Theor. Phys.* **42**, 1098 (1969) and **43**, 1619 (1970).
11. Ovchinnikov, A.: *Soviet Phys. JETP* **30**, 1160 (1970).
12. Shiba, H.: *Phys. Rev.* **B6**, 930 (1972).
13. Coll. C.: to be published.
14. Suezaki, Y.: *Phys. Letters* **38A** 293 (1972).
15. Bernstein, U. and Pincus, P.: *Phys. Rev.*, to be published.
16. Epstein, A. J., Etemad, S., Garito, A. F., and Heeger, A. J.: *Phys. Rev.* **B5**, 952 (1972).
17. Bonner, J. and Fisher, M.: *Phys. Rev.* **135**, A640 (1964).
18. Katsura, S.: *Phys. Rev.* **127**, 1508 (1962).
19. Katsura, S. and Inawashiro, S.: *J. Math. Phys.* **5**, 1091 (1964).
20. Bulaevskii, L. N.: *JETP* **16**, 685 (1963).
21. Silverstein A. J. and Soos, Z.: *J. Chem. Phys.* **53**, 326 (1970).
22. Lieb, E., Schultz, T., and Mattis, D.: *Ann. Phys.* **16**, 407 (1961).
23. des Cloizeaux, J. and Pearson, J. J.: 1962, *Phys. Rev.* **128**, 2131 (1962).
24. Hone, D. and Pincus, P.: *Phys. Rev.*, to be published.
25. Beni, G., Hone, D., and Pincus, P.: *Phys. Rev.*, to be published.
26. Peierls, R. E.: *Quantum Theory of Solids*, Oxford Univ. Press, London, 1955, Chapter V.
27. Adler, D. and Brooks, H.: *Phys. Rev.* **155**, 826 (1967).
28. Hallers, J. J. and Vertogen, G.: *Phys. Rev. Letters* **27**, 404 (1971).
29. Chestnut, D. B.: *J. Chem. Phys.* **45**, 4677 (1966).
30. Pincus, P.: *Solid State Comm.* **9**, 1971 (1971).
31. Beni, G. and Pincus, P.: *J. Chem. Phys.*, to be published.
32. Beni, G.: *J. Chem. Phys.*, to be published.

PART B

ELEMENTARY PARTICLES

PARTONS* **

KERSON HUANG

Laboratory for Nuclear Science and Dept. of Physics, Massachusetts Institute of Technology, Cambridge, Mass. 02139, U.S.A.

1. Introduction

Recent experiments to probe the structure of the nucleon seem to indicate that it is composed of spin $\frac{1}{2}$ constituents, which are 'elementary' in the sense that they act like point particles with respect to electromagnetic interactions. Feynman [1] calls them 'partons'.

In a general sense we have come across partons many times in the history of physics. The first parton was perhaps Dalton's atom, which appears to be 'elementary' in the energy range below a few eV. Then, with Rutherford, as we use α-particles of higher energy to probe into the structure of the atom over smaller distances, we find that the atom itself is made of partons – the nucleus and electrons. In turn, the nucleus appears to be 'elementary' in the energy range below a few MeV, but not when we study it in more detail with probes of higher energy. It too contains partons – protons and neutrons. We are now at the exciting stage of looking deep into the proton and the neutron, and again finding partons, whose nature is still unclear. Nor do we know, of course, whether these partons are the 'ultimate' ones.

In these lectures my purpose is first to present and explain the evidence for partons, and secondly to introduce theoretical ideas that may be relevant to understanding them. Needless to say, there is at present no theory of the structure of nucleons.

The experimental evidence comes from inelastic scattering of electrons from a nucleon. I shall first discuss the analogous case of inelastic neutron scattering from liquid helium, where our intuition is perhaps better. After all, we know that liquid helium is composed of helium atoms. We shall see under what condition the individuality of the atoms emerges from the scattering experiment. This case is very interesting in its own right, because of the hope that it may give direct evidence for the existence of Bose-Einstein condensation in liquid helium.

We then go to the main topic of the so-called deep inelastic scattering of elctrons from a nucleon, and show that a simple notion of partons can account for the experimental results.

Lastly we discuss the interesting fact that deep inelastic scattering takes us into a world confined to the light cone.

Throughout these lectures we use units in which $\hbar = c = 1$.

* This work is supported in part through funds provided by the Atomic Energy Commission under Contract AT(11-1)-3069.

** Center for Theoretical Physics Publication #300

Abecassis de Laredo and Jurisic (eds.), Selected Topics in Phys. Astrophys. and Biophys. 175–213.

2. Neutron Scattering from Liquid Helium

2.1. Liquid helium

For our purpose liquid helium is a bound state of N helium atoms (taken to be non-relativistic point masses), occupying a volume V, in the limit $N \to \infty$, $V \to \infty$, with N/V finite [2]. The Hamiltonian of the system is, in the configuration space representation,

$$H = \sum_{i=1}^{N} \frac{p_i^2}{2M} + \sum_{i<j} v(\mathbf{r}_i - \mathbf{r}_j). \tag{2.1}$$

Its eigenfunctions, denoted by $\Psi_n(\mathbf{r}_1, \ldots, \mathbf{r}_N)$, satisfy

$$H\Psi_n = \varepsilon_n \Psi_n, \tag{2.2}$$

and are orthonormal:

$$(\Psi_n, \Psi_m) \equiv \int \mathrm{d}^3 r_1 \ldots \mathrm{d}^3 r_N \Psi_n^* \Psi_m = \delta_{nm}. \tag{2.3}$$

Alternatively we can use the quantized field description [2] in terms of the field operator $\psi(\mathbf{r})$:

$$\psi(\mathbf{r}) = V^{-1/2} \sum_{\mathbf{k}} e^{i\mathbf{k}\cdot\mathbf{r}} a_k, \tag{2.4}$$

$$[a_k, a_{k'}] = \delta_{kk'}. \tag{2.5}$$

In such a representation the Hamiltonian is given by the operator

$$H = \frac{1}{2M} \int \mathrm{d}^3 r \, |\nabla \psi(\mathbf{r})|^2 + \frac{1}{2} \int \mathrm{d}^3 r_1 \mathrm{d}^3 r_2 \psi_{(2)}^+ \psi_{(1)}^+ v_{12} \psi_{(1)} \psi_{(2)}. \tag{2.6}$$

The eigenvectors $|\Psi_n\rangle$ satisfy

$$H\,|\Psi_n\rangle = \varepsilon_n\,|\Psi_n\rangle, \tag{2.7}$$

$$\langle \Psi_n \mid \Psi_m \rangle = \delta_{nm}. \tag{2.8}$$

The total momentum operator is

$$\mathbf{P}_{\mathrm{op}} = \sum_{\mathbf{k}} \mathbf{k} a_k^+ a_k. \tag{2.9}$$

It commutes with H, and $|\Psi_n\rangle$ can be made a simultaneous eigenfunction of $\mathbf{P}_{\mathrm{op}}$ as well. For calculations it is sometimes convenient to expand any state in terms of a complete set of free particle states. These are labeled by a set of single-particle occupation numbers $\{n_0, \ldots, n_k, \ldots\}$, and are denoted by $|\ldots n_k \ldots\rangle$. The occupation number $n_k = 0, 1, 2, \ldots$. is an eigenvalue of the operator $a_k^+ a_k$.

Although $|\Psi_n\rangle$ is not an eigenfunction of $a_k^+ a_k$, and consequently cannot be described by assigning single-particle quantum states to each atom, the expectation value

$\langle\Psi_n|\, a_k^+ a_k\, |\Psi_n\rangle$ measures the probability of finding an atom of momentum **k**, when the liquid is in the state $|\Psi_n\rangle$. In particular, $\langle\Psi_n|\, a_0^+ a_0\, |\Psi_n\rangle$ is of special significance. The liquid in state $|\Psi_n\rangle$ is said to possess a Bose-Einstein condensate if $N^{-1}\langle\Psi_n|\, a_0^+ a_0|\, \Psi_n\rangle$ approaches a nonzero limit as $N\to\infty$, i.e. if a finite fraction of the particles have zero momentum. If there were no interparicle interactions this fraction would be unity for the ground state. In the presence of interactions this fraction is surely less than one, but still expected to be nonzero.

2.2. Kinematics of Neutron Scattering

We assume that a non-relativistic neutron (of coordinate **R** and momentum operator **P**) interacts with an atom (of coordinate **r**) through a weak potential $U(\mathbf{R}-\mathbf{r})$. Thus the Hamiltonian of liquid helium plus a neutron is

$$\mathscr{H} = H + \frac{\mathbf{P}^2}{2m} + \sum_{i=1}^{N} U(\mathbf{R} - \mathbf{r}_i). \tag{2.10}$$

We treat the scattering of a neutron by liquid helium only to first order in U.

The scattering process we are interested in is

$$\text{neutron} + \text{liquid helium} \to \text{neutron} + \text{anything}, \tag{2.11}$$

where, by virtue of the form of (2.10), 'anything' can only be any eigenstate of H. The

Fig. 1. Kinematics for neutron scattering from liquid helium.

kinematics of the scattering is illustrated in Figure 1. In the laboratory frame, the momentum and energy transferred from the neutron to the liquid are denoted by **q** and ν respectively:

$$\mathbf{q} = \mathbf{k} - \mathbf{k}' \quad \text{(Momentum transfer)} \tag{2.12}$$

$$\nu = E - E' \quad \text{(Energy transfer)} \tag{2.13}$$

It is easily seen that

$$q^2/2m \equiv |\mathbf{q}|^2/2m = (E^{1/2} - E'^{\,1/2})^2 + 4(EE')^{1/2}\sin^2\theta/2, \tag{2.14}$$

$$(E^{1/2} - E'^{\,1/2})^2 \leqslant q^2/2m \leqslant (E^{1/2} + E'^{\,1/2})^2. \tag{2.15}$$

Given the incident energy E, the scattering is kinematically completely determined by two variables, which we may take to be ν and q. The scattering angle θ depends on them through (2.14).

If the momentum and energy transferred by the neutron are completely taken up by the creation of a single excitation in liquid helium, then the scattering cross section is nonvanishing only for correlated values of ν and q, namely they must satisfy the energy-momentum relation characteristic of the excitation. Thus

For phonon creation: $\nu = cq\,(c \approx 240\ \mathrm{m\,s^{-1}})$ (2.16)

For roton creation: $\nu = \Delta + (q - q_0)^2/2\mu$ (2.17)

$(\Delta \approx 8.7\ \mathrm{K},\quad q_0 \approx 2A^{-1},\quad \mu \approx 0.17\ \mathrm{M})$

For particle creation: $\nu = q^2/2M^*$, (2.18)

where particle creation means lifting a particle (of effective mass M^*) from zero momentum to momentum $\mathbf{q}$. If the neutron is scattered by an excitation, then ν and q are not uniquely related. The relation (2.18) furnishes a way to detect the presence of zero momentum atoms in the liquid: Their presence contributes to the cross section a term that depends on ν and q^2 only through the combination q^2/ν, which can only take on values equal to 2 M^*, where M^* is the effective mass of the particle. For a bare helium atom M^* has the unique value M.

2.3. Cross Section

The initial and final states of the scattering process are given as shown as follows:

$$\begin{array}{lll} \text{Initial state:} & V^{-1/2} e^{i\mathbf{k}\cdot\mathbf{R}} & \Psi_n, \\ \text{Final state} & V^{-1/2} e^{i\mathbf{k}\cdot\mathbf{R}} & \Psi_m, \end{array} \tag{2.19}$$

where n is some fixed initial state, and the final state m is to be summed over eventually. For a liquid initially at a given temperature instead of in a given state, the state n is to be averaged over a canonical ensemble.

To first order in U, the cross section $\mathrm{d}\sigma$ per unit volume of the liquid is given by

$$I\,\mathrm{d}\sigma = V^{-1} \sum_m |\langle f|\, U\, |i\rangle|^2\, 2\pi\delta\,(E_f - E_i)\,(2\pi)^{-3}\, V\,\mathrm{d}^3\mathrm{k}' \tag{2.20}$$

$$I = k/mV \tag{2.21}$$

$$\mathrm{d}^3k' = (k')^2\,\mathrm{d}k'\,\mathrm{d}\Omega = mk'\,\mathrm{d}E'\,\mathrm{d}\Omega\,. \tag{2.22}$$

Hence

$$\mathrm{d}\sigma/\mathrm{d}E'\,\mathrm{d}\Omega = \frac{1}{V}\left(\frac{m}{2\pi}\right)^2 \frac{k'}{k} \sum_m \left|\int \mathrm{d}^3R e^{i\mathbf{q}\cdot\mathbf{R}} \left(\Psi_m, \sum_{j=1}^N U(\mathbf{R} - \mathbf{r}_j)\,\Psi_n\right)\right|^2 \times$$
$$\times\, \delta\,(\varepsilon_m - \varepsilon_n - \nu)\,. \tag{2.23}$$

Let

$$U_q \equiv \int \mathrm{d}^3R e^{i\mathbf{q}\cdot\mathbf{R}} U(\mathbf{R})\,. \tag{2.24}$$

Then

$$\int \mathrm{d}^3R e^{i\mathbf{q}\cdot\mathbf{R}} \sum_{j=1}^N U(\mathbf{R} - \mathbf{r}_j) = U_q \varrho_q\,, \tag{2.25}$$

where

$$\varrho_q = \sum_{j=1}^{N} e^{i\mathbf{q}\cdot\mathbf{r}_j} . \tag{2.26}$$

The latter is the qth Fourier transform of the density operator

$$\varrho(\mathbf{r}) = \sum_{j=1}^{N} \delta(\mathbf{r} - \mathbf{r}_j). \tag{2.27}$$

In a quantized field representation they are given by

$$\varrho(\mathbf{r}) = \psi^+(\mathbf{r})\,\psi(\mathbf{r}) \tag{2.28}$$

$$\varrho_q = \sum_{\mathbf{k}} a_{k+q}^{+} a_k, \quad \varrho_q = \varrho_{-q}^{+}. \tag{2.29}$$

Thus we can rewrite (2.23) in the form

$$\frac{d\sigma}{dE'\,d\Omega} = \left(\frac{m}{2\pi}\right)^2 \frac{k'}{k} |U_q|^2 \frac{1}{V} \langle \Psi_n | \varrho_q^+ \delta(H - \varepsilon_n - \nu)\, \varrho_q | \Psi_n \rangle . \tag{2.30}$$

For scattering from a liquid at temperature absolute zero we use (2.30) with $|\Psi_n\rangle = |\Psi_0\rangle$, the ground state. For scattering at finite temperature we average the initial state over a canonical ensemble, obtaining

$$\frac{d\sigma}{dE'\,d\Omega} = \left(\frac{m}{2\pi}\right)^2 \frac{k'}{k} |U_q|^2\, S(\mathbf{q}, \nu) \tag{2.31}$$

$$S(\mathbf{q}, \nu) \equiv \frac{1}{V} \sum_n e^{-\beta\varepsilon_n} \langle \Psi_n | \varrho_q^+ \delta(H - \varepsilon_n - \nu)\, \varrho_q | \Psi_n \rangle \cdot \left[\sum_n e^{-\beta\varepsilon_n}\right]^{-1} . \tag{2.32}$$

If, instead of neutrons, light is scattered from the liquid, then the cross section is still proportional to $S(\mathbf{q}, \nu)$. To first order only the factor in front of it in (2.31) depends on the specific interaction responsible for t he scattering.

2.4. Density correlation function

The function $S(\mathbf{q}, \nu)$ is the Fourier transform of the density correlation function. This can be seen as follows:

$$\begin{aligned} S(\mathbf{q}, \nu) &= \frac{1}{V} \int_{-\infty}^{\infty} \frac{dt}{2\pi} \sum_n e^{-\beta\varepsilon_n} \langle \Psi_n | \varrho_q^+ e^{-it(H - \varepsilon_n - \nu)} \varrho_q | \Psi_n \rangle / \sum_n e^{-\beta\varepsilon_n} = \\ &= \frac{1}{V} \int_{-\infty}^{\infty} \frac{dt}{2\pi} e^{i\nu t} \sum_n e^{-\beta\varepsilon_n} \langle \Psi_n | \varrho_q^+(t)\, \varrho_q(0) | \Psi_n \rangle / \sum_n e^{-\beta\varepsilon_n} = \\ &= \frac{1}{V} \int_{-\infty}^{\infty} \frac{dt}{2\pi} e^{i\nu t} \langle \varrho_q^+(t)\, \varrho_q(0) \rangle , \end{aligned} \tag{2.33}$$

where
$$\langle O\rangle \equiv Tr\,(e^{-\beta H}O)/Tr e^{-\beta H}\,, \tag{2.34}$$
and
$$\varrho_q(t) \equiv e^{itH}\varrho_q e^{-itH}\,. \tag{2.35}$$

In terms of $\varrho(\mathbf{r})$ we can also write

$$S(\mathbf{q},\nu) = \frac{1}{V}\int_{-\infty}^{\infty}\frac{dt}{2\pi}e^{i\nu t}\int d^3r_1\,d^3r_2 e^{-i\mathbf{q}\cdot(\mathbf{r}_1-\mathbf{r}_2)}\langle\varrho(\mathbf{r}_1,t)\,\varrho(\mathbf{r}_2,0)\rangle =$$
$$= \frac{1}{V}\int_{-\infty}^{\infty}\frac{dt}{2\pi}e^{i\nu t}\int d^3r e^{-i\mathbf{q}\cdot\mathbf{r}}\langle\varrho(\mathbf{r},t)\,\varrho(0)\rangle, \tag{2.36}$$

where the last step results from application of translational invariance, and where $\varrho(0)$ is an abbreviation for $\varrho(\mathbf{r}=0,\,t=0)$.

The density correlation function $\langle\varrho(\mathbf{r},t)\,\varrho(0)\rangle$ describes the propagation of a density disturbance in the liquid from $\mathbf{r}=0$, $t=0$ to $\mathbf{r}$, t. Its ν, q Fourier transform picks out the component that propagates like a plane wave of frequency ν and wave vector $\mathbf{q}$.

From our knowledge of liquid helium at low temperatures based on studies of specific heat and superfluid properties, we know that the liquid has collective excitations, which are well-defined when the momentum is relatively small. These are the phonons and rotons. Thus when we scatter neutrons, we expect $S(\mathbf{q},\nu)$ to be peaked at values of ν, $\mathbf{q}$ correlated as in the energy-momentum relation of phonons and rotons. This was in fact borne out by experiments [3], which were in fact the first direct verification of the $\nu-\mathbf{q}$ relation for the collective excitations, as illustrated in Figures 2 and 3. The width of the peak in Figure 2 is inversely proportional to the lifetime of the excitation, and is very small for phonons and rotons. Further experiments [4] reveal that beyond the region covered by Figure 3 there are other types of excitations, which however are highly unstable, as indicated by large widths. This is indicated in Figure 4. Their interpretations are still unclear. We also indicate on Figure 4 the possibility that the upper branch of the excitation curve, which is quite fuzzy in the region of small ν and $\mathbf{q}$, may become narrow as we go to large ν and $\mathbf{q}$, where we expect to see single-atom excitation. The latter has indeed been seen, [7, 8] but the region of overlap has not been covered.

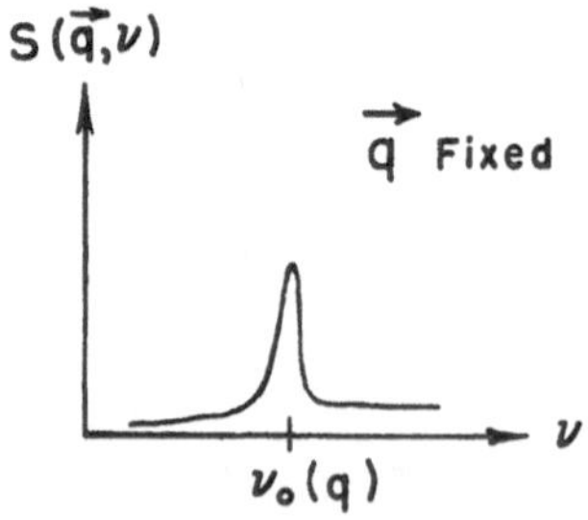

Fig. 2. $S(\mathbf{q},\nu)$ at low $\mathbf{q}$ and ν.

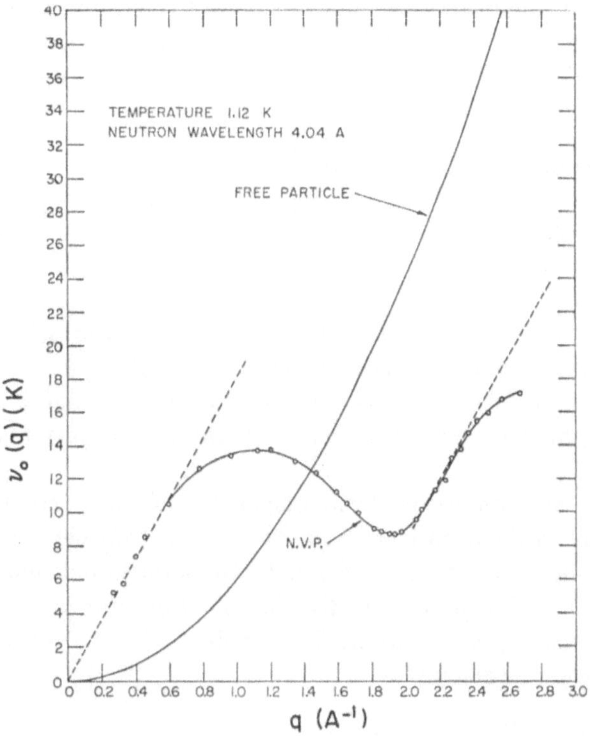

Fig. 3. Energy-momentum relation for a collective excitation in liquid helium.

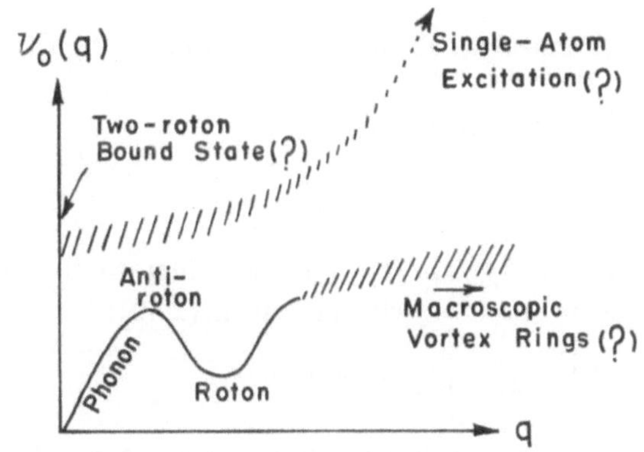

Fig. 4. Other types of excitations.

If we only knew liquid helium through neutron scattering of small ν and $\mathbf{q}$, we would say that it is a system having phonon and roton excitations, but we would not know that it is composed of helium atoms. And it would come as a revelation, when we do the experiment at higher ν and $\mathbf{q}$, to discover that there our neutrons bounce off single atoms.

Indeed it is the single-atom excitation that is of special relevance to us. We want to understand how it comes about that, inspite of interatomic interactions, the neutron is able to see individual atoms if ν, $\mathbf{q}$ are sufficient large. To this question we turn next.

2.5. Deep inelastic scattering [5, 6]

By deep inelastic scattering is meant the kinematic region corresponding to

$$\nu \to \infty, \qquad q^2 \to \infty, \qquad q^2/\nu \quad \text{finite}. \tag{2.37}$$

This idealized limit, convenient for mathematical considerations, in practice means that ν, q^2 are large compared to some characteristic constants of the liquid, which we shall specify later. The main points we want to bring out are that:

(1) As $\nu \to \infty$, neutron sees an individual atom 'frozen' instantaneously. (Impulse approximation).

(2) As $q^2 \to \infty$, neutron scatters from individual atoms incoherently.

The result is that in the deep inelastic region the neutron behaves as if it were scattered by independent free atoms. The initial state of the liquid enters only through the specification of an initial momentum distribution of the atoms.

First the impulse approximation: To calculate $S(\mathbf{q}, \nu)$ from (2.33), we need to know $\varrho_q^+(t)$, which is given by

$$\varrho_q^+(t) = \sum_{\mathbf{k}} a_{k+q}(t)\, a_k^+(t), \tag{2.38}$$

$$a_k(t) = e^{iHt} a_k e^{-iHt}, \tag{2.39}$$

with a_k defined by (2.5). Thus $S(\mathbf{q}, \nu)$ can also be written as

$$S(\mathbf{q}, \nu) = \frac{1}{V} \sum_{\mathbf{k}, \mathbf{p}} \int_{-\infty}^{\infty} \frac{\mathrm{d}t}{2\pi} \langle e^{itH} a_{k+q} a_k^+ e^{-itH} a_{p+q}^+ a_p \rangle\, e^{it\nu}. \tag{2.40}$$

Write $H = H_0 + \Omega$, where $\Omega = \sum_{i<j} v(\mathbf{r}_i - \mathbf{r}_j)$ is the interatomic interaction potential, and consider the following operator, which occurs in (2.40):

$$R = e^{it(H_0 + \Omega + \nu/2)} a_{k+q} a_k^+ e^{-it(H_0 + \Omega - \nu/2)}. \tag{2.41}$$

As $\nu \to \infty$, we neglect Ω, arguing that all its relevant matrix elements will be small compared to ν,* but we do not neglect H_0 because it is unbounded:

$$R \approx e^{it\nu} e^{itH_0} a_{k+q} a_k^+ e^{-itH_0}. \tag{2.42}$$

* This is easy to accept if Ω is a bounded operator. Unfortunately Ω for liquid helium contains a strong repulsive core. The argument relies on the fact that what is important is the comparison between ν and some average of Ω over actual liquid helium wave functions. These wave functions all tend to zero where Ω becomes large, so that the average remains finite.

This is the impulse approximation. It says that the time dependence of $\varrho_q(t)$ is governed by the free particle Hamiltonina H_0. Its validity requires

$$\Omega_{\mathrm{Av}}/\nu \ll 1\,, \tag{2.43}$$

where Ω_{Av} is some average matrix element of Ω. A reasonable estimate might be $\Omega_{\mathrm{Av}} \sim$(depth of interatomic potential), which is about $10\mathrm{K} \sim 10^{-2}$ eV. This turns out to be of the order of the roton energy $\nu_{\mathrm{roton}} = \Delta$:

$$\Omega_{\mathrm{Av}} \sim \nu_{\mathrm{roton}}\,. \tag{2.44}$$

To supply a physical interpretation, we note that liquid helium is a bound state of N atoms, and the attractive part of Ω is responsible for the binding. If Ω were identically zero, then the liquid would be always a collection of free atoms. We may thus interpret Ω_{Av} to be the inverse lifetime of a free particle state, into which the liquid dissociates virtually:

$$\tau_{\mathrm{virtual}} \sim \Omega_{\mathrm{Av}}^{-1}\,. \tag{2.45}$$

On the other hand, the energy transfer ν may be thought of as the inverse collision time:

$$\tau_{\mathrm{coll}} \sim \nu^{-1}\,. \tag{2.46}$$

Hence the validity of the impulse approximation (2.42) requires

$$\tau_{\mathrm{coll}}/\tau_{\mathrm{virtual}} \ll 1\,. \tag{2.47}$$

This is why we want $\nu \to \infty$.

We note

$$\begin{aligned} e^{itH_0} a_k e^{-itH_0} &= e^{-it\varepsilon_k} a_k \\ e^{itH_0} a_k^+ e^{-itH_0} &= e^{it\varepsilon_k} a_k^+ \\ \varepsilon_k &\equiv k^2/2M\,. \end{aligned} \tag{2.48}$$

Hence by (2.42)

$$R \approx e^{it(\nu + \varepsilon_k - \varepsilon_{k+q})}\,. \tag{2.49}$$

Substitution into (2.40) yields

$$S(\mathbf{q}, \nu) \underset{\nu \to \infty}{\to} \sum_{\mathbf{k},\,\mathbf{p}} \delta(\nu + \varepsilon_k - \varepsilon_{k+q}) \langle a_{k+q} a_k^+ a_{p+q}^+ a_p \rangle\,. \tag{2.50}$$

Next we consider $q^2 \to \infty$. In the product $a_{k+q}\, a_k^+\, a_{p+q}^+\, a_p$ in (2.50), the operator a_{p+q}^+ creates a very high momentum atom in the liquid. It is plausible to assume that in a liquid at finite temperature there are no atoms of such high momentum already present. Hence, the operator a_{k+q} must annihilate the very same atom we just created, in order not to yield a vanishing result. Therefore $\mathbf{k} = \mathbf{p}$. This means incoherence between scattering by different atoms. A reasonable estimate of how large q^2 has to be is that $q \gg$ (interatomic distance)$^{-1}$, which is about (4 Å)$^{-1}$, or in momentum units the same order as the roton momentum $q_{\text{roton}} = q_0$:

$$q \gg q_{\text{roton}} . \tag{2.51}$$

Thus

$$S(\mathbf{q}, \nu) \underset{\substack{\nu\to\infty \\ q^2\to\infty}}{\to} \sum_{\mathbf{k}} \delta(\nu + \varepsilon_k - \varepsilon_{k+q}) \langle a_{k+q} a_k^+ a_{k+q}^+ a_k \rangle . \tag{2.52}$$

By use of (2.5), we can write

$$\begin{aligned} \langle a_{k+q} a_k^+ a_{k+q}^+ a_k \rangle &= \langle n_k \rangle + \langle n_{k+q} n_k \rangle \\ n_k &\equiv a_k^+ a_k . \end{aligned} \tag{2.53}$$

The second term can be neglected for large $\mathbf{q}$, for there are few high momentum atoms in the liquid. So finally we have

$$S(\mathbf{q}, \nu) \underset{\substack{\nu\to\infty \\ q^2\to\infty}}{\to} \sum_{\mathbf{k}} \delta\left(\nu - \frac{q^2}{2M} - \frac{\mathbf{q}\cdot\mathbf{k}}{M}\right) \langle n_k \rangle . \tag{2.54}$$

If we truly let $q^2 \to \infty$, $\nu \to \infty$, keeping

$$x = q^2/2M\nu \tag{2.55}$$

finite, then in this 'deep inelastic' limit

$$\nu S(\mathbf{q}, \nu) \to N\delta(1 - x), \tag{2.56}$$

where N is the total number of atoms. The result has a 'scaling' property, in that it depends only on x and not on ν and q^2 separately. The result is a δ-function at $x = 1$ because helium atoms have a unique mass M. If we had a hypothetical gas consisting of atoms with a continuous distribution of effective masses characterized by a distribution function $f(M^*/M)$, then (2.56) would be replaced by

$$\nu S(\mathbf{q}, \nu) \to f(x). \tag{2.57}$$

The integral of (2.56) over all ν:

$$\int_{-\infty}^{\infty} d\nu \nu S(\mathbf{q}, \nu) \to Nq^2/2M \tag{2.58}$$

is just the asymptotic form of the well-known f-sum-rule for $q^2 \to \infty$.

If ν, q^2 are large but not too large, then one can hope to single out $\langle n_0 \rangle$, as follows. We rewrite (2.54) in the form

$$\nu S(\mathbf{q}, \nu) \to \delta(1 - x) \langle n_0 \rangle + \frac{V}{(2\pi)^3} \int d^3k \delta\left(1 - x - \frac{\mathbf{q}\cdot\mathbf{k}}{M\nu}\right) \langle n_k \rangle . \tag{2.59}$$

Assuming that $\langle n_k \rangle$ depends only on $|\mathbf{k}|$, we can perform the angular part of the integration in the second term and get

$$\frac{\nu S(\mathbf{q}, \nu)}{V} \to \frac{\langle n_0 \rangle}{V} \delta(1-x) + \frac{1}{4\pi^2} \int_{\frac{qx}{2(1-x)}}^{\infty} dk k^2 \langle n_k \rangle . \tag{2.60}$$

The second term involves $\langle n_k \rangle$ for $\mathbf{k} \neq 0$. Its form as a function of ν and q^2 may be calculated from models of liquid helium [6, 9], and gives a continuous distribution. For finite ν and q^2, the δ-function in the first term should actually be a peak with finite width, owing to interatomic interaction not completely negligible. One then hopes that this peak, the area under which is the density of the Bose-Einstein condensate, can be distinguished from the background distribution. Unfortunately this does not seem to be the case in present experiments [7, 8, 9], and an estimate of $\langle n_0 \rangle$ can be obtained only by subtraction procedures that are not free of uncertainties. Results are consistent with an independent theoretical estimate of $\langle n_0 \rangle / N \approx 8\%$ near absolute zero.

In the case of neutron scattering from liquid helium, the idealized deep inelastic limit is of course a mathematical fiction, for as we go to higher ν and q^2, what happens in the real liquid is that the neutron begins to look into the helium atom, and then the helium nucleus, and so on. If we want to investigate these inner structures, then neutrons would be a poor probe because it has complicated structures itself. It would be better to use electrons, which as far as we know is a point. This then takes us to the main topic of these lectures.

3. Deep Inelastic Scattering of Electrons from Nucleon

3.1. Kinematics

The process we are interested in is

$$e + N \to e + \text{Anything}. \tag{3.1}$$

To lowest order in the fine structure constant, $\alpha = e^2/4\pi$, the electron interacts with the nucleon through one-photon exchange, as illustrated by the Feynman graph of Figure 5, which also serves to define all relevant 4-momenta. In the laboratory frame the electron energy loss is denoted by ν and the electron scattering angle by θ:

$$\left.\begin{aligned} \nu &= E - E' \\ \theta &= \text{Angle between } \mathbf{k}, \mathbf{k}' \end{aligned}\right\} \text{ in lab. frame}. \tag{3.2}$$

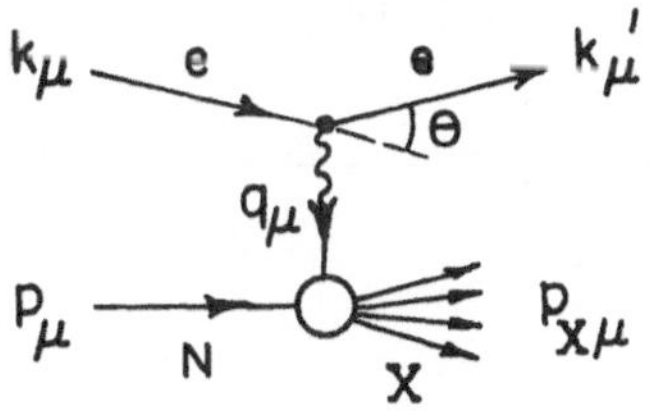

Fig. 5. Feynman graph to lowest order in the fine structure constant.

We also have

$$q^\mu = k^\mu - k'^\mu \text{ (4-momentum transfer)}$$
$$q^0 = \nu \text{ (in lab. frame)} \tag{3.3}$$
$$q^2 = (k - k')^2 \approx -4E'E \sin^2 \theta/2 .$$

The last approximate relation involves neglect of the electron mass m_e, and is stated in terms of laboratory variables. Note that $q^2 < 0$ in the physical region. An invariant expression for ν is

$$\nu = q \cdot P/M . \tag{3.4}$$

For given incident electron energy E, the two invariants ν and $-q^2$ completely specify the kinematics. We also define

$$\frac{1}{\omega} \equiv x \equiv -\frac{q^2}{2M\nu} . \tag{3.5}$$

The final state X is a group of physical hadrons, hence its total invariant mass M_X must be positive:

$$M_x^2 = (P + q)^2 = q^2 + M^2 + 2M\nu \geqslant 0 . \tag{3.6}$$

This leads to the restriction

$$0 \leqslant -q^2 \leqslant 2M\nu + M^2$$

or

$$0 \leqslant x \leqslant 1 + \frac{M}{2\nu} . \tag{3.7}$$

In the $-q^2$, ν plane, the physical region is the shaded region of Figure 6.

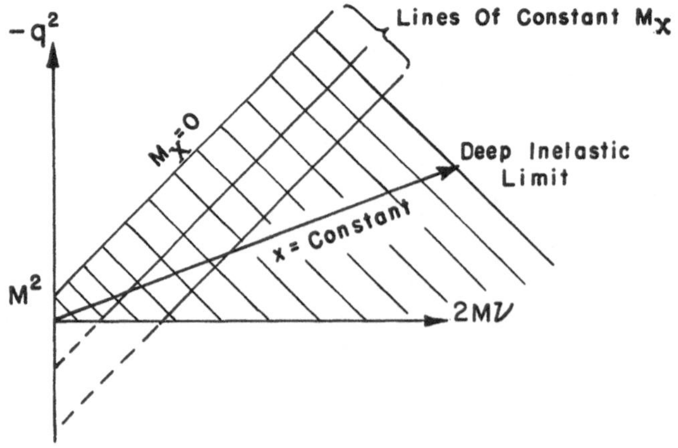

Fig. 6. Kinematics plot. Physical region is the shaded portion.

Creation of a single resonance in the final state X can occur along lines of constant M_X. The deep inelastic limit [11] is defined as

$$\nu \to \infty, \qquad -q^2 \to \infty, \qquad x \text{ finite}, \tag{3.8}$$

and is indicated in Figure 6. In the deep inelastic limit (3.7) becomes $0 \leqslant x \leqslant 1$, or $1 \leqslant \omega < \infty$. The point $x = 1$ (or $\omega = 1$) is the elastic threshold.

3.2. Cross section

We used the convention of Bjorken and Drell [10] for Feynman rules and Dirac matrices. The matrix element m is defined by

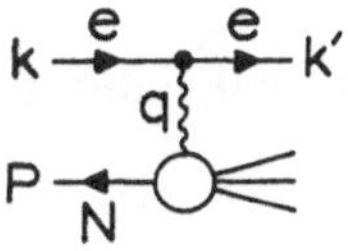

$$\equiv (2\pi)^4\, \delta^4 (k' + P_X - k - P)\, m\,, \tag{3.9}$$

where, for generality at this stage, we consider N to be any particle of mass M and spin s. Thus

$$m = [\bar{u}(\mathbf{k}')\,(-ie\gamma_\mu)\, u(\mathbf{k})]\, \frac{(-ie)}{q^2} \langle X|\, J^\mu\, |N\rangle\,, \tag{3.10}$$

where $J^\mu \equiv J^\mu(0)$ is the current operator evaluated at the origin of space-time coordinates, and is defined by

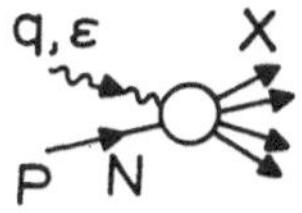

$$\equiv (2\pi)^4\, \delta^4 (P_X - q - P)\, \varepsilon_\mu (-ie) \langle X|\, J^\mu\, |N\rangle\,, \tag{3.11}$$

where q_μ, ε_μ label an off mass shell photon of 4-momentum q_μ and polarization ε_μ. We normalize particle states such that

$$\sum_{\text{1-fermian states}} = \int \frac{\mathrm{d}^3 p}{(p_0/M)\,(2\pi)^3}, \qquad p_0 = \sqrt{\mathbf{p}^2 + M^2}\,,$$

$$\sum_{\text{1-boson states}} = \int \frac{\mathrm{d}^3 p}{(2p_0)\,(2\pi)^3}\,. \tag{3.12}$$

The differential cross section $\mathrm{d}\sigma$ is given by

$$I\, \mathrm{d}\sigma = (2\pi)^4\, \delta^4 (k' + P_X - k - P)\, |m|^2\, d(\text{phase space})\,, \tag{3.13}$$

$$I = v_e (E/m_e) \begin{cases} P_0/M & \text{(fermian target)} \\ 2P_0 & \text{(boson target)}. \end{cases} \tag{3.14}$$

where v_e, the electron velocity, is approximately unity. In the laboratory frame, let

$$a \equiv \begin{cases} P_0/M = 1 & \text{(fermian target)} \\ 2P_0 = 2M & \text{(boson target)}. \end{cases} \tag{3.15}$$

The final phase space element is

$$d\,(\text{phase space}) = (k')^2\, dk'\, d\Omega\,(dP_z) \approx (E')^2\, dE'\,\Omega d\,(dP_x), \tag{3.16}$$

where we neglect the electron mass, and where (dP_X) is the phase space element for the final state X, including all p_0/m or $2p_0$ factors appropriate, and possible factors of $1/n!$ for groups of n identical final particles. Thus in the laboratory frame

$$\frac{d\sigma}{dE'\, d\Omega} = \frac{1}{a\,(E/m_e)\,(E'/m_e)}\frac{(E')^2\,(2\pi)^4}{(2\pi)^3}\sum_X \delta^4(P_x - P - q)\frac{1}{2}\sum_{\text{e spin}}\frac{1}{2s+1}\sum_{\substack{\text{target}\\\text{spin}}}|\mathcal{M}|^2, \tag{3.17}$$

where $\sum_x$ stands for $\int(dP_X)$ and a sum over all possible types of final states X. From (3.10) we have

$$\frac{1}{2}\sum_{\text{e spin}}\frac{1}{2s+1}\sum_{\substack{\text{target}\\\text{spin}}}|\mathcal{M}|^2 = \frac{e^4}{q^4}\, I_{\mu\nu}\,\langle P|\, J^\mu\,|X\rangle\,\langle z|\, J^\nu\,|P\rangle\,, \tag{3.18}$$

where $\langle P|\ldots|P\rangle$ denotes the spin average of $\langle N|\ldots|N\rangle$, and

$$\begin{aligned} I_{\mu\nu} &\equiv \tfrac{1}{2}\sum_{\text{e spin}}\left[\bar{u}(\mathbf{k}')\,\gamma_\mu u\,(\mathbf{k})\right]\left[\bar{u}(\mathbf{k})\,\gamma_\nu u\,(\mathbf{k})\right] = \\ &= \tfrac{1}{2}Tr\left[\gamma_\mu\frac{\not{k} + m_e}{2m_e}\gamma_\nu\frac{\not{k}' + m_e}{2m_e}\right] = \\ &= \frac{1}{2m_e^2}\left[k_\mu k'_\nu + k'_\mu k_\nu + q_{\mu\nu}\,(m_e^2 - k\cdot k')\right]. \end{aligned} \tag{3.19}$$

Let

$$W^{\mu\nu} \equiv \frac{(2\pi)^3}{a}\sum_X \langle P|\, J^\mu\,|X\rangle\,\langle X|\, J^\nu\,|P\rangle\,\delta^4(P_X - P - q)\,. \tag{3.20}$$

Then

$$\frac{d\sigma}{dE'\, d\Omega} = \frac{4\alpha^2 M^2}{q^4}\frac{E'}{E}\, I_{\mu\nu}W^{\mu\nu}, \tag{3.21}$$

where $\alpha = e^2/4\pi = 1/137$.

It is clear from (3.19) that $I_{\mu\nu}$ is dimensionless. The differential cross section $d\sigma/dE'd\Omega$ should be of dimension (length)^3, in units $\hbar = c = 1$. Hence $W^{\mu\nu}$ should be of dimension of length, or equivalently (mass)$^{-1}$.

3.3. Form Factors

The quantity $W^{\mu\nu}$ is a real symmetric tensor that depends only on P^μ and q^μ. Gauge invariance demands that $q_\mu W^{\mu\nu} = 0$. Thus the most general form is

$$W^{\mu\nu} = Ag^{\mu\nu} + Bq^\mu q^\nu + CP^\mu P^\nu + D\,(q^\mu P^\nu + q^\nu P^\mu)\,. \tag{3.22}$$

Gauge invariance requires

$$\begin{aligned} A + Bq^2 + D(q\cdot P) &= 0 \\ C(q\cdot P) + Dq^2 &= 0 . \end{aligned} \tag{3.23}$$

Eliminating B and C, and setting $A = W_1$, $D = -(q\cdot P/q^2)(W_2/M^2)$, we obtain

$$W^{\mu\nu} = -W_1\left(g^{\mu\nu} - \frac{q^\mu q^\nu}{q^2}\right) + \frac{W_2}{M^2}\left(P^\mu - q^\nu\frac{q\cdot P}{q^2}\right)\left(P^\nu - q^\nu\frac{q\cdot P}{q^2}\right), \tag{3.24}$$

where W_1 and W_2 are invariant functions of ν and $-q^2$, and are both of dimension $(\text{mass})^{-1}$. The structure of the target enters exclusively through these two form factors. To find W_1 and W_2 given $W^{\mu\nu}$, let

$$\begin{aligned} C_1 &\equiv W^\mu_\mu \\ C_2 &\equiv \frac{1}{M^2} P_\mu W^{\mu\nu} P_\nu . \end{aligned} \tag{3.25}$$

Then

$$\begin{aligned} W_1 &= \frac{1}{2}\left[\ C_2 - \left(1 - \frac{\nu^2}{q^2}\right) C_1\right]\left(1 - \frac{\nu^2}{q^2}\right)^{-1} \\ W_2 &= \frac{1}{2}\left[3C_2 - \left(1 - \frac{\nu^2}{q^2}\right) C_1\right]\left(1 - \frac{\nu^2}{q^2}\right)^{-2} . \end{aligned} \tag{3.26}$$

A straightforward calculation gives

$$\begin{aligned} I_{\mu\nu}W^{\mu\nu} = &-\frac{W_1}{2m_e^2}\left\{2\left[k\cdot k' - \frac{(q\cdot k)(q\cdot k')}{q^2}\right] + 3(m_e^2 - k'\cdot k)\right\} + \\ &+ \frac{W_2}{2m_e^2 M^2}\left\{2\left[k\cdot P - \frac{(k\cdot q)(P\cdot q)}{q^2}\right]\left[k'\cdot P - \frac{(k'\cdot q)(P\cdot q)}{q^2}\right] + \right. \\ &\left. + (m_e^2 - k\cdot k')\left[M^2 - \frac{(P\cdot q)^2}{q^2}\right]\right\} . \end{aligned} \tag{3.27}$$

This simplifies when we neglect m_e, and evaluate it in the laboratory frame, using

$$\begin{aligned} k\cdot k' &\approx EE'(1 - \cos\theta) \approx -\tfrac{1}{2}q^2 \\ k'\cdot q &\approx k'\cdot(k - k') \approx -\tfrac{1}{2}q^2 \\ k\cdot q &= k\cdot(k - k') \approx \tfrac{1}{2}q^2 \\ k'\cdot P &= ME' \\ k\cdot P &= ME \\ P\cdot q &= M\nu . \end{aligned} \tag{3.28}$$

The final form reads

$$I_{\mu\nu}W^{\mu\nu} = \frac{1}{2m_e^2}\left[-q^2W_1 + \tfrac{1}{2}W_2(q^2 + 4E'E)\right]$$
$$= \frac{EE'}{m_e^2}\left(W_2\cos^2\frac{\theta}{2} + 2W_1\sin^2\frac{\theta}{2}\right). \tag{3.29}$$

Substitution into (3.21) yields

$$\frac{d\sigma}{dE'd\Omega} = \frac{\alpha^2}{4E^2\sin^4\frac{\theta}{2}}\left(W_2\cos^2\frac{\theta}{2} + 2W_1\sin^2\frac{\theta}{2}\right). \tag{3.30}$$

Another form is

$$\frac{d\sigma}{dq^2d\nu} = \frac{4\pi\alpha^2}{q^4}\frac{E'}{E}\left(W_2\cos^2\frac{\theta}{2} + 2W_1\sin^2\frac{\theta}{2}\right). \tag{3.31}$$

We might compare these to the Mott cross section for the scattering of an electron of velocity v by a Coulomb field of charge e: [10]

$$\left(\frac{d\sigma}{dq^2}\right)_{\text{Coulomb}} = \frac{4\pi\alpha^2}{q^4v^4}\left(1 - v^2\sin^2\frac{\theta}{2}\right)\underset{v\to 1}{\longrightarrow}\frac{4\pi\alpha^2}{q^4}\cos^2\frac{\theta}{2} \equiv \sigma_{\text{Mott}}. \tag{3.32}$$

The scattering of an electron by a point Dirac proton is [10]

$$\left(\frac{d\sigma}{dq^2}\right)_{\substack{\text{Dirac}\\\text{proton}}} = \frac{4\pi\alpha^2}{q^4}\frac{E'}{E}\,\frac{\cos^2\frac{\theta}{2} - (q^2/2M^2)\sin^2\frac{\theta}{2}}{1 + (2E/M)\sin^2\frac{\theta}{2}}. \tag{3.33}$$

A convenient way to rewrite (3.31) is

$$\frac{d\sigma}{dq^2d\nu} = \sigma_{\text{Mott}}\left(W_2 + 2W_1\tan^2\frac{\theta}{2}\right). \tag{3.34}$$

3.4. Photo-absorption cross section

The ratio between the form factors W_1 and W_2 can be related to photo-absorption cross sections, defined as the total cross section for

$$\gamma + N \to \text{Anything}, \tag{3.35}$$

for a polarized off-mass-shell photon. The matrix element is given in (3.11). The differential cross section is

$$I d\sigma = (2\pi)^4\,\delta^4(P_x - P - q)\,(dP_x)\,\varepsilon^\mu\varepsilon^\nu\,\langle N|\,J_\mu\,|x\rangle\,\langle x|\,J_\nu\,|N\rangle, \tag{3.36}$$

where $I=(2q_0)a=2\nu a$, with a defined in (3.15). Summing over final states, and using

(3.20), we get

$$\sigma_{\text{tot}} = \frac{4\pi^2\alpha}{\nu}\, \varepsilon^\mu \varepsilon^\nu W_{\mu\nu}, \tag{3.37}$$

for incident photon of polarization ε^μ.

The polarization may be either transverse or longitudinal (Coulomb interaction):

$$\varepsilon^\mu = \begin{cases} \varepsilon_1^\mu, \varepsilon_2^\mu & \text{Transverse (space-like)} \\ \varepsilon_3^\mu & \text{Longitudinal (time-like)} \end{cases} \tag{3.38}$$

They are defined by

$$\begin{aligned} &\varepsilon_i^\mu q_\mu = 0 && (i = 1, 2, 3) \\ &\varepsilon_i^\mu \varepsilon_{i\mu} = \begin{cases} -1 \\ 1 \end{cases} && \begin{matrix} (i = 1, 2) \\ (i = 3) \end{matrix} \end{aligned} \tag{3.39}$$

In the laboratory frame

$$\begin{aligned} &q_\mu = (\nu, 0, 0, |\mathbf{q}|) \\ &|\mathbf{q}| = \nu^2 - q^2 . \end{aligned} \tag{3.40}$$

Using this with (3.39) leads to

$$\begin{aligned} &\varepsilon_1^\mu = (0, 1, 0, 0) \\ &\varepsilon_2^\mu = (0, 0, 1, 0) \\ &\varepsilon_3^\mu = \frac{\nu}{\sqrt{-q^2}} \left(\sqrt{1 - \frac{q^2}{\nu^2}}, 0, 0, 1 \right). \end{aligned} \tag{3.41}$$

The transverse and longitudinal photo-absorption cross sections are then

$$\begin{aligned} \sigma_t \equiv \frac{4\pi^2\alpha}{\nu}\, \tfrac{1}{2} (\varepsilon_1^\mu \varepsilon_1^\nu + \varepsilon_2^\mu \varepsilon_2^\nu)\, W_{\mu\nu} &= \frac{4\pi^2\alpha}{\nu}\, \tfrac{1}{2} (W_{11} + W_{22}) \\ &= \frac{4\pi^2\alpha}{\nu}\, W_1 , \end{aligned} \tag{3.42}$$

$$\begin{aligned} \sigma_l \equiv \frac{4\pi^2\alpha}{\nu}\, \varepsilon_3^\mu \varepsilon_3^\nu W_{\mu\nu} &= \frac{4\pi^2\alpha}{\nu} \left[-W_1 + \frac{W_2}{M^2} (\varepsilon_3 \cdot P)^2 \right] \\ &= \frac{4\pi^2\alpha}{\nu} \left[-W_1 + W_2 \left(1 - \frac{\nu^2}{q^2} \right) \right]. \end{aligned} \tag{3.43}$$

Therefore

$$\frac{\sigma_l}{\sigma_t} = \left(1 - \frac{\nu^2}{q^2} \right) \frac{W_2}{W_1} - 1 , \tag{3.44}$$

or

$$\frac{W_1}{W_2} = \left(1 - \frac{\nu^2}{q^2} \right) \frac{\sigma_t}{\sigma_t + \sigma_l} . \tag{3.45}$$

3.5. Experimental results [12]

We shall merely indicate the salient features of the experimental results:

(1) For proton target $\nu W_2 \to f(x)$ in the deep inelastic limit, where $f(x)$i s shown in Figure 7. That νW_2 approaches a finite limit depending only on x is known as the property of 'scaling'. The fact that $f(0)$ is nonzero will lead us to conclude that if there are partons there would have to be an infinite number.

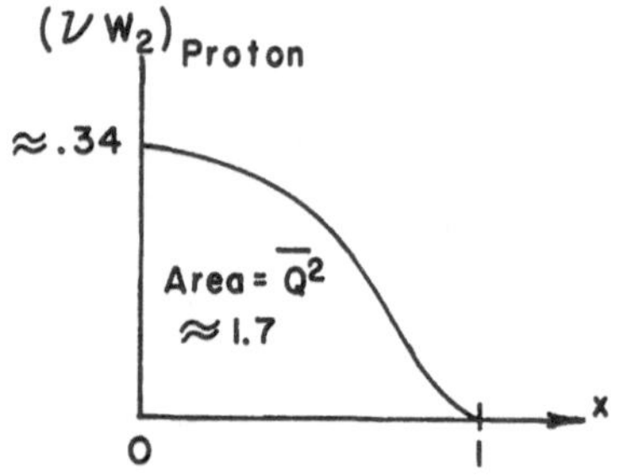

Fig. 7. Qualitative sketch of experimental result of νW_2 in deep inelastic limit.

(2) For proton target W_1/W_2 is, in the deep inelastic limit, consistent with $\sigma_1=0$. This will lead us to conclude that if there are partons they would have to have spin $\frac{1}{2}$.

(3) For neutron target, νW_2 also scales in the deep inelastic limit. The ratio and difference of νW_2 for proton and neutron are shown in Figures 8 and 9.

3.6. Scattering from single point particle

As orientation for understanding the experimental results, let us calculate W_1 and W_2 for a point particle target. In this case X can only be the target particle itself.

First consider a spin 0 target, for which

$$\langle N'| J^\mu |N\rangle = (P + P')^\mu ,$$

where $|N'\rangle$ is the final particle state of 4-momentum P'^μ. We calculate $W^{\mu\nu}$ from (3.20):

$$W^{\mu\nu} = \frac{(2\pi)^3}{2M} \int \frac{d^3 P'}{2P'_0 (2\pi)^3} \delta^4 (P' - P - q) (P + P')^\mu (P + P')^\nu . \tag{3.46}$$

To perform the integration it is convenient to write

$$\int \frac{d^3 P'}{2P'_0} = \int d^4 P' \delta (P'^2 - M^2) \theta (P'_0), \tag{3.47}$$

where

$$\theta (x) = \begin{cases} 1 & \text{if} \quad x > 0 \\ 0 & \text{if} \quad x < 0 . \end{cases} \tag{3.48}$$

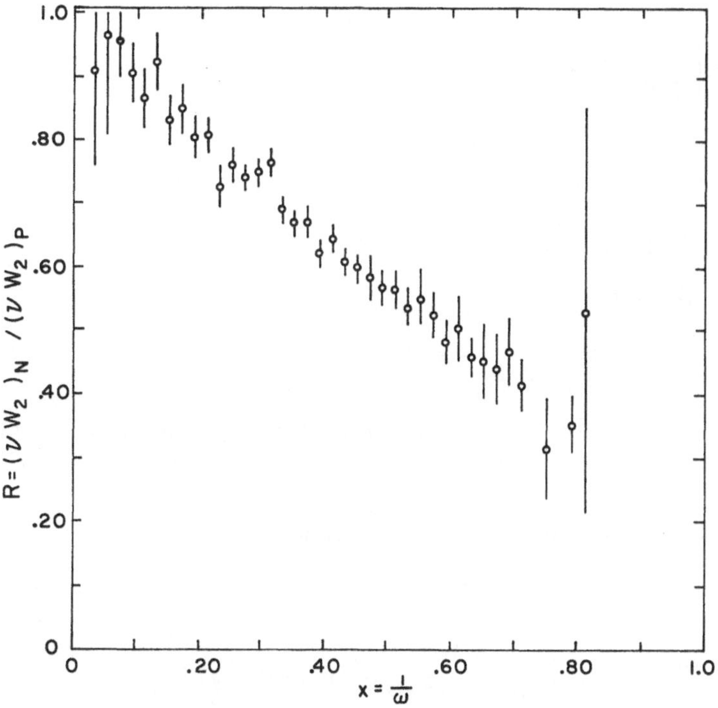

Fig. 8. Ratio of νW_2 of neutron to proton.

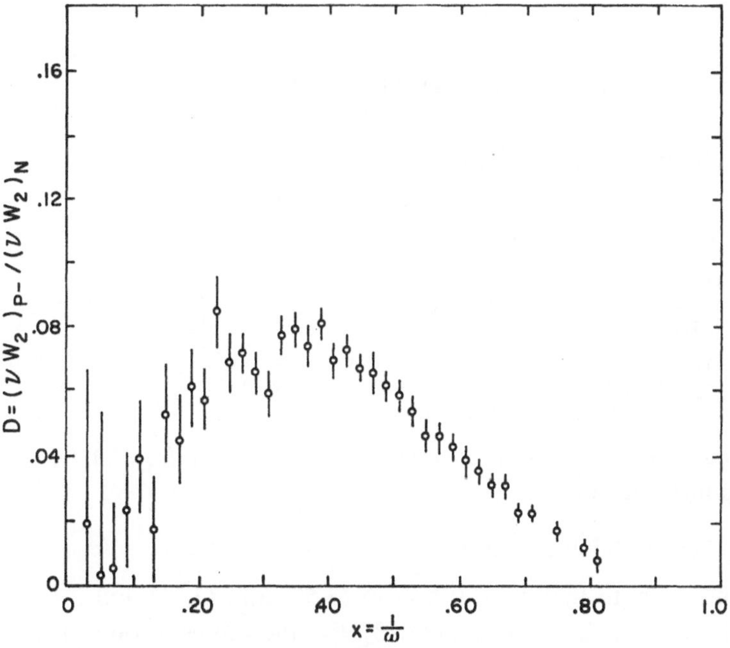

Fig. 9. Difference of νW_2 between proton and neutron.

Thus

$$\int \frac{d^3P'}{2P_0'} \delta^4 (P' - P - q) = \delta (q^2 + 2M\nu), \tag{3.49}$$

and

$$W^{\mu\nu} = \frac{1}{M^2} \delta\left(\nu + \frac{q^2}{2M}\right)\left(P^\mu - q^\mu \frac{q \cdot P}{q^2}\right)\left(P^\nu - q^\nu \frac{q \cdot P}{q^2}\right). \tag{3.50}$$

Comparison with (3.24) shows

$$\begin{aligned} W_1 &= 0 \\ \nu W_2 &= \nu\delta\left(\nu + \frac{q^2}{2M}\right) = \delta (1 - x). \end{aligned} \tag{3.51}$$

Thus we have scaling. By (3.45) the transverse photo-absorption cross section vanishes:

$$\sigma_t = 0 \quad \text{(spin 0 target)}. \tag{3.52}$$

Next we calculate $W^{\mu\nu}$ for a spin $\frac{1}{2}$ target, for which

$$\langle N'| J^\mu |N\rangle = \bar{u}(\mathbf{P}') \gamma^\mu u (\mathbf{P}). \tag{3.53}$$

Hence

$$W^{\mu\nu} = (2\pi)^3 \int \frac{d^3P'}{(P_0'/M)(2\pi)^3} \delta^4 (P' - P - q) I^{\mu\nu}, \tag{3.54}$$

where $I^{\mu\nu}$ is given by (3.19). Doing the integration as before, we obtain

$$W^{\mu\nu} = \delta\left(\nu + \frac{q^2}{2M}\right) I^{\mu\nu}, \tag{3.55}$$

from which W_1 and W_2 can be extracted by (3.26) and (3.25):

$$\begin{aligned} W_1 &= \frac{\nu}{2M} \delta\left(\nu + \frac{q^2}{2M}\right) \\ W_2 &= \delta\left(\nu + \frac{q^2}{2M}\right). \end{aligned} \tag{3.56}$$

or

$$\begin{aligned} \nu W_2 &= \delta (1 - x) \\ \frac{W_1}{W_2} &= \frac{\nu}{2M}. \end{aligned} \tag{3.57}$$

Thus scaling also results. Comparison with (3.45) shows $\sigma_l/\sigma_t = 2M/\nu$, which vanishes in the deep inelastic limit

$$\sigma_l = 0 \quad \text{(spin } \tfrac{1}{2} \text{ target)}. \tag{3.58}$$

In passing we note that substitution of (3.56) into (3.31) and integrating over ν recovers $(d\sigma/dq^2)_{\text{Dirac proton}}$ given in (3.33). For the ν integration q^2 is held fixed, and $\cos^2 \theta/2$, $\sin^2 \theta/2$ must first be rewritten in terms of ν and q^2.

Comparing these results with experiments, we see that the spin $\frac{1}{2}$ case gives closer agreement because $\sigma_l = 0$. However, although νW_2 scales, it does not at all look like the result shown on Figure 7, for it is a δ-function in this calculation. The result does suggest that if we imagine the proton to be composed of a distribution of spin $\frac{1}{2}$ free particles, we might be able to reproduce the experimental results. This leads us to partons.

3.7. Parton Model [1, 13]

We suspect that the salient features of the experimental results, namely scaling and the fact that $\sigma_l = 0$, may be reproduced by a model that regards the proton as a collection of free spin $\frac{1}{2}$ partons. The best we can hope to do is to advance some plausibility arguments, for at present we have no theory of hadron dynamics. Analogy with deep inelastic neutron scattering from liquid helium suggests that the parton picture might be justified by some type of impulse approximation. In the present case, however, we do not know the Hamiltonian, which may not even exist. On the other hand we have Lorentz covariance, which was absent in the liquid helium problem, but which can be exploited to our advantage here.

Imagine the proton to be a bound state of some more 'elementary' partons. It may then be thought of as constantly dissociating into virtual states of free partons and recombining. The lifetime of a virtual state is inversely proportional to the energy difference between the virtual state and the true energy of the proton state. If the collision time, which is of the order of $1/q_0$, is much shorter than this lifetime, then we may argue that the partons can be treated as free particles.

The lifetime of a virtual state depends on the Lorentz frame in which the proton is viewed. If we approach a frame in which the proton is moving infinitely fast (an infinite momentum frame), then this lifetime would become infinite owing to relativistic time dilation: The proton's clock slows down to almost a standstill. That might be long enough for our purpose, except that the collision time may also dilate, and we have to make sure that it remains small relative to the lifetime of virtual states.

To estimate the lifetime of a virtual state, suppose a fast moving proton of momentum $\mathbf{P}$ dissociates virtually into a parton of momentum $y\mathbf{P}$ and a group of parton of total momentum $(1-y)\mathbf{P}$. Then,

$$\begin{aligned} 1/\tau_{\text{virtual}} &\sim \sqrt{(yP)^2 + \mu_1^2} + \sqrt{(1-y)^2 P^2 + \mu_2^2} \\ &\quad - \sqrt{P^2 + M^2} \underset{P\to\infty}{\longrightarrow} \frac{1}{P}\left[\frac{\mu_1^2}{2y} + \frac{\mu_2^2}{2(1-y)} - \frac{M^2}{2}\right] + O\left(\frac{1}{P^2}\right), \end{aligned} \tag{3.59}$$

where $P \equiv |\mathbf{P}|$, and μ_1 and μ_2 are assumed to be finite. The collision time, taken to be $1/q_0$, depends on the specific infinite momentum frame use. We calculate it in the overall CM frame, which is an infinite momentum frame when $E \to \infty$. Consider

$$\begin{aligned} (P+q)^2 &= M^2 + q^2 + 2P\cdot q \approx M^2 + q^2 + 2Pq_0 - 2\mathbf{P}\cdot\mathbf{q} \\ (k-q)^2 &= m_e^2 + q^2 - 2k\cdot q \approx m_e^2 + q^2 - 2Pq_0 - 2\mathbf{P}\cdot\mathbf{q}. \end{aligned}$$

Substracting, we get

$$(P + q)^2 - (k - q)^2 \approx 4Pq_0$$

But $k - q = k'$, and $(k')^2 = m_e^2 \approx 0$. Hence

$$\frac{1}{\tau_{\text{coll}}} \sim q_0 \approx \frac{(P + q)^2}{4P} = \frac{2M\nu + q^2}{4P}, \tag{3.60}$$

and

$$\frac{\tau_{\text{coll}}}{\tau_{\text{virtual}}} \sim \frac{1}{2M\nu + q^2} = \frac{1}{2M\nu} \frac{1}{1 - x}, \tag{3.61}$$

which approaches zero in the deep inelastic limit. Hence, at least in some infinite momentum frame we can view the partons as free. By analogy with the case of neutron scattering from liquid helium, large momentum transfer ($-q^2 \to \infty$) enables us to argue that the partons interact with the electron independently of one another. That is, we add up the cross section from individual partons, instead of the scattering amplitudes. The initial proton state enters only through an initial momentum distribution of the partons. [14]

Note that the arguments above depend on two kinematic conditions: (a) $E \to \infty$, which makes an infinite momentum frame relevant, and (b) the deep inelastic limit, which makes (3.61) small. While (b) necessitates (a), the converse is not true.

The parton model, then, is as follows. Consider the scattering process in some infinite momentum frame, and in the deep inelastic limit. Assume the proton under these conditons appears to the electron (through the virtual photon) as a collection of free partons, which are point particles to the photon. A parton is characterized by its type i, with charge Q_i (in units of e), and its momentum $\mathbf{p}$:

$$\mathbf{p} = (\mathbf{p}_\perp, yP), \qquad (0 < y \leqslant 1), \tag{3.62}$$

where yP is the component parallel to the proton momentum $\mathbf{P}$, with $P \equiv |\mathbf{P}|$, and $\mathbf{p}_\perp$ is the component transverse to it. By restricting the longitudinal fraction y to be between 0 and 1, we are assuming that no parton travels opposite to the proton. A parton distribution function $f_i(y, \mathbf{p}_\perp)$ is defined by

$f_i(y, \mathbf{p}_\perp)\,\mathrm{d}y\,\mathrm{d}^2p_\perp$ = No. of partons of type i, with longitudinal momentum between yP and $(y + \mathrm{d}y)P$, and transverse momentum in $\mathrm{d}^2p_\perp$ about $\mathbf{p}_\perp$. (3.63)

It is assumed that $f_i(y, \mathbf{p}_\perp)$ is zero for $|\mathbf{p}_\perp|$ exceeding a finite bound, or vanishes rapidly beyond that (e.g. exponentially). The normalization condition is

$$\sum_i \int_0^1 \mathrm{d}y \int \mathrm{d}^2p_\perp y f_i(y, \mathbf{p}_\perp) = 1, \tag{3.64}$$

which says that the longitudinal momenta must add up to $\mathbf{P}$. The longitudinal distribution function is

$$n_i(y) = \int d^2 p_\perp f_i(y, \mathbf{p}_\perp),$$
$$\sum_i \int_0^1 dy\, y n_i(y) = 1. \tag{3.65}$$

Since by assumption the physcal cross section is a sum over parton cross sections, and since the cross section is linear in W_1 and W_2, we can just add these form factors from each parton. Since they are Lorentz invariant functions, they can be calculated in any frame, in particular an infinite momentum frame [15, 16]. Assuming the partons to have spin $\frac{1}{2}$, the contribution of one parton to W_2 is, by referring to (3.56), rewriting it as $M\delta(P\cdot q+q^2/2)$, and replacing P^μ by p^μ, and taking into account the fact that the parton charge is Q_i instead of 1:

$$W_2^{1-\text{parton}} = yQ_i^2 M\delta\left(p\cdot q + \tfrac{1}{2}q^2\right). \tag{3.66}$$

The *ad hoc* factor of y in front will be explained later. Let us evaluate $p\cdot q$:

$$p\cdot q = p_0 q_0 - \mathbf{p}\cdot\mathbf{q} = p_0 q_0 - yPq_z - \mathbf{p}_\perp\cdot\mathbf{q}_\perp \underset{P\to\infty}{\longrightarrow} yP(q_0 - q_z) + \\ + \frac{\mathbf{p}_\perp^2 + \mu_i^2}{2y}\frac{q_0}{P} - \mathbf{p}_\perp\cdot\mathbf{q}_\perp + O\left(\frac{1}{P^2}\right), \tag{3.67}$$

where μ_i is the parton mass. Compare this with

$$P\cdot q = P_0 q_0 - Pq_z \underset{P\to\infty}{\longrightarrow} P(q_0 - q_z) + \frac{M^2 q_0}{2P} + O\left(\frac{1}{P^2}\right). \tag{3.68}$$

We see that

$$p\cdot q = yP\cdot q + \left[\left(\frac{\mathbf{p}_\perp^2 + \mu_i^2}{y} - M^2\right)\frac{q_0}{2P} - \mathbf{p}_\perp\cdot\mathbf{q}_\perp\right] + O\left(\frac{1}{P^2}\right). \tag{3.69}$$

We assume that the infinite momentum frame is so chosen that the quantity in square brackets, for finite $y>0$, is small compared with $P\cdot q = M\nu$. This requires q_0/P be finite, or at least small compared with ν as $\nu\to\infty$. It is a restriction on the infinite momentum frame. In particular, if we make a Lorentz transformation from the laboratory frame to the infinite momentum frame, it restricts the direction of that transformation. Carrying out the Lorentz transformation explicitly shows that we are to boost from the laboratory frame in a direction along $\mathbf{q}$ in the laboratory frame, rather than opposed to it. It is important to relaize that the parton model, in this manner, singles out certain Lorentz frames – an unsatisfactory feature which we cannot understand within the parton model itself. In any case, dropping the square bracket in (3.69) gives

$$p\cdot q \approx yP\cdot q = yM\nu. \tag{3.70}$$

Substitution into (3.66) gives

$$W_2^{1-\text{parton}} = Q_i^2 y\delta(y-x) = Q_i^2\delta\left(1-\frac{x}{y}\right). \tag{3.71}$$

This says that, in the deep inelastic limit, when $-q^2/2M\nu$ is held fixed at the value x, the scattering is due only to those partons of longitudinal fraction equal to x.

It is interesting to compare with neutron scattering from liquid helium, where only $x=1$ contributes (because all helium atoms have the same mass.) Here partons behave *as if* they were non-relativistic particles of variable mass yM, and a given value of x picks out the corresponding 'mass'.

The factor y in front of (3.66) is put in to make the one-parton cross section agree with the Mott cross section as $E\to\infty$:

$$\left(\frac{d\sigma}{dq^2 d\nu}\right)_{1-\text{parton}} = \frac{4\pi\alpha^2 Q_i^2}{q^4}\frac{E'}{E}\left(W_2\cos^2\frac{\theta}{2} + 2W_1\sin^2\frac{\theta}{2}\right)\xrightarrow[E\to\infty]{}\frac{4\pi\alpha^2 Q_i^2}{q^4}. \tag{3.72}$$

Without the factor y, the above would have been divided by y, and incorrect. We can attribute the extra y factor to the manner parton states have to be normalized to get the right answer (3.72):

$$\langle xP \mid yP\rangle = y\delta(x-y) = \delta(x/y-1). \tag{3.73}$$

This is invariant under finite Lorentz transformations along the z direction, but would not be if the factor y in front had been left out.

Summing (3.71) over all partons, we obtain

$$\nu W_2 = \sum_i Q_i^2 x n_i(x), \tag{3.74}$$

where $n_i(x)$ is the longitudinal distribution function defined in (3.65). Clearly scaling is reproduced. The fact that νW_2 approaches a finite limit as $x\to 0$ implies that

$$n_i(x)\xrightarrow[x\to 0]{}\frac{a_i}{x}, \tag{3.75}$$

where $a_i \neq 0$ for at least one type i. Hence

$$\sum_i \int_0^1 dx n_i(x) = \infty. \tag{3.76}$$

That is, there has to be an infinite number of partons. Integrating both sides of (3.74) over x, we obtain

$$\sum_i Q_i^2 \int_0^1 dx x n_i(x) = \int_0^1 dx \nu W_2. \tag{3.77}$$

The left side is the mean square charge of the partons, while the right side is the area under the curve in Figure 7:

$$\overline{Q^2} \approx 0.17\,. \tag{3.78}$$

3.8. Quarks as partons [13, 17]

Let us assume that partons are quarks, which are of 6 types p, n, λ add their antiparticles, with quantum numbers listed below:

i	name	Q	B	I	I_z	S
1	p	$\frac{2}{3}$	$\frac{1}{3}$	$\frac{1}{2}$	$\frac{1}{2}$	0
2	n	$-\frac{1}{3}$	$\frac{1}{3}$	$\frac{1}{2}$	$-\frac{1}{2}$	0
3	λ	$-\frac{1}{3}$	$\frac{1}{3}$	0	0	-1
4	$\bar{p}$	$-\frac{2}{3}$	$-\frac{1}{3}$	$\frac{1}{2}$	$-\frac{1}{2}$	0
5	$\bar{n}$	$\frac{1}{3}$	$-\frac{1}{3}$	$\frac{1}{2}$	$\frac{1}{2}$	0
6	λ	$\frac{1}{3}$	$-\frac{1}{3}$	0	0	1

The proton and the neutron then are composed as follows

$$\begin{aligned} P &= (ppn) + \text{'sea'} \\ N &= (pnn) + \text{'sea'}\,, \end{aligned} \tag{3.79}$$

where 'sea' denotes a collection of quark-antiquark pairs, plus possibly mesons, whose presence will not change the results of our model, except the ratio W_1/W_2. Since the experimental result for W_1/W_2 is consistent with partons of spin $\frac{1}{2}$, any mesons present as partons should be of relatively small number. In the following we only calculate νW_2.

By charge symmetry, the 'sea' must be the same for proton and neutron. Furthermore the pp distribution function in the proton must be the same as the nn distributions in the neutron, and the n distribution in the proton must be the same as the p distribution functions in the neutron. Let

$$\begin{aligned} \Delta_2(x) &= \tfrac{1}{2}\,(p \text{ distribution function in proton}) \\ &= \tfrac{1}{2}\,(n \text{ distribution function in neutron}), \\ \Delta_1(x) &= n \text{ distribution function in proton} \\ &= p \text{ distribution function in neutron}, \end{aligned} \tag{3.80}$$

with the normalization conditions

$$\int_0^1 \mathrm{d}x\Delta_2(x) = \int_0^1 \mathrm{d}x\Delta_1(x) = 1\,. \tag{3.81}$$

Let $\sigma_i(x)$ be the distribution function of $i=p, n, \lambda$ in the 'sea'. Let $n'(x)$ be the distribution function of any mesons of charge Q that may be included in the model.

Then the parton distribution functions n_i that enter into (3.74) are as follows for the proton (P) and the neutron (N):

$$\begin{array}{llll} P: & n_p = \sigma_p + 2\Delta_2 & n_{\bar{p}} = \sigma_p & \\ & n_n = \sigma_n + \Delta_1 & n_{\bar{n}} = \sigma_n & \\ & n_\lambda = \sigma_\lambda & n_{\bar{\lambda}} = \sigma_\lambda & \\ & n_{\text{meson}} = n' & & \end{array} \tag{3.82}$$

$$\begin{array}{llll} N: & n_p = \sigma_p + \Delta_1 & n_{\bar{p}} = \sigma_p & \\ & n_n = \sigma_n + 2\Delta_2 & n_{\bar{n}} = \sigma_n & \\ & n_\lambda = \sigma_\lambda & n_{\bar{\lambda}} = \sigma_\lambda & \\ & n_{\text{meson}} = n' & & \end{array} \tag{3.83}$$

By definition all $n_i \geqslant 0$, hence in particular

$$\begin{aligned} \sigma_n + \Delta_1 &\geqslant 0 \\ \sigma_p + \Delta_1 &\geqslant 0 \end{aligned} \tag{3.84}$$

although Δ_1 and Δ_2 need not be positive definite. The normalization condition (3.65) gives separate conditions for P and N. For them to be consistent with each other we must require

$$\int_0^1 \mathrm{d}x\, x (\Delta_1 - \Delta_2) = 0 . \tag{3.85}$$

Substituting (3.82) and (3.83) into (3.74) gives νW_2 for the proton and neutron respectively:

$$\begin{aligned} P(x) &\equiv (\nu W_2)_P = \tfrac{8}{9} x \left[n(x) + \tfrac{1}{8}\Delta_1(x) + \Delta_2(x) \right] \\ N(x) &\equiv (\nu W_2)_N = \tfrac{8}{9} x \left[n(x) + \tfrac{1}{2}\Delta_1(x) + \tfrac{1}{4}\Delta_2(x) \right], \end{aligned} \tag{3.86}$$

where

$$n(x) = \sigma_p(x) + \tfrac{1}{4} \left[\sigma_n(x) + \sigma_\lambda(x) \right] + \tfrac{9}{4} Q^2 n'(x) . \tag{3.87}$$

We introduce the combinations

$$\begin{aligned} D(x) &\equiv P(x) - N(x) = \tfrac{2}{3} x \left(\Delta_2 - \tfrac{1}{2}\Delta_1 \right) \\ R(x) &\equiv \frac{N(x)}{P(x)} = \frac{n + \tfrac{1}{2}\Delta_1 + \tfrac{1}{4}\Delta_2}{n + \tfrac{1}{8}\Delta_1 + \Delta_2} . \end{aligned} \tag{3.88}$$

Applying the conditions (3.85) to $D(x)$ immediately yields the sum rule

$$\int_0^1 \frac{\mathrm{d}x}{x} D(x) = \tfrac{1}{3} . \tag{3.89}$$

The data at present is consistent with this.

Experimentally $D(x)$ is bounded and positive (see Figure 9). Together with (3.88), this requires that Δ_1, Δ_2 be bounded and $\Delta_2 \geqslant \frac{1}{2}\Delta_1$. On the other hand, $n(x)$ is positive definite, and experiments require that it behave like x^{-1} as $x \to 0$.

From (3.89), we can rewrite $R(x)$ as

$$R(x) = \frac{1 + \frac{1}{4}A(x)}{1 + A(x)}, \tag{3.90}$$

where

$$A(x) = \frac{\Delta_2 - \frac{1}{2}\Delta_1}{n + \frac{5}{8}\Delta_1} = \frac{3}{2}\frac{D}{x}\frac{1}{n + \frac{5}{8}\Delta_1}. \tag{3.91}$$

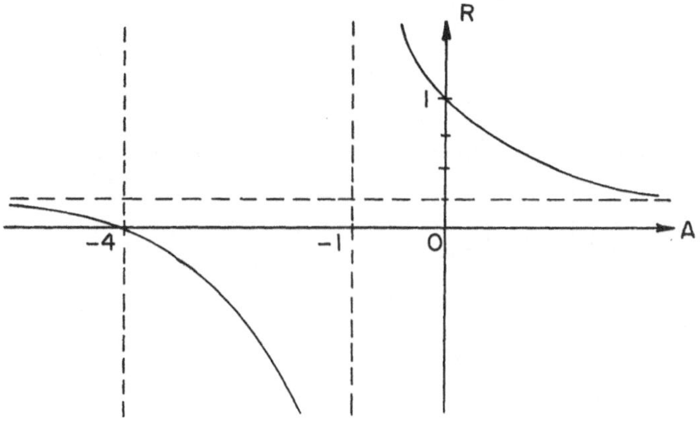

Fig. 10. *R* as a function of *A* [see Equation (3.90)].

A graph of R as a function of A is shown in Figure 10. Since experimentally $0 \leqslant R \leqslant 1$, we conclude that

$$R \geqslant \tfrac{1}{4} \quad \text{if} \quad A \geqslant 0$$
$$R \leqslant \tfrac{1}{4}, \quad \text{if} \quad A \leqslant -4.$$

From (3.84), however, we see that

$$\sigma_p + \tfrac{1}{4}\sigma_n + \tfrac{5}{4}\Delta_1 \geqslant 0.$$

Thus, using (3.87) and the facts that $\sigma_\lambda(x) \geqslant 0$, $n'(x) \geqslant 0$, we conclude

$$n(x) + \tfrac{5}{4}\Delta_1(x) \geqslant 0,$$

which, regardless of the sign of $\Delta_1(x)$, implies

$$n(x) + \tfrac{5}{8}\Delta_1(x) \geqslant 0.$$

By (3.91) this implies, in turn, that $A(x) \geqslant 0$. Therefore

$$R(x) \geqslant \tfrac{1}{4} \text{ for all } x. \tag{3.92}$$

A glance at Figure 8 shows that this is still consistent with present data, but comes dangerously close to being violated. It would be very important to know whether the data is really consistent with this and the sum-rule (3.89). Since these conditions depend only on the assignment of quark charges, a contradiction with experiments will lead to some soul searching.

We now proceed to satisfy all the constraints of the model. Solving for Δ_1 by (3.91) in terms of A, and then expressing A in terms of R by (3.90), we get

$$\Delta_1(x) = \frac{8}{5}\left[\frac{3}{2}\frac{P}{x}(R - \tfrac{1}{4}) - n\right], \tag{3.93}$$

where we have also used the fact

$$P = \frac{D}{1 - R}. \tag{3.94}$$

Since $\Delta_1(x)$ is bounded, while $n \to (\text{constant})/x$ as $x \to 0$, we can identify the constant:

$$n(x) \underset{x\to 0}{\to} \frac{9}{8}\left[\frac{P(x)}{x}\right]_{x=0}. \tag{3.95}$$

Accordingly we put

$$n(x) = \frac{9}{8x} P(x) f(x), \tag{3.96}$$

where $f(x)$ is bounded, positive definite, with $f(0)=1$. Substituting this form into (3.93) yields

$$\Delta_1(x) = \frac{12}{5}\frac{P}{x}(R - \tfrac{1}{4} - \tfrac{3}{4}f). \tag{3.97}$$

The last constraint to satisfy is the normalization condition (3.81):

$$\begin{aligned} 1 &= \frac{12}{8}\int_0^1 \frac{\mathrm{d}x}{x} P(R - \tfrac{1}{4} - \tfrac{3}{4}f) = \\ &= \frac{12}{8}\int_0^1 \frac{\mathrm{d}x}{x}\frac{D}{1-R}\left[\tfrac{3}{4}(1 - f) - (1 - R)\right], \end{aligned} \tag{3.98}$$

which, upon use of (3.89), reduces to

$$\int_0^1 \frac{\mathrm{d}x}{x} P(1 - f) = 1. \tag{3.99}$$

We have now satisfied all constraints, and the arbitrariness of the model is reduced to only one function $f(x)$.

To summarize: Choose a function $f(x)$, arbitrary except for the conditions

$$\begin{aligned}
&\text{(a)} \quad f(x) \text{ is bounded and positive-definite} \\
&\text{(b)} \quad f(0) = 1 \\
&\text{(c)} \quad \int \frac{dx}{x} P(x)\,[1 - f(x)] = 1 .
\end{aligned} \tag{3.100}$$

Then

$$\begin{aligned}
n(x) &= \frac{9}{8x} P(x)\, f(x) \\
\Delta_1(x) &= \frac{12}{5} \frac{P(x)}{x} \left[R(x) - \tfrac{1}{4} - \tfrac{3}{4} f(x)\right] \\
\Delta_2(x) &= \frac{3}{2} \frac{D(x)}{x} + \tfrac{1}{2}\Delta_1(x) .
\end{aligned} \tag{3.101}$$

The functions P, N, D, R are to be taken from experimental data, which are assumed to satisfy

$$\begin{aligned}
&\int_0^1 \frac{dx}{x} D(x) = \tfrac{1}{3} \\
&R(x) \geqslant \tfrac{1}{4} \text{ for all } x ,
\end{aligned} \tag{3.102}$$

otherwise the model is wrong. If they are in fact true, then the function $f(x)$ remains arbitrary apart from (3.100), for substitutions of (3.101) into (3.86) reduces the latter to identities.

4. Light Cone Physics [18]

4.1. Relevance of the Light Cone

We now pursue the subject in a different direction, and ask what scaling implies about the behavior of electromagnetic currents. The answer seems to be that commutators of two currents, notorious in field theory for their singular behavior, have no worse singularities than free-field currents, when the coordinates of the two currents are separated by light-like distances. Thus free-particle behavior again suggests itself, although in a different framework. Since the subject is one under current research, and not clearly understood, the wise thing for me to do is just calculate and talk as little as possible.

To see the relevance of the light cone let us go back to (3.20) and rewrite it as

$$W^{\mu\nu} = \frac{1}{2\pi} \int d^4x\, e^{iq\cdot x} \langle P|\, J^{\mu}(x)\, J^{\nu}(0)\, |P\rangle , \tag{4.1}$$

where

$$J^{\mu}(x) = e^{iP_{op}\cdot x} J^{\mu}(0)\, e^{-iP_{op}\cdot x} . \tag{4.2}$$

The intermediate steps are similar to those we used in arriving at (2.36). In fact there is a formal resemblance between (4.1) and (2.36). We now note that interchanging the order of the two currents in (4.1) would make the integral vanish:

$$\int \mathrm{d}^4 x e^{iq\cdot x} \langle P| J^\nu(0) J^\mu(x) |P\rangle = \\ = (2\pi)^4 \sum_X \delta^4 (P_X - P + q) \langle P| J^\nu(0) |X\rangle \langle X| J^\mu(0) |P\rangle . \tag{4.3}$$

For this to be nonzero, there must exist a physical state $|X\rangle$ for which $P_X = P - q$, or

$$M_X^2 = M^2 + q^2 - 2\nu M . \tag{4.4}$$

Since J^μ conserves baryon number, $|X\rangle$ must have baryon number 1. Since the proton is the lowest state of this type we must have $M_X^2 \geqslant M^2$, and hence by (4.4)

$$q^2 \geqslant 2\nu M . \tag{4.5}$$

This is impossible because $q^2 < 0$, and hence (4.3) must vanish. Therefore we can also write (4.1) as

$$W^{\mu\nu} = \frac{1}{2\pi} \int \mathrm{d}^4 x e^{iq\cdot x} \langle P| [J^\mu(x), J^\nu(0)] |P\rangle . \tag{4.6}$$

Let us now approach the deep inelastic limit. To do this it is conveneint to make the following transformation on all 4-vectors:

$$\begin{aligned} \mathbf{x}^\pm &= \frac{1}{\sqrt{2}} (x^0 \pm x^3) \\ \mathbf{x}_\perp &= (x^1, x^2) . \end{aligned} \tag{4.7}$$

The new metric tensor will be such that

$$\begin{aligned} g^{++} &= g^{--} = 0 \\ g^{+-} &= g^{-+} = 1 \\ g^{ij} &= - \delta_{ij} . \end{aligned} \tag{4.8}$$

The antisymmetric tensor is

$$\begin{aligned} \varepsilon^{+-ij} &= \varepsilon^{ij} \\ \varepsilon^{12} &= - \varepsilon^{21} = 1 \\ \varepsilon^{11} &= \varepsilon^{22} = 0 , \end{aligned} \tag{4.9}$$

The scalar product between two 4-vectors is

$$x\cdot y = x^+ y_+ + x^- y_- + x^i y_i = x^+ y^- + x^- y^+ - \mathbf{x}_\perp \cdot \mathbf{y}_\perp . \tag{4.10}$$

This transformation is not a Lorents transformation, but the limit of a Lorentz transformation with $v \to 1$. It may thus be called an infinite momentum frame. (∞ mom frame).

In terms of the language of the ∞ mom frame:*

$$\begin{aligned} q^2 &= 2q^+q^- - \mathbf{q}_\perp^2 \\ \nu &= \frac{P\cdot q}{M} = \frac{1}{M}(P^+q^- + P^-q^+ - \mathbf{P}_\perp\cdot\mathbf{q}_\perp) \\ \frac{1}{\omega} &= \frac{-q^2}{2M\nu} = \frac{\mathbf{q}_\perp^2 - 2q^+q^-}{2(P^+q^- + P^-q^+ - P_\perp\cdot\mathbf{q}_\perp)}. \end{aligned} \tag{4.11}$$

The deep inelastic limit can be reached by taking

$$q^- \to \infty \text{ all others } (q^+, \mathbf{q}_\perp. P^+, P^-, \mathbf{P}_\perp) \text{ finite}. \tag{4.12}$$

Thus in this limit

$$\begin{aligned} q^2 &\to 2q^+q^- \\ \nu &\to P^+q^-/M \\ \omega &\to -P^+/q^+. \end{aligned} \tag{4.13}$$

In the language of the ∞ mom frame

$$W^{\mu\nu} = \frac{1}{2\pi}\int_{-\infty}^{\infty} dx^+ \int_{-\infty}^{\infty} dx^- \int d^2x_\perp e^{i(q^-x^+ + q^+x^- - q_\perp\cdot x_\perp)} \times$$
$$\times \langle P|\,[J^\mu(x), J^\nu(0)]\,|P\rangle. \tag{4.14}$$

As $q^- \to \infty$, only the neighborhood of $x^+ = 0$ contributes to the integral. We note that by micro-causality the current commutator vanishes for space like x:

$$[J^\mu(x), J^\nu(0)] = 0 \quad \text{for} \quad x^2 < 0. \tag{4.15}$$

But

$$x^2 = 2x^+x^- - \mathbf{x}_\perp^2 \xrightarrow[x^+\to 0]{} -\mathbf{x}_\perp^2. \tag{4.16}$$

Therefore at $x_+ = 0$, $[J^\mu(x), J^\nu(0)]$ is non-vanishing only at $\mathbf{x}_\perp = 0$. The set of space-time points for which $x^+ = 0$, $\mathbf{x}_\perp = 0$, x^- arbitrary, is the light cone, as shown in Figure 11. Therefore, as $q^- \to 0$, only $[J^\mu(x), J^\nu(0)]_{x^2=0}$ is relevant for (4.14).

We may think of x^+ as the new 'time' variable, and think of x^- as defining a new 'space-like' surface, which in the limit $x^+ = \mathbf{x}_\perp = 0$ is the light cone. The current commutator $[J(x), J(0)]$, being an unequal time commutator, generally involves the dynamics of the system whose currents they are. To calculate it generally involves a complete solution of the equations of motion of the system. However, it may be simpler on the light cone, just as it is very simple at equal times.

* From now on we use ω instead of $x = 1/\omega$, in order not to confuse the latter with coordinates

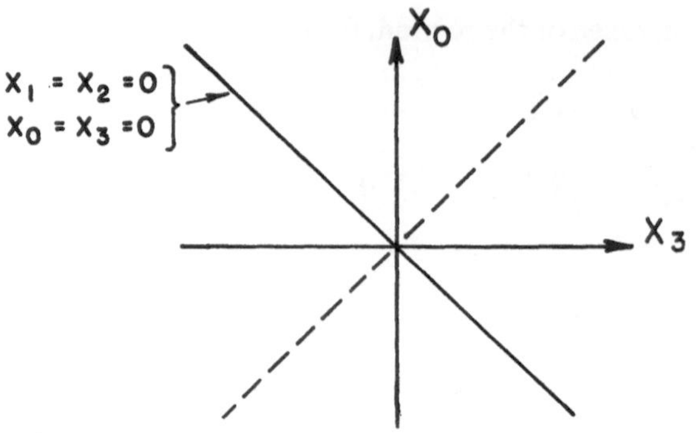

Fig. 11. Light cone.

For future reference we give the components of $W^{\mu\nu}$ in the ∞ mom frame in terms of the invariant form factors W_1 and W_2:

$$
\begin{aligned}
W^{++} &= W_1 \frac{q^+q^+}{q^2} + \frac{W_2}{M^2}\left[P^+ + q^+ \frac{\omega}{2}\right]^2 = \frac{(P^+)^2 f_1}{4M\nu}, \\
W^{--} &= W_1 \frac{q^-q^-}{q^2} + \frac{W_2}{M^2}\left(P^- - q^- \frac{\omega}{2}\right)^2 = \frac{\omega^2 q^-}{4P^+} f_1, \\
W^{-+} &= W^{+-} = - W_1 \left(1 - \frac{q^+q^-}{q^2}\right) + \frac{W_2}{M^2}\left(P^+ + \frac{\omega}{2} q^+\right) \times \\
&\quad \times \left(P^- + \frac{\omega}{2} q^-\right) = \frac{\omega}{4} f_1, \\
W^{+i} &= W_1 \frac{q^+q^i}{q^2} + \frac{W_2}{M^2}\left(P^+ + \frac{\omega}{2} q^+\right)\left(P^i + \frac{\omega}{2} q^i\right) = \\
&= \frac{1}{M\nu}\left[\frac{\omega^2}{4} q^+ q^i f_1 + P^+ \left(\frac{\omega}{2} q^i + P^i\right) \frac{\nu W_2}{M}\right],
\end{aligned}
\tag{4.17}
$$

where

$$
f_1 = \frac{\nu}{M} W_2 - \frac{2}{\omega} W_1 = \frac{2\nu}{\omega} \frac{\sigma_1}{4\pi^2\alpha}, \tag{4.18}
$$

which according to experiments is zero in the deep inelastic limit.

4.2. Quantization on the Light Cone [19]

We want to calculate current commutators on the light cone. One way to do this is to postulate a field theory Lagrangian for the hadronic system, quantize it as usual at equal times, and then solve the theory to get unequal time commutators. The only method we know of to solve such a theory is by use of perturbation theory. Apart from the drawback that this involves extremely hard work, we have no assurance that it gives the correct answer, even if we can calculate to all orders in perturbation theory, for the perturbation series may not converge. The question of convergence might be taken much more seriously here than in the calculation of, say, an S-matrix element. In the latter case the object is to calculate a number, whereas here the object is to calculate a singular operator.

Another way is to quantize our field theory by postulating field commutators on the light cone, instead of at equal times. We can then straightforwardly calculate light-cone current commutators from them. Obviously we have to show that light-cone quantization and conventional equal-time quantization define the same theory. At present this can only be done in a limited way.

Let us illustrate light cone quantization in a simple case: a free scalar field. The classical Lagrangian density and equation of motion are

$$\begin{aligned} L(x) &= \tfrac{1}{2}(\partial_\mu\phi\partial^\mu\phi - \mu^2\phi^2) \\ (\Box + \mu^2)\,\phi &= 0\,. \end{aligned} \tag{4.19}$$

Conventional equal-time quantization proceeds as follows. Define canonical momentum by

$$\pi(x) = \frac{\partial L(x)}{\partial\dot{\phi}(x)} = \dot{\phi}(x)\,, \quad [\dot{\phi}(x) \equiv \partial_0\phi(x)]\,. \tag{4.20}$$

Then postulate

$$\begin{aligned} i[\phi(x), \phi(y)]_{x^0=y^0} &= 0 \\ i[\pi(x), \pi(y)]_{x^0=y^0} &= 0 \\ i[\pi(x), \phi(y)]_{x^0=y^0} &= \delta^3(\mathbf{x}-\mathbf{y})\,. \end{aligned} \tag{4.21}$$

Now let us rewrite (4.19) in the ∞ mom frame:

$$\begin{aligned} L(x) &= \partial_-\phi\partial_+\phi + \tfrac{1}{2}(\partial_i\phi\partial^i\phi - \mu^2\phi^2) \\ (2\partial_+\partial_- + \partial_i\partial^i + \mu^2)\,\phi &= 0 \end{aligned} \tag{4.22}$$

Regarding x^+ as the new 'time', we note that the equation of motion becomes a first order differentiation equation in 'time'. The canonical momentum is, taking $\dot{\phi}(x) \equiv \partial_+\phi(x)$:

$$\pi(x) = \frac{\partial L(x)}{\partial\dot{\phi}(x)} = \partial_-\phi\,, \quad [\dot{\phi}(x) \equiv \partial_+\phi(x)]\,. \tag{4.23}$$

We postulate

$$i[\partial_-\phi(x), \phi(y)]_{x^+=y^+} = \tfrac{1}{2}\delta(x^- - y^-)\,\delta^2(\mathbf{x}_\perp - \mathbf{y}_\perp)\,. \tag{4.24}$$

The factor $\frac{1}{2}$ above comes about as follows. In equal-time commutator, there appears an unknown constant $\hbar$ (which we have set equal to 1), which is determined experimentally. Here similarly an unknown constant appears, which turns out to be $\hbar/2$, by requiring that (4.22) agrees with the Hamiltonian form of the equation of motion, namely

$$i[\mathcal{H}, \pi(x)] = \partial_+\pi(x)\,, \tag{4.25}$$

where

$$\mathscr{H} = \int dx^- \, d^2x_\perp H(x)$$
$$H(x) = \pi(x)\,\partial_+\phi(x) - L(x) = \tfrac{1}{2}(-\partial^i\phi\partial_i\phi + \mu^2\phi^2). \tag{4.26}$$

Integrating both sides of (4.24), we obtain

$$i[\phi(x), \phi(y)]_{x^+=y^+} = \tfrac{1}{4}\varepsilon(x^- - y^-)\,\delta^2(\mathbf{x}_\perp - \mathbf{y}_\perp), \tag{4.27}$$

where

$$\varepsilon(x) = \begin{cases} 1 & x > 0 \\ -1 & x < 0. \end{cases} \tag{4.28}$$

To do the integration we have used

$$\frac{\partial\varepsilon(x)}{\partial x} = 2\delta(x). \tag{4.29}$$

For this case the result of light-cone quantization gives the same theory as the conventional equal-time quantization. To show this note that (4.27) is a c-number, and hence is the same as its vacuum expectation value. Starting from conventional qunatization, we can write the vacuum expectation value of an unequal time commutator in the Lehamann representation [20]

$$i\langle 0|\,[\phi(x), \phi(0)]\,|0\rangle = i\int_0^\infty dm^2\varrho(m^2)\,\Delta(x, m^2), \tag{4.30}$$

where

$$\Delta(x, m^2) = \int \frac{d^4k}{(2\pi)^3}\,\varepsilon(k^0)\,\delta(k^2 - m^2)\,e^{-ik\cdot x}$$
$$\int_0^\infty dm^2\varrho(m^2) = 1. \tag{4.31}$$

Now go to the light cone:

$$\Delta(x, m^2)\,|_{x^+=0} = \tfrac{1}{4}\varepsilon(x^-)\,\delta^2(\mathbf{x}_\perp), \tag{4.32}$$

which follows from (4.31). Substituting this into (4.32) yields the same answer as (4.27).

We now turn to the more relevant case of a free spinor field, for which the classical Lagrangian density and equation of motion are

$$L(x) = \bar{\psi}(x)\,(i\gamma^\mu\partial_\mu - M)\,\psi(x)$$
$$(i\gamma^\mu\partial_\mu - M)\,\psi(x) = 0. \tag{4.33}$$

In the ∞ mom frame they read

$$L(x) = \bar{\psi}(x)\,[i(\gamma^+\partial_+ + \gamma^-\partial_- + \gamma^j\partial_j) - M]\,\psi(x)$$
$$(\gamma^+\partial_+ + \gamma^-\partial_- + \gamma^j\partial_j + iM)\,\psi(x) = 0, \tag{4.34}$$

where

$$\gamma^{\pm} = \frac{1}{\sqrt{2}}(\gamma^0 \pm \gamma^3). \tag{4.35}$$

Some properties of $\gamma^{\pm}$ are

$$\begin{aligned}
(\gamma^{\pm})^2 &= 0 \\
(\gamma^{\pm})^* &= \gamma^{\pm} \\
\{\gamma^{\pm}, \gamma^j\} &= 0 \\
\gamma^+\gamma^- &= 1 + \gamma^3\gamma^0 \\
\gamma^-\gamma^+ &= 1 - \gamma^3\gamma^0,
\end{aligned} \tag{4.36}$$

where* denotes hermitian conjugate. Let

$$\begin{aligned}
P_+ &= \tfrac{1}{2}\gamma^-\gamma^+ = \tfrac{1}{2}(1 - \gamma^3\gamma^0) = \tfrac{1}{2}(1 + \alpha_3) \\
P_- &= \tfrac{1}{2}\gamma^+\gamma^- = \tfrac{1}{2}(1 + \gamma^3\gamma^0) = \tfrac{1}{2}(1 - \alpha_3).
\end{aligned} \tag{4.37}$$

They have the properties

$$\begin{aligned}
(P_{\pm})^2 &= P_{\pm} \\
P_+ + P_- &= 1 \\
P_{\pm}^* &= P_{\pm}
\end{aligned} \tag{4.38}$$

and are therefore projection operators. Accordingly we can make the decomposition

$$\begin{aligned}
\psi(x) &= \psi_+(x) + \psi_-(x) \\
\psi_{\pm}(x) &= P_{\pm}\psi(x).
\end{aligned} \tag{4.39}$$

Similarly

$$\bar{\psi} \equiv \psi^*\gamma^0 = (\psi_+^* + \psi_-^*)\,\gamma^0 = \frac{1}{\sqrt{2}}(\psi_+^*\gamma^- + \psi_-^*\gamma^+). \tag{4.40}$$

The equation of motion can be written as

$$(\gamma^+\partial_+ + \gamma^-\partial_- + \gamma^j\partial_j + iM)\,(\psi_+ + \psi_-) = 0. \tag{4.41}$$

Multiplying through from the left separately by γ^+ and γ^- yields two coupled equations for $\psi_{\pm}$:

$$\begin{aligned}
i\partial_-\psi_- &= \tfrac{1}{2}(i\gamma^j\partial_j + M)\,\gamma^+\psi_+ \\
i\partial_+\psi_+ &= \tfrac{1}{2}(i\gamma^j\partial_j + M)\,\gamma^-\psi_-.
\end{aligned} \tag{4.42}$$

We solve for ψ_- by integrating the first equation:

$$\psi_-(x) = -\frac{i}{2}(i\gamma^j\partial_j + M)\int \mathrm{d}x^-\,\gamma^+\psi_+(x). \tag{4.43}$$

Assuming $\psi_+(x)\to 0$ as $x^-\to\pm\infty$ we can rewrite the indefinite integral above in more explicit form as follows:

$$\int dx^- \psi_+(x) = \tfrac{1}{2}\left(\int_{-\infty}^{x^-} + \int_{\infty}^{x^-}\right) d\xi \psi_+(x^+, \xi, \mathbf{x}_\perp)$$
$$= \tfrac{1}{2} \int_{-\infty}^{\infty} d\xi \varepsilon(x^- - \xi) \psi_+(x^+, \xi, \mathbf{x}_\perp). \tag{4.44}$$

Hence

$$\psi_-(x^+, x^-, \mathbf{x}_\perp) = -\frac{i}{4}(i\gamma^j\partial_j + M) \int_{-\infty}^{\infty} d\xi \varepsilon(x^- - \xi) \gamma^+ \psi_+(x^+, \xi, \mathbf{x}_\perp). \tag{4.45}$$

Substitution of this into the second equation of (4.42) yields

$$\partial_+\psi_+(x^+, x^-, \mathbf{x}_\perp) = \tfrac{1}{4}(i\gamma^j\partial_j + M)(i\gamma^k\partial_k - M)\cdot \cdot \int_{-\infty}^{\infty} d\xi \varepsilon(x^- - \xi) \psi_+(x^+, \xi, \mathbf{x}_\perp) \tag{4.46}$$

which is the equation of motion for evolution in 'time' x^+.

The canonical prescription in ∞ mom frame gives

$$\begin{aligned} \dot{\psi} &\equiv \partial_+\psi \\ \pi &\equiv \frac{\partial L}{\partial \dot{\psi}} = \bar{\psi} i\gamma^+ \\ H &\equiv \pi\dot{\psi} - L = \bar{\psi}[-i(\gamma^-\partial_- + \gamma^j\partial_j) + M]\psi. \end{aligned} \tag{4.47}$$

Decomposing ψ and $\bar{\psi}$ as in (4.39) and (4.40) leads to

$$H = \psi_+^*(x) \Gamma(x) \psi_+(x), \tag{4.48}$$

where $\Gamma(x)$ is the integro-differential operator

$$\Gamma(x) = (i\gamma^j\partial_j + M)(i\gamma^k\partial_k - M)\frac{i}{2\sqrt{2}} \int_{-\infty}^{\infty} d\xi \varepsilon(x^- - \xi). \tag{4.49}$$

We quantize the theory by posutlating the light cone anti-commutators

$$\begin{aligned} \{\psi_+(x), \psi_+(y)\}_{x^+=y^+} &= 0 \\ \{\psi_+(x), \psi_+^*(y)\}_{x^+=y^+} &= \frac{1}{\sqrt{2}} \delta(x^- - y^-) \delta^2(\mathbf{x}_\perp - \mathbf{y}_\perp) P_+. \end{aligned} \tag{4.50}$$

The factor $1/\sqrt{2}$ is determined by requiring that the equation

$$\begin{aligned} i[\mathscr{H}, \psi_+(x)] &= \partial_+\psi_+(x) \\ \mathscr{H} &\equiv \int dx^- d^2x_\perp H \end{aligned} \tag{4.51}$$

shall be identical with (4.46). By (4.50) and (4.45), we have the derived anti-commutator

$$\{\psi_-(x), \psi_+^*(y)\}_{x^+=y^+} = -\frac{i}{4\sqrt{2}}\,\varepsilon(x^- - y^-)\,(i\gamma^i\partial_j + M) \times \\ \times\, \delta^2(\mathbf{x}_\perp - \mathbf{y}_\perp)\,\gamma^+ P_+ . \tag{4.52}$$

The equivalence with conventional quantization can again be shown by a method similar to that for the scalar field described earlier.

4.3. Current commutators [21, 22]

We can now calculate light cone current commutators for a free spinor field. The current operator is

$$J^\mu(x) = \bar{\psi}(x)\,\gamma^\mu\psi(x). \tag{4.53}$$

In ∞ mom frame we use the components $J^\pm$, J^i, defined as

$$\begin{aligned} J^\pm(x) &= \frac{1}{\sqrt{2}}\,[J^0(x) \pm J^3(x)] = \bar{\psi}(x)\,\gamma^\pm\psi(x) \\ J^i(x) &= \bar{\psi}(x)\,\gamma^i\psi(x). \end{aligned} \tag{4.54}$$

Using (4.39) and (4.40) we can write these as

$$\begin{aligned} J^\pm(x) &= \sqrt{2}\psi_\pm^*(x)\,\psi_\pm(x) \\ J^i(x) &= \frac{1}{\sqrt{2}}\,[\psi_-^*(x)\,\gamma^+\gamma^i\psi_+(x) + \psi_+^*(x)\,\gamma^-\gamma^i\psi_-(x)]. \end{aligned} \tag{4.55}$$

Straightforward applications of (4.50) and (4.52) leads to the following

$$[J^+(x), J^+(y)]_{x^+=y^+} = 0. \tag{4.56}$$

$$\begin{aligned} &[J^+(x), J^-(y)]_{x^+=y^+} = \\ &\quad = -\tfrac{1}{2}\partial_-^x\,[\varepsilon(x^- - y^-)\,\delta^2(\mathbf{x}_\perp - \mathbf{y}_\perp)\,\bar{\psi}(x)\,\gamma^-\psi(y)] \\ &\quad - \tfrac{1}{4}\partial_j^x\,\{\varepsilon(x^- - y^-)\,\delta^2(\mathbf{x}_\perp - \mathbf{y}_\perp)\,\bar{\psi}(x)\,[\gamma^i - \varepsilon^{ij}\gamma_j\gamma_5]\,\psi(y)\} \\ &\quad - \text{h.c.} \end{aligned} \tag{4.57}$$

$$\begin{aligned} &[J^+(x), J^i(y)]_{x^+=y^+} = \\ &\quad = -\tfrac{1}{4}\partial_-^x\,\{\varepsilon(x^- - y^-)\,\delta^2(\mathbf{x}_\perp - \mathbf{y}_\perp)\,\bar{\psi}(x)\,[\gamma^i + \varepsilon^{ij}\gamma_j\gamma_5]\,\psi(y)\} \\ &\quad + \tfrac{1}{4}\partial_j^x\,\{\varepsilon(x^- - y^-)\,\delta^2(\mathbf{x}_\perp - \mathbf{y}_\perp)\,\bar{\psi}(x)\,[g^{ij}\gamma^+ + \varepsilon^{ij}\gamma^+\gamma_5]\,\psi(y)\} \\ &\quad - \text{h.c.} \end{aligned} \tag{4.58}$$

The characteristic feature of these commutators is the appearance of operators of the form $\bar{\psi}(x)\,\Gamma\bar{\psi}(y)$, which are called bilocal operators. The fact that in the deep inelastic limit these commutators are relevant for $W^{\mu\nu}$ means that it is in principle possible to measure bilocal operators experimentally. In the approach using light cone quantization these commutators remain unaltered if we couple the spinor field to a quantized scalar or vector field, the latter in a gauge invariant way. However, they are different if one calculates them in conventional perturbation theory. The forms given here leads to scaling, whereas the conventional perturbation theory does not give scaling.

We know in conventional field theory that current commutators are very singular objects, and that a naive calculation such as we have done may miss certain contributions called Schwinger terms [23, 18], whose existence is required by current conservation. The same type of considerations applied here shows that indeed all the commutators above require the addition of Schwinger terms. To see this, consider

$$\langle 0| [J^\mu(x), J^\nu(0)] |0\rangle = (g^{\mu\nu}\Box - \partial^\mu\partial^\nu) \int_0^\infty dm^2 \varrho(m^2)\, \Delta(x, m^2), \qquad (4.59)$$

where $\varrho(m^2)$ and $\Delta(x, m^2)$ are as given by (4.31). The differential operator in front guarantees current conservation, and the integral satisfies micro-causality. From this we deduce

$$\begin{aligned}\langle 0| [J^+(x), J^+(0)]_{x^+=0} |0\rangle &= -\partial_-\partial_- \int_0^\infty dm^2 \varrho(m^2)\, \Delta(x, m^2)\,|_{x^+=0} \\ &= \frac{i}{4}\partial_-\partial_- [\varepsilon(x^-)\,\delta^2(\mathbf{x}_\perp)] \int_0^\infty dm^2 \varrho(m^2) \neq 0. \end{aligned} \qquad (4.60)$$

This does not tell us, however, whether the Schwinger terms are c-numbers or operators. We *assume* they are all c-numbers, in which case they make no contribution to expectation values of current commutators with respect to a nuclear state, for the expectation values are always defined with the vacuum expectation value subtracted.

Jackiw and Cornwall have shown that the light-cone commutators remain the same even if the spin $\frac{1}{2}$ field is coupled to a vector meson field.

We may now use (4.56)–(4.58) to calculate $W^{\mu\nu}$, and hence W_1 and W_2. We obtain immediately by (4.56) and (4.17) $W^{++}=0$, and therefore $f_1=0$, in agreement with experiments. Then, according to (4.17), W_2 can be obtained by calculating W^{+i}, which depends on the light-cone commutator (4.58). The answer involves the expectation value $\langle P| \bar\psi(x)\, \psi(y) |P\rangle$, which depends on possible couplings with other fields. If we take $\psi(x)$ to be a free field, then, not surprisingly, we recover the result (3.56) obtained earlier for a free spin $\frac{1}{2}$ particle.

One can ask the question in reverse: What must the light cone current commutators be in order to reproduce the experimentally observed νW_2? The answer is given by Jackiw *et al.* [24], who wrote down the most general form consistent with scaling and with given νW_2 and W_1 (as functions of ω). The important point of their conclusions are the following:

(1) $\langle P|\,[J^\mu(x), J^\nu(0)]\,|P\rangle$ have singularities on the light cone no worse than those of free-field current commutators.

(2) No known field theory gives the required current commutators.

References

1. Feynman, R. P.: unpublished lectures and *Phys. Rev. Letters* **23**, 1415 (1969).
2. For general reference see Huang, K.: *Statistical Mechanics*, Wiley, New York, 1965.
3. Henshaw, D. G. and Woods, A. D. B.: *Phys. Rev.* **121**, 1266 (1961).
4. Cowley, R. A. and Woods, A. D. B.: *Can. J. Phys.* **49**, 177 (1971).
5. Platzman, P. and Tzoar, N.: *Phys. Rev.* **139**, A140 (1965).
6. Hohenberg, P. C. and Platzman, P. N.: *Phys. Rev.* **152**, 198 (1966).
7. Harling, O.: *Phys. Rev. Letters* **24**, 1046 (1970).
8. Sears, V. F.: *Phys. Rev.* **A1**, 1699 (1970).
9. Puff, R. D. and Tenn, J. S.: *Phys. Rev.* **A1**, 125 (1970).
10. Bjorken, J. D. and Drell, S. D.: *Relativistic Quantum Mechanics*, McGraw-Hill, New York, 1964.
11. Bjorken, J. D.: *Phys. Rev.* **179**, 1547 (1969).
12. Kendall, H. W.: in *Proceedings of the 1971 International Symposium on Electron and Photon Interactions at High Energies*, Cornell University, August, 1971.
13. Bjorken, J. D. and Paschos, E. A.: *Phys. Rev.* **158**, 1975 (1969).
14. See Drell, S. D. and Yan, T. M.: *Phys. Rev. Letters* **177**, 2584 (1969) for an attempt to derive the impulse approximation in field theory.
15. Dirac, P. A. M.: *Rev. Mod. Phys.* **21**, 392 (1949).
16. See Susskind, L.: *Phys. Rev.* **165**, 1535 (1968), for a more detailed discussion of the infinite momentum frame.
17. Kuti, J. and Weisskopf, V. F.: *Phys. Rev.* **D4**, 3418 (1971).
18. For general reviews, see Adler, S. and Dashen, R.: *Current Algebras*, Benjamin, New York, 1968; Treiman, S. B., Jackiw, R., and Gross, D. J.: *Lectures on Current Algebra and Its Applications*, Princeton Univ. Press, Princeton, 1972.
19. Kogut, J. B. and Soper, D. E.: *Phys. Rev.* **D1**, 2901 (1970); Jackiw, R.: in *Springer Tracts in Modern Physics*, No. 62, Springer-Verlag, Berlin, 1972.
20. Lehmann, H.: *Nuovo Cimento* **11**, 342 (1954).
21. Jackiw, R. and Cornwall, J. M.: *Phys. Rev.* **D4**, 367 (1971).
22. Fritsch, H. and Gell-Mann, M.: in *Center for Theoretical Studies, University of Miami Tracts in Mathematics and Natural Science*, Vol. 2, Gordan and Breach, New York, 1971.
23. Schwinger, J.: *Phys. Rev. Letters* **3**, 296 (1959).
24. Jackiw, R., Van Royan, R., and West, G. B.: *Phys. Rev.* **D2**, 2473 (1970).

MULTIPERIPHERAL MODELS

ALBERTO PIGNOTTI
Departamento de Física, Facultad de Ciencias Exactas y Naturales, Universidad de Buenos Aires, Pabellón 1 – Ciudad Universitaria – Núñez, Buenos Aires, Argentina

1. Introduction

This series of lectures is going to deal with strong interactions at high energies. By this we mean energies much larger than characteristic hadronic masses, which are of the order of 1 GeV. Thus high energies will be typically the energies of the present Brookhaven and CERN accelerators (~30 GeV and higher).

In strong interactions it is impossible to isolate particular processes, because all reactions are strongly coupled by unitarity or conservation of probability. Thus, we cannot have an accurate description of any given set of phenomena without having a complete theory, and we do not have such a theory yet. Nonetheless, we are going to use a model – the multiperipheral model – to explore hadronic collisions at high energy. The model is clearly incomplete but we hope to find or guess properties of the real world which are true and more general than the model from which they originate.

What happens in hadronic collisions at high energy? We can classify them in two-body and many-body collisions, depending on the number of final particles that are stable under strong interactions. Two-body processes are either elastic or inelastic. The former have an almost constant cross section at high energy (~10 mb in p-p collisions) whereas the latter tend to zero like some inverse power of the C.M. energy. The contribution of all many-body processes to the total cross section is also approximately constant at high energies (~30 mb in p-p). We conclude that processes in which particles are created are quite important for two different reasons: on the one hand, they constitute the major part of the hadronic cross section; on the other, through the shadow effect they determine important features of the elastic processes at high energy. Because of these reasons, and in spite of the theoretical and experimental difficulties involved, the study of multiparticle production processes has become increasingly active in recent years.

2. Kinematics

2.1. Standard Frames

We describe a process involving a total number N of particles in the initial and final states by a probability amplitude or N-point function depicted in Figure 1. The number of variables on which such an amplitude depends is $3N$-10, where $3N$ comes from the independent components of the four momenta when the mass shell conditions ($p_i^2 = m_i^2$) are taken into account, and 10 is the number of parameters in the Poincaré group. Four of these are associated with space-time translations and invariance under

Abecassis de Laredo and Jurisic (eds.), Selected Topics in Phys. Astrophys. and Biophys. 214–248.

these transformations leads to the four constraints $\sum_{i=1}^{N} p_i = 0$. The remaining six parameters correspond to homogeneous Lorentz transformations, and invariance under this group of transformations makes it possible to choose a particular frame in which six components of the four vectors involved vanish. For example, we can

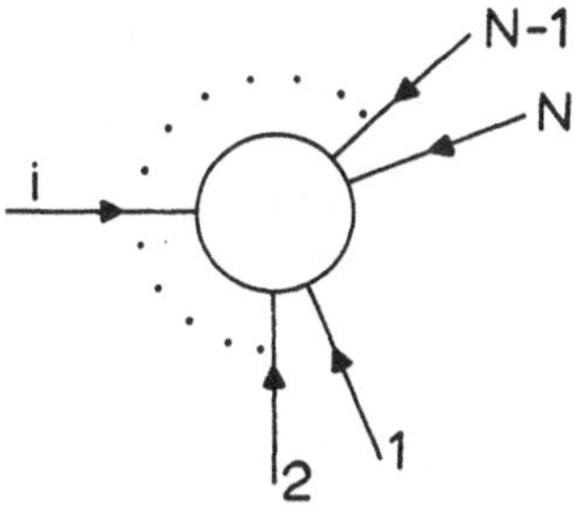

Fig. 1.

choose a frame in which p_1 has only non-vanishing time component, p_2 has non-vanishing time and z component (the latter is chosen to be positive) and p_3 has non-vanishing t, y and z components; namely:

$$\begin{aligned} p_1 &= (m_1, 0; 0; 0) \\ p_2 &= (m_2 \cosh x_2; 0; 0; m_2 \sinh x_2) \\ p_3 &= (m_3 \cosh x_3; 0; m_3 \sinh x_3 \sin \theta_3; m_3 \sinh x_3 \cos \theta_3) \\ &\text{etc.} \end{aligned} \tag{2.1}$$

We call such a frame a STANDARD FRAME with respect to the four vectors p_1, p_2 and p_3 and denote it by $F(p_1, p_2, p_3)$.

2.2. CHANGE OF STANDARD FRAMES

In the following we are going to perform transformations that lead from a given standard frame to a different one.

We discuss two examples in which the momenta involved are not necessarily momenta of external particles but may be a sum over a subset of these. Thus, they may be either time-like or space-like. A typical case is shown in Figure 2 and involves two

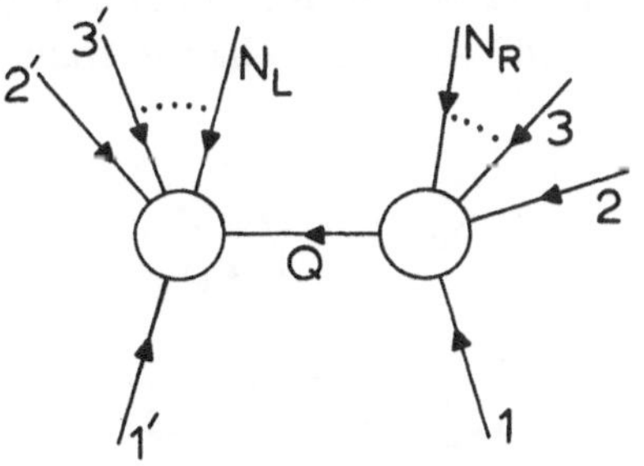

Fig. 2.

initial particles and $N_R + N_L - 2$ final ones. If we have:

$$(p_2 + p_3 + \cdots + p_{N_R})^2 > m_1^2 \quad \text{and} \quad (p'_2 + p'_3 + \cdots + p_{N_L})^2 > m_{1'}^2 \tag{2.2}$$

then, in the physical region for the process

$$1 + 1' \to \bar{2} + \bar{3} + \cdots + \bar{N}_R + \bar{2}' + \cdots + \bar{N}_L$$

Q is a space-like vector, i.e. $Q^2 = t < 0$. Thus, in analogy with the previous definition, we introduce the frame $F(p_1, Q, p_2)$ in which

$$\begin{aligned} p_1 &= (m_1; 0; 0; 0) \\ Q &= (\sqrt{-t}\sinh x; 0; 0; \sqrt{-t}\cosh x) \\ &\text{etc.} \end{aligned} \tag{2.3}$$

In order to change frames to $F(Q, p_1, p_2)$ we have to perform a boost in the z direction of parameter q_{Q_1}. We obtain:

$$\begin{aligned} p_1 &= (m_1 \cosh q_{Q_1}; 0; 0; m_1 \sinh q_{Q_1}) \\ Q &= (\sqrt{-t}\sinh(q_{Q_1} + x); 0; 0; \sqrt{-t}\cosh(q_{Q_1} + x)) \\ &\text{etc.} \end{aligned} \tag{2.4}$$

Thus, the boost parameter that leads to the desired frame is $q_{Q_1} = -x$ and can be written as a function of m_1^2, t and $(p_1 - Q)^2 \equiv S_r$ through the relations:

$$\begin{aligned} \sinh q_{Q_1} &= \frac{m_R^2 - m_1^2 - t}{2m_1\sqrt{-t}} \\ \cosh q_{Q_1} &= \frac{\lambda^{1/2}(m_1^2, m_R^2, t)}{2m_1\sqrt{-t}}, \end{aligned} \tag{2.5}$$

where $\lambda(a, b, c) = a^2 + b^2 + c^2 - 2ab - 2ac - 2bc$.

A second example is one in which the above four vector p_1 is substituted by a space-like momentum transfer Q_1 and is shown in Figure 3. We have $Q_1^2 = t_1 < 0$, $Q_2^2 = t_2 < 0$, $(Q_1 - Q_2)^2 = S > 0$ and want to change frames from $F(Q_1, Q_2, p)$ to $F(Q_2, Q_1, p)$.

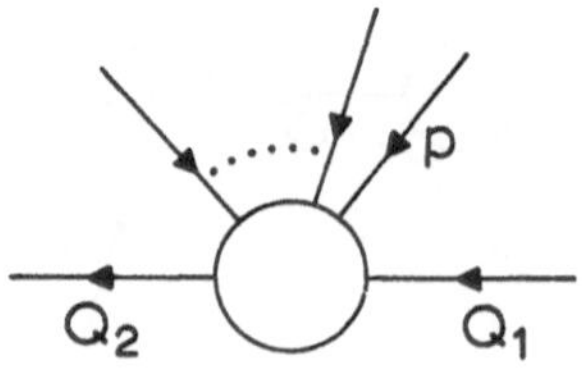

Fig. 3.

Again we will do this by performing a boost in the z direction of parameter $q_{Q_2Q_1}$. In $F(Q_2, Q_1p)$ we have:

$$\begin{aligned} Q_1 &= (\sqrt{-t_1}\,\sinh q_{Q_2Q_1};\, 0;\, 0;\, \sqrt{-t_1}\,\cosh q_{Q_2Q_1}) \\ Q_2 &= (0;\, 0;\, 0;\, \sqrt{-t_2}) \\ &\text{etc.} \end{aligned} \tag{2.6}$$

and

$$S = t_1 + t_2 + 2\sqrt{-t_1}\sqrt{-t_2}\,\cosh q_{Q_2Q_1}$$

i.e.

$$\begin{aligned} \cosh q_{Q_2Q_1} &= \frac{S - t_1 - t_2}{2\sqrt{-t_1}\sqrt{-t_2}} \\ \sinh q_{Q_2Q_1} &= \frac{\lambda^{1/2}(S, t_1, t_2)}{2\sqrt{-t_1}\sqrt{-t_2}}. \end{aligned} \tag{2.7}$$

Observe the similarity but also the difference between these equations and Equations (2.5).

2.3. Generalized partial wave expansions

We consider now a process shown in Figure 4 in which we have N_i incident and N_f final particles, which, for simplicity, are all assumed to be scalar.

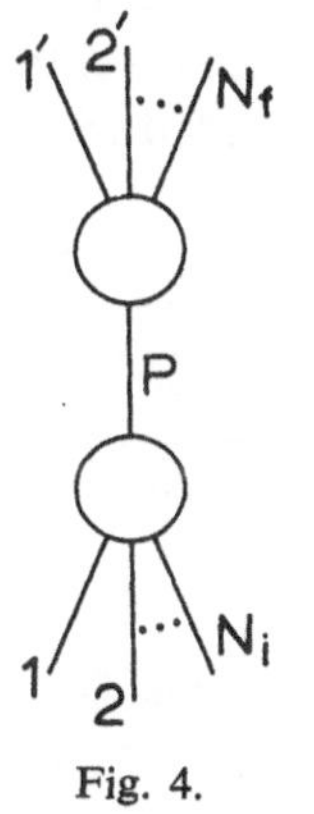

Fig. 4.

We describe separately the initial and final momenta in frames $F(P, p_1, p_2)$ and $F(P, p_1', p_2')$ respectively. We thus introduce $3(N_f+1)-10$ final variables v_f and $3(N_i+1)-10$ initial variables v_i in addition to the C.M. energy squared $P^2 \equiv S$. The description of the process is completed by specifying the relative orientation of frames $F(P, p_1, p_2)$ and $F(P, p_1', p_2')$. In both of these frames P is given by $P = (\sqrt{S}, 0, 0; 0)$, thus, they can differ only by a rotation, i.e., a Lorentz transformation that leaves P invariant, i.e., an element of the little group of the Lorentz group with respect to P. We parametrize this rotation by the usual three Euler angles ψ, θ, φ which we denote collectively by g. Thus the scattering amplitude at fixed S v_i and v_f is a function defined on the rotation group. We define a generalized partial wave amplitude by the expression.

$$A^{j\lambda_f\lambda_i}(v_f, S, v_i) = \int \mathrm{d}g D^{j\lambda_f\lambda_i}(g)\, A(v_f, S, g, v_i) \tag{2.8}$$

and using the unitary property of the rotation matrices $D^{j\lambda\lambda'}$ we write the generalized partial wave expansion.

$$A(v_f, S, g, v_i) = \sum_{j\lambda_f\lambda_i} \frac{(2j+1)}{8\pi^2} A^{j\lambda_f\lambda_i}(v_f, S, v_i)\, D^{j\lambda_i\lambda_f *}(g) \tag{2.9}$$

In the usual case in which $N_i = 2$ the number of initial internal variables in our way of counting is: $3N_i - 7 = -1$. What does this mean? It means that not only no internal variables are needed but in addition that the amplitude does not depend on one of the 'geometrical' variables, i.e., the first Euler angle. This is so because a two-particle initial state defines the frame $F(P, p_1, p_2)$ up to an overall rotation around the z axis. In this case only partial waves with $\lambda_i = 0$ are non-vanishing. If the final state is also a two-body state, the amplitude does not depend on ψ either, and the expansion (2.9) becomes the usual two-body partial wave expansion.

2.4. 0(2, 1) Expansion

We now perform an expansion of the N-point function at fixed momentum transfer Q, defined in Figure 5. We introduce $(3N_R - 7)$ variables v_R associated with the right-hand blob, $(3N_L - 7)$ called v_L and related to the left-hand one, $Q^2 = t$ and the three

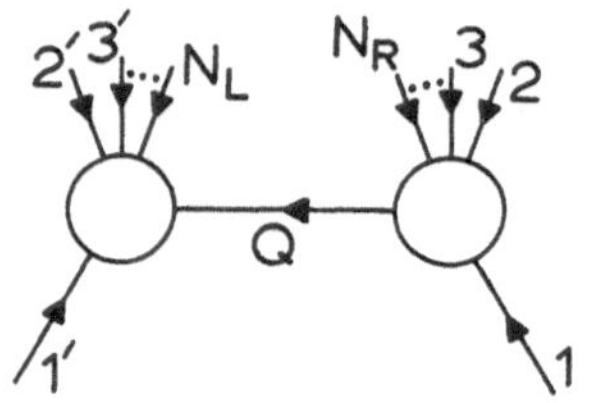

Fig. 5.

group parameters g that describe a transformation from the frame $F(Q, p_1, p_2)$ to $F(Q, p_1', p_2')$. In both of these frames Q has the form

$$Q = (0; 0; 0; \sqrt{-t}). \tag{2.10}$$

Thus they differ by an element of 0(2, 1) which is the little group of the Lorentz Group with respect to the four-vector Q. Therefore, this group contains boosts in the x-y plane and rotations around the z axis. We write the most general 0(2, 1) transformation in a form which is analogous to the three Euler rotations:

$$R_z(\mu)\, B_x(\xi)\, R_z(\nu); \tag{2.11}$$

$R_z(\alpha)$ is a rotation of angle α around the z axis and $B_x(\xi)$ is a boost in the x direction and its effect on a four-vector is given by the matrix

$$B_x(\xi) = \begin{pmatrix} \cosh\xi & \sinh\xi & 0 & 0 \\ \sinh\xi & \cosh\xi & 0 & 0 \\ 0 & 0 & 1 & 0 \\ 0 & 0 & 0 & 1 \end{pmatrix} \tag{2.12}$$

The $0(2, 1)$ group is non-compact, its unitary representations are infinite dimensional and are labelled by a continuous complex parameter J with $\mathrm{Re}J = -\frac{1}{2}$.

Using these representations we can perform the partial wave projection of the production amplitude

$$A^{J\lambda_L\lambda_R}(v_L, t, v_R) = \int dg A(v_L, t, g, v_R)\, D^{J\lambda_L\lambda_R}(g) \tag{2.13}$$

with

$$D^{J\lambda_L\lambda_R}(g) = e^{i\lambda_L\mu} \qquad d^J_{\lambda_L\lambda_R}(\cosh\xi)\, e^{-i\lambda_R\nu} \tag{2.14}$$

The integration over $d(\cosh\xi)$ runs from 1 to ∞, and at large values of $\cosh\xi$ we have:

$$d^J_{\lambda_L\lambda_R} \sim A(\cosh\xi)^J + B(\cosh\xi)^{-J-1} \tag{2.15}$$

thus the integral (2.13) converges provided

$$A(v_L, t, g, v_R) \xrightarrow[(\cosh\xi)\to\infty]{} (\cosh\xi)^\alpha \qquad v_L, v_R, t, \nu, \mu \text{ fixed} \tag{2.16}$$

with $\mathrm{Re}\alpha < -\frac{1}{2}$.

The inverse relation, i.e., the partial wave expansion, is:

$$A(v_L, t, g, v_R) = \int_{-1/2-i\infty}^{-1/2+i\infty} \frac{dJ}{\mathrm{tg}(\pi J)} D^{J\lambda_R\lambda_L^*}(g)\, A^{J\lambda_L\lambda_R}(v_L, t, v_R) \tag{2.17}$$

This is analogous to Equation (2.9), but instead of keeping fixed the total C.M. energy squared S we fix here the momentum transfer squared t. Equation (2.17) is similar to the Sommerfeld Watson transformation used by Regge in the analysis of two-body scattering.

As in the two-body case, we may assume meromorphy of $A^{J\lambda_L\lambda_R}$ as a function of the complex variable J, with poles located at positions $\alpha(t)$ and having factorized residues. As we vary t a pole may come across the path of integration in Equation (2.17). In that case we have to distort the path of integration as shown in Figure 6, and we can separate the pole contribution from that of the background. In the limit cosh $\xi \to \infty$ the pole with the largest Re α dominates, and the amplitude behaves like

$$B_L(v_L, \mu, t)\,(\cosh\xi)^{\alpha(t)}\, B_R(v_R, \nu, t) \tag{2.18}$$

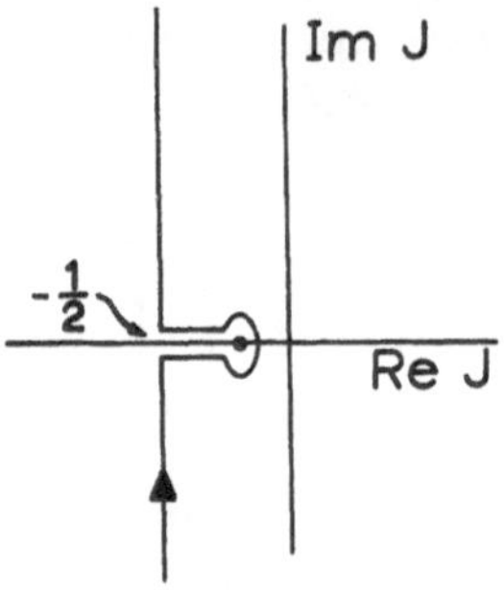

Fig. 6.

What is the limit $(\cosh \xi) \to \infty$? Let us calculate the total energy $S = (p_1 + p_{1'})^2$. For this we calculate p_1 in the rest frame of $1'$, and we have to go through a sequence of frames by performing a chain of transformations:

FRAME: $F(p_{1'}; Q; p_{2'}) \longleftarrow F(Q; p_{1'}; p_{2'}) \longleftarrow F(Q; p_1; p_2) \longleftarrow F(p_1; Q; p_2)$

TRANSFORMATION: $B_z(q_{1'Q})$ $\quad R_z(\mu)\, B_x(\xi)\, R_z(\nu)$ $\quad B_z(q_{1Q})$

$$\frac{p_1}{m_1} = \begin{cases} (1, 0, 0, 0) \\ (\cosh q_{1Q}, 0, 0, \sinh q_{1Q}) \\ (\cosh q_{1Q} \cosh \xi, X, X, \sinh q_{1Q}) \\ (\cosh q_{1'Q} \cosh \xi \cosh q_{1Q} + \sinh q_{1Q} + \sinh q_{1'Q}, X, X, X) \end{cases} \tag{2.19}$$

where X denotes kinematical quantities that we need not specify because the final answer does not depend on them.

If we now call Y the rapidity or boost parameter involved in going from the laboratory to the projectile frame with a single velocity transformation we have:

$$\cosh Y = \cosh q_{1'Q} \cosh \xi \cosh q_{1Q} + \sinh q_{1'Q} \sinh q_{1Q} = \frac{S - m_1^2 - m_{1'}^2}{2m_1 m_{1'}} \tag{2.20}$$

We can see that at fixed values of v_L, v_R, and t; S is linear in $\cosh \xi$. We recall from Equation (2.5)

$$\cosh q_{1Q} = \frac{\lambda^{1/2}(m_1^2, m_R^2, t)}{2m_1 \sqrt{-t}} \tag{2.21}$$

with $m_R^2 = (p_1 - Q)^2$. There are similar expressions for $\sinh q_{1Q}$, $\cosh q_{1Q}$ and $\sinh q_{1'Q}$.

Finally we should point out that whether or not the Regge hypothesis is correct – and we know that strictly speaking it is not – we can always use the group theoretical variables for the study of multiparticle processes.

2.5. Multiple 0(2, 1) analysis. [4]

We are now in a position to go one step forward and perform a multiple 0(2, 1) analysis, i.e., one in which we keep fixed several internal space-like momentum transfers. The analysis is similar to the previous case. The variables are (see Figure 7) $Q_1^2=t_1^2$ and $Q_2^2=t_2^2$, g_1 and g_2 (associated with transformations that leave Q_1 and Q_2 invariant) and the internal variables of each blob. These define the four vectors in special frames; the group theoretical ones describe relative orientations of frames that have in common a given form for a momentum transfer four-vector.

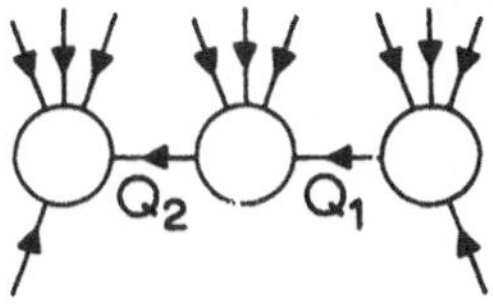

Fig. 7.

As an example we are going to discuss the two-body into three-body amplitude of Figure 8. In particular, we want expressions for the invariants $S=(p_L+p_R)^2$, $S_R=(p_R-Q_L)^2$ and $S_L=(p_L+Q_R)^2$ in terms of the group theoretical variables and $t_R=Q_R^2<0$, $t_L=Q_L^2<0$. The way to obtain these relations is straightforward, though messy. We just have to go through the following succession of frames by means of the transformations shown below where, in order to simplify the notation we ignore the third label of the frames, and we call $q_{Q_R R}\equiv q_R$; $q_{Q_L Q_R}\equiv q_c$ and $q_{LQ_L}\equiv q_L$:

$$\begin{array}{ccccccccccc} F(p_L, Q_L) & \leftarrow & F(Q_L, p_L) & \longleftarrow & F(Q_L, Q_R) & \leftarrow & F(Q_R, Q_L) & \longleftarrow & F(Q_R, p_R) & \leftarrow & F(p_R, Q_R) \\ & \downarrow & & \downarrow & & \downarrow & & \downarrow & & \downarrow & \\ & B_z(q_L) & & R_z(\mu_L)\,B_x(\xi_L)\,R_z(\nu_L); & & B_z(q_c); & & R_z(\mu_R)\,B_x(\xi_R)\,R_z(\nu_R); & & B_z(q_R) & \end{array} \quad (2.22)$$

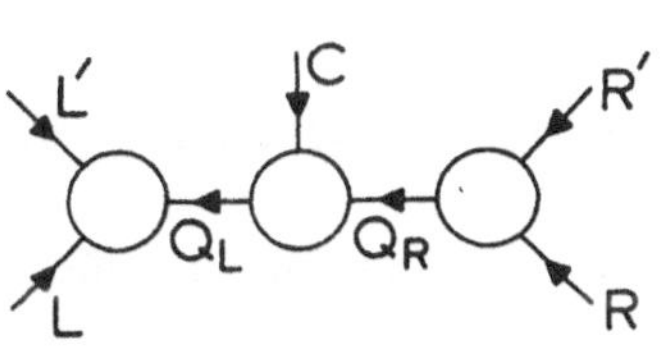

Fig. 8.

If we want to construct the expression of any given four-vector p in the final frame $F(p_L, Q_L, p_L)$, we start from a frame in which p has a simple form, and perform the necessary transformations. We immediately realize that the following simplifications occur, because the vertices in Figure 8 involve only three lines each:

(1) No internal variables are present.

(2) A change in ν_R does not change any four-vector in the final configuration. Thus the amplitude does not depend on it.

(3) A change in μ_L implies a rotation of all the four-vectors involved, and by rotational invariance the amplitude cannot depend on μ_L.

(4) The effects of changing μ_R or ν_L are the same: the amplitude depends in fact on their sum $\omega\equiv\mu_R+\nu_L$. This angle is just the angle between the planes defined by $\mathbf{p}_R$ an $\mathbf{p}'_R$ and $\mathbf{p}_L$ and $\mathbf{p}'_L$, in the rest frame of particle $\bar{c}$.

We are therefore left with the following set of independent variables: t_L, t_R, $\cosh\xi_L$, $\cosh\xi_R$ and ω. In analogy to what was done before in (2.20), we define:

$$\begin{aligned}\cosh Y &= \frac{S - m_R^2 - m_L^2}{2m_R m_L}\\ \sinh Y_R &= \frac{S_R - m_R^2 - t_L}{2m_R\sqrt{-t_L}}\\ \sinh Y_L &= \frac{S_L - m_L^2 - t_R}{2m_L\sqrt{-t_R}}\end{aligned} \tag{2.23}$$

and we obtain:

$$\begin{aligned}\sinh Y_R &= \cosh q_R \sinh q_c \cosh\xi_R + \sinh q_R \cosh q_c\\ \sinh Y_L &= \cosh q_L \sinh q_c \cosh\xi_L + \sinh q_L \cosh q_c\\ \cosh Y &= \cosh q_L \cosh q_R [\cosh q_c \cosh\xi_L \cosh\xi_R +\\ &\quad + \cos\omega \sinh\xi_L \sinh\xi_R] + \sinh q_c [\sinh q_L \cosh q_R \cosh\xi_L +\\ &\quad + \sinh q_R \cosh q_L \cosh\xi_R] + \sinh q_L \cosh q_c \sinh q_R .\end{aligned} \tag{2.24}$$

Observe that if:

$$\begin{aligned}\xi_L &= 0 & Y_L &= (q_L + q_c)\\ \xi_R &= 0 & Y_R &= (q_R + q_c)\\ \xi_L = \xi_R &= 0 & Y &= (q_L + q_c + q_R)\end{aligned} \tag{2.25}$$

as expected from the fact that all the intervening boosts are performed along the same (z) axis.

The double-Regge limit is, instead, the limit in which $\xi_R \to \infty$ $\xi_L \to \infty$ at t_R, t_L and ω fixed. We see that in this limit:

$$\begin{aligned}\cosh Y &\simeq \frac{S}{2m_R m_L} \simeq \cosh q_L \cosh q_R (\cosh q_c + \cos\omega) \cosh\xi_L \cosh\xi_R\\ \sinh Y_R &= \frac{S_R}{2m_R\sqrt{-t_L}} \simeq \cosh q_R \sinh q_c \cosh\xi_R\\ \sinh Y_L &\simeq \frac{S_L}{2m_L\sqrt{-t_R}} \simeq \cosh q_L \sinh q_c \cosh\xi_L\end{aligned} \tag{2.26}$$

and

$$\frac{\cosh Y}{\sinh Y_L \sinh Y_R} \simeq \frac{\dfrac{S}{2m_R m_L}}{\dfrac{S_R}{2m_R\sqrt{-t_L}} \cdot \dfrac{S_L}{2m_L\sqrt{-t_R}}} \simeq \frac{\cosh q_c + \cos\omega}{\sinh^2 q_c} \tag{2.27}$$

then, we obtain:

$$S \simeq S_L S_R \left(\frac{m_c^2 - t_L - t_R + 2\sqrt{-t_L}\sqrt{-t_R}\cos\omega}{\lambda(t_L, t_R, m_c^2)} \right) \tag{2.28}$$

Thus, in the double Regge limit, $S_R \to \infty$, $S_L \to \infty$ and $S/S_R S_L = f(t_R, t_L, \cos\omega)$. Dropping m_c^2 (very frequently the pion mass squared) and taking the extreme cases $\cos\omega = \pm 1$

$$S \simeq S_L S_R \frac{(\sqrt{-t_L} \pm \sqrt{-t_R})^2}{(t_L - t_R)^2} = \frac{S_R S_L}{(\sqrt{-t_L} \mp \sqrt{-t_R})^2}. \tag{2.29}$$

If we have a matrix element that grows with S_R and S_L (i.e., $\alpha > 0$) we see that at *fixed* S the value $\cos\omega = -1$ will allow larger values of S_R and S_L, and then it will be favored. Indeed, distributions in $\cos\omega$ are found experimentally to be peaked around $\cos\omega = -1$. For this value

$$S \simeq \frac{S_L S_R}{4\tau} \qquad \text{with } \tau \simeq \langle -t \rangle_{\text{av.}}$$

$$\frac{S}{4\tau} \simeq \frac{S_L}{4\tau} \times \frac{S_R}{4\tau} \tag{2.30}$$

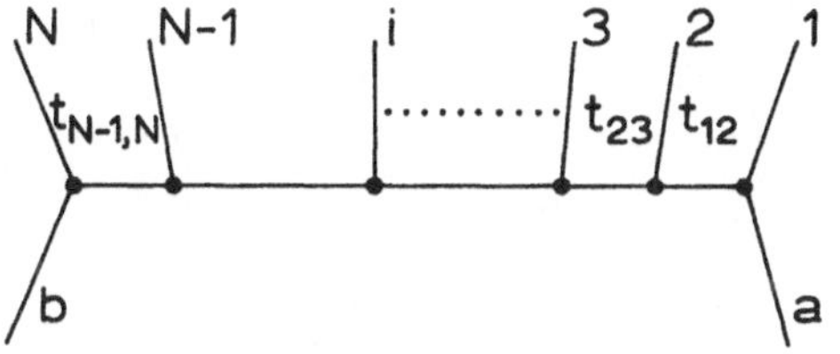

Fig. 9.

Relation (2.30) can be easily generalized to a multiperipheral configuration (see Figure 9)

$$S = \prod_{i=1}^{N-1} S_{i,i+1} \prod_{j=2}^{N-1} f_j \tag{2.31}$$

with

$$f_j = \frac{m_j^2 - t_{j,j-1} - t_{j,j+1} - 2\sqrt{-t_{j,j-1}}\sqrt{-t_{j,j+1}}\cos\omega_j}{\lambda(t_{j,j-1}, t_{j,j+1}, m_j^2)},$$

then

$$\frac{S}{4\tau} \simeq \prod_{i=1}^{N-1} \left(\frac{S_{i,i+1}}{4\tau} \right) \tag{2.32}$$

The multi-Regge behavior for the production amplitude is then:

$$A \simeq \beta(t_{N-1,N})\,\beta(t_{1,2}) \prod_{i=1}^{N-1} (\cosh \xi_{i,i+1})^{\alpha_{i,i+1}} \prod_{j=2}^{N-1} \beta(t_{j-1,j}, \omega_j, t_{j,j+1})$$

$$\omega_j;\ t_{i,i+1} \text{ fixed}$$

$$\xi_{i,i+1} \to \infty \tag{2.33}$$

3. Peripheralism

It is well known that two-body processes are dominantly peripheral at high energy. By this we mean that they are dominated by low momentum transfer events. An example of a peripheral model is the Regge model, in which the amplitude for the process $ab \to \bar{1}\,\bar{2}$ at high energy is of the form:

$$\begin{aligned} A &\sim G_{a1}(t)\,G_{b2}(t)\,(\cosh \xi)^{\alpha_t(t)} \quad \text{at fixed } t = (p_a + p_1)^2 \\ &\sim G_{a2}(u)\,G_{b1}(u)\,(\cosh \xi)^{\alpha_u(u)} \quad \text{at fixed } u = (p_a + p_2)^2, \end{aligned} \tag{3.1}$$

where $\alpha_{t,u}$ depend on the quantum numbers of the systems $a1$ and $a2$, respectively. Peripherality is injected through the residue functions G_{ij} which decrease fast as $|t|$ or $|u|$ grow. As a consequence we can draw figure 10. The figures in the right-hand side

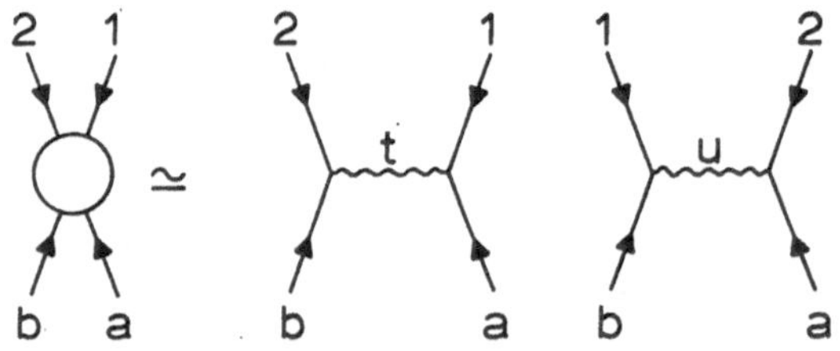

Fig. 10.

are not just a kinematic definition of variables, but also carry the dynamical implication of peripherality as defined above. As a consequence, in each diagram we have:

(1) A well-defined longitudinal ordering of the final particles: in the C.M., for the first diagram $p_1^l \gg p_2^l$, where p_i^l is the component of the momentum of the ith final particle in the direction of p_a in the c.m.s. Of course for the second diagram $p_2^l \gg p_1^l$, and therefore both diagrams do not interfere.

(2) The average transverse momenta of the final particles are limited, i.e., do not grow at high energy.

4. Multiperipheralism

4.1. Basic assumptions

The extension of the concepts contained in Section 3 to multiple production processes is called multiperipheralism.

In the case of Figure 9 in which we have N final particles we can draw $N!$ diagrams. We call them multiperipheral (MP) diagrams. Each one defines a complete set of variables:

$$\begin{aligned} &\cosh \xi_{i,\,i+1}, i = 1, \ldots, N-1 \\ &t_{i,\,i+1}, \qquad i = 1, \ldots, N-1 \\ &\omega_i, \qquad\quad i = 2, \ldots, N-1 \end{aligned}$$

The first statement of multiperipheralism says that as we keep N fixed and let S grow, the production amplitude may be important only in regions of phase space in which all t_i's of some of the $N!$ graphs are small. This is indeed a small fraction of phase space because, as we increase the energy, the kinematically allowed region for t_i grows, whereas we are saying that the dynamically relevant region for t_i does not grow with s.

The second hypothesis of multiperipheralism is that the multiparticle amplitude factorizes, i.e., it can be written as a product of factors, each of which depends on the momenta of a limited number of neighbouring lines in the corresponding MP diagram. We see that the Multi-Regge model is an example of such a model. Indeed, $\cosh \xi_{i,\,i+1}$ can be written in terms of the five four-momenta shown in Figure 11, whereas we need two more momenta to define ω_i (as indicated in Figure 12).

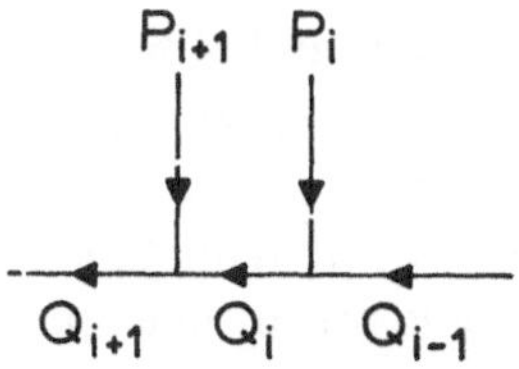

Fig. 11.

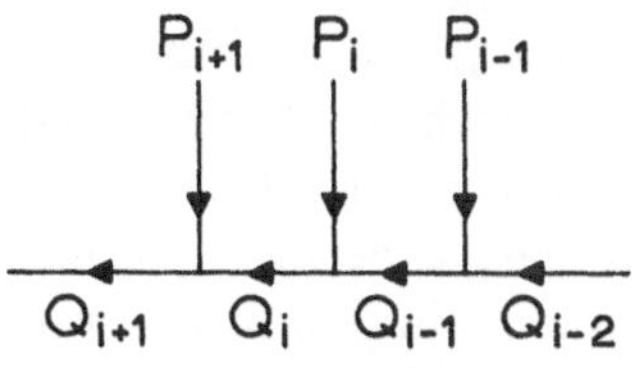

Fig. 12.

Again the residue functions are supposed to provide the suppression of the matrix element at large values of t. We should remark that although peripheralism at high energy leads us to the Regge region, multiperipheralism does not necessarily lead us to the Multi-Regge region.

This region is such that all $\cosh \xi_{i,i+1}$ grow, instead we can stay in the MP region (i.e., the region of limited $t_{i,i+1}$'s) by letting only one of the subenergies grow. Therefore, the MR limit may not be quite relevant to a large part of the production amplitude at high energy.

Again in the MR region we find that the final particles have a characteristic configuration: we are going to show that:

(1) they are longitudinally ordered, i.e., $p_1^l \gg p_2^l \gg \cdots \gg p_N^l$.

(2) their transverse momenta are limited.

In order to do this, we write the expression for the energy and longitudinal momentum of particle i in the cms in an invariant form (for $m_a = m_b = m$)

$$E_i = p_i \cdot \frac{(p_a + p_b)}{\sqrt{S}}$$
$$p_i^l = - p_i \cdot \frac{(p_a - p_b)\sqrt{S}}{\lambda^{1/2}(S, m^2, m^2)} \tag{4.1}$$
$$p_i^{\perp^2} = E_i^2 - p_i^{l^2} - m_i^2 \simeq 4m_i^2 \frac{(p_i \cdot p_a)^2 + (p_i \cdot p_b)^2}{S^2} + \\ + \frac{4(p_i \cdot p_a)(p_i \cdot p_b)}{S} - m_i^2 .$$

where, again, X denotes unspecified kinematical quantities.

We now evaluate the above expressions in the rest frame of particle i, i.e., in $F(p_i, Q_{i-1}, Q_i)$. We only need the time components of p_a and p_b in this frame, and the same analysis that led us to the expression for S in this limit now gives (also in the rest frame of i)

$$p_a \sim (\cosh \xi_{1,2} \cosh \xi_{2,3} \cdots \cosh \xi_{i-1,i}; X; X; X) \\ p_b \sim (\cosh \xi_{i,i+1} \cosh \xi_{i+1,i+2} \cdots \cosh \xi_{N-1,N}; X; X; X). \tag{4.2}$$

Thus, when we go from the rest frame of particle i, to that of particle $i+1$, the time component of p_b decreases, whereas that of p_a increases. Hence remembering that $S \sim \prod_{i=1}^{N-1} \cosh \xi_{i,\,i+1}$ we immediately see that $p_i^{\perp} \simeq$ constant and $p_i^l \gg p_{i+1}^l$.

In the case in which all t's are limited, but only *one* $\cosh \xi_{j,\,j+1}$ grows, we can devide the final particles into two groups: those with index $i \leqslant j$ and those with $i \geqslant j+1$. The longitudinal ordering holds only for pairs of particles belonging to different sets. However, the limitation of the transverse momenta holds for all final particles also in this case. Thus we conclude that limitation of the transverse momenta of the produced particles is a consequence of multiperipheralism. This means that all the sequence of transformations that we have perfomed after all lead us to a configuration in momentum space in which the distribution of particles is roughly longitudinal. The approximately constant ratio between longitudinal components of the momenta of the final particles suggests a description in terms of a variable that is roughly the logarithm of the longitudinal momentum. Thus we are led naturally to the introduction of rapidity y used in Section 2 [5].

Instead of performing boosts in all directions as we have done up to now, we just boost along the longitudinal direction. Thus, we do not go through a frame in which particle i is at rest, but through one in which particle i has only transverse motion. The boost parameter that takes us from the frame in which a is at rest to the frame in which $p_i^l = 0$ is the rapidity of particle i:

$$p_i^l = \sqrt{p_i^{\perp^2} + m_i^2} \sinh y_i \equiv w_i \sinh y_i$$

and

$$E_i = w_i \cosh y_i \tag{4.3}$$

Observe that $d^3p_i/E = d^2p_i^{\perp}\,dy_i$ is a Lorentz invariant i.e., differences of rapidities are invariant under boosts in the longitudinal direction that leave $\mathbf{p}^{\perp}$ invariant.

In terms of the rapidity $-\,\mathbf{p}^{\perp}$ varibales we have:

$$\begin{aligned} S_{i,i+1} &= m_i^2 + m_{i+1}^2 + 2w_i w_{i+1} \cosh(y_{i+1} - y_i) - 2\mathbf{p}_i^{\perp}\mathbf{p}_{i+1} \\ t_{i,i+1} &= -\left(\sum_{j=1}^{i} w_j e^{-y_j} - m_a e^{-y_a}\right)\left(\sum_{j=i+1}^{N} w_j e^{y_j} - m_b e^{y_b}\right) - \\ &\quad -\left(\sum_{j=1}^{i} \mathbf{p}_i^{\perp}\right)^2 \simeq -w_i w_{i+1} e^{y_{i+1}-y_i} - \left(\sum_{j=1}^{i} p_i^{\perp}\right)^2 . \end{aligned} \tag{4.4}$$

The limitation of $|\mathbf{p}_i^{\perp}|$, implies that only a small fraction of phase space is populated at high energies. This limitation is a well-known experimental fact but we can ask ourselves why it is so. We do not have a completely satisfactory answer, but we can argue like this: Suppose that the production amplitude is just 1. The N-particle production cross section would then be:

$$\sigma_{ab\to N} = \int \frac{|1|^2\,dR_N}{N!\,I} \tag{4.5}$$

with

$$I = 4\sqrt{S}q_{cN} \approx 2\,S$$

and

$$dR_N = \prod_{i=1}^{N} \frac{d^3p_i}{(2\pi)^3\,2E_i}\,\delta^{(4)}\left(p_a + p_b - \sum_{j=1}^{N} p_j\right)(2\pi)^4$$

then

$$\int dR_N \sim S^{N-2} \tag{4.6}$$

from where we conclude that

$$\sigma_{ab\to N} \sim S^{N-3} \tag{4.7}$$

and already for $N=4$ this violates the Froissart bound (which says that the total cross section cannot grow faster than $\log^2 s$ if probability is conserved). There is no unique way out of this difficulty. One, for instance, would be for the production amplitude to behave as some inverse power of S, with a power that grows with the multiplicity. Such an amplitude would not obey the factorization hypothesis. The way chosen by Nature to conserve probability is instead not to populate the phase space region of large $p^{\perp}$. Thus, addition of one more particle in the final state does not introduce an extra power of S, and this resolves the difficulty.

4.2. Simplified MP model [5, 6]

We construct now the simplest possible version of the MP model. For this we assume that:

(1) All produced particles are identical.
(2) We can ignore spin and isospin.

(3) We have a universal exchange. Because the t_i's are limited we ignore the variation of $\alpha(t)$ with t and set $\alpha(t)=\bar{\alpha}$.

(4) We can use Regge behaviour at all energies.

(5) We can perform kinematic approximations that are true only in the MR limit.

(6) The final particles 1 and N have the internal quantum numbers of a and b, respectively.

Thus we write

$$S_{i,\,i+1} \simeq w_i w_{i+1} e^{Z_{i,\,i+1}} \tag{4.8}$$

$$t_{i,\,i+1} \simeq -\, w_i w_{i+1} e^{-Z_{i,\,i+1}} - \mathbf{Q}_i^{\perp 2}, \tag{4.9}$$

where

$$Z_{i,\,i+1} \equiv y_i - y_{i+1}$$

and

$$\mathbf{Q}_i^{\perp} = \sum_{j=1}^{i} \mathbf{p}_j^{\perp}$$

If we use an exponential damping in t, i.e., e^{bt}, we can write:

$$A_{ab\to N} \simeq G_a G_b g^{N-2} \exp\left[\bar{\alpha} \sum_{i=1}^{N-1} Z_{i,\,i+1}\right] \prod_{i=1}^{N-1} \exp[-\,\mathrm{b}e^{-Z_{i,\,i+1}}] \exp[-\,b\mathbf{Q}_i^{\perp 2}] \tag{4.10}$$

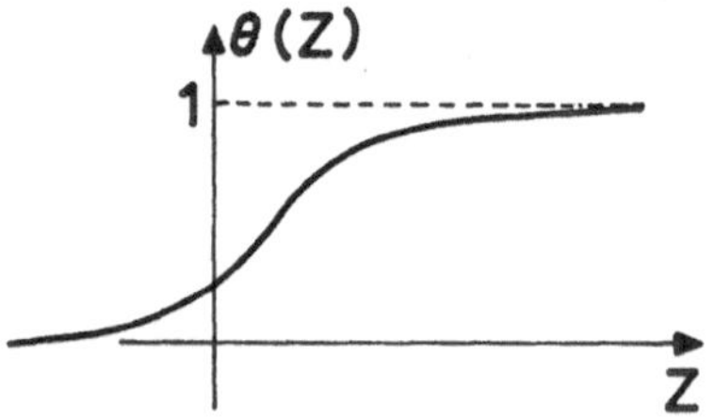

Fig. 13.

The function $\theta(Z)\equiv e^{-be^{-z}}$ is of the form indicated in Figure 13, where a change in b amounts to a shift of the origin. We therefore see that the effect of damping in t not only produces damping in the transverse momenta, but also orders the longitudinal components, i.e.:

$$y_1 \gtrsim y_2 \gtrsim y_3 \gtrsim \cdots \tag{4.11}$$

Thus

$$\sigma_{ab\to N} = \frac{1}{N!}\frac{1}{I}\int |A_{ab\to N}|^2 \,\mathrm{d}R_N = \frac{1}{2S}\int G_a^2 G_b^2 g^{2N-4}$$
$$\times\, e^{2\bar{\alpha}(y_1-y_N)}\theta(Z_{1,\,2})\ldots\theta(Z_{N-1,\,N})\,\mathrm{d}^2p_1^{\perp}\ldots\mathrm{d}^2p_N^{\perp}\,\mathrm{d}y_1\ldots\mathrm{d}y_N$$
$$\times\,\delta^{(4)}(P_{\text{final}} - P_{\text{initial}}), \tag{4.12}$$

where θ is some kind of step function producing the effect described in Figure 13. In this equation the $N!$ has been cancelled by the contribution of the $N!$ different graphs that do not interfere and differ from each other by permutation of the final particles.

We re arrange the energy and longitudinal components in the argument of the δ-function the following way.

$$\begin{aligned}(E+p^l)_{\text{initial}} &= m_a e^Y + m_b \simeq m_a e^Y \\ (E-p^l)_{\text{initial}} &= m_a e^{-Y} + m_b \simeq m_b \\ (E \pm p^l)_{\text{final}} &= \sum_{i=1}^{N} w_i exp(\pm y_i)\end{aligned} \tag{4.13}$$

thus

$$\delta^{(4)}(P_{\text{final}} - P_{\text{initial}}) = \tfrac{1}{2}\delta^{(2)}\left(\sum_{i=1}^{N} \mathbf{p}_i^{\perp}\right)\delta\left(\sum_{i=1}^{N} w_i e^{y_i} - m_a e^{y}\right)\times$$
$$\times\left(\sum_{i=1}^{N} w_i e^{-y_i} - m_b\right) \simeq \tfrac{1}{2}\delta^{(2)}(\mathbf{Q}_N^{\perp})\,\delta(w_1 e^{y_1} - m_a e^Y)\,\delta(w_N e^{-y_N} - m_b) \simeq$$
$$\simeq \frac{1}{2m_a m_b e^Y}\,\delta^{(2)}(\mathbf{Q}_N^{\perp})\,\delta(y_1 - Y)\,\delta(y_N) \tag{4.14}$$

Thus, particles 1 and N have roughly the same rapidities as a and b, respectively. We can then write, setting $N-2=n$ and absorbing the result of integrating over the transverse momenta in the coupling constants:

$$\sigma_{ab\to N} \sim G_a^2 G_b^2 e^{(2\bar{\alpha}-2)Y} g^{2n} \int_0^Y \mathrm{d}y_2 \ldots \int_0^Y \mathrm{d}y_{N-1}\,\theta(Z_{1,2})\ldots\theta(Z_{n-1,N}) \tag{4.15}$$

In (4.15) the rapidities of the final particles are ordered by the θ functions. If we assume for simplicity that these are step functions at the origin we have:

$$I_n(Y) \equiv \frac{g^{2n}}{n\,!}\int_0^Y \mathrm{d}y_2 \ldots \int_0^{y_{N-2}} \mathrm{d}y_{N-1} = \frac{(g^2 Y)^n}{n\,!}, \tag{4.16}$$

thus

$$\sigma_n(Y) = G_a^2 G_b^2 \frac{(g^2 Y)^n}{n\,!} e^{-(2-2\bar{\alpha})Y} \tag{4.17}$$

The total cross section is then

$$\sigma_{\text{TOT}}(Y) = \sum_{n=0}^{\infty} \sigma_n(Y) = G_a^2 G_b^2 e^{(g^2+2\bar{\alpha}-2)Y} \tag{4.18}$$

If we want σ_{TOT} to be asymptotically constant we need to set $g^2=2-2\bar{\alpha}$. We also find:

$$\bar{n} \equiv \sum_{n=0}^{\infty} n \frac{\sigma_n}{\sigma_{\text{TOT}}} = g^2 Y \tag{4.19}$$

For the multiplicity distribution we have:

$$P_n \equiv \frac{\sigma_n}{\sigma_{\text{TOT}}} = \frac{\bar{n}^n e^{-\bar{n}}}{n\,!} \tag{4.20}$$

which is a Poisson distribution.

From the experimental values of the average multiplicity we fix:

$$g^2 \simeq 1.5 \quad \bar{\alpha} = 2 - g^2/2 \simeq 0.25. \tag{4.21}$$

Let us summarize the main features of this very simple model:

(1) All partial cross sections go to zero at high the energy (see (4.17))

(2) The total cross section is constant in energy.

(3) The average multiplicity grows like log s (see (4.19)).

(4) The multiplicity distribution is Poisson-like.

Let us calculate the average subenergy. At fixed n, because of (2.32) we have (in units of 4τ):

$$\overline{\log S_i}\,|_n = \frac{\log S}{n+1} \tag{4.22}$$

Now we average over n, counting $n+1$ subenergies at each n:

$$\overline{\overline{\ln S_i}} = \frac{\sum_{n=0}^{\infty} P_n\,(n+1)\,\overline{\log S_i}|_n}{\sum_{n=0}^{\infty} P_n\,(n+1)} = \frac{\log S}{n+1} \simeq \frac{1}{g^2} \tag{4.23}$$

Then

$$\frac{S_i}{4\tau} \simeq e^{1/g^2} \simeq 2, \text{i.e.,} \quad S_i \simeq 8\,\tau. \tag{4.24}$$

Taking $\tau \simeq \langle \mathbf{p}_\perp^2 \rangle \simeq \cdot 11$ GeV^2 we find $S_i \simeq 0.9$ GeV^2 i.e., an energy between the ϱ and f^0.

Thus as we increase the energy, it goes into creating more particles, rather than increasing the subenergies. The latter stay constant, and if g^2 is very small they will be in average fairly large, and the assumptions used may become more justifiable. That is why we say that the model is good in the weak coupling limit, but, unfortunately the actual value of the coupling is not weak, but of order 1.

5. Generalizations

5.1. Exchange of Quantum Numbers

In order to extend the applications of the MP model beyond the case of the total cross section we recall that the S-matrix is related to the scattering amplitude A by the expression:

$$S_{fi} = \delta_{fi} + i\,(2\pi)^4\,\delta^4\,(P_f - P_i)\,A_{fi}. \tag{5.1}$$

Unitarity of the S-matrix implies a quadratic relation for the scattering amplitude

$$\operatorname{Im} A_{fi} = \tfrac{1}{2} \sum_{N=2}^{\infty} \int \mathrm{d}R_N A_{Nf}^{*} A_{Ni}. \tag{5.2}$$

We restrict ourselves here to consider a 2-body initial state i with momenta p_a and p_b, a 2-body final state with momenta p'_a and p'_b, and an N body intermediate state in which particles 1 and N carry the quantum numbers of a and b, respectively, and n additional pions of momenta $p_2 \ldots p_{N-1}$ are present. If we choose $p_a = p'_a$ and $p_b = p'_b$ we find the optical theorem:

$$\operatorname{Im} A_{ii} = \lambda^{1/2} (S m_a^2 m_b^2)\, \sigma_i^{\mathrm{TOT}} \simeq S \sigma_i^{\mathrm{TOT}} \tag{5.3}$$

We have used a model of particle production to calculate σ^{TOT} or, equivalently, the imaginary part of the elastic amplitude in the forward direction. The unitarity relation shows that we can also use it, at least in principle, to calculate the imaginary part of any 2-body amplitude, both at or away from the forward direction.

Let us discuss the case of a forward 2-body amplitude in which quantum numbers are exchanged [7]. For concreteness, let us consider a model in which we have only pions produced by exchange of $I=1$ objects (I: isospin), like the ϱ and A_2; for the amplitude with $I_t=0$, we obtained

$$\operatorname{Im} A^{I_t=0} \sim e^{(2\bar{\alpha}-1+g^2 I_{t=0})} \tag{5.4}$$

We have just added an index $I_t=0$ to the coupling constant because we construct our amplitude by considering diagrams of the type shown in Figure 14, in which we have to project out the $I_t=0$ component. For each pion in the intermediate state we there-

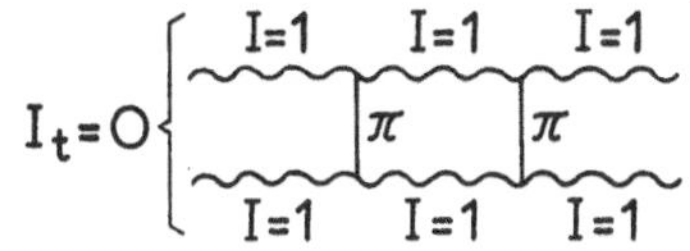

Fig. 14.

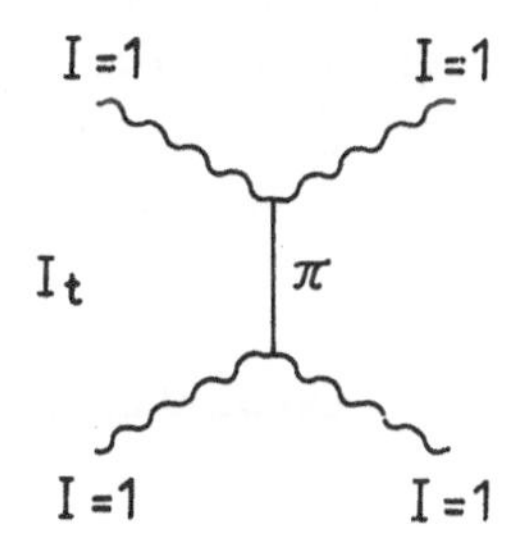

Fig. 15.

fore have a factor $g_{I_t}^2$ associated with the projection of the diagram of Figure 15, on a well defined I_t. Indeed, this diagram has a well-defined isospin in the S-channel (the pion isospin, i.e., $I_S=1$) and it is well known that such an amplitude is a linear combination of amplitudes of well defined I_t. The inverse relation is

$$A^{I_t} = \sum_{I_s=0}^{2} \beta^{I_t I_s} A^{I_s} \tag{5.5}$$

with

$$\beta^{I_t I_s} = \begin{pmatrix} \frac{1}{3} & 1 & \frac{5}{3} \\ \frac{1}{3} & \frac{1}{2} & -\frac{5}{6} \\ \frac{1}{3} & -\frac{1}{2} & \frac{1}{6} \end{pmatrix} \tag{5.6}$$

Therefore if we have an amplitude of $I_s=1$, the relative contributions to $I_t=0, 1, 2$ are given by the central column in the matrix, i.e.,

$$A^{I_t=0} : A^{I_t=1} : A^{I_t=2} = 1 : \tfrac{1}{2} : -\tfrac{1}{2} \tag{5.7}$$

or

$$g^2_{I_t=0} : g^2_{I_t=1} : g^2_{I_t=2} = g^2 : \tfrac{1}{2}g^2 : -\tfrac{1}{2}g^2$$

Therefore, writing

$$\operatorname{Im} A^{I_t}(s, t=0) \sim \exp(\alpha^{I_t}_{\text{out}} Y) \tag{5.8}$$

we obtain

$$\begin{aligned} \alpha^{I_t=0}_{\text{out}} &= 2\bar{\alpha} - 1 + g^2 \\ \alpha^{I_t=1}_{\text{out}} &= 2\bar{\alpha} - 1 + g^2/2 \\ \alpha^{I_t=2}_{\text{out}} &= 2\bar{\alpha} - 1 - g^2/2 \end{aligned} \tag{5.9}$$

with $\bar{\alpha}=0.25$ and $g^2=1.5$ we get

$$\begin{aligned} \alpha^{I_t=0}_{\text{out}} &= 1 \\ \alpha^{I_t=1}_{\text{out}} &= 0.25 \\ \alpha^{I_t=2}_{\text{out}} &= -1.25\,. \end{aligned} \tag{5.10}$$

The strong repression of the $I_t=2$ pole is welcome, because no particles with $I=2$ are known. The $\alpha^{I_t=1}_{\text{out}}=0.25$ is consistent with the input value $\bar{\alpha}=0.25$. This is an example of how bootstrap conditions can be imposed. In our example, the condition of $\alpha^{I_t=0}_{\text{out}}=1$ and the condition $\alpha^{I_t=1}_{\text{out}}=\bar{\alpha}$ are identical and that is why the second one was automatically satisfied after the first one was imposed.

5.2. Integral equation [1, 8, 5]

We can exhibit the (slightly redefined) external couplings in Im $A_{ab,\,ab}$: the n-particle intermediate state contribution to

$$\operatorname{Im} A_{ab,\,ab} = G_a^2 B(Y)\, G_a^2 = G_b^2 \sum_{n=0}^{\infty} B_n(Y)\, G_b^2\,, \tag{5.11}$$

where

$$B_n(Y) = g^{2n} \int_0^Y K(z_{1,2})\, \mathrm{d}z_{1,2} \ldots \int_0^Y K(z_{n-1,n})\, \mathrm{d}z_{n-1,n} \times \delta\left(\sum_{i=1}^{n-1} z_{i,i+1} - Y\right) \tag{5.12}$$

with

$$K(z) = \theta(z)\, e^{(2\bar{\alpha}-1)z}\,. \tag{5.13}$$

In particular

$$B_0(Y) = K(Y)$$

$$B_1(Y) = g^2 \int K(z_{1,2})\,\mathrm{d}z_{1,2} \int K(z_{2,3})\,\delta(z_{1,2} + z_{2,3} - Y)\,\mathrm{d}z_{2,3}$$

$$\left(\text{see that} \int K(z_{2,3})\,\delta(z_{1,2} + z_{2,3} - Y)\,\mathrm{d}z_{2,3} = B_0(Y - z_{1,2})\right)$$

$$B_n(Y) = g^2 \int_0^Y K(z)\,B_{n-1}(Y - z)\,\mathrm{d}z = g^2 \int_0^Y B_{n-1}(Y')\,K(Y - Y')\,\mathrm{d}Y', \tag{5.14}$$

Summation over n gives:

$$B(Y) = K(Y) + g^2 \int_0^Y B(Y')\,K(Y - Y')\,\mathrm{d}Y' \tag{5.15}$$

this is pictorially shown in Figure 16.

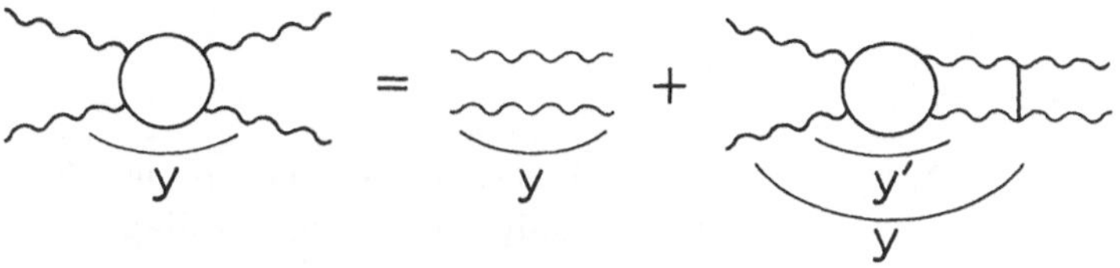

Fig. 16.

Equation (5.15) is in the form of a convolution. If we take the Laplace transform of it we write

$$b(j) = k(j) + g^2 b(j)\,k(j) \tag{5.16}$$

with

$$b(j) = \int e^{-yj} B(y)\,\mathrm{d}y, \text{ etc}.$$

Equation (5.16) can now be solved algebraically and the solution for (5.15) is

$$B(y) = \frac{1}{2\pi i} \int_{c-i\infty}^{c+i\infty} e^{jy} b(j)\,\mathrm{d}j = \frac{1}{2\pi i} \int_{c-i\infty}^{c+i\infty} e^{jy} \frac{k(j)}{1 - g^2 k(j)}\,\mathrm{d}j, \tag{5.17}$$

where the contour of integration is taken to the right of the rightmost singularity of $b(j)$

If we take a step function for $\theta(z)$ we have

$$k(j) = \frac{1}{j + 1 - 2\bar{\alpha}}; \quad b(j) = \frac{1}{j + 1 - 2\bar{\alpha} - g^2} \tag{5.18}$$

and consequently

$$B(y) = e^{(2\bar{\alpha} - 1 + g^2)y} \tag{5.19}$$

We see that the effect of iterating the integral equation has been to displace the pole that in the Born term was at $j = 2\bar{\alpha} - 1$ to $j = 2\bar{\alpha} - 1 + g^2$. We have thus reproduced the result of Equation (4.17), in a somewhat fancier language.

5.3. The Two-Channel Problem [9]

We now generalize the previous treatment to be able to incorporate more than one exchange mechanism. For concreteness, we consider the possibility of pomeron and meson exchanges. Thus, B becomes a matrix, and G_a^2 and G_b^2 become column vectors, where the row and the column indices label the nature of the exchanges.

The absorptive part of the elastic amplitude is given by

$$\tilde{G}_a^2 B G_a^2 \equiv G_{aP}^2 \left[B_{PP} G_{bP}^2 + B_{PM} G_{bM}^2\right] + G_{aM}^2 \left[B_{MP} G_{bP}^2 + B_{MM} G_{bM}^2\right]. \tag{5.20}$$

The coupling constant g^2 is replaced by the matrix

$$\begin{pmatrix} 0 & g_{PM}^2 \\ g_{PM}^2 & g_{MM}^2 \end{pmatrix} \tag{5.21}$$

where the g_{PP}^2 element is zero because we assume that only pions are produced and they do not couple to two pomerons. Finally, the kernel is a diagonal matrix:

$$K(z) = \begin{pmatrix} K_P(z) & 0 \\ 0 & K_M(z) \end{pmatrix} \tag{5.22}$$

with

$$K_{P,M}(z) = \theta(z)\, e^{(2\alpha_{P,M} - 1)z}$$

The integral equation is now written in matrix form:

$$B(y) = K(y) + \int B(y')\, g^2 K(y - y')\, dy'. \tag{5.23}$$

Again we perform a Laplace transformation that reduces the equation to an algebraic matrix equation:

$$b = k + b g^2 k$$

then

$$b = (k^{-1} - g^2)^{-1} \tag{5.24}$$

Now

$$(k^{-1} - g^2)^{-1} = \frac{1}{\Delta} \begin{pmatrix} \dfrac{1}{k_M} - g_{MM}^2 & g_{PM}^2 \\ g_{PM}^2 & \dfrac{1}{k_P} \end{pmatrix} \tag{5.25}$$

with

$$\Delta = \frac{1}{k_P k_M} - \frac{g_{MM}^2}{k_P} - g_{PM}^4 \tag{5.26}$$

If we use a step function for $\theta(z)$ we get

$$b(j) = \begin{pmatrix} j - f_M & g_{PM}^2 \\ g_{PM}^2 & j - f_P \end{pmatrix} \times \frac{1}{(j - f_P)(j - f_M) - g_{PM}^4} \tag{5.27}$$

with

$$\begin{aligned} f_M &= 2\alpha_M + g_{MM}^2 - 1 \\ f_P &= 2\alpha_P - 1 . \end{aligned} \tag{5.28}$$

We therefore find two poles at the solutions of the equation

$$j^2 - j(f_P + f_M) + f_P f_M - g_{PM}^4 = 0$$

i.e.

$$j_\pm = \tfrac{1}{2}\left[f_P + f_M \pm \sqrt{(f_P - f_M)^2 + 4g_{PM}^4}\right] \tag{5.29}$$

We observe $g_{PM}^2 = \sqrt{(j_\pm - f_P)}\,\sqrt{(j_\pm - f_M)}$ and therefore the residues in b factorize: i.e., near $j = j_+$

$$b(j) \simeq \frac{1}{(j - j_+)} \begin{pmatrix} \sqrt{\dfrac{j_+ - f_M}{j_+ - j_-}} \\ \sqrt{\dfrac{j_+ - f_P}{j_+ - j_-}} \end{pmatrix} \begin{pmatrix} \sqrt{\dfrac{j_+ - f_M}{j_+ - j_-}} & \sqrt{\dfrac{j_+ - f_P}{j_+ - j_-}} \end{pmatrix} \tag{5.30}$$

and similarly near $j = j_-$.
Therefore

$$a(j) \equiv \tilde{G}_a^2 b(j) G_b^2 = \frac{\gamma_a^+ \gamma_b^+}{j - j_+} + \frac{\gamma_a^- \gamma_b^-}{j - j_-} \tag{5.31}$$

with

$$\gamma_a^\pm = \begin{pmatrix} G_{aP}^2 & G_{aM}^2 \end{pmatrix} \begin{pmatrix} \sqrt{\dfrac{j_\pm - f_M}{j_\pm - j_\mp}} \\ \sqrt{\dfrac{j_\pm - f_P}{j_\pm - j_\mp}} \end{pmatrix} \tag{5.32}$$

and

$$\gamma_b^\pm = \begin{pmatrix} \sqrt{\dfrac{j^\pm - f_M}{j_\pm - j_\mp}} & \sqrt{\dfrac{j^\pm - f_P}{j_\pm - j_\mp}} \end{pmatrix} \begin{pmatrix} G_{bP}^2 \\ G_{bM}^2 \end{pmatrix} \tag{5.33}$$

If the two channels decouple, the output poles tend to:

$$J_\pm \begin{matrix} \nearrow f_P \\ \searrow f_M \end{matrix} \tag{5.34}$$

and

$$a(j) = \frac{G_{aP}^2 G_{bP}^2}{j - f_P} + \frac{G_{aM}^2 G_{bM}^2}{j - f_M}, \tag{5.35}$$

where the first term is related to the elastic and the second one to the inelastic cross section. In general we can write for the total cross section:

$$\sigma_{ab} \sim \gamma_a^+ \gamma_b^+ e^{(j_+ - 1) Y} + \gamma_a^- \gamma_b^- e^{(j_- - 1) Y}. \tag{5.36}$$

We may ask how often a pomeron is exchanged in the MP chain. To answer this question we compute the average number of MM and PM vertices in the chain and for this purpose we expand in powers of g_{PM}^2:

$$\sigma_{ab} = \sum_k \sum_i (g_{MM}^2)^i (g_{PM}^2)^k \sigma_{ab}^{ik} \tag{5.37}$$

then:

$$\bar{n}_{MM} = \sum_k \sum_i i (g_{MM}^2)^i (g_{PM}^2)^k \sigma_{ab}^{ik} / \sigma_{ab} =$$

$$= g_{MM}^2 \frac{\partial \sigma_{ab}}{\partial g_{MM}^2} \bigg/ \sigma_{ab} = Y g_{MM}^2 \frac{\partial j_+}{\partial g_{MM}^2} + \text{constant}. \tag{5.38}$$

and

$$\bar{n}_{PM} = g_{PM}^2 \frac{\partial \sigma_{ab}}{\partial g_{PM}^2} \bigg/ \sigma_{ab} = Y g_{PM}^2 \frac{\partial j_+}{\partial g_{PM}^2} + \text{constant} \tag{5.39}$$

Now

$$\frac{\partial j_+}{\partial g_{MM}^2} = \frac{1}{2}\left[1 + \frac{f_M - f_P}{\sqrt{(f_M - f_P)^2 + 4 g_{MP}^4}}\right] \tag{5.40}$$

$$\frac{\partial j_+}{\partial g_{PM}^2} = \frac{1}{2} \frac{4 g_{PM}^2}{\sqrt{(f_M - f_P)^2 + 4 g_{PM}^4}} \tag{5.41}$$

Let us define $\alpha_P = 1 - \varepsilon$ and set $\varepsilon = 0.05$, $j_+ = 1$ and $j_- = 0.5$. We therefore find $f_P = 0.9$, $f_M = 0.6$ and $g_{PM}^2 = 0.2$. If we choose as in the one-channel case $\alpha_M = 0.25$ we get $g_{MM}^2 = 1.1$. Therefore:

$$\begin{aligned} \bar{n}_{MM} &= 0.22\, Y + \text{constant} \\ \bar{n}_{PM} &= 0.16\, Y + \text{constant} \\ \bar{n} = n_{MM} + n_{PM} &= 0.38\, Y + \text{constant}. \end{aligned} \tag{5.42}$$

This value is too low and this probably indicates that the iterated pomeron exchange is much less important than it appears in this model, and that other dynamical mechanisms should be included. Keeping this in mind, we complete the description of the model because it is sufficiently simple and instructive. We first note that $\bar{n}_{MM}/\bar{n}_{PM} \approx_{y \to \infty} 1.5$. Thus, we may say that, with this choice of parameters, typical components of the MP chain are of the type shown in Figure 17.

Self-consistency is not too satisfactory, we have an input P at 0.95, an output P at 1, an input M at 0.25 and an output M at 0.5. The reason for the need of a lower value for α_M input than for α_M output may be that π exchange is important in the input whereas in the output it is negligible compared to contributions of poles at $\alpha \approx 0.5$. In order to determine two secondary trajectories by self-consistency requirements, one should work out a three-channel model.

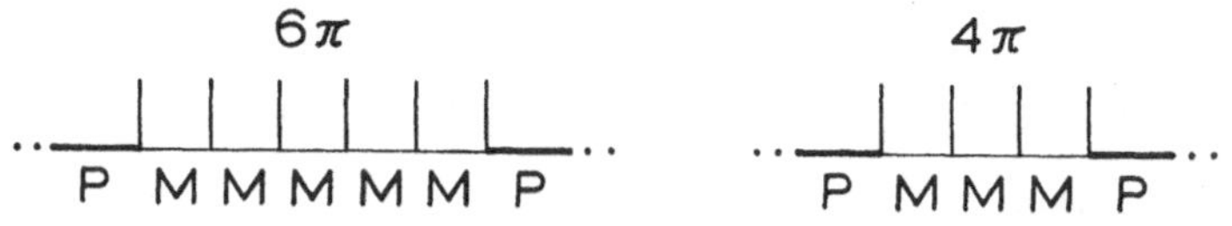

Fig. 17.

In summary, the features of the two-channel model are:

(1) As we increase the energy partial cross sections σ_{ab}^N grow, reach a maximum, decrease and flatten out, with an almost constant asymptotic behaviour.

(2) The total cross section is constant.

(3) The average multiplicity is asymptotically linear in log s, but shows some curvature at accelerator energies.

(4) The distribution of multiplicities is not Poisson any more.

We can use the many-channel formalism to perform a comparison between the pp and p$\bar{\text{p}}$ case [10]. In the pp case, we have up to now neglected the N$\bar{\text{N}}$ and $\bar{\text{N}}$N channels, which give rise to diagrams with N$\bar{\text{N}}$ loops. We based this in the fact that mostly pions are produced at high energies. Thus, we have a three-channel problem (MM, N$\bar{\text{N}}$ and $\bar{\text{N}}$N) of which only one channel is important, and in which the coupling between channels is weak. In spite of this the N$\bar{\text{N}}$ channel is important in the p$\bar{\text{p}}$ case, because it couples strongly to the external particles. This coupling is possible because the π is a bound state of p$\bar{\text{p}}$. Iteration of this diagram gives rise to the p$\bar{\text{p}}$ annihilation cross section. Thus we find:

$$\sigma_{\text{p}\bar{\text{p}}}^{\text{TOT}} = \sigma_{\text{pp}}^{\text{TOT}} + \sigma_{\text{p}\bar{\text{p}}}^{\text{Annih.}} \tag{5.43}$$

$\sigma_{\text{p}\bar{\text{p}}}^{\text{Annih}}$ goes to zero at $Y \to \infty$, and the parameters in the model can be adjusted to reproduce this effect. Thus, we find a Regge secondary contribution to $\sigma_{\text{p}\bar{\text{p}}}^{\text{TOT}}$ which arises because of the presence of a low-energy bound state in this channel. This is in agreement with the Harari-Freund hypothesis [11], but it should not be pushed too far. We understand that, due to the presence of a p$\bar{\text{p}}$ bound state, there is an additional contribution in this non-exotic channel. We do not understand why Regge secondaries are absent or cancel each other in exotic channels.

5.4. ABSFT MODEL [1]

The integral equation approach has been pushed to a high degree of sophistication. For instance, detailed calculations in the spirit of the original ABSFT model have been done [12]. In these calculations the kernel that is iterated is just the elastic $\pi\pi$ cross section. Its strength is not an adjustable parameter any more: the elastic $\pi\pi$ cross section is approximately known, and the ϱ contribution to it has a well-known width and cannot exceed the unitarity bound. As a consequence, these calculations do not succeed in building up a realistic total cross section. The reason may be that there is more than pion exchange and $\pi\pi$ elastic scattering in the guts of the inelastic cross section at high energy.

In general, we may say that the factorization property of the MP amplitude makes it possible to write an integral equation and that this generates a leading singularity in J which implies Regge behaviour and logarithmic multiplicity.

5.5. Shadow Scattering

Using the unitarity relation, we may calculate the imaginary part of the elastic scattering amplitude also away from the forward direction if we have a model of multiparticle production,

$$\operatorname{Im} A_{fi}(s,t) = \tfrac{1}{2}\sum_N \int \mathrm{d}R_N A^*_{Nf} A_{Ni} = \sum_N \operatorname{Im} A^N_{fi}(s,t) \tag{5.44}$$

where the $N!$ in the denominator has been cancelled by the contribution of $N!$ different diagrams. We see that for $f \neq i$, the integrand in Equation (5.44) is not positive definite, and that the result is sensitive to the phase of A_{Ni}, which is not the case for the forward elastic amplitude or, equivalently, for the multiparticle cross section.

The importance of the phases was shown quite clearly in some Monte Carlo calculations of Im A_{fi} (s, t) [13], in which the model of Chan Łoskiewicz and Allison [14] was used for the production amplitude. This model has multi-Regge behaviour in the multi-Regge limit, and approaches phase space at lower values of the subenergies. It contains several parameters that have been adjusted to fit particle production data.

When used to calculate Im $A^N_{fi}(s,t)$ at $t \neq 0$, if no phases are included in the model, the contributions of the various multiplicities to the diffraction peak are much too wide: the model clearly does not work. But the simple inclusion of Regge phases of the type $\exp(-i\pi\alpha(t)/2)$ makes a dramatic difference and sharpens the diffraction peak considerably (maybe too much). The results of these calculations are shown in Figure 18 taken from ref. [13].

If we analyze the behaviour of these contributions we observe the following fact [15]: the larger N, the sharper is the t dependence of Im $A^N_{fi}(s,t)$ at fixed s. This implies that if we expand Im $A^N_{fi}(s,t)$ in partial waves, we find a larger average value of J the higher is N. In the forward direction this expansion is just equivalent to the angular momentum expansion of the cross section for the process in which N particles are produced. Thus the model predicts that at a given energy, the higher the angular momentum, i.e., the higher the impact parameter, the higher the *average* number of particles produced. Of course, as the impact parameter becomes large the cross section decreases and the number of particles decreases, only the average number of produced particles increases. Some people find this pictures anti-intuitive: a frontal collision should produce, in average, more fragments than a glancing collision. It is not clear how much we should trust this kind of intuition, but we would like to make a couple of comments.

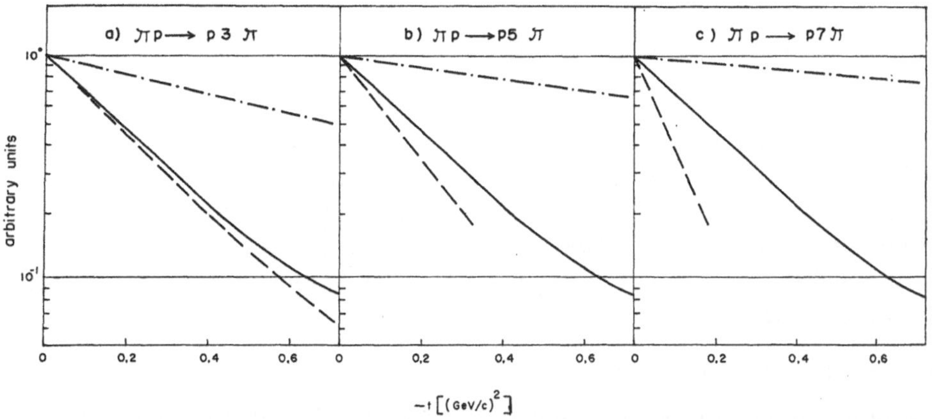

Fig. 18. Imaginary part of elastic-scattering amplitude calculated as the shadow of the three 8 GeV/c $\pi+p$ inelastic channels in the multi-Regge exchange model without phases (—·—·—) and with Regge phases (— — —). For comparison the amplitude from the measurement of elastic scattering is shown with full lines.

Let us assume that the width in t of Im $A^N(s, t)$ is energy independent. As we increase the energy, higher values of N become more important and Im $A(s, t)$ is going to shrink if higher values of N are associated with sharper t-dependence. Such a shrinking effect has been found up to ISR energies [16]. The picture is somewhat obscured by the fact that quite likely, Im $A^N(s, t)$ also shrinks, but probably this shrinkage is not strong enough. Indeed, let us consider the case in which this shrinkage occurs, because the subenergy associated to a pomeron exchange increases. Thus, it originates from an integral of the type (in the linear approximation $\alpha_p(t) = \alpha_p(0) + \alpha'_p t$

$$I(s) = \int s^{\alpha'_p(t_1+t_2)} \, dt_1 \, dt_2 \,, \tag{5.45}$$

where $t_1 = Q_1^2$, $t_2 = Q_2^2$, $t = Q^2 = (Q_1 + Q_2)^2$ (see Figure 19). Thus, $I(s)$ behaves like

$$I(s) \sim s^{\alpha'_p(t_1+t_2)_{\text{MIN}}} \,, \tag{5.46}$$

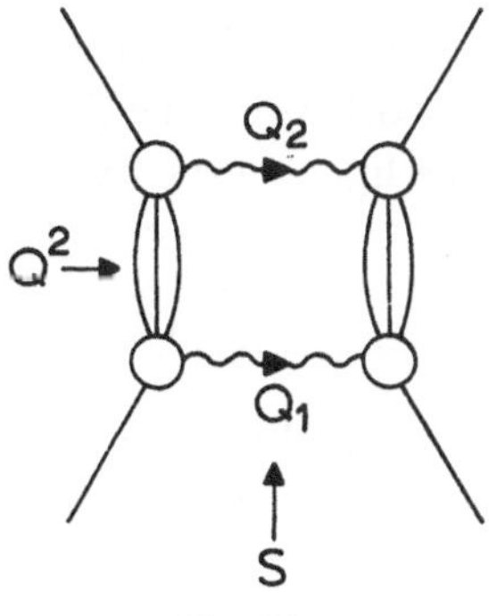

Fig. 19.

and

$$(t_1 + t_2)_{\text{MIN}} = (Q/2)^2 + (Q/2)^2 = t/2$$
$$I(s) = s^{\alpha'_{\text{P}}(t/2)} = e^{(\alpha'_{\text{P}} t/2) \ln s} \tag{5.47}$$

We call $\alpha'_{\text{p}}/2$ the rate of logarithmic shrinkage, and we conclude that the rate of shrinkage of Im $A^N(s, t)$ is $\frac{1}{2}$ the value associated to a pure pomeron pole, i.e., $\frac{1}{2}$ smaller than in the case of Im $A(s, t)$. A similar argument can be given for the case of more than one pomeron. Therefore, we conclude that an additional shrinkage is needed and this may very well originate from the 'anti-intuitive' picture in the way described above.

5.6. Impact Parameter Picture

I want to reproduce here an argument due to Gribov [17] that throws some light on the connection between the impact parameter and the multiplicity. We introduce N two-dimensional impact parameter vectors $\mathbf{b}_i$ which are the Fourier variables conjugate to $\mathbf{p}_i^{\perp}$. We neglect for simplicity the constraint $\mathbf{Q}_N^{\perp} = \sum_{i=1}^{N} \mathbf{p}_i^{\perp} = 0$ and we recall that the expression for $t_{i,i+1}$ contains a term $-(\mathbf{Q}_i^{\perp})^2$. Thus if we perform the transformation to the impact parameter space, we may consider an integral of the type

$$\int e^{i(\mathbf{p}_1^{\perp}\cdot\mathbf{b}_1 + \cdots + \mathbf{p}_N^{\perp}\cdot\mathbf{b}_N)} a((\mathbf{Q}_1^{\perp})^2) \cdots a((\mathbf{Q}_{N-1}^{\perp})^2) =$$
$$= \int e^{i[\mathbf{Q}_1^{\perp}\cdot(\mathbf{b}_1 - \mathbf{b}_2) + \mathbf{Q}_2^{\perp}\cdot(\mathbf{b}_2 - \mathbf{b}_3) + \cdots]} a((\mathbf{Q}_1^{\perp})^2)\, a((\mathbf{Q}_2^{\perp})^2) \ldots \equiv \tilde{a}(b_{12})\, \tilde{a}(b_{23}) \cdots,$$

where $\mathbf{b}_{ij} = \mathbf{b}_i - \mathbf{b}_j$ is the relative impact parameter associated to particles i and j. The usual exponential dependence on $t_{i,i+1}$ leads to a gaussian in $|\mathbf{Q}_i^{\perp}|$ and in the Fourier transform variable $b_{i,i+1}$. Thus, to each step in the MP chain we can associate a step in the impact parameter plane, of limited size. After n such steps we expect to have travelled a distance in impact parameter space proportional to $\sqrt{n}$. In this fashion larger multiplicities are associated in the MP model with larger values of the impact parameter.

6. Inclusive Processes

6.1. Definitions

The difficulties involved in both the experimental and theoretical study of multiparticle processes have led the workers in the field to concentrate on simpler experiments, which still throw some light on the dynamics of strong interactions.

The so-called inclusive processes are of the type [18]

$$a + b \to c_1 + \cdots + c_n + \text{anything else} \tag{6.1}$$

in which some given features of the final state are specified, and all processes sharing these features are included. The simplest case is that of the single particle distributions of the type

$$a + b \to c + \text{anything} \tag{6.2}$$

which we describe in terms of the 'spectrum'

$$\varrho_{ab}^{c}(p^{l}, p^{\perp}, s) = \frac{\mathrm{d}n_{ab}^{c}}{\dfrac{\mathrm{d}^3 p_c}{E_c}} = \frac{E_c}{\sigma_{ab}} \frac{\mathrm{d}\sigma_{ab}^{c}}{\mathrm{d}^3 p_c}. \tag{6.3}$$

Because when n particles of kind c are produced in an event each one is counted separately, we have

$$\int \varrho_{ab}^{c} \frac{\mathrm{d}^3 p_c}{E_c} = \bar{n}_{ab}^{c}(S) \tag{6.4}$$

We define the c.m. inelasticity η_{ab}^{c} as the fraction of the incident c.m. energy that is carried by particles of kind c in the final state. Thus

$$\eta_{ab}^{c} = \frac{1}{\sqrt{s}} \int \varrho_{ab}^{c} \, \mathrm{d}^3 p. \tag{6.5}$$

Energy conservation requires

$$\sum_{c} \eta_{ab}^{c} = 1. \tag{6.6}$$

6.2. Single particle distribution in the MP model [5, 19]

How do we calculate inclusive spectra in the MP model? Let us go back to our example in which only pions are produced at the internal vertices, and calculate the pion spectrum. For simplicity we ignore transverse momenta. For the distribution of the ith pion

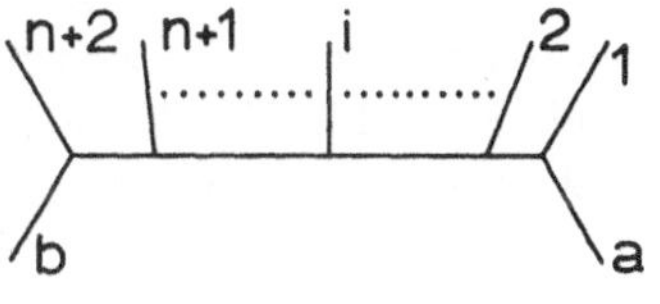

Fig. 20.

when n such particles are produced we have (see Figure 20)

$$\frac{\mathrm{d}\sigma_{ab}^{n,i}}{\mathrm{d}y} = \int \mathrm{d}\sigma^{n} \delta(y^{i} - y), \tag{6.7}$$

where

$$\mathrm{d}\sigma_{ab}^{n} = e^{-Y} G_{a}^{Y2} K_a(y_a - y_2)\, g^2 K(y_2 - y_3)\, g^2 \cdots g^2 K_b(y_{n+1} - y_b). \tag{6.8}$$

Here we have allowed the kernels adjacent to external particles to depend on the nature of these particles. Thus, remembering that

$$B_{ab}(y_a - y_b) = K_{ab}(y_a - y_b) + \sum_{n=1}^{\infty} \int K_a(y_a - y_2)\, g^2 \cdots \\ \cdots g^2 K_b(y_b - y_{n+1})\, \mathrm{d}y_2 \cdots \mathrm{d}y_{n+1} \tag{6.9}$$

we can write setting $i=i_r$, $i_l=n+3-i$

$$\frac{\mathrm{d}\sigma_{ab}^{\pi}}{\mathrm{d}y}=\sum_{n=1}^{\infty}\sum_{i=2}^{n+1}\frac{\mathrm{d}\sigma_{ab}^{n,i}}{\mathrm{d}y}=e^{-Y}G_a^2B_a(y_a-y)\,g^2B_b(y-y_b)\,g_b^2 \tag{6.10}$$

$$\bar{\varrho}_{ab}^{\pi}\equiv\frac{1}{\sigma_{ab}}\frac{\mathrm{d}\sigma_{ab}^{\pi}}{\mathrm{d}y}=\frac{B_{a\pi}(y_a-y)\,g^2B_{b\pi}(y-y_b)}{B_{ab}(y_a-y_b)}. \tag{6.11}$$

In the crudest model in which $B(y)=\theta(y)\,e^{jy}$ with θ a step function we have

$$\bar{\varrho}_{ab}^{\pi}(y, y_a, y_b)=\theta(y_a-y)\,g^2\theta(y-y_b) \tag{6.12}$$

i.e., a rectangular spectrum. In more refined versions $B_{ik}(y)$ has additional y-dependence at lower values of y, but still behaves like $C_i\,C_k\,e^{jy}$ as $y\to\infty$. Thus, we can distinguish three limits:

(1) fragmentation of particle a: in this limit we keep y_a-y fixed and let $y-y_b$ (and therefore, y_a-y_b) grow. We find

$$\bar{\varrho}_{ab}^{\pi}(y, y_a, y_b)=B_{a\pi}(y_a-y)\,g^2\frac{C_\pi}{C_a}e^{j(y-y_a)}=f_a^{\pi}(y_a-y) \tag{6.13}$$

This property is called limiting fragmentation of particle a into c [19].

(2) fragmentation of particle b, which is analogous to case 1 with $a\leftrightarrow b$.

(3) pionization. In this limit we let y_a-y and $y-y_b$ grow. We find

$$\bar{\varrho}_{ab}^{\pi}=g^2C_\pi^2 \tag{6.14}$$

In a graphical form, this behaviour is shown in Figure 21 in which we have chosen $y_b=0$. As y_a increases the curve stretches out and develops a plateau. This behaviour has been observed at the CERN ISR [21]. We should remark that the height of the plateau is just the coefficient of $Y=y_b-y_a$ in the expression for the pionic average multiplicity.

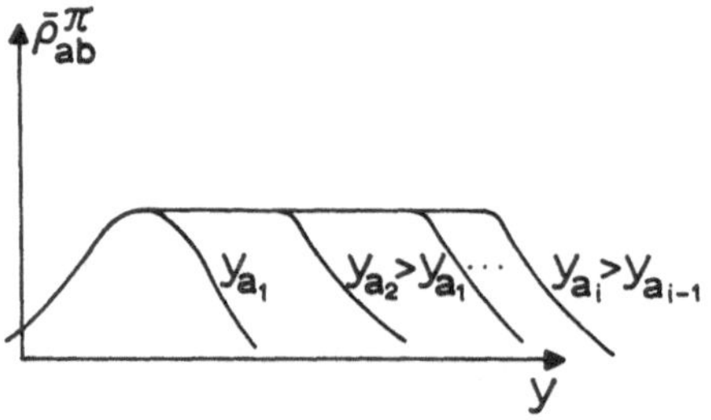

Fig. 21.

6.3. Scaling

The above properties are most conveniently summarized using the variable $x=2p_{cm}^l/\sqrt{S}$ introduced by Feynman [18]. Clearly, the a-fragmentation limit corresponds to fixed $x>0$ and $S\to\infty$, the b fragmentation limit to fixed $x<0$ and $S\to\infty$, and the pionization limit, to $x=0$ and $S\to\infty$. Thus, we write

$$\varrho_{ab}^{c}(p^{l},p^{\perp},S)\underset{S\to\infty}{\longrightarrow}\begin{cases}\varrho_{a}^{c}(x,p^{\perp}) & x>0\\ \varrho^{c}(p^{\perp}) & x=0\\ \varrho_{b}^{c}(x,p^{\perp}) & x<0\end{cases} \tag{6.15}$$

and we call this the scaling hypothesis for inclusive hadronic processes. Observe that

$$\eta_{ab}^{c}=\frac{1}{\sqrt{S}}\int\varrho\mathrm{d}^{3}p^{c}=\tfrac{1}{2}^{c}\int_{-1}^{1}\varrho\mathrm{d}^{2}p_{\perp}\,\mathrm{d}x=\text{constant}. \tag{6.16}$$

6.4. End of Spectrum

We may investigate the behaviour of $f(x)$ as x approaches the kinematic boundary, i.e., $x\to\pm1$. Clearly, the spectrum will be dominated by the diagram in which the observed particle is emitted at positions either 1 or N in the chain. Thus, we need only consider the diagram of Figure 22. We write the Regge propagator in the form

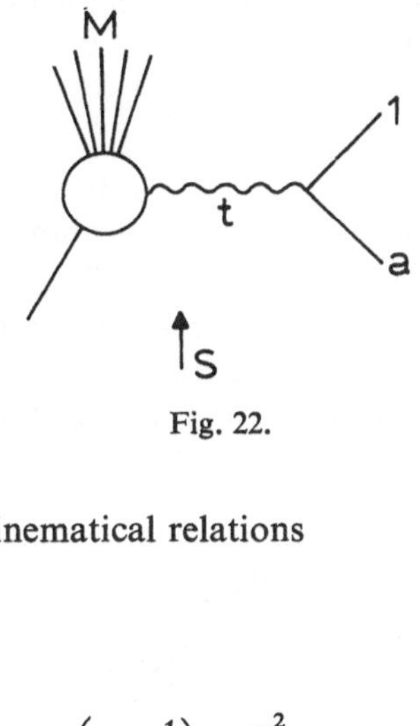

Fig. 22.

$(S/M^{2})^{2\alpha(t)}$ and using the kinematical relations

$$\frac{S}{M^{2}}\simeq\frac{1}{1-x} \tag{6.17}$$

$$t=M_{a}^{2}(1-x)+M_{1}^{2}\left(1-\frac{1}{x}\right)-\frac{p_{\perp}^{2}}{x} \tag{6.18}$$

we obtain, if we approach the kinematic boundary at constant t [18, 22]

$$\varrho_{ab}^{c}\sim\beta_{a}^{c}(t)\left(\frac{1}{1-x}\right)^{2\alpha(t)-1}. \tag{6.19}$$

Thus, the behaviour of ϱ near the boundary is controlled by the value of the highest Regge trajectory which couples to the incident particle a and the produced particle 1.

We observe that the double limit $x\to1$, $S\to\infty$ is such that both M^{2} and $S/M^{2}\to\infty$.

Therefore, reggeization of the wavy line in Figure 23 is justified. Furthermore, it is kinematically justified to use the Regge behaviour for the 'cross section' reggeon-particle → anything. Pictorially we represent this cross section near the kinematic boundary by the diagram of Figure 23 which defines the three reggeon coupling [23].

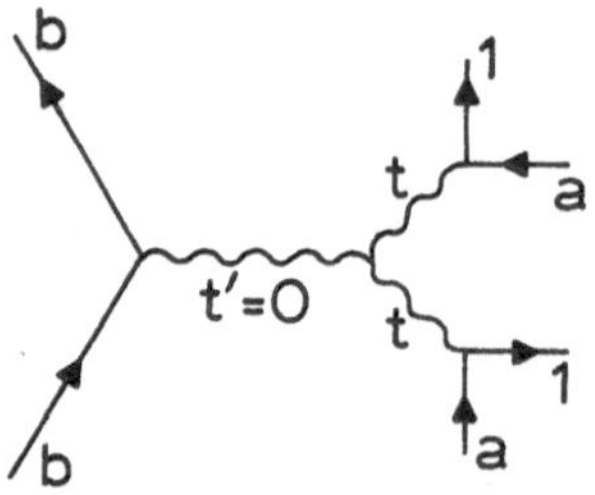

Fig. 23.

In the case in which particles 1 and a have the same quantum numbers, and the process can go via a pomeron exchange at that vertex, we are dealing with diffraction dissociation of particle b into large masses, – and we encounter the three-pomeron vertex. This vertex has to be zero if the pomeron goes through one with a nonvanishing residue, as we have already discussed.

6.5. Generalized Optical Theorem

We can write the optical theorem for the forward three-body elastic scattering amplitude. It just relates the absorptive part of this amplitude to the total cross section for three-body into anything. But if we analytically continue this expression, we can relate a discontinuity of the forward three-body amplitude to an inclusive cross section. This is shown in diagrams in Figure 24.

Mueller [24] showed that with the assumption of Regge behaviour with factorized

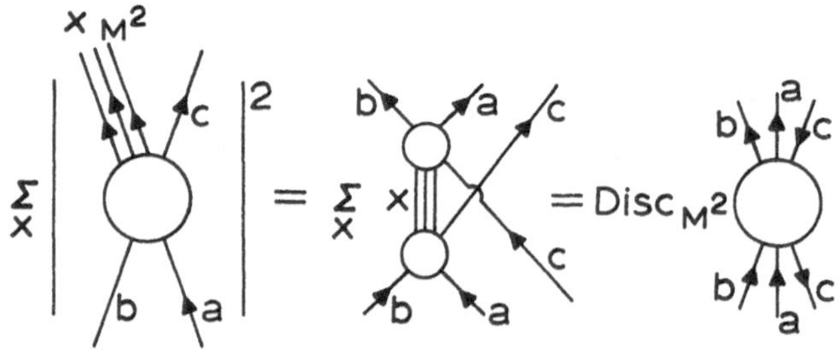

Fig. 24.

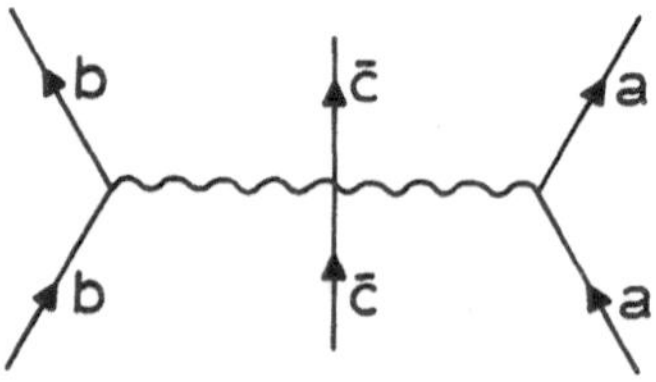

Fig. 25.

residues for this discontinuity we obtain the full content of scaling. For instance, the pionization limit is dominated by the diagram in Figure 25. Here we are making use of the fact that a Regge pole dominates not only at fixed t and $S_{a\bar{c}}=(p_a+p_{\bar{c}})^2\to\infty$ butalso as $S_{a\bar{c}}\to-\infty$. In the inclusive processes within the pionization region we indeed have $(p_a+p_{\bar{c}})^2=(p_a-p_c)^2\to-\infty$ and $(p_b-p_c)^2\to-\infty$.

6.6. A COUNTEREXAMPLE

In order to clarify the meaning of scaling, I believe that it is instructive to examine a model that does not scale. One such model is the CKP model [25], which has been used for many years. That model was based on three basic features

(1) limitation of $p^{\perp}$,

(2) constant inelasticity, and

(3) slowly increasing average multiplicity.

In it the inclusive pion cross section is

$$\frac{\mathrm{d}\sigma}{\mathrm{d}p^l\mathrm{d}^2p^{\perp}}\simeq Ae^{-Bp_l^{cm}/s^{1/4}}e^{-Cp_{\perp}}. \tag{6.20}$$

We find

$$\bar{n}^{\pi}\sim S^{1/4}\int e^{-Bu}\,du\sim S^{1/4} \tag{6.21}$$

$$\eta^{\pi}\sim\int ue^{-Bu}\,du=\text{constant}. \tag{6.22}$$

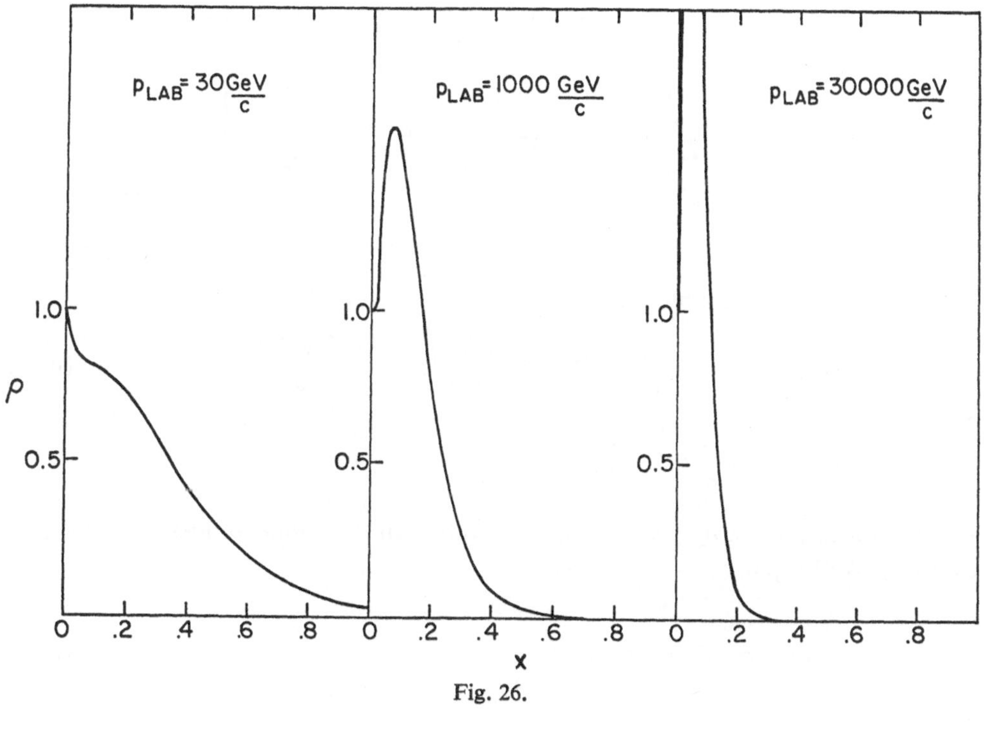

Fig. 26.

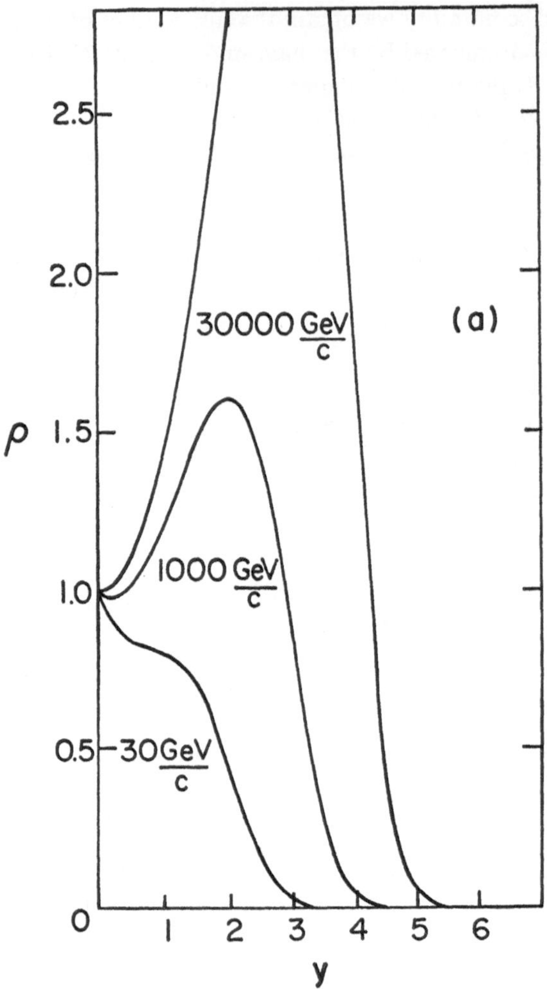

Fig. 27.

In terms of the x variable we have

$$\varrho(x, p_\perp) = \frac{1}{\sigma} A e^{-B(\sqrt{s}x)/2} e^{-cp^\perp} \frac{\sqrt{S}}{2} \sqrt{x^2 + \frac{p^{\perp 2} + M^2}{S/4}}. \tag{6.23}$$

Thus

$$\varrho(0, p_\perp) = \frac{A}{\sigma} e^{-cp^\perp} \sqrt{p_\perp^2 + M^2}$$

whereas for any $x \neq 0$, $\varrho(x, p_\perp) \to_{s\to\infty} 0$. Therefore in this model ϱ does not scale (except at $x=0$) in spite of the fact that

$$\int_{-1}^{1} \varrho(x, p_\perp)\, \mathrm{d}x = \text{constant} \tag{6.24}$$

Figures 26 and 27 show some examples of ϱ in this model as functions of x and y at various incident incident energies. Recent experiments at ISR [21] have definitely ruled out this kind of behaviour and as we have already mentioned, they are in good agreement with the scaling prediction.

7. Conclusion

The aim of these lectures has been to show a model in, perhaps, an oversimplified fashion, but such that can give some quick answer to questions that arise in the complex field of multiparticle production. The model has some basic limitations, and several of its features are probably not quite correct. For instance, the interplay of poles and cuts in the angular momentum plane, and the full content of unitarity are difficult to incorporate into the model. However, the MP model provides a guideline: when we are aware of its predictions we can ask more subtle questions and gain deeper insight in the study of hadronic processes at high energy.

References

1. Bertocchi, L., Fubini, S., and Tonin, M.: *Nuovo Cim.* **25**, 626 (1962); and Amati, D., Stanghellini, A., and Fubini, S.: *Nuovo Cim.* **26**, 896 (1962).
2. Werle, J.: *Relativistic Theory of Reactions*, Wiley, New York, 1966.
3. Toller, M.: *Nuovo Cim.* **37**, 631 (1965).
4. Bali, N., Chew, G., and Pignotti, A.: *Phys. Rev.* **163**, 1572 (1967).
5. De Tar, C.: *Phys. Rev.* **D3**, 128 (1971).
6. Chew, G. and Pignotti, A.: *Phys. Rev.* **176**, 2112 (1968).
7. Caneschi, L. and Pignotti, A.: *Phys. Rev.* **180**, 1525 (1969).
8. Chew, G., Goldberger, M., and Low, F.: *Phys. Rev. Letters* **22**, 208 (1969).
9. Chew, G. and Snider, D.: *Phys. Rev.* **D1**, 3453 (1970).
10. Ting, P.: *Phys. Rev.* **181**, 1942 (1969).
11. Freund, P.: *Phys. Rev. Letters* **20**, 235 (1968); and Harari, H.: *Phys. Rev. Letters* **20**, 1385 (1968).
12. Tow, D.: *Phys. Rev.* **D2**, 154 (1970).
13. Michejola, L., Turnau, J., and Biatas, A.: *Nuovo Cim.* **56A** 241 (1968).
14. Chan, H. M., Łoskiewicz, J., and Allison, W.: *Nuovo Cim.* **57A**, 93 (1968).
15. Michejda, L.: *Fortschr. Physik* **16**, 707 (1968).
16. Barbiellini, G. *et al. Phys. Letters* **39B**, 663 (1972).
17. Gribov, V.: *Yadern. Fiz.* **9**, 640 (1969) (*Soviet J. Nucl. Phys.* **9**, 369 (1969)).
18. Feynman, R.: *Phys. Rev. Letters* **23**, 1415 (1969); in *High. Energy Collisions*, Gordon and Breach, New York, 1969, p. 237.
19. Ripa, P.: *Nucl. Phys.* **B45**, 217 (1972).
20. Benecke, J., Chou, T. T., Yang, C. N., and Yen, E.: *Phys. Rev.* **188**, 2159 (1969).
21. Sens, J. C.: Invited paper presented at the Fourth International Conference on High Energy Collisions, Oxford, April 1972.
22. Caneschi, L. and Pignotti, A.: *Phys. Rev. Letters* **22**, 1219 (1969).
23. De Tar, C., Jones, C., Low, F., Weis, J., Young, J., and Tan, C. I.: *Phys. Rev. Letters* **26**, 675 (1971).
24. Mueller, A.: *Phys. Rev.* **D2**, 2963 (1970).
25. Cocconi, G., Koester, L. J., and Perkins, H. D.: UCRL-10022, p. 167 (1961), unpublished; and Cocconi, G.: CERN preprint, Fieb. 1971.

PART C

NUCLEAR PHYSICS

THREE NUCLEON SYSTEMS

EDWARD HARMS
Dept. of Physics, Fairfield University, Conn., U.S.A.

1. Introduction

The subject of these lectures is the study of three-body problem with emphasis upon the three-nucleon system. The lectures are divided into three sections. In Section 2 we examine two-body system introducing our notation and obtaining results which will be used in the later sections. Section 3 develops the more formal aspects of the three-body problem, introducing the Faddeev equations and developing the formalism used for calculating three-body scattering processes. In Section 4 the three-nucleon problem is discussed. This section is subdivided into one section on the bound state system and another on the scattering states. Very good discussion of work through 1969 can be found in reference [1] to [5].

I would like to express my thanks to Dr L. Laroze for his assistance in the preparation of these notes.

2. Two-Body Problem

We begin our study of the three-body problem with a review of the two-body problem. Our starting point is the Schroedinger equation

$$(E - K - V)\,|\psi\rangle = 0. \tag{2.1}$$

A traditional way of handling Equation (2.1) in practical problems has been to use the coordinate representation and solve the resulting differential equation for $\psi(\mathbf{r})$. For example the adjusting of phenomenological nucleon-nucleon potentials is usually done in this manner. Much of the recent work in the three-body problem, however, is done using the momentum representation, and solving integral equations for transition operators rather than wave functions. This is the form in which we will study the two-body problem. We first rewrite Equation (2.1) as an integral equation in the form

$$|\psi_{\mathbf{k},\alpha}\rangle = |\mathbf{k},\alpha\rangle + G_0(E + i\varepsilon)\,|\psi_{\mathbf{k},\alpha}\rangle. \tag{2.2}$$

Here $|\mathbf{k},\alpha\rangle$ is a plane wave state of momentum $\mathbf{k}$ satisfying $(E-K)\,|\mathbf{k},\alpha\rangle=0$ and $\langle\mathbf{k}',\alpha\mid\mathbf{k},\alpha\rangle=\delta_{\alpha\alpha'}\delta^3(\mathbf{k}-\mathbf{k}')$. The free Green's function $G_0(s)$ is given by $G_0(s)=(s-K)^{-1}$, the $i\varepsilon(\varepsilon>0)$ guarantees outgoing spherical waves. We designate spin and isospin variables by α and note that E has the value k^2.

We use units in which $\hbar=M=1$ and lengths are measured in Fermis (F). One energy unit then equals 41.47 MeV.

Strictly speaking to obtain an integral equation for Equation (2.2), we must choose a particular representation of the operators and states present. In the coordinate

Abecassis de Laredo and Jurisic (eds.), Selected Topics in Phys. Astrophys. and Biophys. 251–306.

representation, we have

$$|\psi_{\mathbf{k},\alpha}(\mathbf{r})\rangle = \frac{1}{(2\pi)^{3/2}} e^{i\mathbf{k}\cdot\mathbf{r}} |\alpha\rangle + \\ + \iint G_0^+(\mathbf{r}, \mathbf{r}'; k)\, V(\mathbf{r}', \mathbf{r}'')\, |\psi_{\mathbf{k},\alpha}(\mathbf{r}'')\rangle\, \mathrm{d}^3 r'\, \mathrm{d}^3 r'', \tag{2.2a}$$

where

$$G_0^+(\mathbf{r}, \mathbf{r}'; k) = -\frac{1}{4\pi} \frac{e^{ik|\mathbf{r}-\mathbf{r}'|}}{|\mathbf{r} - \mathbf{r}'|}$$

and $|\alpha\rangle$ is the initial spin-isospin state. The potential as written here has its most general form. In many cases of interest the potential $V(\mathbf{r}', \mathbf{r}'')$ takes on the simple form of a local or static potential

$$V(\mathbf{r}', \mathbf{r}'') = V(\mathbf{r}')\, \delta^3(\mathbf{r}' - \mathbf{r}''). \tag{2.3}$$

Another simple form for $V(\mathbf{r}', \mathbf{r}'')$ that has played an important role in the three-body problem is the separable form

$$V(\mathbf{r}', \mathbf{r}'') = \sum_n \lambda_n^{-1} g_n(\mathbf{r}')\, h_n(\mathbf{r}''). \tag{2.4}$$

As we shall see, for this choice of potential, Equation (2.2a) reduces to quadratures.

In the momentum representation Equation (2.2) takes the form

$$|\psi_{\mathbf{k},\alpha}(\mathbf{p})\rangle = \delta^3(\mathbf{p} - \mathbf{k})\, |\alpha\rangle + \frac{1}{k^2 + i\varepsilon - p^2} \int V(\mathbf{p}, \mathbf{p}')\, |\psi_{\mathbf{k},\alpha}(\mathbf{p}')\rangle\, \mathrm{d}^3 p', \tag{2.2b}$$

where the Green's function is now diagonal. The long range oscillatory behavior of the kernel of Equation (2.2a) becomes a pole at $p^2 = k^2$ in Equation (2.2b). A similar situation will develop in the three-body problem. Numerically it is usually easier to treat poles such as those of Equation (2.2b) rather than long range oscillatory behavior, due to the difficulty in integrating such functions. This will be especially true in the three-body problem.

In the two-nucleon problem, the potential takes on a very complicated form. We can quite generally write it in the coordinate representation as [6]

$$V(\mathbf{r}, \mathbf{r}') = \sum_{\alpha = JST} \mathscr{Y}_{JLS}^{M}(\Omega_{\mathbf{r}})\, V_{L,L'}^{\alpha}(r, r')\, \mathscr{Y}_{JL'S}^{M}(\Omega_{\mathbf{r}'})^{+}\, \mathbb{P}_T, \tag{2.5}$$

where the $\mathscr{Y}^{LL'M}_{\ \ 'S}$ are total angular momentum states

$$\mathscr{Y}_{JLS}^{M}(\Omega) = \sum_{\mu} \langle JM \mid L\mu, S, M - \mu\rangle\, Y_L^{\mu}(\Omega)\, |S, M - \mu\rangle \tag{2.6}$$

and $\mathbb{P}_T$ is the isospin projection operator

$$\mathbb{P}_T = \sum_I |T, I\rangle\!\rangle \langle\!\langle T, I|\,. \tag{2.7}$$

A similar form holds for $V(\mathbf{p}, \mathbf{q})$ with the substitution

$$V^{\alpha}_{L,L'}(p, q) \to V^{\alpha}_{L,L'}(r', r). \tag{2.8}$$

The relationship between the two is

$$V^{\alpha}_{L,L'}(p, q) = \frac{2i^{L-L'}}{\pi} \iint j_L(pr)\, r^2\, \mathrm{d}r V^{\alpha}_{L,L'}(r, r')\, r'^2\, \mathrm{d}r' j_{L'}(qr'). \tag{2.9}$$

An advantage of the integral Equation (2.2) over the Schroedinger equation is that the boundary conditions on ψ (incident plane wave of momentum $\mathbf{k}$, outgoing scattered wave) are already contained in Equation (2.2) while they must be imposed as subsidiary conditions if Equation (2.1) is solved in coordinate space.

If E is negative we obtain the homogeneous version of Equation (2.2)

$$|\psi_B\rangle = G_0(E_B)\, V\, |\psi_B\rangle. \tag{2.10}$$

This has as its solutions the bound state eigenfunctions and energies of the potential V.

We now introduce the idea of a *transition operator* by looking for a solution of Equation (2.2) of the form (we drop the labels $\mathbf{k}$ and α unless necessary).

$$|\psi\rangle = |\mathbf{k}\rangle + G_0(E + i\varepsilon)\, T(E + i\varepsilon)\, |\mathbf{k}\rangle. \tag{2.11}$$

Putting this expression in Equation (2.2), we find that $T(s)$ satisfies the operator equation

$$T(s) = V + VG_0(s)\, T(s) \tag{2.12}$$

called the Lippmann-Schwinger equation.

This two-particle T-matrix will play an important role in what follows. The most general form of $T(s)$ is given by Equation (2.12) in which s enters parametrically and may take on any complex value. If T appears in the form $T(k^2+i\varepsilon)\,|\mathbf{k}, \alpha\rangle$, we then refer to T as being half-off-the-energy-shell. We see from Equation (2.4) that the wave function is determined by the half-off-shell T-matrix. The form $\langle \mathbf{k}', \alpha'|\, T(k^2+i\varepsilon)\, |\mathbf{k}, \alpha\rangle$ with $k'^2 = k^2$ is called the fully-on-shell T-matrix. The two-particle differential scattering cross section is given in terms of the on-shell T-matrix by

$$\frac{\mathrm{d}T_{\alpha\alpha'}(\mathbf{k}_f, \mathbf{k}_i)}{\mathrm{d}\Omega} = 4\pi^4\, |\langle \mathbf{k}_f \alpha|\, T(k^2 + i\varepsilon)\, |\mathbf{k}_i \alpha'\rangle|^2 \tag{2.13}$$

with

$$E = k^2 = k_i^2 = k_f^2.$$

Thus elastic scattering experiments give us information about the on-shell T-matrix only.

If the potential has the form of Equation (2.5), a similar form holds also for T with the replacement

$$T^{\alpha}_{L,L'}(p, q, s) \to V^{\alpha}_{L,L'}(r, r'). \tag{2.14}$$

The T-matrix elements are obtained from the partial wave Lippmann-Schwinger equation

$$T^{\alpha}_{L,L'}(p,q,s) = V^{\alpha}_{L,L'}(p,q) + \sum_{L''} \int V^{\alpha}_{L,L''}(p,k') \frac{k'^2\,\mathrm{d}k'}{s-k'^2} T^{\alpha}_{L'',L'}(k',q'\,s). \tag{2.15}$$

It is in this form that one usually solves for T.

Since numerical methods play an important role in the three-body problem, we will illustrate how we would solve Equation (2.15) numerically. A common method for solving integral equations such as (2.15) is to replace the integration by a numerical quadrature formula such that

$$\int_0^{\infty} f(p)\,\mathrm{d}p \approx \sum_{i=1}^{N} f(p_i)\,w_i,$$

where the p_i and w_i are the quadrature points and weights. Neglecting possible couplings between different L, we then consider Equation (2.15) with p and q taking on the values of the quadrature points. We obtain a matrix equation for the value of T at these points of the form

$$T(p_i,p_j,s) = V(p_i,p_j) + \sum_k \frac{V(p_i,p_k)\,p_k^2 w_k T(p_k,p_j,s)}{s-p_k^2}$$

or, in an abbreviated form $\mathsf{T}=\mathsf{V}+\mathsf{V}\mathsf{G}_0\mathsf{T}$. As long as s is not real and positive (in particular if s is negative), we can solve this system of equations on a computer with little difficulty using standard matrix inversion procedures.

If s is positive, however, a pole appears in the kernel of Equation (2.15). To solve for T in this case we can rely, for example, upon a method developed by Kowalski [7] and Noyes [8].

We consider Equation (2.15) for a central potential. Setting for the moment $s=k^2+i\varepsilon$ and $q=k$ we must solve the integral equation

$$T(p,k,k^2+i\varepsilon) = V(p,k) + \int V(p,k') \frac{k'^2\,\mathrm{d}k'}{k^2+i\varepsilon-k'^2} T(k',k;k^2+i\varepsilon). \tag{2.15a}$$

We now set $p=k$, multiply the equation by $V(p,k)/V(k,k)$, and subtract the result from the above equation. After arranging terms, and defining $T(k)=T(k,k;k^2+i\varepsilon)$, we find

$$T(p,k;k^2+i\varepsilon) = \frac{V(p,k)}{V(k,k)} T(k) + \\ + \int \Lambda(p,k';k) \frac{k'^2\,\mathrm{d}k'}{k^2+i\varepsilon-k'^2} T(k',k;k^2+i\varepsilon)$$

with

$$\Lambda(p,q;k) = V(p,q) - \frac{1}{V(k,k)} V(p,k)\,V(k,q).$$

Dividing by $T(k)$ gives

$$\frac{T(p,k;k^2+i\varepsilon)}{T(k)} \equiv f(p,k) = \frac{V(p,k)}{V(k,k)} + \int \frac{\Lambda(p,k';k)\,k'^2\,\mathrm{d}k' f(k',k)}{k^2-k'^2}. \tag{2.16}$$

where we have defined the half-off-shell function $f(p,k)$. The new kernel has no singularity at $k'=k$ since Λ vanishes here and the equation is well suited for numerical treatments using the quadrature method. Having solved for the half-off-shell function, we may obtain the on-shell T-matrix from (see Equation (2.15a))

$$T(k) = V(k,k)\left\{1 + \int \frac{V(k,k')\,k'^2\,\mathrm{d}k' f(k',k)}{k^2+i\varepsilon-k'^2}\right\}^{-1} \tag{2.17}$$

and the half-off-shell T-matrix from

$$T(p,k;k^2+i\varepsilon) = f(p,k)\,T(k). \tag{2.18}$$

Performing a similar analysis on the full T-matrix we find that

$$T(p,q;k^2+i\varepsilon) = f(p,k)\,T(k)\,f(q,k) + R(p,q;k), \tag{2.19}$$

where R is a real function satisfying

$$R(p,q;k) = \Lambda(p,q;k) + \int \Lambda(p,q';k)\,\frac{q'^2\,\mathrm{d}q'}{k^2-q'^2}\,R(q',q;k) \tag{2.20}$$

or $(\mathsf{R}=\mathsf{\Lambda}+\mathsf{\Lambda}\mathsf{G}_0\mathsf{R})$ and vanishes if $p=k$ or $q=k$. Once again, the quadrature method is well suited for solving this equation. Thus we have a method for solving for T which is numerically easy to handle. We may regard the two-body problem as solvable to any desired accuracy.

We now approach the solution of Equation (2.1) from another point of view. Considerable use will be made of the identity

$$\frac{1}{A} - \frac{1}{B} = \frac{1}{B}(B-A)\frac{1}{A} = \frac{1}{A}(B-A)\frac{1}{B} \tag{2.21}$$

valid for any operators A and B which have inverses.

The full two-particle Green's function is defined as the inverse of the Schroedinger operator of Equation (2.1)

$$G(s) = (s-K-V)^{-1} = (s-H)^{-1}. \tag{2.22}$$

Setting $A^{-1}=G(s)$ and $B^{-1}=G_0(s)$ in the identity (2.21.) gives an operator equation for $G(s)$:

$$\begin{aligned} G(s) - G_0(s) &= G_0(s)\,VG(s) \\ &= G(s)\,VG_0(s). \end{aligned} \tag{2.23}$$

This operator or the essentially equivalent operator $T(s)$ provide a complete solution to the two-body problem. For example, given $G(s)$ we can construct $|\psi_{\mathbf{k}}\rangle$ from

$$|\psi_{\mathbf{k}}\rangle = \lim_{\varepsilon\to 0} \{i\varepsilon G(k^2 + i\varepsilon) |\mathbf{k}\rangle\}. \tag{2.24}$$

Proof: Multiply the first of Equation (2.23) on the left by $i\varepsilon$ and operate to the right on $|\mathbf{k}\rangle$. It is easy to show that

$$i\varepsilon G_0(k^2 + i\varepsilon) |\mathbf{k}\rangle = |\mathbf{k}\rangle \tag{2.25}$$

and that the right side of Equation (2.24) therefore satisfies Equation (2.2).

Alternatively if we perform the same operations using the second of Equation (2.23), we find another expression for $|\psi_{\mathbf{k}}\rangle$:

$$|\psi_{\mathbf{k}}\rangle = |\mathbf{k}\rangle + G(k^2 + i\varepsilon) V|\mathbf{k}\rangle \tag{2.26}$$

which along with Equation (2.11) implies that

$$G_0(s) T(s) = G(s) V \tag{2.27}$$

Using the result alternately in the Lippmann-Schwinger equation and the integral equation for G we obtain relations between G and T such as

$$T(s) = V + VG(s) V \tag{2.28}$$

and

$$G(s) = G_0(s) + G_0(s) T(s) G_0(s) \tag{2.29}$$

displaying the equivalence of these operators.

Since knowledge of T gives the two-body wave function it will also provide the amplitudes for inelastic processes such as the photo-effect or nucleon-nucleon Bremsstrahlung (providing we assume that we know the electromagnetic interaction). Conversely studying such processes gives additional information about the T-matrix. For the deuteron photoeffect, for example, we can write the amplitudes as

$$\mathscr{F} = \langle\mathbf{k}_f| \{T(E_f + i\varepsilon) G_0(E_f + i\varepsilon) + 1\} H^{EM} |\psi_D\rangle \tag{2.30}$$

which depends upon the half-off-shell T-matrix.

The Green's function also contains the information about the bound states. Using a complete set of states formed from the eigenfunctions of H, we have

$$1 = \sum_{\substack{\text{bound}\\\text{states}}} |\psi_i\rangle\langle\psi_i| + \int |\psi_{\mathbf{k}}\rangle\, d^3k\, \langle\psi_{\mathbf{k}}| \tag{2.31}$$

and

$$\frac{1}{s - H} = \sum \frac{|\psi_i\rangle\langle\psi_i|}{s - E_i} + \int \frac{|\psi_{\mathbf{k}}\rangle\, d^3k\, \langle\psi_{\mathbf{k}}|}{s - k^2}. \tag{2.32}$$

Thus the poles of $G(s)$ have the bound state wave functions as their residue. Conversely if we know the bound state wave function and scattering state wave function (half-off-shell T-matrix) we can construct $G(s)$ and hence $T(s)$ from the above equation.

It might appear from Equation (2.28) that in order to get T we need the potential. However, from Equation (2.29) we see that

$$T(s) = (s - k)\left[G(s) - G_0(s)\right](s - K).$$

Similarly the bound state poles in T appear as

$$\frac{V\,|\psi_i\rangle\,\langle\psi_i|\,V}{s - E_i} \rightarrow \frac{(E_i - p^2)\,|\psi_i\rangle\,\langle\psi_i|\,(E_i - p^2)}{s - E_i} \tag{2.33}$$

which is true by the Schroedinger equation. Once we know $T(s)$, V can be obtained from $V = \lim_{s\to\infty}\{T(s)\}$ or by solving the Lippmann-Schwinger equation backwards to get V from T. Of course for a *local* potential we can obtain V from Equation (2.1) and the wave function at any one energy.

Another important property of T, is that of unitary. From the relationship between T and G we have

$$T(E + i\varepsilon) - T^+(E + i\varepsilon) = V\left[G(E + i\varepsilon) - G^+(E + i\varepsilon)\right]V.$$

From the operator identity (2.21) with $A^{-1} = G(E + i\varepsilon)$ and $B^{-1} = G(E - i\varepsilon)$, we find

$$G(E + i\varepsilon) - G(E - i\varepsilon) = -\,2i\varepsilon G(E + i\varepsilon)\,G(E - i\varepsilon)$$

and

$$T - T^+ = -\,2i\varepsilon VGG^+V = -\,2i\varepsilon TG_0G_0^+T^+$$

or

$$T(E + i\varepsilon) - T(E - i\varepsilon) = -\,2\pi iT(E + i\varepsilon)\,\delta(E - K)\,T(E - i\varepsilon) \tag{2.34}$$

since

$$\lim_{\varepsilon\to 0}\left\{\frac{\varepsilon}{(E - K)^2 + \varepsilon^2}\right\} = \pi\delta(E - K).$$

The unitary property is a generalization of the well-known optical theorem which expresses the conservation of probability in quantum mechanical processes. The Kowalsky-Noyes representation of the T-matrix (Equation (2.19)) gives an explicit representation of unitary since (2.34) will be satisfied for any real symmetric function R which vanishes half on-shell and any f such that $f(k, k) = 1$.

We have spent a considerable amount of time discussing the T-matrix since in current formulations of the three-body problem, the T-matrix rather than the potential is used to supply the two-body information.

A number of authors [18–22] have indicated how to construct a T-matrix that incorporates two-body on-shell information directly without going through a potential and have indicated the arbitrariness present in this construction.

However, the usual procedure for constructing the T-matrix is to fit a potential to the phase shift and deuteron data and then solve the Lippmann-Schwinger equation for T. The arbitrariness in T then arises from our freedom of choice in the form for V. The form most usually chosen is that of an approximately local potential. (A completely local form will not suffice since it is now well established that the two-nucleon potential has some non-locality, i.e. momentum dependence. In particular the fitting of the spin zero, even parity phase shifts requires a more attractive potential in the S-states than in the higher angular momentum states) [17, 23]. A current example is the potential of Reid [17], who has fit the data with a form in which $V^{\alpha}_{L,L'}(r, r')$ is taken to be local but different for each partial wave. Other phenomenological potentials achieve the same result by introducing explicitly various spin, isospin and angular momentum operators into the potential form.

One reason for choosing a local potential for the two-nucleon interaction, aside from the greater familiarity with such forms, is that the one pion exchange potential which gives the long range tail of the two-nucleon interaction is local. Such a potential form may therefore be expected to be valid in high partial waves. For S-waves however where the nucleons come very close to one another this may not be true. In fact general quantum mechanical principles would indicate that the true potential contains some non-locality [23, 24]. The restriction to local interactions in some not clearly understood manner will limit the type of off-shell extrapolation of the T-matrix that we obtain. Recent work in the three-body problem is concerned with determining the effects of different choices for the two-body potential form.

One point that is not stressed often but which is very relevant for the three-body problem is that the T-matrix is usually quite different from the potential from which it is derived. To see this we now consider separable potentials. A separable potential is of the form

$$V(\mathbf{p}, \mathbf{q}) = \sum_{n=1}^{N} \lambda_n^{-1} g_n(\mathbf{p})\, h_n(\mathbf{q}).$$

While at first glance this potential form seems to have little resemblance to the more familiar local potentials, we want to show that a separable potential can provide a very good approximation to the two-body T-matrix. The reason for this is quite simple. If in a particular partial wave the two-body system has a bound state, then for s near the bound state energy $(-B)$, $T(s)$ has the form [5] (see Equation (2.33))

$$T(s) \underset{s \to -B}{\sim} \frac{V\,|B\rangle\,\langle B|\,V}{s + B} = \frac{(p^2 + B)\,|B\rangle\,\langle B|\,(p^2 + B)}{s + B}$$

which is separable.

We might be tempted to use this pole term as an approximation to T. However, since it is real, it would not provide a unitary T-matrix and therefore we would not conserve probability. To obtain the pole structure in T and yet satisfy unitarity, we derive another approximate form for T called the Unitary Pole Approximation (UPA) [9–13]. Taking a separable potential of the form

$$V_u = -\lambda_u^{-1} V\,|B\rangle\,\langle B|\,V$$

and using it in the Lippmann-Schwinger equation, we find

$$\begin{aligned} T_u &= -\lambda_u^{-1} V\,|B\rangle\,\langle B|\,V - \lambda_u^{-1} V|B\rangle\,\langle B|\,V G_0 T \\ &= -\lambda_u^{-1} V\,|B\rangle\,\langle A|\,. \end{aligned}$$

Solving for $\langle A|$ gives the UPA T-matrix

$$T_u(s) = \frac{-V|B\rangle\langle B|V}{\lambda_u + \langle B|VG_0(s)V|B\rangle}$$

We now choose λ_u to have the value $-\langle B|VG_0(-B)V|B\rangle$ so that the pole in T_u occurs at the correct energy. Having guaranteed the correct pole position it is easy to see that the UPA also has the same residue as T since

$$\frac{\mathrm{d}}{\mathrm{d}s}\{-(\lambda_u + \langle B|VG_0(s)V|B\rangle\}|_{s=-B} = \langle B|VG_0^2(-B)V|B\rangle =$$
$$= \langle B \mid B\rangle = 1,$$

and $|B\rangle$ satisfies the Schroedinger equation and is assumed normalized. Thus the unitary pole approximation gives a T-matrix which agrees with the actual T-matrix at the bound state and satisfies the unitarity condition.

The usefulness of such an approximation depends, of course, upon the range of energies for which it provides a good approximation to the true T-matrix.

As an example of the type of agreement which is obtainable, we show in Figure 1 the UPA to the T-matrix of a local potential of the form

$$V(r) = -\lambda_a \frac{e^{-\mu_a r}}{r} + \lambda_b \frac{e^{-\mu_b r}}{r}$$

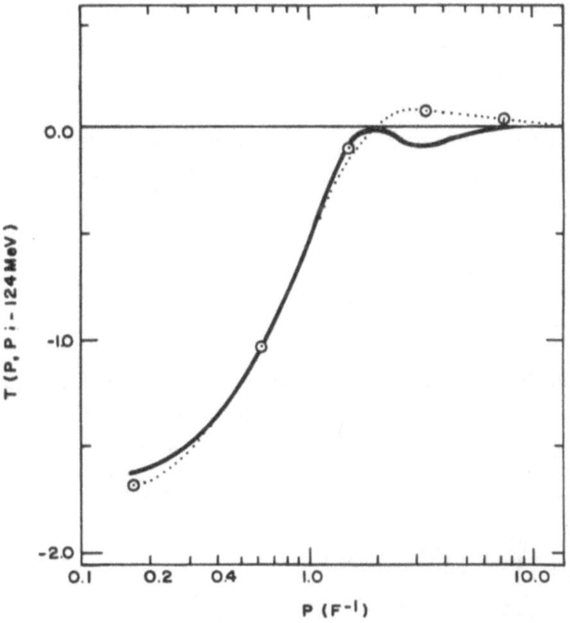

Fig. 1. Diagonal T-matrix for the MT potential VI (ref. [14]). The exact T-matrix (⊙) is compared with the UPA to it (——) and a three term UPE (....) retaining in addition to the UPA term one repulsive and another attractive term in the expansion. The energy is −120 MeV. The figure is from ref. [12].

found by averaging phenomenological singlet and triplet two-nucleon potentials of a similar shape, having soft core repulsion. The energy here is -120 MeV in the center of mass system. The agreement is very good even though we are very far from the bound state pole.

In the two-nucleon system, the triplet channel contains a bound state so this discussion is directly applicable. In the singlet channel, the low energy T-matrix is also dominated by a pole, but this pole occurs on the second Riemann sheet and is an antibound state pole. We can develop an approximation for this channel and also indicate how to add correction terms to the UPA by considering separable expansions of the T-matrix.

Our starting point [15, 16] is the homogeneous form of the Lippmann-Schwinger equation

$$|\psi_n(s)\rangle = \lambda_n(s)\, VG_0(s)\, |\psi_n(s)\rangle .$$

The eigenvalues and eigenfunctions of this equation have a simple physical interpretation. Setting s at negative value, we see that the functions $|\phi_n(s)\rangle = G_0(s)|\psi_n(s)\rangle$ satisfy the equation

$$|\phi_n(s)\rangle = G_0(s)\, \lambda_n(s)\, V\, |\phi_n(s)\rangle$$

that is, they are bound states of energy s of the potentials $\lambda_n(s)\, V$. For a purely attractive local potential, there are an infinite number of these functions and a corresponding infinite number of eigenvalues. The eigenvalues approach infinity asympototically as n^2 and the coordinate space eigenfunctions have $(n-1)$ nodes. For a potential with a short range repulsion and long range attraction, there are two series of eigenfunctions and eigenvalues corresponding to bound states supported by the attractive part of the potential $(\lambda_n^A > 0)$ and by the repulsive part of the potential $(\lambda_n^R < 0)$. Given the solutions of this eigenvalue problem, it is easy to see that V can be formally expanded (we assume the normalization $\langle\psi_n(s)|\, G_0(s)\, |\psi_m(s)\rangle = -\delta_{n,m}$) as

$$V = \sum_{n=1}^{\infty} \frac{|\psi_n(s)\rangle\, \langle\psi_n(s)|}{-\lambda_n(s)} .$$

For practical calculations we of course truncate this expansion at some point. To get an approximation for T we can proceed in at least two different directions. If we associate s with the energy parameter in $T(s)$ and insert the expansion of V into the Lippmann-Schwinger equation we find the Weinberg [15] series

$$T_W^N(s) = \sum_{n=1}^{N} \frac{|\psi_n(s)\rangle\, \langle\psi_n(s)|}{-\lambda_n + 1} .$$

Note that if there is a bound state at energy s, then one eigenvalue will have the value 1 and $T_W^N(s)$ has the correct pole. Unfortunately if s is positive the $|\psi_n(s)\rangle$ become complex and $T_W^N(s)$ is not unitary. In addition we must solve for the λ_n and $|\psi_n\rangle$ for every value of s desired, and for large s the eigenvalues and eigenfunctions become poorly behaved.

A more convenient method, called the Unitary Pole Expansion (UPE) [16], is obtained by regarding s as fixed at some non-positive value $-B$ in the expansion for V. Solving for the T-matrix we find

$$T_u(s) = \sum_{nm} |\psi_n(-B)\rangle \, \Delta_{n,m}(s) \, \langle \psi_m(-B)|$$

with

$$[-\Delta^{-1}(s)]_{n,m} = \lambda_n(-B)\,\delta_{n,m} + \langle \psi_n(-B)|\, G_0(s)\, |\psi_m(-B)\rangle .$$

This form for T will be unitary and has the computational advantage that once we have calculated the λ_n and ψ_n the T-matrix at any energy is obtained by quadratures. If a partial wave has a bound state, choosing B as the binding energy guarantees that T will have the correct bound state pole. In fact, for $N=1$, we reproduce the UPA so that the UPE provides a systematic method for improving upon the UPA. For a partial wave without a bound state, we may take B at any convenient value, for example zero.

In Figure 1, the dotted curve shows the UPE T-matrix when one additional attractive and repulsive term beyond the UPA are retained. We see that the three term UPE and the actual T-matrix are in very good agreement. Figure 2 shows a three term UPE to the Reid [17] singlet potential. The choice $B=0.0$ has been used here. Once again the agreement is very good.

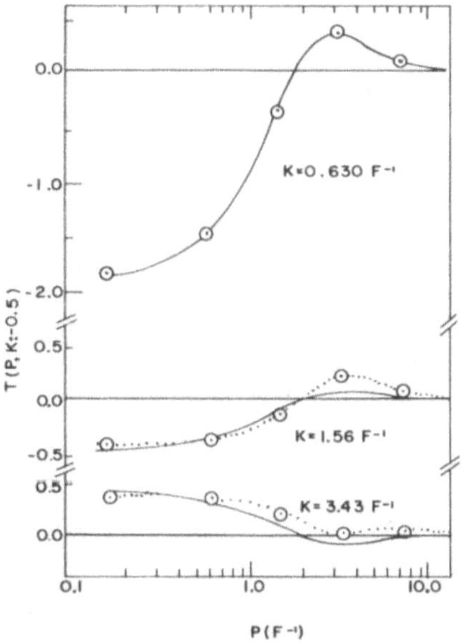

Fig. 2. Off-diagonal T-matrices for the Reid $^1S_0$ potential. The notation is the same as for Figure 1. The energy is -20 MeV. The figure is from reference [16].

We would like to draw a few observations from the results presented here. As we remarked previously, the T-matrix is very different from the potential which is used to calculate it from the Lippmann-Schwinger equation. Of course it contains the information about the potential, but the functional form is quite different. Even though the potential in these examples is local, the T-matrix is dominated by the separable pole term and the difference between the pole term and the true T-matrix can be made up by the addition of a few additional separable terms. Thus although we have started with a local potential to extrapolate off-the-energy shell, it is the pole term which provides the major extrapolation to negative energies. Jackson and Lande [25] have argued that the rapid convergence of the additional separable terms is due to the increase in the eigenvalues with n, but also to the increased number of nodes in the eigenfunctions. This latter factor should be especially effective for the repulsive terms since they are essentially confined to the repulsive core of the potential. These authors find that the UPA plus an additional repulsive term accurately reproduces the two-body defect wave function (Figure 3).

Another point that arises from this discussion is that the T-matrix drastically reduces the effects of the repulsive terms in the potential. This is easily seen from the Weinberg expansion where the $-\lambda_n+1$ denominator accents the positive eigenvalues and strongly suppresses the negative eigenvalues which may have values of -0.06. [16]. Thus although the core may be a dominant feature in the potential (For example, the $^1S_0$ Reid potential in momentum space is everywhere positive), it is drastically reduced in the T-matrix and approximately separable.

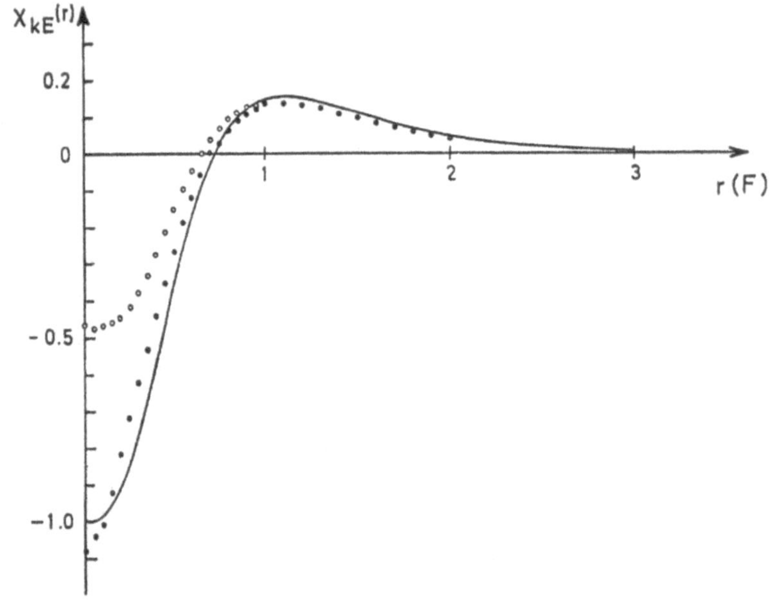

Fig. 3. Defect wave function for the Reid $^1S_0$ potential. The exact function (——) is compared with the UPA (...) and a UPE (○○○) retaining the UPA and one additional repulsive term at an energy of -100 MeV. The figure is taken from reference [25].

The approximate separability of the T-matrix has been displayed by other authors taking somewhat different approaches. Van Wageningen *et al.* [26] have shown that a local delta function potential

$$V(\mathbf{r}, \mathbf{r}') = V(r)\,\delta^3(\mathbf{r} - \mathbf{r}') \quad \text{with} \quad V(r) = \delta(r - a)$$

which is separable, can be fit very accurately to the two-nucleon phase shifts in the S-states. Doing this, they then solve for the properties of the trinucleon finding results similar to that obtained with more realistic local potentials. In addition they find that the form factor for their separable potential and the required strength are very close to those obtained from the UPA. They then conclude that any potential which reproduces the two-nucleon S-wave phase shifts up to a few hundred MeV and also the bound-state three-nucleon properties, is strong at intermediate distances and approximately separable. Another paper in this area is by Fiedeldey and Erens [27] who find that a large number of more or less realistic local potentials are what they term 'nearly separable' in that the half-off-shell function $f(p, k)$ is nearly independent of the strength of the two-body interaction. This is exactly true for separable potentials.

A number of other separable expansions have been developed. In particular Fuda [28] and Osborn [29] use Kowalski's representation of the T-matrix, Equation (2.19), and expand the non-separable term R in eigenfunctions of the kernel ΛG_0. This expansion has the advantage of giving the correct half-off-shell T-matrix in all orders, but has difficulties when s is negative. All the expansions appear to have about the same rates of convergence, with the UPE being somewhat faster than the others. The UPA however appears to be the best one term approximation by a substantial margin. For practical computations in the three-body problems, energy independent expansions such as the UPE are to be favored over energy dependent expansions since the amount of work necessary to solve the two-body problems is then much smaller.

The simplicity gained from the use of separable potentials has led a number of authors to try to obtain fits of the two-body data starting with a separable form for the potential. The first extensive of these fits was by Tabakin [30] who chose rank one or two separable potentials for each of the functions $V^{\alpha}_{L,L'}(p, q)$. Unfortunately Tabakin's fits suffer from a number of errors, giving an incorrect value for the two-body binding energy and deuteron quadrupole moment. A more extensive set of fits was attempted by Mongan [31]. Mongan's fits also suffer from a number of drawbacks. They all have a very small percent D-state in the deuteron (about 1%) and apparently have unrealistic effective ranges in the singlet S-states. I know of no separable potential fits to the two-body data which achieve nearly the accuracy of the local potential fits.

To discuss the two-body equations we may also use [5] some results from the theory of integral equations. In particular an integral operator K, with the property that Trace $[KK^+]$ is finite (or for which some power of the kernel has this property) is called and $\mathfrak{L}^2$-operator. For the two-body kernel it can easily be shown that this trace is proportional to $[I\text{mag}\{\sqrt{s}\}]^{-1}$. Thus the kernel is $\mathfrak{L}^2$ for s not on the positive real axis. By appropriate manipulation [5] we can show that the kernel is effectively $\mathfrak{L}^2$ even on the real axis.

An $\mathfrak{L}^2$ kernel has a number of useful properties. In particular we have (i) it can be approximated as accurately as desired by a kernel of finite rank and is amenable to numerical treatment, (ii) if the kernel depends analytically upon some parameter s, then the solutions of the inhomogeneous equation are unique except for a possible finite number of discrete values of s at which the homogeneous equation has a solution.

3. The Three-Body Problem

We will study the more formal aspects of the three-body problem by trying to proceed in a manner similar to the two-body problem and seeing where the procedure breaks down. We will then discuss the methods that have been developed to produce a workable system of three-body equations.

The three-body Schroedinger equation has the form

$$(E - K - V) |\psi\rangle = 0 \tag{3.1}$$

with

$$V = \sum_{\alpha=1}^{3} V_\alpha = V^0 \tag{3.2}$$

and

$$K = \tfrac{1}{2}(p_1^2 + p_2^2 + p_3^2). \tag{3.3}$$

Here for example V_1 is the potential between particles 2 and 3. We will also be interested in the channel potential defined by

$$V^\alpha = \sum_{\beta \neq \alpha} V_\alpha = V - V_\alpha \tag{3.4}$$

and the Hamiltonians for a pair of interacting particles with the third one free, defined by

$$H_\alpha = K + V_\alpha. \tag{3.5}$$

It is often convenient to extend this notation defining

$$V_0 = 0$$

if only two-body forces exist. The sum above will then be from 0 to 3. If we then wish to consider three-body forces, most of our results go right through by defining V_0 to be the three-body potential energy. In what follows, however, we will consider only two-body forces. This notation is also convenient when dealing with the break-up reaction.

For our coordinates we choose the coordinate of the center of mass $\mathbb{R} = \frac{1}{3}(\mathbf{r}_1 + \mathbf{r}_2 + \mathbf{r}_3)$, the separation of any particle pair $\mathbf{r}_{\beta\gamma} = \mathbf{r}_\beta - \mathbf{r}_\gamma$ and the separation between the third particle on the center of mass of the other two, $\boldsymbol{\varrho}_\alpha = \mathbf{r}_\alpha - \frac{1}{2}(\mathbf{r}_\beta + \mathbf{r}_\gamma)$. The corresponding conjugate momenta are the total momentum

$$\mathbb{P} = \mathbf{p}_1 + \mathbf{p}_2 + \mathbf{p}_3, \tag{3.6}$$

the relative momentum of β and γ

$$\mathbf{q}_\alpha = \tfrac{1}{2}(\mathbf{p}_\beta - \mathbf{p}_\gamma) \tag{3.7}$$

and the momentum

$$\tilde{\mathbf{p}}_\alpha = \tfrac{2}{3}[\mathbf{p}_\alpha - \tfrac{1}{2}(\mathbf{p}_\beta + \mathbf{p}_\gamma)] \tag{3.8}$$

In the center of mass system ($\mathbb{P}=0$), in which we will always work, $\tilde{\mathbf{p}}_\alpha$ has the value $\mathbf{p}_\alpha$ so that our basic momentum variables, will be $\mathbf{p}_\alpha$ and $\mathbf{q}_\alpha$. The possibilities for $\alpha=1$, 2 or 3 give us three sets of momenta related, for example, by

$$\begin{aligned} \mathbf{q}_1 &= -\tfrac{1}{2}\mathbf{q}_2 + \tfrac{3}{4}\mathbf{p}_2 = -\tfrac{1}{2}\mathbf{q}_3 - \tfrac{3}{4}\mathbf{p}_3 \\ \mathbf{p}_1 &= -\mathbf{q}_2 - \tfrac{1}{2}\mathbf{p}_2 = \mathbf{q}_3 - \tfrac{1}{2}\mathbf{p}_3 . \end{aligned} \tag{3.9}$$

The kinetic energy has the value

$$K = \tfrac{3}{4}p_\alpha^2 + q_\alpha^2 = p_\alpha^2 + p_\beta^2 + \mathbf{p}_\alpha\cdot\mathbf{p}_\beta . \tag{3.10}$$

For the Hamiltonians H_0, H_γ and H we will have the associated Green's functions

$$\begin{aligned} G_\gamma(s) &= (s - H_\gamma)^{-1}, \quad \gamma = 0, 1, 2, 3 \\ G(s) &= (s - H)^{-1} \end{aligned} \tag{3.11a}$$

and T-matrices

$$\begin{aligned} T_\gamma(s) &= V_\gamma + V_\gamma G_\gamma(s)\, V_\gamma, \quad \gamma = 0, 1, 2, 3 \\ T(s) &= V + VG(s)\, V . \end{aligned} \tag{3.11b}$$

The two-particle Green's function in the three-body space $G_\alpha(s)$ has the representation

$$\langle \mathbf{p}_\alpha \mathbf{q}_\alpha |\, G_\alpha(s)\, |\mathbf{p}'_\alpha \mathbf{q}'_\alpha\rangle = \delta^3(\mathbf{p}_\alpha - \mathbf{p}'_\alpha)\, \langle q_\alpha |\, \hat{G}_\alpha(s - \tfrac{3}{4}p_\alpha^2)\, | q'_\alpha\rangle , \tag{3.12}$$

where $\hat{G}_\alpha(s-\tfrac{3}{4}p_\alpha^2)$ is the normal two-body Green's function. Thus the Green's function in the three-body space is essentially the same as in the two-body space but evaluated at the three-body energy minus the relative kinetic energy of the third particle and the two-body subsystem. (If necessary, we designate two-body operators by putting hats on them).

Using the identity (2.21) we can obtain many relations between the various Green's function and T-matrices. In particular we have that

$$\begin{aligned} G(s) - G_\gamma(s) &= G_\gamma(s)\, V^\gamma G(s) \\ &= G(s)\, V^\gamma G_\gamma(s)', \quad \gamma = 0, 1, 2, 3 \end{aligned} \tag{3.13}$$

We now try to write down an integral equation for the three-body wave function. We have a number of different possibilities for the wave function depending upon the preparation of the scattering experiment. For the initial states we may choose (i) a free particle α incident upon a bound state of particles β and γ, $\alpha=1,2$ or 3. This state, denoted by $\Phi_\alpha=|\phi_\alpha, p_\alpha\rangle$, satisfies $H_\alpha\Phi_\alpha=E_\alpha\Phi_\alpha$ with $E_\alpha=-B_\alpha+\tfrac{3}{4}p_\alpha^2$.

(ii) three free particles ($\alpha=0$). This state, denoted by $\Phi_0=|\mathbf{p}, \mathbf{q}\rangle$, satisfies $K\Phi_0= =E_0\Phi_0$ with $E_0=q^2+\tfrac{3}{4}p^2$.

The corresponding wave function is then obtained from

$$\Psi_\alpha^+ = \lim \{i\varepsilon G(E_\alpha + i\varepsilon)\, \Phi_\alpha\}\,, \tag{3.14}$$

where $G(s)$ is the full three-body Green's function. To relate G to G_0 we can use the first of the Equation (3.13) with $\gamma=0$, giving $G=G_0+G_0VG$, then we find that

$$\Psi_\alpha^+ = \lim_{\varepsilon\to 0} \{i\varepsilon G_0(E_\alpha + i\varepsilon)\, \Phi_\alpha\} + G_0(E_\alpha + i0)\, V\Psi_\alpha^+ \tag{3.15}$$

If $\alpha=0$ we have

$$i\varepsilon G_0(E_0 + i\varepsilon)\, \Phi_0 = \frac{i\varepsilon}{E_0 + i\varepsilon - K}\, \Phi_0 = \frac{i\varepsilon}{E_0 + i\varepsilon - E_0}\, \Phi_0 = \Phi_0$$

then

$$\Psi_0^+ = \Phi_0 + G_0(E_0 + i0)\, V\Psi_0^+\,. \tag{3.16}$$

However, if we choose $\alpha\neq 0$ then

$$i\varepsilon G_0(E_\alpha + i\varepsilon)\, \Phi_\alpha = \frac{i\varepsilon}{E_\alpha + i\varepsilon - K}\, \Phi_\alpha = \frac{i\varepsilon}{-B_\alpha - q_\alpha^2 + i\varepsilon}\, \Phi_\alpha \to 0$$

since the denominator has no pole. Hence

$$\Psi_\alpha^+ = G_0(E_\alpha + i0)\, V\Phi_\alpha^+\,. \tag{3.17}$$

and Ψ_α^+, $\alpha\neq 0$, is a homogeneous solution of the equation

$$F^+ = \Phi_0 + G_0(E + i0)\, F^+ \tag{3.18}$$

and hence this Lippmann-Schwinger equation does not have a unique solution and therefore does not fulfill the role of combining the associated differential equation and boundary conditions into one equation.

The difficulties of the Lippmann-Schwinger formulation of the three-body problem can be also expressed in terms of the non-$\mathfrak{L}^2$ nature of the kernel. The kernel $K=G_0V$ has terms of the form

$$\frac{V_\alpha(q_\alpha, q_\alpha')\, \delta^3(\mathbf{p}_\alpha - \mathbf{p}_\alpha')}{s - \frac{3}{4}p_\alpha^2 - q_\alpha^2}\,.$$

Taking Trace$[KK^+]$ we find integrals over $[\delta^3(\mathbf{p}-\mathbf{p}')]^2$ which leads to a divergent integral. All power of the kernel will have terms in which iterations of only one two-body potential will occur. These so called 'disconnected' terms lead to the divergence of the $\mathfrak{L}^2$-norm of K.

The Russian mathematician L. D. Faddeev [32], has shown how to write down a set of equations with a well behaved kernel for three-body systems. His starting point for deriving the equations is the formal iterative solution of the Lippmann-Schwinger equation for the three-body T-matrix

$$\begin{aligned} T &= V + VG_0T = V + VG_0V + VG_0VG_0V + \cdots \\ &= V_1 + V_2 + V_3 + (V_1 + V_2 + V_3)\, G_0(V_1 + V_2 + V_3) + \cdots \end{aligned} \tag{3.19}$$

We rearrange the terms in this series as follows

$$T = V_1 + V_1 G_0 V_1 + V_1 G_0 V_1 G_0 V_1 + \cdots + V_2 + V_2 G_0 V_2 + \cdots + \\ + (V_1 + V_1 G_0 V_1 + \cdots)\, G_0 \,(V_2 + V_2 G_0 V_2 + \cdots) + \cdots$$

hence

$$T = T_1 + T_2 + T_3 + T_1 G_0 T_2 + T_1 G_0 T_3 + \cdots,$$

where iteration of the two-body potentials have been summed to give the two-body T-matrices. There is no single integral equation for which this last series represents the perturbation expansion. However, if we define

$$T = T^1 + T^2 + T^3 \tag{3.20}$$

then the coupled system of equations

$$\begin{pmatrix} T^1 \\ T^2 \\ T^3 \end{pmatrix} = \begin{pmatrix} T_1 \\ T_2 \\ T_3 \end{pmatrix} + \begin{pmatrix} 0 & T_1 & T_1 \\ T_2 & 0 & T_2 \\ T_3 & T_3 & 0 \end{pmatrix} G_0 \begin{pmatrix} T^1 \\ T^2 \\ T^3 \end{pmatrix} \tag{3.21}$$

when iterated gives the expansion for T.

The kernel of these equation still contains disconnected terms and its trace is infinite. However, if the equations are iterated once, the iterated kernel contains only terms of the form

$$T_\alpha G_0 T_\beta G_0\,, \quad \alpha \neq \beta\,.$$

These terms will give a finite trace for s not on the positive real axis, and hence the square of the Faddeev kernel is an $\mathfrak{L}^2$ operator for s off the real axis. The analysis can be carried further to show [32] that under very weak restrictions on the potentials, the fifth power of the kernel is a compact operator in a certain Banach space even for real s. If the potential is a superposition of Yukawas, then this result can be shown to be true for the once iterated kernel also [33].

Thus we have much useful information about the nature of the Faddeev equations. In particular this solution is unique except for certain isolated values of s which occur at the three-body bound states. In addition we know that the Faddeev equations are amenable to a numerical solution, at least in principle. This appears to be true in practice also.

We consider now the calculation of the various three-body scattering cross-sections. In the two-body case, the scattering amplitudes were obtained by taking appropriate matrix elements of the two-body T-matrix. Three-body systems have a number of distinct channels associated with the elastic, exchange or break-up reactions. The corresponding scattering amplitudes may be obtained by the following procedure.

The amplitude $T_{\alpha\beta}$ for the transition from the state $|\phi_\beta, \mathbf{p}_\beta\rangle$ representing particle β incident upon a bound state of particles γ and α to the state $|\phi_\alpha, \mathbf{p}_\alpha\rangle$ which has α as the free particle is given by

$$T_{\alpha\beta} = \langle \phi_\alpha, \mathbf{p}_\alpha|\, V^\alpha \,|\psi_\beta^+\rangle\,, \tag{3.22}$$

where $|\psi_\beta^+\rangle$ is the full three-body wave function corresponding to the initial conditions represented by $|\phi_\beta, \mathbf{p}_\beta\rangle$, i.e.

$$(E - K - V)\,|\psi_\beta^+\rangle = 0 \tag{3.23}$$

and

$$|\psi_\beta^+\rangle = |\phi_\beta, \mathbf{p}_\beta\rangle + \text{outgoing scattering waves}.$$

Using the identity (2.21) we can obtain a representation for $|\psi_\beta^+\rangle$ analogous to that found in the two-body case

$$|\psi_\beta^+\rangle = [1 + G(E + i\varepsilon)\,V^\beta]\,|\phi_\beta, \mathbf{p}_\beta\rangle. \tag{3.24}$$

Thus

$$T_{\alpha\beta} = \langle\phi_\alpha, \mathbf{p}_\alpha|\, U_{\alpha\beta}^+(E + i\varepsilon)\,|\phi_\beta, \mathbf{p}_\beta\rangle \tag{3.25}$$

with

$$U_{\alpha\beta}^+(s) = V^\alpha + V^\alpha G(s)\,V^\beta. \tag{3.26}$$

Here E has the on-shell value

$$E = -\,B_\alpha + \tfrac{3}{4}p_\alpha^2 = E_\alpha = -\,B_\beta + \tfrac{3}{4}p_\beta^2 = E_\beta. \tag{3.27}$$

For the break-up reaction, the same equations are valid with α set equal to 0, the final state taken as $|\psi_0\rangle = |\mathbf{p}_\alpha, \mathbf{q}_\alpha\rangle$ and with $E = \frac{3}{4}p_\alpha^2 + q_\alpha^2$. Lovelace [9] has developed Faddeev type equations for the operator $U_{\alpha\beta}^+$. Numerous other operators differing slightly from $U_{\alpha\beta}^+$ but having the same on-shell matrix elements exist in the literature. By methods to be shown below, we can develop Faddeev type equations for any of these operators. For example the operators of Alt *et al.* [34] (AGS) defined by

$$U_{\alpha\beta}(s) = (1 - \delta_{\alpha\beta})\,(s - K) + V^0 - V_\alpha - V_\beta + \delta_{\beta\alpha}V_\alpha + V^\alpha G(s)\,V^\beta \tag{3.28}$$

have the same on shell matrix elements as $U_{\alpha\ \beta}^{+}$ and satisfy the equations

$$U_{\alpha\beta}(s) = (1 - \delta_{\alpha\beta})\,(s - K) + \sum_{\gamma\neq\alpha} T_\gamma(s)\,G_0(s)\,U_{\gamma\beta}(s) =$$
$$= (1 - \delta_{\alpha\beta})\,(s - K) + \sum_{\delta\neq\beta} U_{\alpha\beta}G_0(s)\,T_\delta(s). \tag{3.29}$$

Rather than dealing with operators such as $U_{\alpha\beta}$ which give the scattering amplitudes directly, we consider the operators

$$\tilde{U}_{\alpha\beta}(s) = V_\alpha + \delta_{\alpha\beta}(s - K - V_\alpha) + V_\alpha G(s)\,V^\beta. \tag{3.30}$$

The operator $\delta_{\alpha\beta}(s-K-V_\alpha)$ will vanish on-the-energy-shell. It is included in the definition of $\tilde{U}_{\alpha\beta}$ to obtain a simple form for the final equations.

It is easy to show that

$$T_{\alpha\beta} = \sum_{\gamma\neq\alpha} \langle\phi_\alpha, \mathbf{p}_\alpha|\, \tilde{U}_{\gamma\beta}(E + i\varepsilon)\,|\phi_\beta, \mathbf{p}_\beta\rangle. \tag{3.31}$$

While it is necessary to take a sum of the $\tilde{U}_{\alpha\beta}$ to obtain any matrix element, these operators satisfy a much simpler set of integral equations than do the $U_{\alpha\beta}$ of AGS. In practice they would appear to be the easiest to deal with.

The main ingredient to derive the equations satisfied by these operators is the resolvent identity

$$G(s) = G_\gamma(s) + G_\gamma(s)\, V^\gamma G(s).$$

Inserting this in the definition for $\tilde{U}_{\gamma\beta}$ we find

$$\begin{aligned}\tilde{U}_{\gamma\beta} &= V_\gamma + \delta_{\gamma\beta}(s - K - V_\gamma) + V_\gamma G_\gamma [1 + V^\gamma G(s)]\, V^\beta = \\ &= V_\gamma + \delta_{\gamma\beta}(s - K - V_\gamma) + T_\gamma G_0 V^\gamma + T_\gamma G_0 \sum_{\delta \neq \gamma} V_\delta G V^\beta .\end{aligned}$$

Using the definitions of $\tilde{U}_{\delta\beta}$ to replace $V_\delta G V^\beta$, expanding the resulting expression and using the Lippmann-Schwinger equation to combine terms, we arrive at the final result

$$\tilde{U}_{\gamma\beta}(s) = \delta_{\gamma\beta}(s - K) + T_\gamma G_0 \sum_{\delta \neq \gamma} \tilde{U}_{\delta\beta} . \tag{3.32}$$

Written in matrix form we find, for example.

$$\begin{pmatrix} \tilde{U}_{11} \\ \tilde{U}_{21} \\ \tilde{U}_{31} \end{pmatrix} = \begin{pmatrix} s - K \\ 0 \\ 0 \end{pmatrix} + \begin{pmatrix} 0 & T_1 & T_1 \\ T_2 & 0 & T_2 \\ T_3 & T_3 & 0 \end{pmatrix} G_0 \begin{pmatrix} \tilde{U}_{11} \\ \tilde{U}_{21} \\ \tilde{U}_{31} \end{pmatrix} . \tag{3.33}$$

These equations have the same kernel as the Faddeev equations and hence the same desirable properties.

The equations given here, and those developed before, have the interesting feature that the two-body information is contained in the two-body T-matrices. The potentials never occur in the final equations. We recall that it is the T-matrix which is most closely related to experiment. However, as we shall see, these T-matrices are required fully off-shell and we need two body information which is not measurable in two-body experiments.

For the break-up reaction all of the above equations still hold with $\alpha = 0$ and we have

$$T_{0\beta} = \sum_{\gamma \neq 0} \langle \phi_0 |\, \tilde{U}_{\gamma\beta}\, | \phi_\beta, \mathbf{p}_\beta \rangle \tag{3.34}$$

which expresses the break-up amplitudes in terms of the operators for elastic scattering. For break-up reactions, however, the $\tilde{U}_{\alpha\beta}$ are needed at different momenta than for elastic scattering.

Another form for the break-up amplitude is useful. If we define the operators

$$U_{\alpha\beta} = \sum_{\gamma \neq \alpha} \tilde{U}_{\gamma\beta} \tag{3.35}$$

and do this sum in Equation (3.30) satisfied by $\tilde{U}_{\gamma\beta}$ we find

$$\begin{aligned} U_{\alpha\beta} &= \sum_{\gamma \neq \alpha} \delta_{\gamma\beta}(s - K) + \sum_{\gamma \neq \alpha} \{T_\gamma G_0 \sum_{\delta \neq \gamma} \tilde{U}_{\delta\beta}\} = \\ &= (1 - \delta_{\alpha\beta})(s - K) + \sum_{\gamma \neq \alpha} T_\gamma G_0 U_{\gamma\beta} . \end{aligned} \tag{3.36}$$

Thus these $U_{\alpha\beta}$ are just the AGS operators. The AGS break-up operator $U_{0\beta}$ is given in terms of the $U_{\gamma\beta}$ by

$$U_{0\beta} = (s - K) + \sum_{\gamma} T_\gamma G_0 U_{\gamma\beta} \tag{3.37}$$

and in terms of the $\tilde{U}_{\gamma\beta}$ by

$$U_{0\beta} = \sum_{\gamma} \tilde{U}_{\gamma\beta} \tag{3.38}$$

in agreement with our previous result. We note that on-the-energy shell $(s-K)$ vanishes.

Using Equation (3.29) satisfied by the $U_{\alpha\beta}$ we can obtain another form for $U_{0\beta}$ (dropping an $(s-K)$ term)

$$U_{0\beta} = [1 + T_\alpha G_0]\, U_{\alpha\beta}, \quad \alpha = 1, 2 \text{ or } 3 \tag{3.39}$$

This form is useful for studying the effects of final state interactions in the break up reactions. To see this we consider the on-shell matrix element

$$T_{0\beta} = \langle \mathbf{p}_\alpha \mathbf{q}_\alpha |\, [1 + T_\alpha (E + i\varepsilon)\, G_0 (E + i\varepsilon)]\, U_{\alpha\beta} (E + i\varepsilon)\, |\phi_\beta, \mathbf{p}_\beta\rangle . \tag{3.40}$$

We remember that T_α is the two-body T-matrix in the three-body space. Since here the momenta have values such that $E = \frac{3}{4}p_\alpha^2 + q_\alpha^2$, the two-body T-matrix $\hat{T}_\alpha$ will be evaluated at $q_\alpha^2 + i\varepsilon$ (the three-body energy minus the kinetic energy of the relative motion of α and the (β, γ) subsystem. Thus

$$\langle \mathbf{q}_\alpha, \mathbf{p}_\alpha |\, [1 + T_\alpha (E + i\varepsilon)\, G_0 (E + i\varepsilon)] = \langle \chi_\alpha^{(-)}|\, \langle p_\alpha | ,$$

where $\langle \chi_\alpha^- |$ is the full 2-particle scattering state (with incoming spherical waves) of energy q_α^2.

Thus

$$T_{0\beta}(s) = \langle \chi_\alpha^- |\, \langle \mathbf{p}_\alpha |\, \tilde{U}_{\alpha\beta}(s)\, |\phi_\beta, \mathbf{p}_\beta\rangle . \tag{3.41}$$

We may therefore expect that the behavior of β and γ in the break-up reaction could be related to the two-body wave function. Later we will discuss the extent to which this is true in the three-nucleon system.

The next step in our formal study of the three-body problem is to reduce Equation (3.30) for the $\tilde{U}_{\alpha\beta}$ to a workable system of integral equations. To do this we will first consider the simplifications which result when we take the identity of the scattering particles into account. Then we will take the resultant equations into the momentum representation and perform an angular momentum reduction.

The rearrangement amplitude for the scattering of three identical particles is given by

$$T = \sum_{\alpha, \beta = 1}^{3} \frac{1}{\sqrt{3}} \langle \phi_\alpha, \mathbf{p}_\alpha |\, \sum_{\gamma \neq \alpha} U_{\gamma\beta}\, |\phi_\beta, \mathbf{p}_\beta\rangle \frac{1}{\sqrt{3}}, \tag{3.42}$$

where we assume that $|\phi_\alpha\, \mathbf{p}_\alpha\rangle$ is antisymmetric in particles β and γ. For example

$$(12)\,|\phi_3, \mathbf{p}_3\rangle = -\,|\phi_3, \mathbf{p}_3\rangle,$$

where (12) is the permutation operator for particles 1 and 2. We define $|\chi_\gamma\rangle$ by

$$|\chi_\gamma\rangle = \sum_\beta \tilde{U}_{\gamma\beta}\,|\phi_\beta, \mathbf{p}_\beta\rangle. \tag{3.43}$$

Then

$$T = \tfrac{1}{3} \sum_{\alpha=1}^{3} \sum_{\gamma\neq\alpha} \langle\phi_\alpha, \mathbf{p}_\alpha \mid \chi_\gamma\rangle.$$

Multiplying the equation for $\tilde{U}_{\gamma\beta}$ on the right by $|\phi_\beta, \mathbf{p}_\beta\rangle$ and summing over β, we find the equation

$$|\chi_\gamma\rangle = \sum_\beta \delta_{\gamma\beta}(E-K)\,|\phi_\beta, \mathbf{p}_\beta\rangle + T_\gamma G_0 \sum_{\delta\neq\gamma} \sum_\beta \tilde{U}_{\delta\beta}\,|\phi_\beta, \mathbf{p}_\beta\rangle$$

or,

$$|\chi_\gamma\rangle = V_\gamma\,|\phi_\gamma, \mathbf{p}_\gamma\rangle + T_\gamma G_0 \sum_{\delta\neq\gamma} |\chi_\delta\rangle. \tag{3.44}$$

From the symmetry of these equations it is clear that

$$|\chi_2\rangle = (132)\,|\chi_3\rangle \quad \text{and} \quad |\chi_1\rangle = (123)\,|\chi_3\rangle,$$

where (123) and (132) are particle permutation operators. Thus

$$|\chi_3\rangle = V_3\,|\phi_3, \mathbf{p}_3\rangle + T_3 G_0\,[(123) + (132)]\,|\chi_3\rangle.$$

Since we are dealing with identical fermions we have from our explicit representation for T_3 (Equations (2.5) and (2.14)) that

$$(12)\,T_3 = -\,T_3 = T_3\,(12)$$

and similarly for V_3. These imply that

$$(12)\,|\chi_3\rangle = -\,|\chi_3\rangle.$$

Thus

$$\begin{aligned} T_3 G_0 (132)\,|\chi_3\rangle &= [-\,T_3\,(12)]\,G_0\,(132)\,[-\,(12)\,|\chi_3\rangle] = \\ &= T_3 G_0\,(12)\,(132)\,(12)\,|\chi_3\rangle = T_3 G_0\,(123)\,|\chi_3\rangle \end{aligned}$$

and

$$|\chi_3\rangle = V_3\,|\phi_3, \mathbf{p}_3\rangle + 2T_3 G_0\,(123)\,|\chi_3\rangle. \tag{3.45}$$

This is the equation which is eventually to be solved.

We can use the identity of the particles to further simplify the calculation of the scattering amplitude. All terms in the sum are clearly identical. Hence

$$\begin{aligned} T = \sum_{\gamma\neq 3} \langle\phi_3, \mathbf{p}_3 \mid \chi_\gamma\rangle &= \langle\phi_3, \mathbf{p}_3|\,(123) + (132)\,|\chi_3\rangle = \\ &= 2\langle\phi_3, \mathbf{p}_3|\,(123)\,|\chi_3\rangle. \end{aligned} \tag{3.46}$$

To develop the integral equation of Equation (3.45) we generalize the representation of the two-body T-matrix to include the presence of the third particle. The desired expression is

$$T_3(s) = \sum_{\substack{JSTM \\ LL' \\ jml}} |\mathscr{Y}^M_{JLS}\rangle\, |jml\rangle\, \langle jml|\, \langle \mathscr{Y}^M_{JL'S}| \otimes$$

$$\otimes \sum_{i_z=\pm 1/2} |i_z\rangle\!\rangle\, \mathbb{P}_T \langle\!\langle i_z| \otimes \int |p_3\rangle\, p_3^2\, \mathrm{d}p_3 T^{\alpha}_{LL'}(s - \tfrac{3}{4}p_3^2)\, \langle p_3|\,.$$

We have decomposed the unit operator of the third particle into angular momentum states and isospin states and indicated the form of the dependence of the T-matrix on the momentum of the third particle. The next step is to rearrange the terms in the sums to obtain states of total three-body angular momentum, isospin and parity. For example the first factor becomes

$$\sum_{\mathscr{J}\mathscr{M},\, jl} |\mathscr{J}\mathscr{M}, J(LS); j(l)\rangle\, \langle \mathscr{J}\mathscr{M}, J(L'S), j(l)|$$

while for the isospin

$$\sum_{\mathscr{T}\mathscr{I}} |\mathscr{T}\mathscr{I}(T)\rangle\!\rangle\, \langle\!\langle \mathscr{T}\mathscr{I}(T)|$$

Denoting all these quantum numbers by ξ, we write

$$|\xi\rangle = |\mathscr{J}\mathscr{M}, J(LS), jl\rangle\, |\mathscr{T}\mathscr{I}(T)\rangle\!\rangle\,. \tag{3.47}$$

The T-matrix takes the form

$$T_3(s) = \sum_{\xi\xi'} |\xi\rangle \left[\int |p_3\rangle\, p_3^2\, \mathrm{d}p_3 T^{\alpha}_{LL'}(s - \tfrac{3}{4}p_3^2)\, \langle p_3|\right] \langle \xi'|\,. \tag{3.48}$$

The quantum numbers in ξ' are the same as for ξ with possible exception of L'.

If we put this into Equation (3.45), with a similar expansion for V_3, it is clear that $|\chi_3\rangle$ has the form

$$|\chi_3\rangle = \sum_{\xi} |\xi\rangle\, |\chi^{\xi}\rangle\,. \tag{3.49}$$

Using these expansions and going to momentum space we find (dropping the subscript 3 on all momenta)

$$\langle pq \mid \chi^{\xi}\rangle = \langle pq|\, \langle \xi|\, V_3\, |\phi_3, p_3^{(i)}\rangle + 2 \sum_{\xi'\xi''} \int q'^2\, \mathrm{d}q'\, p''^2\, \mathrm{d}p''\, q''^2\, \mathrm{d}q''$$

$$\frac{\langle q|\, T^{\alpha}_{LL'}(E - \tfrac{3}{4}p^2)\, |q'\rangle\, \langle pq'|\, \langle \xi'|\, (123)\, |\xi''\rangle\, |p''q''\rangle\, \langle p''q'' \mid \chi^{\xi''}\rangle}{E - \tfrac{3}{4}p^2 - q'^2}\,. \tag{3.50}$$

We will not discuss the general treatment of the matrix element $\langle p'q'|\langle \xi'|(123)|\xi''\rangle|p''q''\rangle$. This has been discussed by a number of authors [35]. However it is clear that the operation of (123) leaves the total three-body quantum numbers unchanged and has non-diagonal matrix elements only between different two-body subsystems. Therefore the equations break up into blocks each characterized by a particular value of $\mathscr{J}\mathscr{M}\mathscr{T}\mathscr{I}$ and π. The number of coupled equations in each block is equal to the number of different ways a third particle may be coupled to the two-body T-matrix partial waves in order to obtain these quantum numbers. For example if we consider the $\mathscr{J}^\pi=\frac{1}{2}^+$ state of the three-nucleon system, keeping the ${}^1S_0$ and ${}^3S_1+{}^3D_1$ two-body interactions, the different terms in the integral equation will correspond to the following table

J	L	S	j	l
1	0	1	$\frac{1}{2}$	0
1	0	1	$\frac{3}{2}$	2
1	2	1	$\frac{1}{2}$	0
1	2	1	$\frac{3}{2}$	2
0	0	0	$\frac{1}{2}$	0

Therefore for this state we will have to solve five coupled two-dimensional integral equations.

We will consider the form that Equations (3.50) takes for a simple case in which the two-body forces act only in the singlet and triplet two-body S-states. For this interaction the quantum numbers for total three-body orbital angular momentum and spin are also conserved and we can rearrange the coupling in the T-matrix to reflect this. For the $\mathfrak{L}=0$ state, we find that the only states which contribute are the following

J	L	S	j	l
1	0	1	$\frac{1}{2}$	0
0	0	0	$\frac{1}{2}$	0

For the doublet ($\mathscr{S}=\frac{1}{2}$) $\mathscr{L}=0$ state we have

$$\begin{aligned}\langle pq|\,\langle\xi|\,(123)\,|\xi'\rangle\,|pq'\rangle &= \langle\mathscr{S}=\tfrac{1}{2}(S_{12})|\,(123)\,|\mathscr{S}=\tfrac{1}{2}(S'_{12})\rangle\times\\ &\times\langle\!\langle\mathscr{T}=\tfrac{1}{2}(T_{12})\,|(123)|\,\mathscr{T}=\tfrac{1}{2}(T'_{12})\rangle\!\rangle\times\\ &\times\langle pq|\,\langle\mathscr{L}=0(l=0)\,|(123)|\,\mathscr{L}=0(l=0)\rangle\,|p'q'\rangle, \qquad (3.51)\end{aligned}$$

where the product of spin and isospin matrix elements has the value

S_{12}, T_{12} \ S'_{12}, T'_{12}:	1,0	0,1
1,0	$\frac{1}{4}$	$-\frac{3}{4}$
0,1	$-\frac{3}{4}$	$\frac{1}{4}$

For the quartet ($\mathscr{S}=\frac{3}{2}$) $\mathscr{L}=0$ state on the other hand, only the triplet two-body interaction enters and we have

$$\langle \mathscr{S}=\tfrac{3}{2}(1)|\,(123)\,|\mathscr{S}=\tfrac{3}{2}(1)\rangle \langle\!\langle \mathscr{T}=\tfrac{1}{2}(0)|\,(123)\,|\mathscr{T}=\tfrac{1}{2}(0)\rangle\!\rangle = -\tfrac{1}{2}. \tag{3.52}$$

To evaluate

$$\mathscr{P} \equiv \langle pq|\,\langle \mathscr{L}=0(l=0)|\,(123)\,|\mathscr{L}=0(l=0)\rangle \tag{3.53}$$

we insert complete momentum states and use the relation

$$\langle \mathbf{p}''\mathbf{q}''|\,(123)\,|\mathbf{p}'''\mathbf{q}'''\rangle = \delta^3(\mathbf{q}''-[\mathbf{p}'''+\tfrac{1}{2}\mathbf{p}''])\,\delta^3(\mathbf{q}'''+[\mathbf{p}''+\tfrac{1}{2}\mathbf{p}'''])$$

which follows from the relations between the various momenta. The external momenta fix the length of the internal momenta, and we find

$$\mathscr{P} = \int \langle \mathscr{L}=0(l=0)\,|\,\hat{\mathbf{p}}''\hat{\mathbf{q}}''\rangle\,\mathrm{d}\hat{p}''\,\mathrm{d}\hat{q}''\,\delta^3(\mathbf{q}''-[\mathbf{p}'''+\tfrac{1}{2}\mathbf{p}''])\times \\ \times\,\delta^3(\mathbf{q}'''+[\mathbf{p}''+\tfrac{1}{2}\mathbf{p}'''])\,\mathrm{d}\hat{p}'''\,\mathrm{d}\hat{q}'''\,\langle \hat{\mathbf{p}}'''\hat{\mathbf{q}}'''\,|\,\mathscr{L}=0(l=0)\rangle.$$

Doing the integrations over $\hat{\mathbf{q}}''$ and $\hat{\mathbf{q}}'''$ the delta functions become delta functions in the length of the vectors $\mathbf{q}''$ and $\mathbf{q}'''$. Since these lengths are fixed, they are effectively delta functions in the angle between $\mathbf{p}''$ and $\mathbf{p}'''$. Doing all the other angular integrals gives a factor of $8\pi^2$. Since $\langle \mathfrak{L}=0(l=0)\,|\,\hat{p}\hat{q}\rangle = 1/4\pi$ we find

$$\mathscr{P} = \tfrac{1}{2}\int_{-1}^{1} \mathrm{d}\cos\theta_{p''p'''}\,\frac{\delta(q-|\mathbf{p}'''+\tfrac{1}{2}\mathbf{p}''|)}{q^2}\,\frac{\delta(q'-|\mathbf{p}''+\tfrac{1}{2}\mathbf{p}'''|)}{q'^2}. \tag{3.54}$$

Converting the q' delta function to one in $\cos\theta_{p''p'''}$ gives

$$\delta(q'-|\mathbf{p}''+\tfrac{1}{2}\mathbf{p}'''|) = \frac{2q'}{pp'}\,\delta\left(\cos\theta_{p''p'''}-\frac{q'^2-p^2-\frac{1}{4}p'^2}{pp'}\right) \tag{3.55}$$

and

$$\delta(q-|\mathbf{p}'''+\tfrac{1}{2}\mathbf{p}''|) = 2q\delta([q^2+\tfrac{3}{4}p^2]-[q'^2+\tfrac{3}{4}p'^2]). \tag{3.56}$$

In addition since $|\cos\theta|\leqslant 1$ we have

$$\left|\frac{q'^2-p^2-\frac{1}{4}p'^2}{pp'}\right| \leqslant 1$$

or

$$(p-\tfrac{1}{2}p')^2 \leqslant q'^2 \leqslant (p+\tfrac{1}{2}p')^2. \tag{3.57}$$

Putting this all back into equation (3.50) we find

$$\chi_n(pq) = \delta_{n,\,\mathrm{triplet}}\Phi_n(p,q) + 2\sum_{n'}\varphi_{n,n'}\times \\ \times\left\{\int_0^\infty \frac{p'^2\,\mathrm{d}p'}{p}\int_{(p-1/2p')^2}^{(p+1/2p')^2}\frac{\mathrm{d}q'^2}{2}\,\frac{t_n(q,[q'^2+\frac{3}{4}p'^2-\frac{3}{4}p^2]^{1/2};s-\frac{3}{4}p^2)}{s-\frac{3}{4}p'^2-q'^2}\,\chi_{n'}(p'q')\right\}, \tag{3.58}$$

where n and n' run over the two-body channels and the $\varphi_{nn'}$ are the spin-isospin coefficients. As we see, in addition to being a two-dimensional equation, the kernel of the three-body equations have singularities arising from the T-matrix and the Green's function and the limits on the q' integration are variable. This makes the equations particularly poorly suited for a solution by the matrix method. Current progress has been achieved by using an iterative technique to solve these equations.

The T-matrix enters into the three-body equations completely off the energy shell. Thus we need information about the two-body interaction which is not obtainable from two-body experiments alone.

If the two-body T-matrix is separable, the three-body equations reduce considerable in complexity. Assuming

$$T_3 = |g\rangle \, \Delta \, \langle g| \tag{3.59}$$

we have that Equation (3.45) becomes

$$|\chi_3\rangle = |\chi^{(i)}\rangle + 2\,|g\rangle \, \Delta \, \langle g| \, G_0(123) \, |\chi_3\rangle \,. \tag{3.60}$$

Writing

$$|\chi_3\rangle = |\chi^{(i)}\rangle + |g\rangle \, \Delta \, |\Gamma\rangle \tag{3.61}$$

which defines $|\Gamma\rangle$, we obtain

$$|\Gamma\rangle = \langle g| \, G_0(123) \, |\chi_3\rangle = \langle g| \, G_0(123) \, |\chi^{(i)}\rangle + \langle g| \, G_0(123) \, |g\rangle \, \Delta \, |\Gamma\rangle \,. \tag{3.62}$$

Since $|g\rangle$ determines the dependence upon the relative coordinates. The 'spectator function $|\Gamma\rangle$ depends only on the coordinates of the third particle. After taking this equation for $|\Gamma\rangle$ into the momentum representation and performing an angular momentum reduction we are left with a system of one-dimensional integral equations. The number of equations depends, among other things, upon the number of separable terms used to represent the T-matrix. These equations are much easier to solve than those with local interaction and a large number of investigations have been done using separable potentials.

4. The Three-Nucleon System

We now move from our formal considerations on the three-body problem to a discussion of the three-nucleon system. Several excellent reviews covering work up through 1969 are available [1–3].

The three years since these reviews were written have seen considerable progress made in this field. In these lectures we will review some of the well established aspects of the three-nucleon system and then focus attention upon recent developments.

The primary reason for studying the three-nucleon system is to learn about the interactions between nucleons. As we have seen the two-nucleon system is somewhat restrictive in this area since for elastic scattering it gives us information about the on-shell T-matrix only. Measurements on the deuteron give information about the important deuteron pole in T, but even here information is still missing.

The properties of the three-nucleon system however depend upon the fully-off-shell

T-matrix, and we hope to exploit this dependence to learn about the off-shell behavior of T. The sensitivity of the three-nucleon system to our choice for the two-nucleon interaction will be an important focus for these lectures. We may expect that explicit three-body forces and additional relativistic effects will also enter. The question of how to distinguish between different unknown features of the nuclear interactions is of considerable current interest. In fact to a large extent it is the uncertainty in the three-body force which is currently the main impediment to the use of the three-nucleon system for the testing of two-nucleon potentials.

In principle we might hope to learn about the off-shell T-matrix from studying the interaction of the two-nucleon system with another weakly interacting particle as in nucleon-nucleon bremsstrahlung or deuteron photodisintegration. Unfortunately this has not been the case and in many instances the same theoretical uncertainties which inhibit our understanding of the two-nucleon interaction also make our extraction of information from such experiments subject to doubt [2–3].

4.1. The three-nucleon bound state

Experimentally the triton energy, E_T, has the value -8.48 MeV. For a given two-body potential form, this is the easiest three-body quantity to calculate. Fortunately, along with the doublet scattering length, it is also one of the quantities most sensitive to the form of the two-body potential and is therefore of major importance.

Calculations on the triton have a long history, some details of which may be found in the review articles. Here we will content ourselves with indicating what is well established in the field, taking examples from recent calculations, and considering areas of current interest.

Since the three-body problem is to a large extent a problem of numerics, we first briefly review the calculational methods used. Of the important methods, the oldest and most frequently used is the variational method. An early application (1935) of this method by Thomas [39] showed that a zero range two-nucleon force (at that time consistent with experiment) would give infinite binding in the triton. Probably the most famous and certainly the most extensive series of calculations in the three-nucleon field is the work of Blatt, Delves and co-workers [40, 41] on the Hamada-Johnston potential. We will review the latest version [40] of this calculation as an indication of the difficulty engendered in the use of a modern phenomenological potential in the three-nucleon system.

In principle the variational method is simple, one chooses a trial wave function containing a number of variable parameters, calculates the expectation value of the Hamiltonian and then varies the parameters to minimize this expectation value obtaining an upper bound to the energy and an estimate for the wave function. In practice, for realistic potentials, it is quite difficult. This is in large part due to the complexity of the two-nucleon interaction. Any such interaction produces strongly correlated wave functions due to the strong, short range repulsions. The existence of tensor forces, spin-orbit forces and quadratic spin-orbit forces requires the inclusion of all possible wave function configurations. The recent progress with the Hamada-

Johnston potential has come about due to the inclusion in the trial function of a 'core' wave function which is similar to the deuteron and to the emphasis on linear variational parameters. The former helps to incorporate the strong correlation induced by the interaction while the latter reduces the amount of computational effort necessary and assures sufficient flexibility of the basis set to guarantee eventual convergence. A maximum of 59 linear parameters was used in the latest calculation. Throughout careful checks are made on the numerical accuracy of the results. After 10 yr of effort the calculation has converged to an energy of -6.5 ± 0.2 MeV. The uncertainties are numerical.

Another recent variational calculation is that of Jackson *et al.* [42] who use a trial function expanded in three-body harmonic oscillator states to calculate E_T for the Reid potential. They use as many as 478 linear parameters and the oscillator length is a non-linear parameter.

More recent than the variational method is the use of separable potentials. As we have seen, by specifying the functional form of the two-body subsystems, the separable approximation reduces the three-body problem to an equivalent two-body problem and the two-dimensional integral equations reduce to coupled one-dimensional integral equations. The solution to the boundstate problem using separable potentials is usually considered to be exact in the sense that the answer can be calculated to an accuracy which is much greater than is presently needed for comparison with experiments. Not all calculations have achieved this accuracy but good numerical methods are available and very extensive calculations have been performed with apparently reliable results even for scattering problems where the kernels develop singularities.

Most recently, methods have been developed for the solutions of the Faddeev equations using local potentials. One class of such methods involving separable expansions of the local potential T-matrix has already been discussed. These methods reduce the local problem to an equivalent multi-rank separable potential problem, with the corresponding simplification to coupled one-dimensional equations.

Methods have also been developed for solving directly the two-dimensional Faddeev equations, the most important being the iteration method developed by Malfliet and Tjon [14, 43, 44]. This method has been applied to evaluate E_T for a number of realistic local potentials. The theoretical framework can be found in ref. [43]. In brief, if we consider the homogeneous Faddeev equations with the addition of a linear eigenvalue

$$\phi = \lambda K_F(E)\, \phi$$

then the function

$$\phi_n = K_F^n(E)\, \phi_0$$

in the limit as $n \to 0$ approaches the eigenfunction associated with the eigenvalue of smallest absolute value λ_0. The ratio

$$\frac{\phi_n}{\phi_{n+1}} \to \lambda_0 .$$

In practice this eigenvalue is close to 1. By making successive choices for E we can eventually find an E such that $\lambda_0(E)=1$. This energy and the resulting eigenfunction are the solutions to the bound state problem. (The fact that the smallest eigenvalue is the desired one is an indication of the reduction in the T-matrix of the repulsive part of the two-body potential. For a realistic local potential the smallest eigenvalue of the Lippmann-Schwinger equation is about -0.06 and there are four or five negative eigenvalues less than 1 in absolute value [16]).

The great usefulness of an iterative method of solution is that we do not have to invert, diagonalize, or store large matrices. We are thus able to handle the solution on a computer of reasonable size and can also be confident of obtaining a reasonably

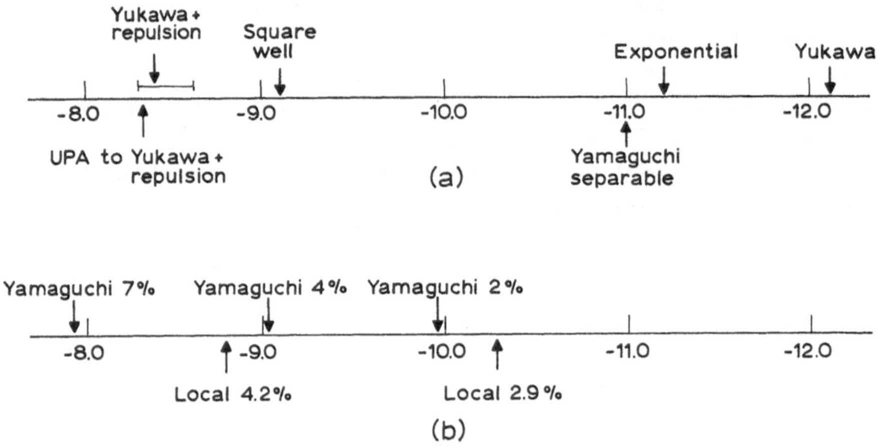

Fig. 4. Some typical triton energies obtained from spin dependent central interactions (4a) and purely attractive potentials containing a tensor force in the triplet state (4b).

accurate solution. For the Faddeev equations, the complicated domains of integration which occur make a matrix method of solution particularly difficult. As we shall see the application of iterative methods in scattering problems has also provided a useful computational scheme.

As is to be expected the form we choose for the two-body interaction has a large effect upon E_T. Purely attractive central potentials overbind the triton.

In Figure 4a, we show the energies obtained from typical examples of such potentials adjusted to fit the singlet and triplet effective range parameters. Choosing different forms for the potential produces large variations in E_T, and we conclude that E_T is sensitive to other aspects of the two-nucleon interaction in addition to the low energy data.

Further investigation of the sensitivity of E_T to the two-body input were made by Kok *et al.* [45], who calculated the binding energies of three-particles interacting by means of three different spin-independent central potentials. Some results are shown in Table I.

TABLE I

Triton energies for purely attractive potentials

Potential	E_T (MeV)
Local Hulthen	−14.6
Yamaguchi	−12.5
Local Bargmann	−10.9

The Yamaguchi potential is a separable potential of the form $V(p, q) = -\lambda^{-1} g(p) g(q)$ with $g(p) = (\beta^2 + p^2)^{-1}$ and is the UPA to the Hulthen potential. Hence it has the same bound state wave function. The Bargmann potential was constructed to reproduce the Yamaguchi phase shifts. These results indicate that for the potentials considered, neither the phase shifts nor the bound state wave function are by themselves sufficient input to determine E_T.

That this may *not* be the case for more realistic potentials is indicated by considering central potentials with a short range repulsion which are fit to the nucleon-nucleon data up to 300 MeV. The results of calculations by Malfliet and Tjon [14] and Erens [46] using such potentials are also shown in Figure 4a. More detailed numbers are given in Table II along with the values for E_T calculated with the UPA to these local potentials.

TABLE II

Triton energies for potentials containing short range repulsion

	MT I and III		MT I and IV		MT II and IV
MT [14]	8.3 ± 0.1		8.4 ± 0.1		12.1 ± 0.1
Erens [46]	8.5	(8.8)	8.4	(8.7)	12.1
UPA [13]	8.5		8.4		10.4

Potentials I and II are singlet potentials, III and IV are triplet potentials, and I and III contain repulsive cores. On line 2 the numbers in parentheses are Eren's calculated values which include contributions from all even L partial waves while lines 1 and 3 use only S waves. The numbers without parentheses have been corrected for these additional contributions and should be used for comparison.

The values for E_T obtained with UPA agree well with those obtained from the local potentials which contain repulsion. However for the last column, in which the potentials are purely attractive, the discrepancies are similar to those found by Kok *et al.* (Table I). These results support the conclusions of Section 1 where the pole dominance was displayed by examining the two-body T-matrix.

Additional confirmation of pole dominance and an indication of the corrections to be expected from the neglected parts of T come from studying the convergence of separable expansion calculations. In Table III we show the results of a UPE calculation [16] for spinless bosons acting through the average of potentials I and III of Table II.

TABLE III

Convergence of a unitary pole expansion calculation of the triton energy

Terms		E_T (MeV)
1 A (UPA)		7.4
2 A + 1 R		−7.49
3 A + 2 R		−7.51
1 A + ∞ R		−7.35
∞ A		−7.64
∞ A + ∞ R		−7.55
	Malfliet-Tjon [14]	−7.3 ± 0.1
	Erens [46]	−7.5 (7.8)

The 2 $A+1$ R result for example is obtained by keeping the UPA and in addition one attractive and one repulsive correction term in the UPE. The ∞A or R results are obtained by an extrapolation. All the repulsive terms contribute only 0.9 MeV to E_T while the non-pole attractive terms contributes −0.2 MeV, indicating that the accuracy of the UPA energy reflects the dominance of the pole and *not* the near cancellation of large attractive and repulsive corrections. Similar results have been found using the UPE for the Reid $^1S_0$ potential in conjunction with a separable triplet tensor potentials [16]. We conclude that for the more realistic potentials considered here, E_T is determined mainly by the T-matrix poles, and the results of Kok *et al.* are characteristic of purely attractive potentials. We will return to the question of pole dominance at various points in these lectures.

Some effects on E_T of the introduction of tensor forces are shown in Figure 4b. The triton binding energy decreases as the strength of the tensor potential (as measured by P_D, the percentage of D-state in the deuteron) increases. This is to be expected. Tensor forces depend upon the orientation of the relative position vector with respect to the total spin favoring situation (a) of Figure 5 as compared to (b).

(a) $\uparrow \boldsymbol{\sigma}_1$ / $\uparrow \boldsymbol{\sigma}_2$ (b) $\boldsymbol{\sigma}_1 \uparrow \uparrow \boldsymbol{\sigma}_2$

Fig. 5.

In the deuteron the nucleons can arrange themselves to take advantage of this preference. In the triton, however, this is not possible to as large an extent and the tensor force loses some of its effectiveness in binding.

From the values shown, we see that local and separable results are reasonably similar. However, one should be careful in drawing any conclusions. For separable potentials a wave function with small P_D (all fit the quadrupole moment) appears to be very long range and unrealistic. For the local potential, however, the low energy data could not be fit for P_D values greater than 4.2% indicating perhaps a basic difference between the two types of interaction. Replacing the local potential by a Yamaguchi separable potential, Malfliet and Tjon [43] find the latter gives about 0.3 MeV more binding.

We see that E_T depends quite strongly upon P_D. Experimentally this quantity is not well known. We know the deuteron has a D-wave component since it has a quadrupole moment of about 0.28 F^2. This number alone cannot fix P_D however, since it depends upon the radial form of the wave function. An estimate of P_D can be made from the magnetic moment of the deuteron. The difference of 0.22 μ_n between experiment and the sum of the neutron and proton magnetic moments can be explained if we assume $P_D \approx 4\%$. However, we realize that the magnetic moment is also sensitive to various 'interaction' effects such as meson exchange or non-localities in the two-nucleon potential. Since we don't know how to calculate the contribution of these effects, P_D is still undetermined.

Instead of using the magnetic moment to fix P_D, we might try to predict P_D using potentials that fit the other two-body data. If we choose a *local* potential having an OPEP tail, fitting the deuteron and low-energy data and having a hard or soft-core to fit the high energy phases, we find $P_D = 7\%$. On the other hand, a *non-local* description of the core region by means of a boundary condition model can fit the same data with only 4.5% D state. (If we are willing to accept a separable parametrization of the interaction, then we can get a P_D as low as 1%). The choice between different forms for the short range nucleon-nucleon potential cannot be made for much the same reasons that we cannot calculate 'interaction' effects.

Levinger [37] has suggested a possible means of directly measuring P_D using elastic electron-deuteron scattering and observing the tensor polarization of the recoil deuterons. In Figure 6 we show polarization results obtained by Brady [38] for different values of P_D. A rough measurement would serve to rule out the low values of P_D

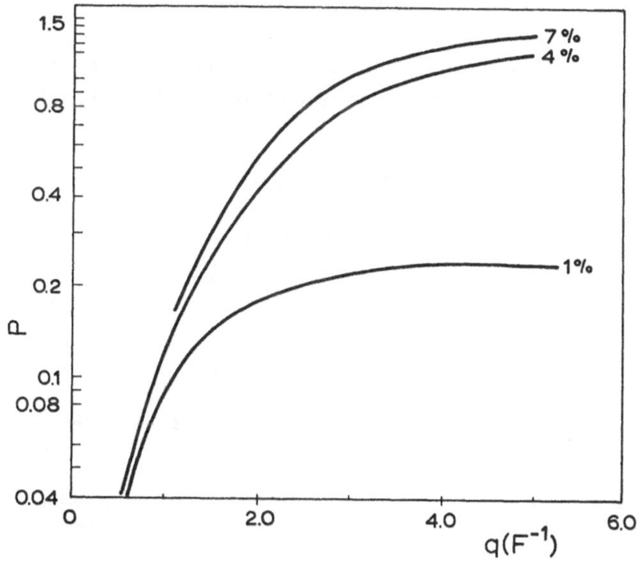

Fig. 6. Tensor polarization of recoil deutron in elastic electron-deuteron scattering as a function of the deuteron percent D-state and the momentum transfer. The figure is taken from ref. [37].

(already highly unlikely from other considerations). A precise measurement would be needed to distinguish P_D's of 4 to 7%. The extent of interaction effects in such an experiment are unknown but may be small since the magnetic and electric scattering can be separated.

From the central potentials with repulsion, we obtain values of E_T which are close to experiment. Since the inclusion of tensor forces further reduces the triton binding, we may expect that modern phenomenological potentials which contain both of these components will underbind the triton. This is seen in Figure 7, where we show triton energies obtained from such potentials.

The most elaborate calculations using separable potentials have been performed by Brady [11, 47] who has calculated E_T using Mongan's separable potentials and also a modified version of Tabakin's potentials. Unfortunately all of these potentials possess some drawback due to very low values for P_D or inaccurate fits to the deuteron or low energy data. In Figure 7 we have attempted to adjust Brady's published numbers to account for the errors in the Mongan effective ranges. The Schrenk-Mitra value is obtained from a purely attractive separable triplet potential and a rank-two separable singlet potential that fits the two-nucleon scattering data. The value shown is that recalculated by Brady [11] and gives 0.7 MeV less binding than the value published by Schrenk and Mitra [48]. The UPA to Reid result [49] will be discussed in more detail later. It is obtained by using the UPA to both the singlet and triplet Reid soft core potentials. The value for E_T is similar to other values obtained [11] for rank-3 separable potentials that have $P_D=7\%$ and repulsion in the triplet and singlet states.

The values for E_T from the model local tensor potential (⊗ in the figure) have been shifted a very large amount by the addition of repulsion in the singlet (about 2.8 MeV for $P_D=2.9\%$ and 1.7 MeV for $P_D=4\%$) as compared to similar changes in the separable potentials.

For the realistic local potentials, we see definite indication of underbinding of the triton. Just how much underbinding is not clear however. The Hamada-Johnston

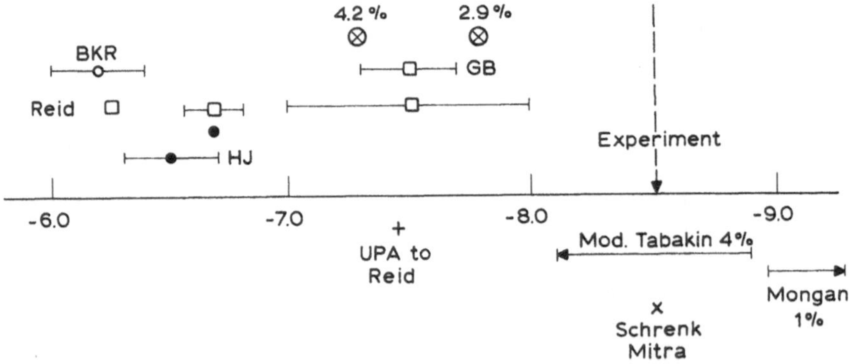

Fig. 7. Triton energies from calculations with realistic nucleon-nucleon potentials. All results contain only $^3S_1$-$^3D_1$ and $^1S_0$ interaction. Calculated results which included other partial waves have been increased by 0.25 MeV. References for the values given are included in the text

potential result [40] is shown along with a confirming value of 6.7 MeV found by Hu [50] and this value is at the present time well established.

The situation for the Reid potential is much more confused. Calculated triton energies with the Reid potential are shown in the table below. The variational calculations include two-body partial waves other than $^3S_1$–$^3D_1$ and $^1S_0$ and the contribution to E_T from the other partial waves has been subtracted out from the published numbers, which are in parentheses (Table IV).

TABLE IV

Comparison of triton energy calculations with the Reid potential

	Full calculation for $^3S_1$–$^3D_1$ and $^1S_0$	Truncated wave function	S-wave T-matrix only	Method
Jackson *et al.* [42]	$-6.25(-6.50)$	–	–	Var.
Tjon *et al.* [43, 51]	–	-6.5 ± 0.5	-6.8 ± 0.5	Fadd.
Harper *et al.* [52]	-6.7 ± 0.07	-6.4 ± 0.06	-6.7 ± 0.05	Fadd.
Hennel and Delves [53]	-7.5 ± 0.5 $-(7.75)$	–	–	Var.
UPA [49]	-7.45	-6.8	-7.0	Sep. Pot.

The second and third columns show results obtained in the Faddeev approach by neglecting that part of ψ_3 which has the third particle in a D-state, and by retaining only the S-wave part of the two-body T-matrix respectively. The Faddeev results are consistent with one another but neither of the three local potential results overlap. The differences between the calculations are large when compared to the discrepancies with experiment.

Rather than deal with (apparently) approximate numerical methods, we can look at an approximate but exactly solvable problem and use the UPA to calculate E_T. Results are shown in row five of Table IV. We see that for the S-wave T-matrices, the UPA result is in good agreement with the local potential calculations (rows 2 and 3). This reflects the accuracy with which the UPA reproduces the S-wave part of the singlet and triplet T-matrices [16, 49]. Using a truncated wave function, we find somewhat poorer agreement. The largest discrepancy in behavior arises when going from the truncated to the full wave-function. The UPA value for E_T decreases by 0.65 MeV while Harper *et al.* find a decrease of only 0.3 MeV indicating different behavior between a rank-2 separable and a local tensor potential. That the UPA might also underestimate the repulsive effects of the tensor force is indicated by the comparison of the phase shifts for the Reid and UPA potentials in Figure 8.

This disagreement in E_T (if in fact it does exist) should not necessarily be thought of as an off-shell effect. We would expect such effects to be contained primarily in the S-wave T-matrices, where the UPA works quite well indicating deuteron dominance. It seems more likely that the discrepancies would result from poor reproduction of the on-shell T-matrix by the UPA. Whether or not the inclusion of the correct on-shell behavior

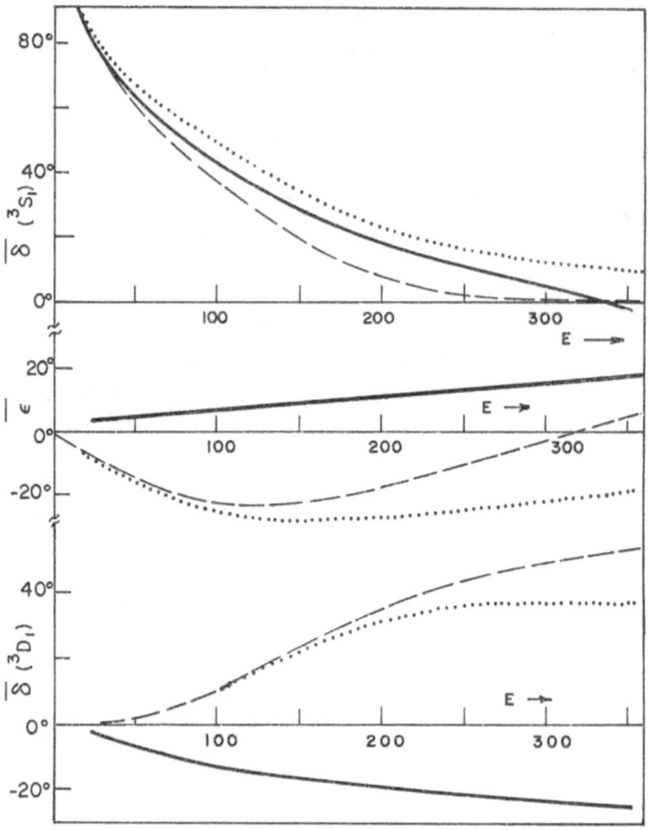

Fig. 8. Phase shifts and mixing parameter for the Reid $^3S_1$–$^3D_1$ potential (——), for the UPA (----), and for a separable Yamaguchi tensor potential (.....). The figure is from ref. [37].

would bring the UPA result into agreement with the Faddeev calculations is not known. It is of course in excellent agreement with the Hennell and Delves result.

The other values shown in Figure 7 are a calculation by Hennell and Delves [53] on the Gammel-Brueckner potential ($E_T = -7.5 \pm 0.2$ MeV) and by Malfliet and Tjon [44] using the BKR potential ($E_T = -6.2 \pm 0.2$ MeV). Since the latter calculation uses a truncated triton wave function this result should probably be increased by 0.3 MeV.

From the results shown here, we can conclude that phenomenological potentials of the type for which calculations have been per formed (i.e. with a local parametrization of the short range repulsion and $P_D \approx 7\%$) underbind the triton by 1 to 2 MeV. The additional binding energy must therefore be accounted for in terms of three-body forces, relativistic effects or inadequacies in the form chosen for the two-body potential.

The computational aspects of the different calculations with local potentials are interesting. Delves and Hennell use a core wave function and 59 linear parameters to obtain the Hamada-Johnston energy. The build up of numerical uncertainties limits the use of more terms in this calculation and also in the Reid calculation. Jackson *et al.* use 487 linear parameters and report no difficulties with the build up of numerical errors. The Faddeev calculations would appear to be the simplest of all in terms of programming and computer time. The uncertainties reported in the latest calculations are quite small. It may be that the appearance of the T-matrix rather than the potential produces numerical simplifications. The approximate separability of T will lead to a comparitively simple wave function and the suppression in T of the large momentum components in the potential should make their effects easier to treat.

We now turn attention to the effects of the unknown properties of the T-matrix upon the triton energy. In particular we want to see to what extent E_T is determined by the two-body scattering data and the bound and antibound state poles. Since it is a well known fact that the bound states and phase shifts uniquely determine a *local* potential, we are forced to consider the effects of non-localities to answer this question. More precisely since the two-body phase shifts are measured only to some finite energy, there is still some ambiguity in the two-body input. Fiedeldey [54] has investigated the dependence of E_T upon the high energy two-body phase shifts and found little effect from varying the high energy phases. In a non-relativistic model, these phases are related to the parametrization of the short range repulsion. Folk and Bonnem [55] and Afnan and Tang [56] have found that velocity dependent, hard-core or soft-core representations of the repulsion give essentially equivalent results for E_T. Related results have been obtained by Kharchenko *et al.* [57]. They use square well potentials containing a repulsive square well cores of different ranges but fit to approximately the same phase shifts. They find that E_T is essentially independent of the size of the core. Their calculation is done using the Weinberg expansion of the T-matrix. In all cases the 1 term result (analogous to the UPA) is quite accurate.

Explicitly off-shell effects have been investigated by a number of authors. Fiedeldey [58] has shown that if one uses a two term separable potential to fit a phase shift which has one sign change, then one of the separable form factors may be taken arbitrarily. He uses this freedom to investigate E_T keeping the on-shell T-matrix constant and finds large variations. This procedure however does not leave the wave function constant. If we apply the requirement of constant phase shift and wave function to Fiedeldey's work then the two-term potential is uniquely specified and no off-shell variation is possible.

In Figure 9 we show the results for E_T found by Afnan and Serduke [59]. These authors and others construct potentials that lead to identical on-shell but different off-shell T-matrices by applying to the Hamiltonian a unitary transformation of the form $U=1-2\Lambda$ with $\Lambda+\Lambda^+=2\Lambda\Lambda^+$. If $\langle r|\,\Lambda\,|r\rangle\to 0$ faster than $1/r$ as $r\to\infty$ then two Hamiltonians H and $\tilde{H}=UHU^+$ will have identical phase shifts, but the transformed potential (which is non-local $\tilde{V}=\tilde{H}-K$ will give rise to different off-shell behavior. These authors apply such transformations to a spin dependent central separable potential. In the singlet state, they place no restriction upon the variations and find large changes in E_T even for very short range variations. In the triplet state, they require that the deuteron be unchanged by the transformation and find much

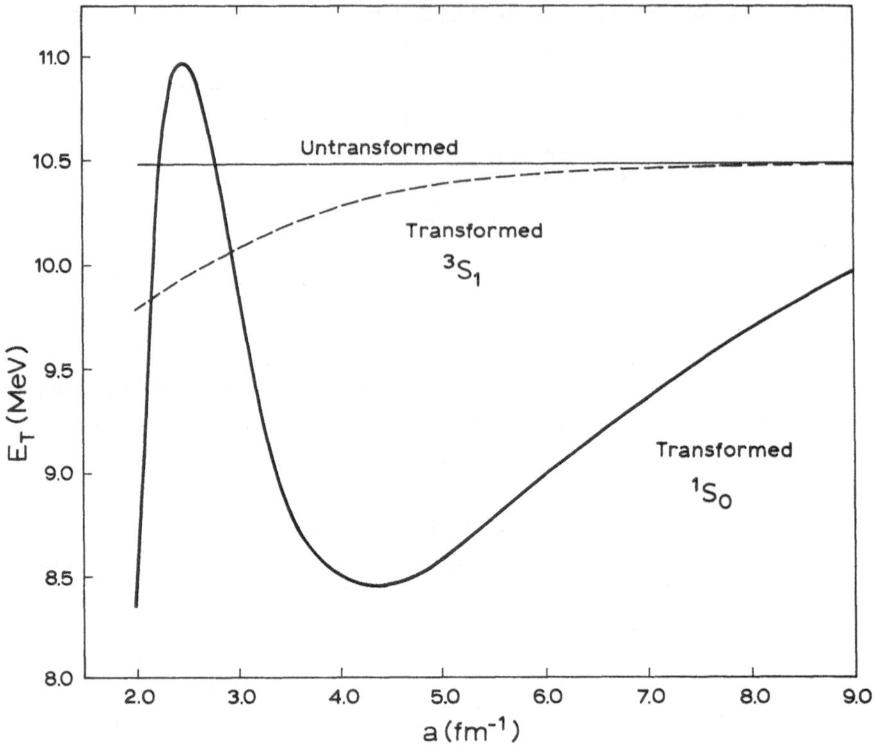

Fig. 9. Triton energies obtained by making a unitary transformation on the two-body interaction. The inverse range of the transformation is a. The figure is from ref. [59].

smaller variation which disappear rapidly as the range of Λ decreases. Similar results have been found by Hadjimichael and Jackson [60] and by Haftel [61]. The former use the same transformation method as Afnan and Serduke applying it to the Reid potential. For variation in the singlet, they find changes in E_T similar to Figure 9. In the triplet, if the deuteron is unchanged, E_T changes only slightly but if the wave function (as determined by the electron form factors) changes, E_T changes appreciably also. Haftel finds similar results using the Faddeev equations for the local spin independent potential of MT (Table III). If he considers free variations of the off-shell matrix elements he finds -3.6 MeV$\leqslant E_T \leqslant -9.7$. Fixing the deuteron, however, he finds $-6.5 \leqslant E_T \leqslant -7.6$ MeV. Using similar transformations in nuclear matter, the binding energy per nucleon ranges between 6 and 33 MeV.

Thus it would seem that given the deuteron and on-shell T-matrix, the triton will have little sensitivity to off-shell variations in the triplet channel. In the singlet channel we cannot hope to measure the anti-bound state form factor directly. If, however, we have a theory which agrees with the experimental observations on the deuteron, we may hope that a similar theory will give us a good extrapolation in the singlet state. As Levinger has said [37], the deuteron plays a crucial role.

Assuming that the postulated insensitivity of the triton to off-shell effects is true, it may be possible to combine the triton and nuclear matter to disentangle the effects of off-shell behavior and three-body forces. Haftel and Tabakin [62] and others have shown that nuclear matter is very sensitive to such off-shell effects. The procedure we envision is using the triton to fix the contribution of three-nucleon forces and then using nuclear matter to evaluate possible off-shell effects. Of course, when we learn how to solve the four-body problem, we will have another testing ground.

The difference between the binding energy of H^3 and He^3 is 0.76 MeV. If we assume the charge symmetry of nuclear forces, this difference must be due to the Coulomb repulsion between the protons in He^3. Nearly all estimates of the Coulomb energy fall short of this value by about 0.1 MeV. For example Delves and Hennell [40] find that for the Hamada-Johnston wave function, the Coulomb energy is 0.56 MeV. Since the triton is underbound by this potential and is therefore more spread out, we can expect a larger value for the Coulomb energy from a wave function which fits the experimental energy. However the addition of a three-body force to bring the energy to -8.5 MeV increases the Coulomb energy only to 0.61 MeV [40]. In order to obtain the correct Coulomb energy by this method, it is necessary to bind the triton by about 16 MeV. It should be noted that in this calculation, the Coulomb interaction has been added to the nuclear force and then used in the variational calculation. This should eliminate the difficulty that wave functions obtained in variational energy calculations are not always reliable for the calculation of other expectation values. An extensive study of the Coulomb energy problem has been made by Okamoto and Lucas [63] who found that the best they could do for the Coulomb energy was 0.65 MeV. Similar results have been found by Erens [46] who also includes the Coulomb interaction in the potential used in a variational calculational. Finally de la Rippelle [64] has used symmetry considerations to relate the Coulomb energy to the electron scattering form factors. This model independent estimate of E_C also falls 0.1 MeV short of experiment. These results would seem to indicate the presence of charge asymmetry in the nucleon-nucleon interaction. An increase of about 1% in the contribution of the n-n interaction to the triton binding as compared to the p-p interaction would explain this discrepancy.

In addition to obtaining energies from the various calculations of E_T we also obtain a three-body wave function. The triton is a $\mathscr{J}^\pi = \frac{1}{2}^+$ system and, aside from very small admixtures of $\mathscr{T} = \frac{3}{2}$ state, has $\mathscr{T} = \frac{1}{2}$. The identification of all states that have these quantum numbers has been performed. In L-S coupling the possible states are ${}^2S_{1/2}$, ${}^2P_{1/2}$, ${}^4P_{1/2}$ and ${}^4D_{1/2}$. These states can be further classified according to symmetry properties under the permutations of the three-particles. Only the S-states however are obtained accurately enough from calculations or distinguished sufficiently in experiments to warrant discussion of separate states.

The permutation group of three objects has three different irreducible representations, symmetric, antisymmetric and mixed. For spin or isopin states we can have only the symmetric ($\mathscr{S}$ or $\mathscr{T} = \frac{3}{2}$) or mixed ($\mathscr{S}$ or $\mathscr{T} = \frac{1}{2}$, S_{12} or $T_{12} = 0$ or 1). The total spin-isopin states can have all three representations however. These states must then be combined with appropriate functions of the coordinates to get a totally antisymmetric state. For the S-states all the permutation types can occur. The symmetric component is the most important forming roughly 90% of the wave function. The antisymmetric S-state is very unlikely. The mixed symmetry S-state (commonly referred as S' state) is not negligible and is of interest due to its importance in a number of reactions.

Values for S, S' and D components of the wave function from recent calculations give in Table V.

TABLE V

Different wave function components in the triton from recent calculations

Potential	$P(S)$	$P(S')$	$P(D)$
Hamada-Johnston [40]	89	1.8	9.0
HJ + three-body force [40]	89	1.0	10.0
Reid-HD [53]	90	≈ 1.0	9.5
HKT [52]	90	1.7	8.6
JLS [42]	90	0.5	9.0
MT [43]	91	2.5	6.5
BKR [44]	91	1.4	7.8

The percentage of P states is very small. With a few exceptions we see that $P(S) \approx 90\%$, $P(D) \approx 9\%$ and $P(S') \approx 1$–2%. These results are also in agreement with those obtained from separable potential calculations.

In addition to the percentages of various states in the triton, we are interested in the functional form that the wave function has. Forgetting about spin for the moment, the wave function may be regarded as a function of the length of two vectors, $\mathbf{r}_{12}$ and $\boldsymbol{\varrho}_3$, and the angle between them. Rather than discussing a function of three-variables we will consider the individual dependences upon the lengths.

The dependences upon r_{12} is contained in the two-body correlation function $g(x)$ defined as

$$g(x) = \int \mathrm{d}\Omega_x \int \langle \psi_t \mid \mathbf{r}_{12}, \boldsymbol{\varrho}_3 \rangle \, \delta^3 (\mathbf{r}_{12} - \mathbf{x}) \, \mathrm{d}^3\mathbf{r}_{12} \, \mathrm{d}^3\boldsymbol{\varrho}_3 \, \langle \mathbf{r}_{12}, \boldsymbol{\varrho}_3 \mid \psi_t \rangle$$

and giving the probability distribution in r_{12} independent of the other variables. There is considerable interest in this function, but it is very difficult to measure experimentally. Given a three-body wave function however it is a simple matter to calculate. This has been done by Hadjimichael [65] for the Reid wave function of ref. [42]. As is to be expected the correlation function looks very similar to the deuteron at small inter-particle separations. This behavior is also retained in the UPA to the Reid potential wave function [66] as seen in Figure 10.

The single particle correlation function (i.e. with respect to $\boldsymbol{\varrho}_3$) is easier to measure experimentally, for example by electron scattering, and is of considerable interest since the measurements by McCarthy *et al.* [67] showed the existence of a minimum in the charge from factor of He^3 at a momentum transfer squared of 11.6 F^{-2}. This minimum is believed to be related to the short range repulsion in the two-nucleon interaction. As a result it is of interest to see if wave functions obtained from current

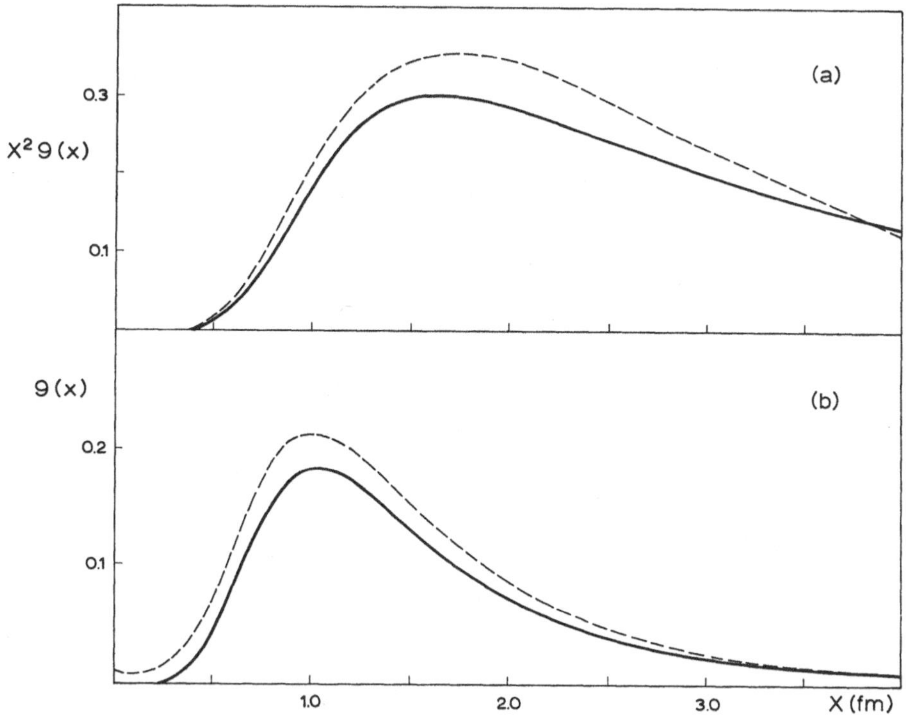

Fig. 10. The two-body correlation function in the triton (dashed curve) obtained from the UPA to Reid wave function. The figure is from ref. [66]. The solid curve is the deuteron correlation function.

TABLE VI
Charge radii and the position of the minimum in the H^3 charge form factor from recent calculations

Potential	R_{ch}(He)	R_{ch}(H)	Position of minimum
Hamada-Johnston	1.90	1.85	12.5
+ three-body		1.70	14
Reid-HKT	1.96		15.5
Reid-YJ		1.81	12.9
Reid-TGI	2.05	1.80	17.0
Reid-UPA	1.97	1.76	17.0
Gammel-Brueckner	1.74	1.65	15.0
Experiment	1.87	1.70	11.6

two-body force models can fit the observed form factors. In Table VI we show some results from recent calculations.

A number of results appear, in particular the potentials shown all give a minimum at too large a value of q^2. This is particularly relevant since all the local potentials underbind the triton by about 2 MeV and the work on the Hamada-Johnston potential shows that the addition of a three-body force to bring the energy into better agreement with experiment also increases the value of q^2 at the minimum. The two Faddeev

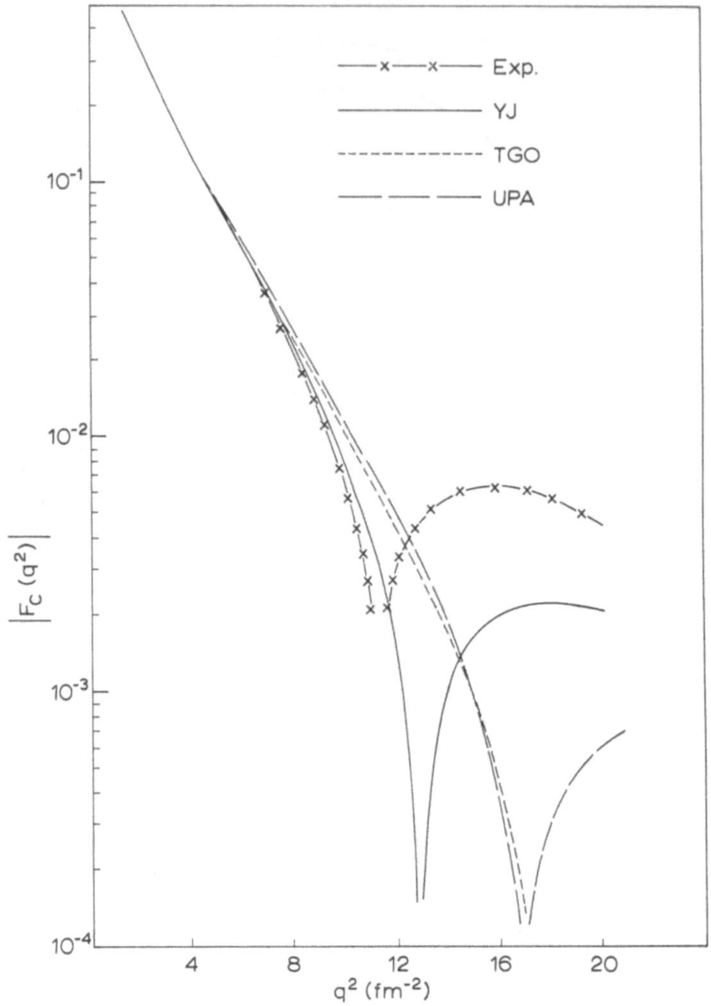

Fig. 11. Some examples of theoretical calculation of the charge form factor of He^3. The experimental values are from ref. [67]. The calculations are YJ (ref. [61]), TGO (ref. [51]) and UPA (ref. [66]). The figure is from ref. [66].

calculations and the UPA result [66] are in agreement with one another for the Reid potential, while that of Yang and Jackson [68] gives a much smaller value. The charge and magnetic radii are in reasonable agreement with experiment and should be quite close if the correct binding energy of the triton is obtained. An example of the over all fit to the form factors is shown in Figure 11. In addition to having difficulty in fitting the location of the minimum, the recent calculations are unable to obtain a large enough secondary maximum. From their studies with wave functions of different size cores, Hennel and Delves find that a wave function that fits the triton energy and also the minimum in the form will have a core region 0.54 F in size.

4.2. The scattering of three-nucleons

Some of the most successful applications of current three-body theory have come in the study of the scattering of three-nucleons. It now appears that the main features can be represented using simple forms for the two-body interaction in the framework of an exact three-particle theory such as the Faddeev equations.

Neglecting non-central forces for the moment, nucleon-deuteron scattering occurs in either the quartet ($\mathscr{S}=\frac{3}{2}$) or doublet ($\mathscr{S}=\frac{1}{2}$) spin state but since the deuteron is an isosinglet, we have $\mathscr{T}=\frac{1}{2}$ only. Because the quartet spin state is completely symmetric under permutation of the particles and the isospin state is necessarily of mixed symmetry, the coordinate state must also be of mixed symmetry. This has the effect of keeping the nucleons apart and the quartet is quite insensitive to the form used for the nucleon-nucleon interaction. The doublet state, which contains the triton is more sensitive to the form of the interaction. However, as we shall see, the predominant scattering occurs in the quartet state and the sensitivity of the doublet is largely masked.

To begin our study, we consider the multiple scattering series obtained by iterating Equation (II) for $U_{\beta\alpha}$. We find

$$U_{\beta\alpha} = (1 - \delta_{\beta\alpha})(E - K) + \sum_{\beta\neq\gamma\neq\alpha} T_\gamma + \sum_{\beta\neq\gamma\neq\delta\neq\alpha} T_\gamma G_0 T_\delta + \cdots$$

and refer to the different terms according to the number of T-matrix factors which enter. Thus, the first term is the zeroth-order while the impulse terms are the first order. Written out to first order we find for the elastic transition operator,

$$U_{33} = T_{23} + T_{13} + \cdots$$

Here we have written T_{23} instead of T_1 to make the physical content of the equations clearer. For the rearrangement operators we find

$$U_{13} = (E - K) + T_{13} + \cdots$$
$$U_{23} = (E - K) + T_{23} + \cdots$$

and for break-up

$$U_{03} = T_{13} + T_{23} + \cdots$$

We have used the vanishing of $(E-K)$ on-the-energy-shell for the break-up reaction. A diagrammatic representation of these operators is given in Figure 12.

The zeroth order term plays a very important role in nucleon-deuteron scattering. Calculating its matrix elements we find (using Equation 2.10)

$$\langle\phi_1, \mathbf{p}_1| (E - K) |\phi_3, \mathbf{p}_3\rangle = \langle\mathbf{p}_1| \langle\phi_1| V_1 \frac{1}{E - K} V_3 |\phi_3\rangle |\mathbf{p}_3\rangle .$$

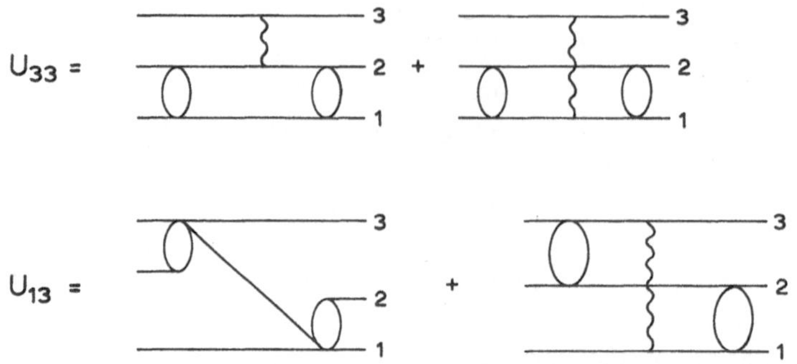

Fig. 12. Diagrammatic representation of the terms in the multi-scattering series. Circles correspond to the deuteron wave function and wavy lines to the T-matrix.

The ket $V_3|\phi_3\rangle = |\psi_3\rangle$ is the residue at the T-matrix pole and has a non-singular behavior. Evaluating this matrix element gives

$$\langle \phi_1, \mathbf{p}_1 | (E - K) | \phi_3, \mathbf{p}_3 \rangle = \frac{\psi_1(\mathbf{p}_3 + \frac{1}{2}\mathbf{p}_1)\,\psi_3(-\mathbf{p}_1 - \frac{1}{2}\mathbf{p}_3)}{E - (p_1^2 + p_3^2 + \mathbf{p}_1 \cdot \mathbf{p}_3)} \times \varphi_{13}$$

As shown in Figure 12 (U_{13}) this exchange term is associated with the transfer of particle 2 between the bound state of 1 and 2 and the bound state of 2 and 3.

The constant φ_{13} is the overlap of the initial and final spin and isospin states

$$\varphi_{13} = \langle\!\langle \mathcal{T}(T_{23}) | \langle \mathcal{S}(S_{23}) \mid \mathcal{S}(S_{12}) \rangle | \mathcal{T}(T_{12}) \rangle\!\rangle = \begin{cases} \frac{1}{4}, & \mathcal{S} = \frac{1}{2} \\ -\frac{1}{2}, & \mathcal{S} = \frac{3}{2} \end{cases}.$$

The exchange term is therefore attractive for the doublet state and repulsive in the quartet. Physically we expect this term to be peaked in the backward direction, which also follows from the above expression.

Due to the large size of the deuteron the exchange process has a very long range. As a result partial waves with $\mathfrak{L}$ greater than 0 become apparent at very low energies. In addition this produces singularities in the scattering amplitude that are very close to threshold. Recalling that $E = -B + \frac{3}{4}p_1^2 = -B + \frac{3}{4}p_3^2 = -B + \frac{3}{4}p^2$ we see that the partial wave amplitudes associated with the exchange term have cuts from $p^2 = -4B$ to $p^2 = -\frac{4}{9}B$. Since the incoming nucleon is indistinguishable from the constituent nucleons, the physical elastic scattering amplitudes will always contain these cuts. As we shall see they play an important role in behavior of the S-wave phase shifts.

The terms in the above expansions which contain the two-body T-matrices have the obvious interpretation of representing the incident particle being scattered by one of the constituent nucleons in the deuteron. For the rearrangement collisions the interaction ejects one of the originally bound particles leaving the incident particle bound in the deuteron. The impulse term will be peaked in the forward direction.

Additional terms in $U_{\alpha\beta}$ correspond to multiple scatterings. It has been argued [69, 70] that these terms should become increasingly isotropic since the particles soon lose knowledge of the initial direction. This is born out by calculations [70]. As a result, with the exception of the *S*- and *P*- waves, nucleon-deuteron scattering is dominated by the exchange and impulse terms. We now proceed to investigate some of these features in more detail.

4.2.1. *Very Low Energy n-d Scattering*

For n-d scattering near zero energy only $\mathscr{L}=0$ states are important and the scattering is given by the doublet and quartet scattering lengths, a_2 and a_4. The determination of the scattering lengths has had an interesting history. For some time there were two sets of scattering lengths consistent with experiment

Set A $\quad a_2 = 0.7 \pm 0.3$ F $\qquad a_4 = 6.38 \pm 0.6$ F

or

Set B $\quad a_2 = 8.26 \pm 0.12$ F $\qquad a_4 = 2.6 \pm 0.2$ F

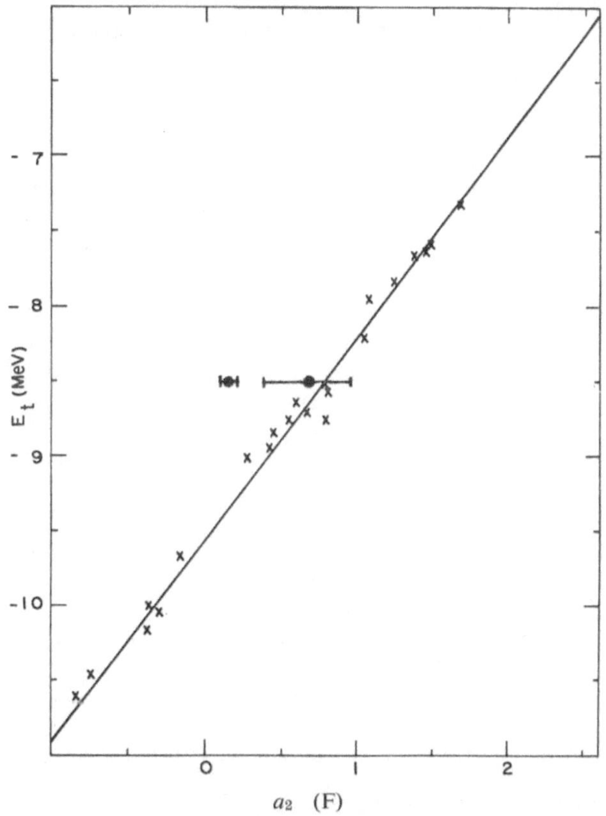

Fig. 13. Doublet scattering length and triton energies obtained from a number of calculations with separable potentials. Also shown are the old (set A) and new (set A′) values for a_2. The figure is from ref. [75].

Their value was obtained from the Fermi-Marshall [71] value for the 'free' n-d total cross section $\sigma_{\text{free}}=4\pi(\frac{2}{3}a_4^2+\frac{1}{3}a_2^2)=3.44\pm0.06b$ and the Hurst and Alcock [72] measurements of the ratio a_2/a_4 or its inverse. This ambiguity was resolved by the experiments of Alfimenkov *et al.* [73] by measuring the cross section for polarized neutrons scattering from polarized deuterons. Their results definitely establish that $a_4>a_2$. Therefore low energy scattering is clearly dominated by the quartet state.

More recently, however, Van Oers and Seagrave [74] obtained new values for the scattering lengths:

$$\text{Set A}' \quad a_2 = 0.15 \pm 0.05\ \text{F} \qquad a_4 = 6.13 \pm 0.04\ \text{F}$$

by examining new data on coherent and incoherent n-d scattering.

The calculation of the scattering lengths is fairly straightforward and most bound state programs can be easily modified to perform the calculation. The A′ set is particularly inconvenient from a theoretical point of view as we show in Figure 13. The graph is taken from a paper [75] by the Rensselaer group.

The plotted points give the value for the triton binding energy and doublet scattering length obtained for a large number and variety of separable potentials. This type of

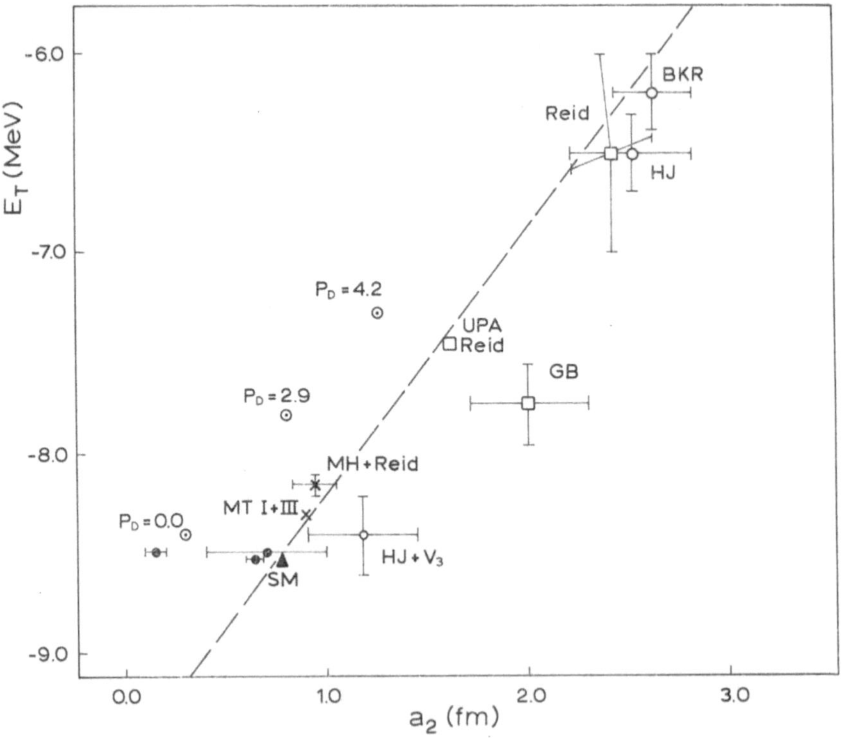

Fig. 14. Doublet scattering length and triton energies obtained from realistic nucleon-nucleon interactions.

graph was first drawn by Phillips [76] who found a similar result, i.e. that these two quantities satisfy a nearly linear relationship for a wide range of potentials. A least square fit to the data shown here gives

$$[E_T + 8.5]\ (\text{MeV}) = \tfrac{4}{3}[a_2 - 0.75]\ (\text{F})$$

Two things are apparent. First that the data for separable potentials clearly favor the old doublet scattering length of set A. Second, that at least for these potentials, E_T and a_2 are very closely related quantities. In Figure 14 we show that this is true for more realistic interactions also. The values plotted here are given in Table VII. All calculations except for local potentials with purely attractive triplet states and the Gammel-Brueckner potential lie on or very near the Phillips line.

The HJ + V_3 values are a collection of the values obtained by Delves and Hennell [40] by adding a three-body potential energy to the interaction in order to bind the triton correctly. All three of their potential choices gave essentially the same value for a_2. The error bars on the energy are assumed to be the same as for the Hamada-Johnston potential alone and those for a_2 are a sum of the computational error bars and the spread in the extrapolation.

The discrepancy between the A′ value for a_2 and theoretical calculations may have been removed by recent experiments performed by Dilg *et al.* [77]. These authors measure the coherent scattering length and free n-d cross section and obtain the values

Set A $\quad a_2 = 0.65 \pm 0.04$ F $\qquad a_4 = 6.35 \pm 0.02$ F

We have redefined set A to be the newer and presumably more accurate values, since this new set A is consistent with the old. Because theory and experiment are now in

TABLE VII

Triton energies and doublet scattering lengths obtained in recent calculations with realistic potentials

Potential	a_2(F)	E_T(MeV)
Bressel *et al.* (BKR) [44]	2.6 ± 0.2	−6.2 ± 0.2
Hamada-Johnston (HJ) [40]	2.5 ± 0.3	−6.5 ± 0.2
and three-body force (HJ + V_3) [40]	1.15 ± 0.3	−8.4 ± 0.2
Reid [43]	2.4 ± 0.2	−6.5 ± 0.5
Reid Alternate (not shown) [43]	2.3 ± 0.2	−6.5 ± 0.5
Gammel-Brueckner (GB) [53]	2.0 ± 0.3	−7.75 ± 0.2
UPA to Reid	1.6	−7.45
Modified Hulthen Triplet [90]		
and Local Reid Singlet (MH + Reid)	0.93 ± 0.10	−8.14 ± 0.05
Schrenk-Mitra (SM) [75]	0.77	−8.51
Malfliet and Tjon Soft Core [43]		
Singlet +		
Local tensor $P_D = 4.2\,\%$	1.25	−7.3
Local tensor $P_D = 2.9\,\%$	0.8	−7.8
Attractive Yukawa $P_D = 0.0\,\%$ (not shown)	0.3	−8.4
Yukawa and Core $P_D = 0.0\,\%$ (MT I + III)	0.9	−8.3

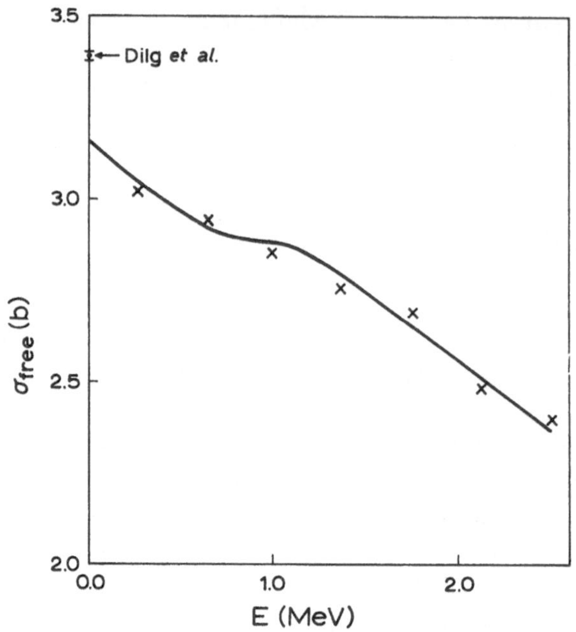

Fig. 15. Total cross section for n-d scattering. The experimental points and solid curve are from ref. [74]. The Dilg *et al.* value is from ref. [77].

agreement, it would seem to indicate that the scattering length cannot be used as an independent check upon nuclear interaction models for some time to come. This is particularly true since the Delves and Hennell calculation shows that for a three-body force of the form chosen by them, the triton energy and doublet scattering length extrapolate along the Phillips line. An explanation for the Phillips line is still to be given.

The acceptance of the new set A and the corresponding $\sigma_{free}=3.39\pm0.012\, b$ implies a very rapid variation in σ_{free} near zero energy. This is seen in Figure 15 where we have plotted the n-d total scattering cross sections as given by Seagrave [74], along with his spline fit. Error bars on the data are less than 50 mb (Figure 18 of ref. [74]). At least in this data the necessary rapid rise has not begun to appear. (Note added in proof: Recent measurements and calculations do find this rise. See P. Stoler *et al.*, Phys. Rev. Letters **29** (1972), 1745.)

In Figure 16 we show the behavior of ${}^{2\mathscr{S}+1}K=k\cot\{{}^{2\mathscr{S}+1}\delta\}$ for laboratory energies less than 20 MeV.

The quartet phase shifts can apparently be represented by an effective range expansion. For the doublet phases however a fit linear in k^2 will clearly have a very limited validity. In fact the data strongly suggests a pole in ${}^2K$ at a small negative energy. A pole in ${}^2K$ will correspond to a zero in the T-matrix.

Computations with spin independent interactions predict a three-body excited state. For the three-nucleon system, we might suppose that this second bound state

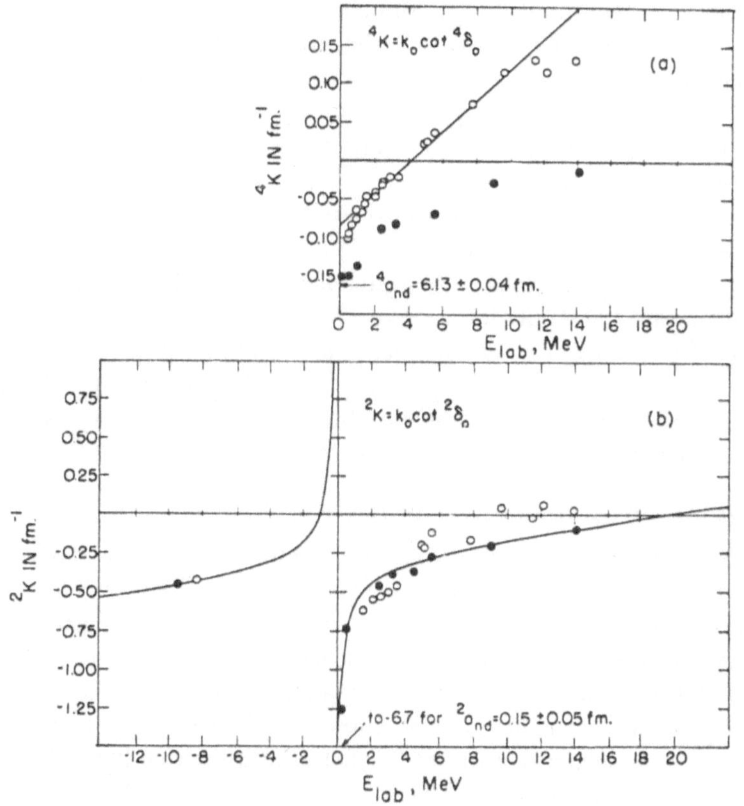

Fig. 16. Values of ${}^{2\mathscr{S}+1}K$ for doublet and quartet n-d (solid circles) and p-d (open circles) scattering. The figure is from ref. [74].

has moved onto the second energy sheet and become an antibound state. The zero in the T-matrix might then be ascribed to the necessary zero between the bound and antibound state poles. That this simple argument is incorrect can be seen as follows. We consider the scattering length as a function of the strength of the interaction and assume the behavior is the same as for potential scattering. As the potential strength increases from zero, the scattering length approaches minus infinity then becoming positively infinite when the first bound state appears. By increasing the potential strength further, the scattering length goes through zero and becomes negative again. Thus if the zero in T is to occur on the first energy sheet, a_2 must be negative. This simple argument therefore predicts the wrong sign for a_2.

Phillips and Barton [78] resolve this discrepancy, pointing out that near threshold the cut from the zeroth order exchange amplitude has an important influence on the behavior of the scattering amplitude. Fitting the quartet scattering by effective range theory, they find a pole at $p^2 = -\frac{4}{3}B$ and one considerably further away. The near pole lies on the exchange cut and presumably simulates its effects.

In the doublet state, due to the spin-isospin coefficient in the exchange term, the exchange amplitude has a cut of one half the strength of the quartet and opposite in sign. To include the effects of this cut, Phillips and Barton introduce a pole at the same position as in the quartet but with a residue opposite in sign and of one half the magnitude. Including in addition the triton pole and adjusting a far pole to fit a_2, they reproduce the rapid variation in ^2K.

The same authors [79] have performed an N/D calculation using as input the exchange cut from the zeroth order term. For the quartet they obtain a scattering length of 6.3 F. In the doublet state, after introducing the doublet scattering length as a subtraction constant, the rapid variation of ^2K is obtained. Furthermore the shape is independent of the value of a_2 except very near the elastic threshold. Thus it would appear that except for the doublet scattering length and triton energy, the low energy n-d system is fairly model independent since the exchange term is so well known. This model independence carries on to higher energies also.

4.2.2. *Elastic Scattering at Moderate Energies*

Seagrave [74] has tabulated n-d scattering cross sections over a wide range of energies. In the region below 50 MeV the results are very consistent. Partial waves other than $\mathscr{L}=0$ are apparent even at low energies as a result of the long range exchange term. A minimum in the differential cross-section develops at low energies, becomes pronounced by 10 MeV and moves toward back angles as the energy increases.

Separable potential calculations, even without the inclusion of tensor forces, are able to reproduce the major features of this behavior quite well [80, 81]. Comparison between theory and experiment, at several energies, is given in Figure 17. As can be seen except for small angles and the highest energies, the agreement is quite good. That it can be expected to be fairly model independent can be seen from work by Sloan [70] who investigates the convergence of the multiple scattering series for a separable model. He finds that at 14 MeV, for example, the series diverges for the *S*-waves (doublet and quartet). For the *P*-waves however it converges by third order in the doublet and by second order in the quartet. For *D* or higher waves, the expansion to first order is sufficient. Thus the scattering is to a large extent associated with the impulse and exchange terms which will depend mostly upon the on shell *T*-matrix and can be expected to be very insensitive to off-shell effects. The forward maximum in the cross section comes from the impulse term, and the back peak from the exchange diagram. The doublet *S*-waves are probably not given correctly by a simple separable model, but once we have the triton energy and doublet scattering length there may not be too much model dependence left. Also the quartet state dominates the scattering and as we have seen it is very insensitive to the interaction. Note that due to the rapid convergence of the multiple scattering series for large $\mathscr{L}$-partial waves, higher order terms in this series contribute mostly to the *S*-wave scattering. However even up to 100 MeV the *P*-waves require terms to second order.

The model independence of elastic scattering can also be seen by comparing a separable potential calculation with the recent local potential calculation of Kloet and

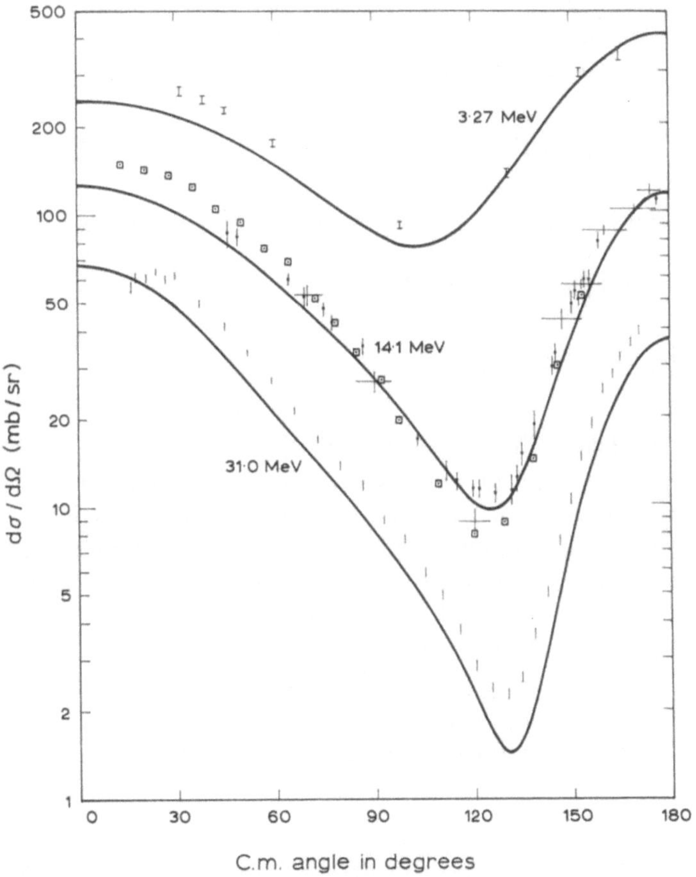

Fig. 17. Comparison of experimental n-d differential cross section with separable potential calculations. The figure is from ref. [81].

Tjon [82]. In Table VIII we compare the phase shifts obtained by these authors with those of a separable potential calculation [83].

Kloet and Tjon used local potentials with soft cores. The separable potentials are purely attractive. Considering the differences between the interactions the agreement is remarkable.

TABLE VIII

Comparison of n-d phase shifts obtained using local and separable potentials

Angular momentum ($\mathfrak{L}$)	Doublet				Quartet			
	Separable		Local		Separable		Local	
	$\bar{\delta}$	η	$\bar{\delta}$	η	$\bar{\delta}$	η	$\bar{\delta}$	η
0	126	0.46	106	0.50	71.9	0.98	73.2	0.94
1	12.9	0.69	14.4	0.69	29.4	0.92	32.9	0.41
2	6.4	0.95	6.9	0.95	−8.6	0.98	−8.9	0.98
3	−1.2	0.99	−1.3	0.99	3.0	1.0	3.1	1.0

The real part of the phase shift is $\bar{\delta}$, $\eta = \exp(-2\, I\mathrm{mag}\{\delta\})$.

To obtain their results, Kloet and Tjon used Pade approximants to sum the multiple scattering series. For the Faddeev equations with a linear parameter, the multiple scattering series gives an expression of the form

$$T^N = \sum_{n=0}^{N} \lambda^n M_n,$$

where M_n contains the nth order terms. We retain $2N+1$ terms in the series and write

$$T(\lambda) = T_{[N,N]}(\lambda) + O(\lambda^{2N+1})$$

$$T_{[N,N]}(\lambda) = \frac{Q_N(\lambda)}{R_N(\lambda)}$$

adjusting the coefficients in the polynomials Q and R to agree with the calculated terms in the power series expansion. These terms are obtained by numerically iterating the Faddeev equations. It can be shown that the Pade approximants converge to the correct result. In the calculations of Kloet and Tjon a [6,6] approximant was needed for $\mathfrak{L}=0$, but the order rapidly decreased with increasing $\mathfrak{L}$. Similar results for the convergence of the Pade approximants were obtained with separable potentials by Brady and Sloan [83]. Thus Sloan's results on the convergence of the multiple scattering series for separable potential are probably valid also for local potentials. It should also be noted that the local potentials give a larger cross section at the forward angles than the separable potentials thereby improving the agreement with experiment.

4.2.3. *The Break-Up Reaction*

For positive energies, in addition to elastic scattering there is the possibility of break-up of the deuteron. The most reliable calculations of this process are by Cahill and Sloan [84]. In this energy region, the scattering calculations become very difficult since the kernels develop logarithmic singularities that lie on the path of integration. In current calculations these singularities are avoided by rotating the path of integration away from the real axis and solving the integral equation along the rotated contour. The amplitude along the real axis is then obtained by using the integral equation again, with one momentum real but integrating over the complex momenta. For elastic scattering this poses no difficulty, but for break-up care must be taken to treat appropriately the singularities of the integrand. A careful discussion is given in ref. [84].

In Figure 18 we show the results of the calculation by Cahill and Sloan at 14.1 MeV. The rich structure of the break-up reaction is apparent. The most prominent feature is the large enhancement at the minimum proton energies and small angles. This peak is associated with the final state interaction (FSI) of the n-n pair.

To see the origin of the peak, we consider Equation (3.40) for the break-up amplitude. Assuming that particle 3 is the proton, and 1 the incoming neutron we may write

$$T_{01} = \langle \mathbf{p}_3, \mathbf{q}_3 | V_{31} | \Phi_1, \mathbf{p}_1 \rangle + \\ + \int \frac{\hat{T}_3 (q_3, q_3'; q_3^2 + i\varepsilon)\, q_3'^2\, dq_3' \langle \mathbf{p}_3, \mathbf{q}_3' | U_{31}(E) | \Phi_3, p_1 \rangle}{q_3^2 - q_3'^2 + i\varepsilon}.$$

For the maximum proton energies $q_3^2 \approx 0$ and we are therefore very near the pole in T_3.

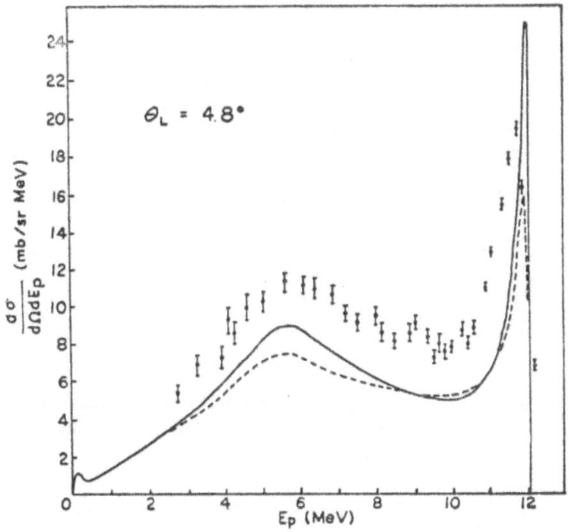

18a.

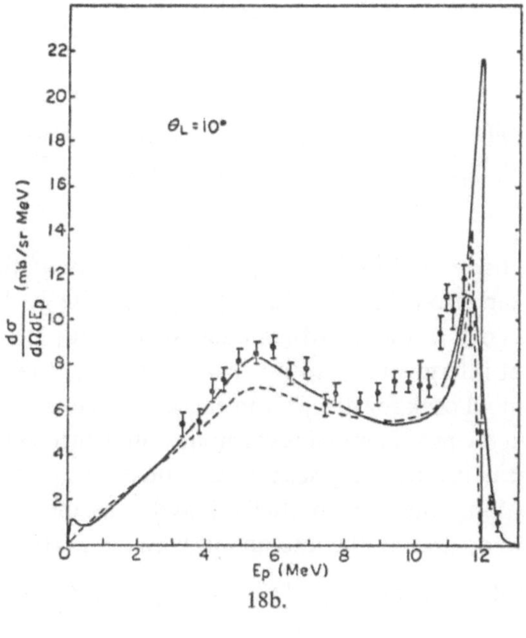

18b.

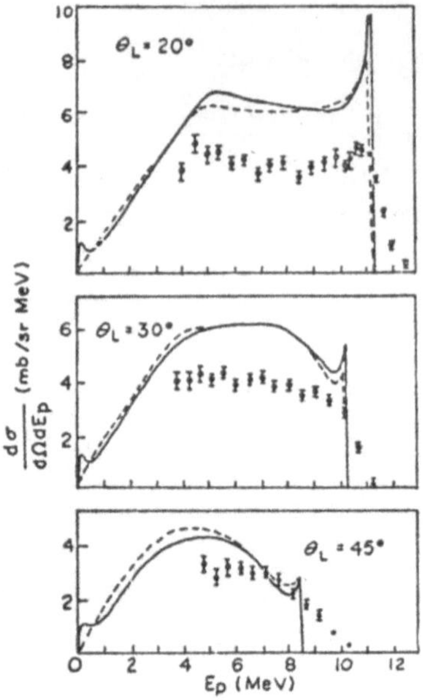

18c.

Fig. 18a, b, c. Comparison of experimental results for the n-d breakup reaction at 14.1 MeV with the results from a separable potential model. The figures are from ref. [84].

It is this pole which produces the rapid variation in the cross section. Since near the pole $\hat{T}_3$ has the form

$$\hat{T}_3(q_3, q_3'; q_3^2 + i\varepsilon) \propto \frac{f(q_3^2)\, f(q_3'^2)}{\alpha + iq_3}$$

We may try to write

$$T_{01} \sim \frac{Cf(q_3^2)}{\alpha + iq_3} \int \frac{f(q_3'^2)\, q_3'^2\, \mathrm{d}q_3' \langle \mathbf{p}_3, \mathbf{q}_3' | \, U_{31}(E) \, |\phi_1, q_1\rangle}{q_3^2 - q_3'^2 + i\varepsilon} = \\ = \hat{T}_3(q_3, q_3; q_3^2 + i\varepsilon)\, C'$$

in which the three-body amplitude is expressed in terms of the on-shell two-body amplitude which near zero energy varies very rapidly. This is the spirit of final state interaction theories such as that of Migdal and Watson [6].

The broad peak at intermediate proton energies in Figure 18 is associated with the n-p FSI peak. The small peak for low proton energies is the quasi free scattering (QFS) peak produced when the proton remains essentially undisturbed by the n-n interaction. The origin of this peak can be easily seen if we evaluate the break-up amplitude in the impulse approximation. Choosing particles 1 and 2 as the neutrons with 1 as the incident particle and neglecting the interaction between particles 1 and 3 we find

$$T_{01} = \langle \mathbf{p}_3^f, \mathbf{q}_3^f | \, T_3(E) \, |\phi_1, \mathbf{p}_1^i\rangle = T_3(\mathbf{q}_3^f, \mathbf{q}_1^i; q_3^{f2} + i\varepsilon)\, \phi_1(-\varepsilon),$$

where $\boldsymbol{\varepsilon}$ is the final momentum of the proton in the lab system, $\mathbf{q}_3^f$ is the final n-n relative momentum and $\mathbf{q}_3^i = \frac{1}{2}(\mathbf{p}_{\mathrm{lab}} + \boldsymbol{\varepsilon})$ is the initial relative momentum.

For reasonably high energies and small $\boldsymbol{\varepsilon}$, the T-matrix is almost on-shell. The deuteron wave function has a sharp maximum for $\boldsymbol{\varepsilon} = 0$ which produces the QFS peak.

As can be seen in Figure 18, the calculation fits the data reasonably well. The cross section is too small at forward angles and too large at back angles indicating perhaps difficulties with the separable approximation since similar discrepancies in elastic scattering appear to be largely removed by the use of a local potential.

In the break-up reaction the doublet state is very important. Both FSI peaks are primarily associated with scattering in the singlet two-body state and hence they show up in the doublet three-body channel. This large doublet contribution is also indicated by the calculations of elastic scattering where large inelasticities were observed in the doublet phase shifts. One of the checks performed on the break-up calculation was to see that the integrated break-up cross section agreed with the predictions from the imaginary parts of the scattering phase shifts.

The FSI and QFS peaks have recently received a considerable amount of attention [86] in the hope that they enable us to extract information about the n-n scattering parameters. The situation is quite complex. Cahill and Sloan and others [85] have found that those parts of the amplitude ending in the interaction of the two neutrons gives a peak which is broader than is observed. However when all terms in the amplitude are added together, the peak is much closer to that expected from final state interaction theories such as that of Migdal and Watson [6].

Cahill and Sloan have also investigated the effects of using charge dependent forces on the shape of the FSI peak. Varying the n-n scattering length they find that the shape of the peak does in fact change but that the difference is much less than experimental uncertainties. It may therefore be difficult to carry out a study of the n-n interaction using experiments of the type in which only one of the particles is detected.

This may not be the case however for kinematically complete reactions. Ebenhöh [87] has used separable potential model to study the FSI peak for such experiments, and finds that the Migdal-Watson theory gives good results. For an input n-n scattering length of -16F, his largest reported discrepancy using Migdal-Watson theory is 0.6 F and most are much less. His program has been used to analyse a kinematically complete reaction with 18.4 MeV neutrons [88]. The data was analysed to obtain the n-n scattering length (primarily from variations of the n-n FSI peak). The results give $a_{nn} = -14.5 \pm 0.8$ and $r_{nn} = 2.7 \pm 0.5$F. The former value is smaller by about 1.5 F than results obtained from applying FSI theories to other such reactions. The reliability of these calculations is still open to question since the separable model may not be sufficient to extract such detailed results from experiment. However this type of analysis is certainly of interest and will no doubt play an important role in the future.

It is of interest to ask if iterative methods may also be applied to the break-up problem so that local potentials can be used. To calculate elastic scattering only one number per partial wave is required. In the break-up reaction however we need each partial wave amplitude as a function of two momentum variables. The Pade method may be applied for each momentum point desired, but a simpler approach is available in the method of moments [89].

This method essentially involves using iterates of the integral equation $\chi = \phi_0 + K\chi$ to expand its solution. Schematically, we define

$$\chi_0 = \phi_0$$

and

$$\chi_n = \phi_0 + K\chi_{n-1}$$

then writing

$$\chi^N = \sum_{n=0}^{N} \beta_n^N \chi_n$$

we obtain a simple set of algebraic equations for the β_n^N by inserting this expansion into the integral equation and contracting on the left with χ_m. Once again this method uses only iterates of the scattering equations. It has been shown to converge rapidly for separable potential calculations of the wave functions and is guaranteed to converge for the local potential case as well.

References

1. Amado, R. D.: *Ann. Rev. Nucl. Sci.* **19**, 61 (1969).
2. Delves, L. M. and Phillips, A. C.: *Rev. Mod. Phys.* **41**, 497 (1969).
3. Mitra, A. N.: *Adv. Nucl. Phys.* **3**, 1 (1969).
4. McKee, J. S. C. and Rolph, P. M. (eds.): *Three Body Problem in Nuclear and Particle Physics*, Amer. Elsevier Publ. Comp., New York 1969.
5. Watson, K. M. and Nuttal, J.: *Topics in Several Particle Physics*, Holden Day, San Francisco 1967.
6. Goldberger, M. L. and Watson, K. M.: *Collision Theory*, Wiley, New York 1969.
7. Kowalski, K. L.: *Phys. Rev. Letters* **20**, 798 (1965).
8. Noyes, H. P.: *Phys. Rev. Letters* **15**, 538 (1965).
9. Lovelace, C.: *Phys. Rev.* **135**, B1225 (1964).
10. Fuda, M. G.: *Nucl. Phys.* **A116**, 83 (1968).
11. Brady, T., Fuda, M., Harms, E., Levinger, J. S., and Stagat, R.: *Phys. Rev.* **186**, 1069 (1969).
12. Harms, E. and Levinger, J. S.: *Phys. Letters* **30B**, 449 (1969).
13. Harms, E. and Newton, V.: *Phys. Rev.* **C2**, 1214 (1970).
14. Malfliet, R. and Tjon, J. A.: *Nucl. Phys.* **A127**, 161 (1969).
15. Weinberg, S.: *Phys. Rev.* **131**, 440 (1963).
16. Harms, E.: *Phys. Rev.* **C1**, 1667 (1970); Harms, E. and Laroze, L.: *Nucl. Phys.* **A160**, 499 (1971).
17. Reid, R. V.: *Ann. Phys.* **50**, 411 (1968).
18. Amado, R. D.: *Phys. Rev.* **C6**, 2439 (1970).
19. Baranger, M., Giraud, B., Mukhopadhyay, S. K., and Sauer, P. U.: *Nucl. Phys.* **A138**, 1 (1969).
20. Haftel, M.: *Phys. Rev. Letters* **25**, 120 (1970).
21. Kowalski, K. L., Monahan, J., Shakin, C., and Thaler, M.: *Phys. Rev. C* **3**, 1146 (1971).
22. Sauer, P.: *Nucl. Phys.* **A170**, 497 (1971).
23. Noyes, H. P.: in ref. [4].
24. Feshbach, H. and Kerman, A.: *Comm. Nucl. Part. Phys.* **2**, 22 (1968); **2**, 78 (1968).
25. Jackson, A. and Landé, A.: preprint.

26. Van Wageningen, R., Bakker, B. L. G., Bruinsma, J., Erens, G., and Struivenberg, J. H.: *Symp. Nucl. Three-Body Probl.*, Budapest, Hungary, 1971.
27. Fiedeldey, H. and Erens, G.: *Phys. Letters* **38B**, 15 (1972).
28. Fuda, M. G.: *Phys. Rev.* **186**, 1078 (1969).
29. Osborn, T. A.: *Nucl. Phys.* **A138**, 305 (1969).
30. Tabakin, F.: *Ann. Phys.* **30**, 51 (1964).
31. Mongan, T. R.: *Phys. Rev.* **175**, 1260 (1968); **178**, 1597 (1969).
32. Faddeev, L. D.: *Soviet Phys. JETP* **12**, 1014 (1961).
33. Rubin, M. H., Sugar, R. L., and Tiktopoulos, G.: *Phys. Rev.* **146**, 1130 (1966); **159**, 1348 (1967) and **162**, 1555 (1967).
34. Alt, E. O., Grassberger, P. and Sandhas, W.: *Nucl. Phys.* **B2**, 167 (1967).
35. Stingl, M. and Rinat, A. S. (Reiner): *Nucl. Phys.* **A154**, 613 (190) and ref. therein.
36. Harms, E.: *Nucl. Phys.* **A159** 545, (1970) and ref. therein.
37. Levinger, J. S.: *Symp. Nucl. Three-Body Probl.*, Budapest, Hungary, 1971.
38. Brady, T.: private communication to J. S. Levinger.
39. Thomas, L. H.: *Phys. Rev.* **47**, 903 (1935).
40. Delves, L. and Hennell, M.: *Nucl. Phys.* **A168**, 347 (1971).
41. Delves, L., Blatt, J., Pask, C. and Davies, B.: *Phys. Letters* **28B**, 472 (1969).
42. Jackson, A., Landé, A., and Sauer, P.: *Phys. Letters* **35B**, 365 (1971).
43. Malfliet, R. A. and Tjon, J. A.: *Ann. Phys.* **61** 425, (1970).
44. Malfliet, R. A. and Tjon, J. A.: *Phys. Letters* **35B** 487, (1971).
45. Kok, L. P., Erens, G., and Van Wageningen, R.: *Nucl. Phys.* **A122**, 684 (1968).
46. Erens, G.: Thesis, Univ. of Amsterdam, 1970.
47. Brady, T.: *Phys. Letters* **32B** 85, (1970).
48. Schrenk, G. L. and Mitra, A. N.: *Phys. Rev. Letters* **19**, 530 (1967).
49. Bhatt, S. C., Levinger, J. S., and Harms, E.: *Phys. Letters* **40B**, 23 (1972).
50. Hu, Chi-yu: *Phys. Rev.* **C3**, 2151 (1971).
51. Tjon, J. A., Gibson, B. F., and O'Connell, J. S.: *Phys. Rev. Letters* **25**, 540 (1970).
52. Harper, E., Kim, V., and Tubis, A.: *Phys. Rev. Letters* **28**, 1533 (1972).
53. Hennell, M. and Delves, L.: *Phys. Letters* **40B** 20, (1972).
54. Fiedeldey, H.: *Phys. Letters*, **35B** 195 (1971), *Nucl. Phys.*: **A156**, 242 (1970).
55. Folk, R. and Bonnem, E.: *Nucl. Phys.* **63**, 513 (1965).
56. Afnan, I. R. and Tung, Y. C.: *Phys. Rev.* **175**, 1337 (1968).
57. Kharchenko, V., Shadchin, S., and Storozhenko, S.: *Phys. Letters* **37B**, 131 (1971).
58. Fiedeldey, H.: *Nucl. Phys.* **A135**, 533 (1969); *Phys. Letters* **30B**, 603 (1969).
59. Afnan, I. and Serduke, J.: preprint.
60. Hadjimichael, E. and Jackson, A.: *Nucl. Phys.* **A180**, 217 (1972).
61. Haftel, M. I.: *Bull. Am. Phys. Soc.* **17**, 439 (1972).
62. Haftel, M. I. and Tabakin, F.: *Phys. Rev.* **C3**, 921 (1971).
63. Okamoto, K. and Lucas, C.: *Nucl. Phys.* **B2**, 347 (1967).
64. Fabre de la Ripelle, M.: *Fizika* **4**, 1 (1972).
65. Hadjimichael, E. and Brown, G. E.: *Phys. Letters* **39B**, (1972).
66. Hadjimichael, E., Harms, E., and Newton, V.: *Phys. Letters* **40B**, 61 (1972).
67. McCarthy, J. S., Sick, I., Whitney, R., an Yearian, M.: *Phys. Rev. Letters* **25**, 884 (1970).
68. Yang, S. N. and Jackson, A. D.: *Phys. Letters* **36B**, 1 (1971).
69. Queen, N. M.: *Nucl. Phys.* **55**, 177 (1964).
70. Sloan, I. H.: *Phys. Rev.* **185**, 1361 (1969).
71. Fermi, E. and Marshall, L.: *Phys. Rev.* **75**, 578 (1949).
72. Hurst, D. and Alcock, J.: *Can. J. Phys.* **29**, 36 (1951).
73. Alfimenkov, V., Lushchikov, V., Nikolenko, V., Taran, Yu., and Shapiro, F.: *Phys. Letters* **24B**, 151 (1967).
74. Seagrave, J.: in ref. [4].
75. Brady, T., Harms, E., Laroze, L., and Levinger, J. S.: *Phys. Rev.* **C2**, 59 (1970).
76. Phillips, A. C.: *Nucl. Phys.* **A107**, 209 (1968).
77. Dilg, W., Koester, L., and Nistler, W.: *Phys. Letters* **36B**, 208 (1971).
78. Phillips, A. C. and Barton, G.: *Phys. Letters* **28B**, 378 (1969).
79. Barton, G. and Phillips, A. C.: *Nucl. Phys.* **A132**, 97 (1969).

80. Aaron, R., Amado, R. D., and Yam, Y.: *Phys. Rev.* **140**, B1291 (1965).
81. Sloan, I. H.: *Nucl. Phys.* **A168**, 211 (1971).
82. Kloet, W. and Tjon, J. A.: *Phys. Letters* **37B**, 460 (1971).
83. Brady, T. and Sloan, I. H.: *Phys. Letters* **40B**, 55 (1972).
84. Cahill, R. T. and Sloan, I. H.: *Nucl. Phys.* **A165**, 161 (1971).
85. Aaron, R. and Amado, R. D.: *Phys. Rev.* **150**, 857 (1966).
86. Slaus, I., in ref. [4].
87. Ebenhoh, W.: *Nucl. Phys.* **A191**, 97 (1972).
88. Zeitnitz, B., Maschuw, R., Suhr, P., and Ebenhöh, W.: *Phys. Rev. Letters* **28**, 1656 (1972).
89. Harms, E.: *Phys. Letters* **41B**, 26 (1972).

THE AVERAGE FIELD IN THE NUCLEUS

PHILIP J. SIEMENS

The Niels Bohr Institute, Copenhagen, Denmark

1. The Concept of the Average Field

1.1. The nuclear interaction

Of the many sources of information about the forces between nucleons, the most easily interpreted are the measurements of the scattering of two nucleons. These experiments show that two nucleons colliding gently with each other experience attractive forces. But two nucleons approaching each other with high relative velocity (greater than about $\frac{1}{2}$c) experience predominantly repulsive forces, if they hit each other head-on (relative angular momentum$=0$). These repulsive forces seem to predominate when the nucleons are separated by less than about 0.7 fm, but within this region they are very strong.

The force between two nucleons has a rather strong dependence on their relative angular momentum, and on the relation between their spins and the angular momentum of their relative motion. On the other hand, the nuclear interaction depends only slightly on the isospin of the nucleons. The angular momentum characteristics of the nuclear force are most pronounced in its long-range components, which are generally agreed to result from the exchange of π mesons (ref. [1]). The angular momentum dependence of the nuclear force introduces a great deal of complexity into calculations.

Though the main features of the nuclear force can be deduced from nucleon-nucleon scattering measurements, the information obtained this way is far from sufficient to permit the deduction of nuclear properties from two-nucleon scattering, for several reasons:

(a) the imperfect precision of scattering measurements. This is especially serious in the deduction of the strength of the tensor force, which permits transfer of angular momentum between the internal-spin and relative-motion degrees of freedom. Another place where experimental inaccuracies are especially large is the measurement of the force between nucleons moving with large relative angular momentum.

(b) Present methods of parametrizing the nuclear force do not permit the use of information derived from the scattering of two nucleons whose relative kinetic energy is so large that pions are produced. Thus the analysis is limited to experiments with nucleon beams at less than about 300 to 400 MeV kinetic energy in the lab. The relative momentum of the nucleons in these experiments is not great enough for them to probe the nuclear force in detail at small separations.

(c) Nucleons in nuclei are different from free nucleons because the attractive

Abecassis de Laredo and Jurisic (eds.), Selected Topics in Phys., Astrophys. and Biophys., 307–324.

force acting on them decrease their rest mass, among other reasons. Thus the forces between nucleons in nuclei may be different from the forces between free nucleons. This effect cannot possibly be measured in two-nucleon elastic scattering. However, as progress is made in understanding the nuclear force, it may be possible to use information from other experiments such as π-nucleon scattering. Preliminary work [2] suggests that the nuclear force is not very different from a potential, with non-locality limited mainly to angular-momentum dependence.

1.2. Evidence for independent-particle nuclear structure

Although a nucleon in a nucleus interacts with each of the other nucleons with very strong attractive forces, and even stronger repulsive forces should the two chance to approach too near each other, most nuclear physicists spend their time explaining nuclear phenomena in terms of a model adapted from atomic physics, where the forces between two electrons in an atom is rather weak. This model is called the shell model. The philosophy behind the shell model is that each nucleon moves freely throughout the nucleus until it reaches the edges of the nucleus, where a lessening of the attractive potential due to the other nucleons prevents it from leaving the nucleus. The average attractive force of the other nucleons produces an average potential U which is the most important effect determining the nucleon's motion. Mayer and Jensen showed that this picture can explain the angular momenta of all nuclear ground states, if the potential U is assumed to depend on the relative orientation of the nucleon's spin and orbital angular momenta. The picture also explains the existence of 'magic' nuclei, which are especially tightly bound, in terms of the degeneracies of the eigenvalue spectrum of U. In the last two decades, many phenomena have been explained in terms of this picture; especially impressive are:

(a) Many excited states of nuclei with one nucleon more or fewer than a 'magic' nucleus can be interpreted in terms of a nucleon moving in an eigenstate of U. Not only can the energies and spins of these states be correlated in this way; many features of the cross-sections for production of these states in stripping and pickup reactions can be explained.

(b) Many other nuclear excited states can be explained if the independent motion of the nucleons is correlated by a mild interaction between them [3].

We will not dwell on the success of the shell-model picture, which can be demonstrated by reference to any volume of *Physical Review* (or *Nuclear Physics A*). Our goal is rather to reconcile the shell model's success with the characteristics of the nuclear force.

1.3. Conventional Hartree-Fock theory

In atomic physics, the shell model can be deduced from Hartree-Fock theory, so we might try the same approach in the case of the nucleus. Since the aim of the theory is a picture in which each nucleon moves independently, we conjecture that the wave-function of a nucleus with A nucleons might resemble the product of A single-particle wavefunctions,

$$\psi \simeq \prod_{i=1}^{A} \phi_i(r_i, s_i),$$

where r_i and s_i are the position and spin of the ith nucleon. After this we will write only r_i, but remember that this is a short notation for all the coordinates of the ith nucleon. Since nucleons are fermions, we can't be satisfied with a simple product wavefunction, but must make it antisymmetric by taking a Slater determinant of the single-particle wavefunctions:

$$\psi_{\text{HF}} = \frac{1}{A} \sum_{P} (-1)^P \prod_{i=1}^{A} \phi_i(r_{P(i)}), \tag{1.1}$$

where $P(i)$ is a permutation of the integers from 1 to A, and $(-1)^P$ is ± 1 according to whether the permutation is even or odd. One often sees Equation 1.1 written in a short notation

$$\psi_{\text{HF}} = \mathscr{A} \prod_{i=1}^{A} \phi_i, \tag{1.1a}$$

where the $\mathscr{A}$ means 'antisymmetrized'.

In conventional Hartree-Fock theory, the wavefunction (1.1) is used as a trial wavefunction for the Hamiltonian operator

$$H = \sum_i \left(\frac{-h^2}{2m}\right) \nabla_i^2 + \prod_{i=1}^{A} \sum_{j=1}^{i-1} v_{ij}. \tag{1.2}$$

The 2-body interaction v_{ij} is the (not necessarily local) interaction deduced from nucleon-nucleon scattering. (In atomic physics, v_{ij} would be the Coulomb interaction between the electrons i and j, and there would be an additional term in H representing the Coulomb field produced by the nucleus). The condition that the expectation value of H be a minimum,

$$\langle \psi_{\text{HF}} | H | \psi_{\text{HF}} \rangle = \text{minimum}, \tag{1.3}$$

leads to a set of coupled, non-linear equations for the ϕ_i:

$$E_i \phi_i(r) = -\frac{h^2}{2m} \nabla^2 \phi_i(r) + U_D(r)\, \phi_i(r) - \int \mathrm{d}^3 r' U_x(r, r')\, \phi_i(r'), \tag{1.4}$$

where – if v is local in coordinate space –

$$U_D(r) = \int \mathrm{d}^3 r' \sum_j \phi_j(r')\, \phi_j(r')\, v(r - r')$$

$$U_x(r, i) = \sum_j \phi_j(r)\, v(r - r')\, \phi_j(r'). \tag{1.5}$$

The Equations (1.4 and (1.5) can be solved on an electronic computer by an iterative technique in which the ϕ_i are guessed, U_D and U_x calculated from them, new ϕ_i calculated from U_D and U_x, the potentials recalculated, etc.

The trouble with the Hartree-Fock procedure is that, when v is a reasonable nucleon-nucleon interaction, it contains so much repulsion that the potentials U_D and U_x are repulsive, or at best only slightly attractive, in any case, the expectation value of H is not nearly attractive enough. We do not have to look far for the reason for this trouble. Our trial wavefunction $|\psi_{HF}\rangle$ is not nearly flexible enough. It does not permit two nucleons to avoid coming within the range of their repulsive interaction. In the nucleus, we expect that two nucleons on a collision course will be pushed apart by the repulsive interaction, and thus will avoid feeling its full effects. Somehow, we must make a nuclear theory in which nucleons avoid approaching too close to each other.

1.4. Healing

From the lack of success of Hartree-Fock theory, it would appear that the motion of pairs of nucleons in a nucleus is correlated in such a way that they avoid coming too close to one another. This conclusion could be tested (there is definite evidence in favour of it, in the case of the deuteron, from inelastic electron scattering, and only experimental imprecision stands in the way of similar tests in heavier nuclei [4]). But if the motion of nucleons is strongly correlated, then how can we explain all the shell-model evidence for independent-particle motion in nuclei? The answer is in the concept of 'healing' [5]. While two nucleons coming very close to each other – say within a distance d – may disturb each others' motion, they can return to their original motion after the collision, when they are again separated by a distance greater than d. If the distance d, outside of which the effects of their interaction are small, is less than the average spacing between nucleons, then each nucleon can move like an independent particle for most purposes: you will have to look at it just when it is within a distance d of another nucleon, if you want to see any disturbance in its motion.

Why should two interacting nucleons return to their original motion after colliding in a nucleus? (They do not do that if they collide outside a nucleus, for example in a scattering experiment!) This question is perhaps best answered by considering the other possibilities open to them. If they do not return to their original motion, they must be scattered into some other states of independent-particle motion. It is here that the presence of the other nucleons comes into play. All the low-energy independent-particle states are already occupied by other nucleons; being fermions, our scattered nucleons are excluded from taking the same motion as the other nucleons. Thus, if they are scattered, they must go to independent-particle states of large energy. This they cannot do, except while the potential of their interaction with each other is strong enough to help them to do it. When they separate, they no longer have much potential energy from the two-body force v, so they have to go back to their original states. The distance d is kept small by the fact that the available single particle states not only have large kinetic energy, but also less potential energy, because a fast-moving nucleon feels more of the effect of the repulsive part of the nucleon-nucleon force.

Already, with what we know of the shell model, we can see that the above argument will not always hold for all nucleons in all nuclei. Valence nucleons in a non-magic nucleus will usually be able to choose among several independent-particle states of nearly the same energy. Thus they will be able to scatter from each other and from the other nucleons in the nucleus, and we can expect long-range correlations to appear. But we are already familiar with ways to treat these correlations, within the framework of the shell model.

1.5. METHODS FOR OBTAINING THE AVERAGE FIELD

Several methods have been proposed to build the healing property into a theoretical model, leading to equations for the average field. Of these methods, the one which has received the most attention is the Brueckner theory, which is consequently in a more advanced stage of development than any other nuclear many-body theory. But Brueckner theory is not the only interesting method of deriving the average field. We will mention two other methods, Jastrow theory and unrenormalized perturbation theory.

The Jastrow theory is perhaps the most transparent way of introducing short-range correlations into the average field [6]. The inadequacy of the Hartree-Fock variational wavefunction $|\psi_{\mathrm{HF}}\rangle$ are remedied by using a slightly more flexible trial wavefunction $|\psi_{\mathrm{J}}\rangle$:

$$|\psi_{\mathrm{J}}\rangle = |\psi_{\mathrm{HF}}\rangle \prod_{i=1}^{A} \prod_{j=1}^{i-1} f(r_i - r_j) \tag{1.6}$$

The function f should approach a value 1 when $r_i - r_j$ is greater than the healing distance d, but become small when $r_i - r_j$ is small. In this way, the wavefunction $|\psi_{\mathrm{J}}\rangle$ is capable of including the effects we think are important.

Thus $|\psi_{\mathrm{J}}\rangle$ is a good candidate for a variational wavefunction. The task is to find, not only the single-particle eigenstates ϕ_i, but also the function f such that $\langle\psi| H |\psi_{\mathrm{J}}\rangle$ is a minimum. Nothing could be more elegant.

Unfortunately, there are severe technical difficulties in carrying out the minimization in the Jastrow method. Although the wavefunction $|\psi_{\mathrm{J}}\rangle$ is very simple and easy to understand, the expression for the expectation value $\langle\psi_{\mathrm{J}}| H |\psi_{\mathrm{J}}\rangle$ is quite complicated. In practice, it is necessary to resort to a perturbation expansion to determine $\langle\psi_{\mathrm{J}}| H |\psi_{\mathrm{J}}\rangle$. Then subsidiary conditions must be imposed on f due to the antisymmetry of $|\psi_{\mathrm{J}}\rangle$, and to try to keep the perturbation series convergent. A good deal of progress has been made in the application of Jastrow theory, and it is likely that further development will make it an increasingly useful method for studying nuclear correlations. Already, the theory has shown itself to be especially well adapted to the study of the pairing correlation.

Another method for calculating the correlations induced by the nuclear force is a conventional perturbation theory, in which the expression (1.3) is considered to be the first term in a perturbation expansion of the nuclear energy in powers of the two-body interaction v. Obviously, the crucial question is: How rapidly does the perturbation series converge? For the conventional parametrizations of v, there is no sign whatever of convergence in the first terms; in fact, as we mentioned above, the first term even has the wrong sign. However, the uncertainties in deducing v from two-body data are such that the two-nucleon scattering does not exclude a v with rather small repulsion at short distances [7]. Such a v has been used in perturbation theory, and reasonable results are obtained with inclusion of the second-order term – though there is some problem of obtaining the correct nuclear radii, which tend to come out too small for heavy nuclei. The main objection to this theory is that the nuclear force may well have too much repulsion for perturbation theory to be useful. This question must be settled by experiments on short-range correlations in nuclei, and by a theoretical understanding of the force's origin. The saturation difficulties encountered with 'soft' forces hint that nature may not be so kind. In any case, Brueckner calculations are not much more difficult than a second-order perturbation calculation.

Finally, we come to Brueckner theory, which is based on perturbation theory, but says that the lowest-order term of Equation (1.3) is not very much like the energy of a nucleus, because the potentials (1.4) and (1.5) do not much resemble the forces acting on a nucleon in a nucleus. The force on a nucleon is, instead, approximated by the sum of a class of terms in perturbation theory. This class of terms is chosen to include the effects discussed in Section 1.4, in a way which is actually rather similar to what is done in Jastrow's wavefunction (1.6). The Brueckner classification of perturbation theory forms the subject of the next section.

2. Brueckner Theory

2.1. Motivation

The theory of Brueckner is intented to be a way of allowing for the ability of two nucleons to avoid approaching too close to each other, while retaining the concept of an average field. The idea of the method is to exploit the 'healing' property discussed in Section 1.4 above. Since the correlation of the motion of two nucleons disappears when they are far apart, the main effect of their correlation is to reduce the effect of the repulsive force between them. Thus, it should be possible to define an *effective interaction* g between two nucleons in a nucleus, such that the energy of interaction of two uncorrelated nucleons, interacting by way of the effective interaction g, is the same as the energy of interaction of two correlated nucleons interacting by way of the true nuclear interaction v.

The formal tool for defining the effective interaction is perturbation theory. The perturbation theory is performed in a basis constructed from independent-particle wavefunctions, and a certain class of terms is summed to infinitely many orders by means of an integral equation.

2.2. Basis for perturbation theory

The perturbation theory of Brueckner is performed in a basis of wavefunctions which resemble the Hartree-Fock wavefunction $|\psi_{HF}\rangle$ of Equation (1). To construct the basis, an auxiliary Hamiltonian H_0 is introduced. H_0 is taken to be the sum of A single-nucleon operators.

$$H_0 = \sum_{i=1}^{A} H_1(r_i), \tag{2.1}$$

where the one-body operator H_1 is a sum of kinetic and potential terms

$$H_1 = -\frac{\hbar^2}{2m}\nabla^2 + U. \tag{2.2}$$

The intention, of course, is to make U resemble the average field, but formally U can be any Hermitian operator, provided the perturbation series is convergent. We will return later to discuss the choice of U, in Section 2.4. But for the moment, the main reason for introducing H_1 is to make use of its eigenfunctions ϕ_i

$$H_1\phi_i = E_i\phi_i. \tag{2.3}$$

The basis in which we perform our perturbation theory is not the eigenfunctions ϕ_i, but the eigenfunctions of H_0, since we are interested in the wavefunction of the whole nucleus. The eigenfunctions of H_0 are simply antisymmetric products of the eigenfunctions of H_1:

$$|\psi_n\rangle = \mathscr{A} \prod_{i=1}^{A} \phi_{i,n}. \tag{2.4}$$

The wavefunction $|\psi_n\rangle$ can be specified in terms of which single-nucleon eigenfunctions $\phi_{i,n}$ have been used to construct it. The eigenvalue of ψ_n is just the sum of the eigenvalues of its $\phi_{i,n}$:

$$H_0|\psi_n\rangle = \left(\sum_{i=1}^{A} E_{i,n}\right)|\psi_n\rangle = \mathscr{E}_n|\psi_n\rangle. \tag{2.5}$$

We want to look for the nuclear ground state $|\psi\rangle$ by means of perturbation theory. The unperturbed wave function we use as a starting point is called $|\psi_1\rangle$, and is the product of the A single-nucleon eigenfunctions $\phi_{i,1}$ whose eigenvalues $E_{i,1}$ are lowest. We assume that there is one unique $|\psi_1\rangle$ of lowest energy (otherwise we will not get the healing property, as we remarked in Section 1.4). To perform a perturbation calculation, we need matrix elements

$$\langle\psi_n| H - H_0 |\psi_m\rangle = \langle\psi_n| \tfrac{1}{2}\sum_{i,j} v(r_i - r_j) - \sum_i U(r_i) |\psi_m\rangle. \tag{2.6}$$

Each matrix element is an integral over the coordinates of A nucleons. Most of these integrals are trivial, however, because the variables of integration do not appear in $H - H_0$. These integrals just lead to δ-functions

$$\langle\phi_{i,n} | \phi_{i,m}\rangle = \delta_{(i,n)(i,m)}.$$

Thus the matrix element is a sum (because of the antisymmetrization) of relatively simple terms:

$$\langle\psi_n|\, H - H_0\, |\psi_m\rangle = \sum_i \sum_{j<i} S\{\langle\phi_{i,n}\phi_{j,n}|\, v\, |\phi_{i,m}\phi_{j,m}\rangle - \\ - \langle\phi_{j,n}\phi_{i,m}|\, v\, |\phi_{i,m}\phi_{j,m}\rangle\} - \sum_i S\langle\phi_{i,n}|\, U\, |\phi_{i,m}\rangle\,. \tag{2.7}$$

The S is just a kind of selection rule arising from the δ-functions: S is 1 if all the ϕ's in $|\psi_n\rangle$ are the same as those in $|\psi_m\rangle$, except for the ϕ's which appear explicitly in the matrix element afterwards. If any of these other ϕ's are different in $|\psi_n\rangle$ and $|\psi_m\rangle$, then S is 0.

Now to perturbation theory. We need to evaluate

$$\langle\psi|\, H\, |\psi\rangle = \frac{\langle\psi_1|\, H\, |\psi\rangle}{\langle\psi_1 \mid \psi\rangle}\,, \tag{2.8}$$

where

$$\langle\psi_1|\, H |\psi\rangle = \langle\psi_1|\, H |\psi_1\rangle - \\ - \sum_n \langle\psi_1|\, H - H_0 |\psi_n\rangle \frac{1}{\mathscr{E}_n - \mathscr{E}_1} \langle\psi_n|\, H - H_0 |\psi_1\rangle + \\ + \sum_n \sum_m \langle\psi_1|\, H - H_0 |\psi_n\rangle \frac{1}{\mathscr{E}_n - \mathscr{E}_1} \langle\psi_n|\, H - H_0 |\psi_m\rangle \frac{1}{\mathscr{E}_m - \mathscr{E}_1} \times \\ \times \langle\psi_m|\, H - H_0 |\psi_1\rangle + \cdots\,. \tag{2.9}$$

We can see that the perturbation series is quite complicated, especially when we remember that each matrix element in (2.9) is the sum of many terms as in (2.7).

In order to make it easier to think about the perturbation series (2.9), we can introduce a pictorial, or 'diagram', way of writing the terms in the series. Because of the 'selection rule' S of Equation (2.7), we realize that the wavefunctions appearing in (at least) the first terms in the perturbation series will not be very different from $|\psi_1\rangle$, in the sense that most of their constituent ϕ's will be the same. Thus, when we want to indicate a wavefunction $|\psi_n\rangle$, we will only mention how it differs from $|\psi_1\rangle$. A term in perturbation theory, written from right to left, is 'drawn' in a diagram from bottom to top as follows:

(1) A line with an arrow going up (called a particle line in the literature) means a

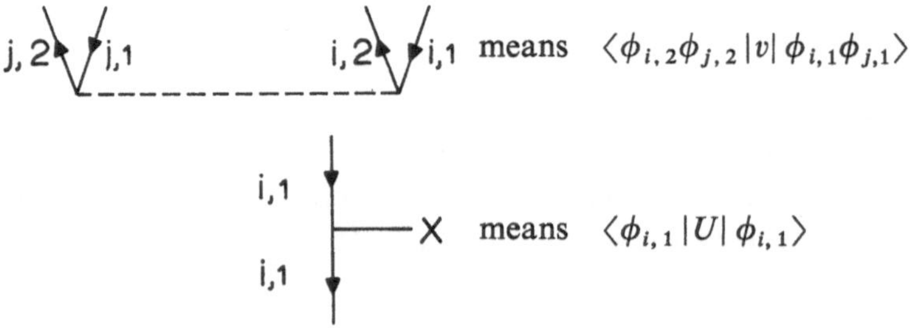

Fig. 1.

ϕ in $|\psi_n\rangle$ which isn't in $|\psi_1\rangle$. (2) A line with an arrow going down (called a hole line in the literature) means a ϕ in $|\psi_1\rangle$ which is missing in $|\psi_n\rangle$. (3) A horizontal dotted line means an interaction v. (4) A horizontal line with an x means an interaction U. These are illustrated in Figure 1. The sums over $|\psi_n\rangle$ can be converted to sums over the constituents $\phi_{i,2}$ and $\phi_{i,1}$. For example,

$$\sum_n \langle\psi_1|\, H - H_0\, |\psi_n\rangle \frac{1}{\mathscr{E}_n - \mathscr{E}_1} \langle\psi_n|\, H - H_0\, |\psi_1\rangle =$$

$$= \sum_{ij} \sum_{i'j'} \langle\phi_{i'}\phi_{j'}|\, v\, |\phi_i\phi_j - \phi_j\phi_i\rangle^2 \frac{1}{E_i' + E_j' - E_i - E_j} +$$

$$+ \sum_i \sum_{i'} \langle\phi_{i'}|\, U\, |\phi_i\rangle^2 \frac{1}{E_{i'} - E_i} -$$

$$- 2\sum_{ij} \sum_{i'} \langle\phi_{i'}|\, U\, |\phi_i\rangle \frac{1}{E_{i'} - E_i} \langle\phi_{i'}\phi_j|\, v\, |\phi_i\phi_j\rangle, \tag{2.10}$$

where the sums over i' and j' extend only over states $\phi_{i'}, \phi_{j'}$ which do not appear in the determinant $|\psi_1\rangle$. To remind ourselves that the sum over i' and j' is limited by S, we write $Q/(E_{i'} + E_{j'} - E_i - E_j)$ in the expressions in (2.10), and say that Q 'excludes' all the single-nuclear states ϕ_i in $|\psi_1\rangle$ from the sum.

$$\langle\psi_1|\, H - H_0\, |\psi_n\rangle \frac{1}{\mathscr{E}_n - \mathscr{E}_1} \langle\psi_n|\, H - H_0\, |\psi_1\rangle =$$
$$- \langle\phi_i\phi_j|\, v\, |\phi_{i'}\phi_{j'} - \phi_{j'}\phi_{i'}\rangle\, Q/(E_{i'} + E_{j'} - E_i - E_j) \times$$
$$\times \langle\phi_{i'}\phi_{j'}|\, v\, |\phi_i\phi_j - \phi_j\phi_i\rangle + \langle\phi_i|\, U\, |\phi_{i'}\rangle\, Q/(E_{i'} - E_i)\, \langle\phi_{i'}|\, U\, |\phi_i\rangle -$$
$$- 2\langle\phi_i\phi_j|\, v\, |\phi_{i'}\phi_j - \phi_j\phi_{i'}\rangle\, Q/(E_{i'} - E_i)\, \langle\phi_{i'}|\, U\, |\phi_i\rangle. \tag{2.11}$$

The diagram representation of (2.10), or (2.11) is shown in Figure 2.

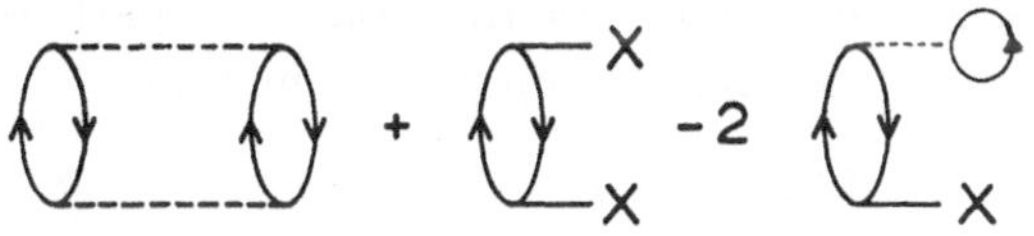

Fig. 2.

2.3. Partial sums of perturbation series

If the idea of 'healing' is correct, then the most important terms in the perturbation series will be those which represent two nucleons scattering from each other, while the other nucleons are passive.

The sum of these terms is given a special notation. Sometimes it is called K, sometimes t, sometimes G. We will call it g. The first few terms contributing to g are

$$\langle\phi_i\phi_j|\, g\, |\phi_i\phi_j\rangle = \langle\phi_i\phi_j|\, v\, |\phi_i\phi_j\rangle -$$
$$- \langle\phi_i\phi_j|\, v\, |\phi_{i'}\phi_{j'}\rangle\, Q/(E_{i'} + E_{j'} - E_i - E_j)\, \langle\phi_{i'}\phi_{j'}|\, v\, |\phi_i\phi_j\rangle +$$
$$+ \langle\phi_i\phi_j|\, v\, |\phi_{i''}\phi_{j''}\rangle\, Q/(E_{i''} + E_{j''} - E_i - E_j)\, \langle\phi_{i''}\phi_{j''}|\, v\, |\phi_{i'}\phi_{j'}\rangle \times$$
$$\times Q/(E_{i'} + E_{j'} - E_i - E_j)\, \langle\phi_{i'}\phi_{j'}|\, v\, |\phi_i\phi_j\rangle + \cdots \tag{2.12}$$

In Figure 3, g is denoted by a wavy line, so that we can write

Fig. 3.

This series of terms which makes up g is rather like the perturbation series for the scattering of two free particles. Just as in that case, we can derive an integral equation for the sum of the series, provided we are willing to define not only $\langle\phi_i\phi_j|\, g\, |\phi_i\phi_j\rangle$, the 'diagonal' matrix elements but also $\langle\phi_{i'}\phi_{j'}|\, g\, |\phi_i\phi_j\rangle$, the 'off-diagonal' matrix elements. The integral equation is called the Bethe-Goldstone equation,

$$\langle\phi_{i'}\phi_{j'}|\, g\, |\phi_i\phi_j\rangle = \langle\phi_{i'}\phi_{j'}|\, v\, |\phi_i\phi_j\rangle - \\ - \langle\phi_{i'}\phi_{j'}|\, v\, |\phi_{i''}\phi_{j''}\rangle\, Q/(E_{i''} + E_{j''} - E_i - E_j)\, \langle\phi_{i''}\phi_{j''}|\, g\, |\phi_i\phi_j\rangle\,. \tag{2.13}$$

It is easy to verify that iterating Equation (2.13) produces Equation (2.12). Similar integral equations can be derived to sum other classes of diagrams in which two nucleons interact repeatedly by way of v, but the arguments here are rather involved and we will not go into them. The result of the arguments is that, practically speaking, g can be thought of as an effective interaction, which is used in a new perturbation series. The structure of the terms in this new perturbation series is rather more involved than the old series, and each term is more complicated to calculate. But the increased complication is compensated by a much more rapid convergence of the series in g. At least, that is what is hoped – we will say a little more about this in Section 2.5. There are complete discussions of this topic in refs. [8] and [9].

There is an alternative way of viewing the result of summing the perturbation terms representing the repeated interaction of the same two particles: we can calculate the wavefunction in the same perturbation approximation

$$|\psi_B\rangle = |\psi_1\rangle - |\zeta\rangle\,, \tag{2.14}$$

where

$$|\zeta\rangle = \sum_{n\neq 1} |\psi_n\rangle \frac{1}{\mathscr{E}_n - \mathscr{E}_1} \langle\psi_n|\, g\, |\psi_1\rangle\,. \tag{2.15}$$

The equivalence between these two viewpoints is most easily seen from the equation

$$g\, |\psi_1\rangle = v\, |\psi_B\rangle\,. \tag{2.16}$$

The wavefunction $|\zeta\rangle$, called the 'defect wavefunction', is the correlation induced by the interaction of two nucleons at a time; it represents the probability density for two nucleons moving differently from their independent-particle behavior ϕ_i and ϕ_j, due to the influence of the force v between them.

Its norm,

$$K = \langle\zeta\,|\,\zeta\rangle\,, \tag{2.16a}$$

is the total probability of correlation, in this approximation. In nuclei, calculations show that K is about 0.14 (for the Reid force); roughly speaking, this means that the independent-particle picture is about 85% correct. Of course, some nuclear properties are more sensitive to K than others. A particularly sensitive property is the saturation of nuclear densities, i.e. the tendency of all nuclei to have nearly the same densities in their interior regions.

2.4. The Average Potential

We are now in a better position to discuss the choice of the single-nucleon operator U, which is used to define the basis functions for our perturbation theory. Naturally, we would like our wavefunction $|\psi_B\rangle$ to be as much like the nuclear ground state as we can make it. We know experimentally that the nuclear ground state is rather like an independent-particle determinant, and we even know something about what the independent-particle wavefunctions are like: they are like the shell-model wavefunctions deduced from nuclear reactions. Of course, no nuclear state is a pure single-particle state, so the single-particle wavefunctions are in a sense a fiction introduced in the shell model. Correspondingly, there is an uncertainty introduced in deducing them from experimental states (to say nothing of the uncertainties associated with our imperfect understanding of reaction mechanisms). But let us suppose for the moment that we know how to extract the shell model states ϕ_i^{sm} and their energies E_i^{sm}. It is then very tempting to identify our perturbation-theory basis with these measurable quantities, i.e. to try to make

$$\begin{aligned} \phi_i &= \phi_i^{\mathrm{sm}} \\ E_i &= E_i^{\mathrm{sm}} . \end{aligned} \tag{2.17}$$

How can we construct the operator U which makes possible the identification 2.17? The answer (ref. [10]) turns out to be quite simple: we must take very seriously the phrase 'the average potential felt by a single nucleon, due to all the other nucleons'. That is, our definition must be

$$\langle\phi_k|\, U\, |\phi_k\rangle = \sum_{i=1}^{\infty} \langle\phi_k\phi_i|\, g\, |\phi_k\phi_i\rangle\, P(i), \tag{2.18}$$

where $P(i)$ is the probability that there is a nucleon in the single-nucleon state ϕ_i. When the probability $P(i)$ is expanded in a perturbation series in g, the first terms are as shown in Figure 4.

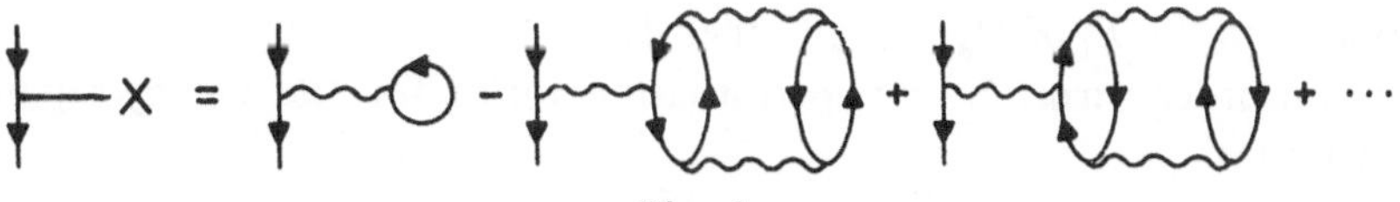

Fig. 4.

The first term is the sum of the potential energies contributed by each nucleon in the states ϕ_i which make up $|\psi_1\rangle$. The second and third terms represent corrections for the fact that some of the time, the nucleons are not in these ϕ_i's, but are in other single-particle states due to the correlating effect of the two-body interaction. (Numerically, the third term turns out to be small and is often omitted from computations).

2.5. Convergence of the Partially-Summed Series

Although we know experimentally that independent-particle motion is a pretty good approximation to the nuclear wavefunction, it would be very nice to have some quantitative understanding of this in terms of the speed of convergence of our expansion in powers of g. Unfortunately, this problem has not yet been solved; it has not even been shown that the expansion is convergent! What *is* known is that, if there is a rapidly-convergent expansion, it is *not* an expansion in powers of g. Instead, we must group together all terms in which the same 3 nucleons interact repeatedly by means of g, and solve a set of coupled integral equations (the Bethe-Fadeeyev equations) analogous in spirit to the one we obtained for the successive interaction of two nucleons. Probably this procedure must be repeated for the repeated interaction of four nucleons. Brandow has classified the terms in the series according to the power of the wavefunction $|\zeta\rangle$ which appears in them. This classification provides a sort of series in powers of K, which is a small parameter. But since the properties of the weighting functions of the various orders of $|\zeta\rangle$ are not clear, there is no proof of the convergence of the expansion [11].

A more intuitive approach would be to look at the goodness of the healing criterion, that the defect wavefunction $|\zeta\rangle$ should be small if all the pairs of nucleons are separated by a distance d which is less than the average distance between them. This criterion is well satisfied for the calculated $|\zeta\rangle$, except for one component: the $J=1$, $T=0$ tensor correlation between the spins and angular momenta of neutron-proton pairs, where the correlation function has an uncomfortably long range due to the great strength of the π-meson exchange force in this channel. It is conceivable that there may be a long-range, many-body tensor correlation in nuclei, though there is no experimental evidence to support or detract from this possibility.

One thing that can be said with certainty about the convergence of the revised perturbation series is that it is strongly affected by the choice of U. One good way of seeing this is to consider the g-matrices resulting from computations using two different potentials U_A, and U_B, chosen to give the same basis wave functions ϕ_i, but with different eigenvalue spectrums. Suppose the potentials are constructed to give the same E_i for all the ϕ_i in ψ_i, but that the E_i of U_B are shifted by an amount ΔU compared to the E_i of U_A, for all other ϕ.

The lowest-order terms in the perturbation series for the energy (apart from kinetic energy) is

$$\mathscr{E}^{(1)} = \sum_i \sum_{j<i} \langle \phi_i\phi_j | \, g \, | \phi_i\phi_j - \phi_j\phi_i \rangle . \tag{2.19}$$

It can be shown that the potentials U_A and U_B lead to lowest-order energies which differ by

$$\mathscr{E}_A^{(1)} - \mathscr{E}_B^{(1)} = K\,\Delta U \tag{2.20}$$

if ΔU is small. But since the sum of the perturbation series should be independent of the basis, this means that the sum of the higher-order terms will also be different in the two cases, by an amount $-K\Delta U$.

The realization that the convergence of the perturbation expansion can be manipulated by adjusting U suggests a different way of choosing U than the shell-model potential, Equation (2.18). Why not try to adjust U to minimize the effect of higher-order correlations on the system's energy? The trouble is, we would *also* like to have our single-particle wavefunctions ϕ_i and energies E_i resemble their shell-model counterparts. Fortunately, it seems to be possible to have our cake and eat it too. This is because the short-range correlations built into $|\psi_B\rangle$ by Equation (2.15) are due mostly to states ϕ_i with very high kinetic energy, which are not interesting in the context of the shell model. Thus we can use the shell-model potential (2.18) for states of low kinetic energy, but modify the definition for states of higher kinetic energy so as to make the lowest-order energy (2.19) a good approximation to the sum of all the terms in the series.

The only case where extensive numerical studies have been made of the higher-order terms in the perturbation series in g is the hypothetical problem called nuclear matter, an infinitely large system of nucleons in which the Coulomb force has been turned off. Because of translational invariance, the single-nucleon wavefunctions ϕ_i in this system are plane waves, which immensely simplifies the computations. Here, the terms involving successive interaction of 3 and 4 different nucleons have been studied (ref. [12]). These studies have shown that the potential energy U of states ϕ with large kinetic energy should be fairly small if $\mathscr{E}^{(1)}$ is to be a good approximation to the energy of the system's ground state. A fairly good approximation is to set $E_i =$ = kinetic energy for all the states ϕ_i which do not appear in $|\psi_1\rangle$. (This approximation also introduces computational simplifications.) Then the higher-order terms are estimated to be in the neighbourhood of -4 MeV (ref. [9]) while $\mathscr{E}^{(1)}$ is about -35 MeV per nucleon. The choice E_i = kinetic energy for states not in $|\psi_1\rangle$ is very different from the shell-model U, which is attractive for states whose kinetic energy is a little larger than that of the states in $|\psi_1\rangle$, but repulsive (about half as large as the kinetic energy) for states of more than about 100 MeV kinetic energy.

3. Computations in Magic Nuclei

3.1. Harmonic oscillatorre presentation

The numerical problem of solving the Bethe-Goldstone equation is complicated by the appearance of ϕ, a single-nucleon operator, and v, a two-nucleon operator, successively in Equation (2.13). The potential v is a function of the relative coordinate of the two interacting nucleons, and both its angular and radial dependences on this coordinate are so strong and so complicated that it is necessary to introduce the relative coordinate explicitly in the representation used for solving Equation (2.13). Thus the numerical solution would be greatly facilitated if the product $\phi_i\phi_j$ could be rewritten in a simple way as a function of the relative and center-of-mass coordinates of the two nucleons. One way to do this is to express ϕ_i and ϕ_j in a representation $|n\rangle$,

$$\phi_i = \sum_n c_{in} |n\rangle . \tag{3.1}$$

The basis n should be such that the product of two of the basis states factors into a product of functions of the relative and center-of-mass coordinates of the two-nucleon system. Then $|\phi_i\phi_j\rangle$ will be the sum of simple terms, integrals can be evaluated quickly and precisely, and the Bethe-Goldstone equation expressed in terms of the c_{in}, integrals of v, and Q. There are just two kinds of functions $|n\rangle$ which have the property of separating relative and center-of-mass motion: the plane-wave states, and the harmonic-oscillator eigenfunctions.

Both representations have been applied in the numerical solution of the Bethe-Goldstone equation. In principle, either representation should be equally satisfactory in any given application, since they are completely equivalent. In practice, approximations are necessary, because of computer limitations, so that it is necessary to choose the representation to fit the problem. The harmonic-oscillator representation is especially suited to lighter nuclei, because there the shell-model single-particle wavefunctions ϕ_i can be quite similar to the harmonic-oscillator wavefunctions $|n\rangle$, if the oscillator parameter $\hbar\omega$ is suitably chosen.

The main approximation of harmonic-oscillator-representation computations is in the operator Q. Instead of using the correct expression,

$$Q = \left(1 - \sum_{i=1}^{A} \sum_{j=1}^{\infty} |\phi_i\phi_j\rangle \langle\phi_i\phi_j|\right) \left(1 - \sum_{i=1}^{\infty} \sum_{j=1}^{A} |\phi_i\phi_j\rangle \langle\phi_i\phi_j|\right), \tag{3.2}$$

one uses the easier-to-compute approximation

$$Q^{\mathrm{H,O}} = \left(1 - \sum_{n=1}^{A} \sum_{m=1}^{\infty} |nm\rangle \langle nm|\right) \left(1 - \sum_{n=1}^{\infty} \sum_{m=1}^{A} |nm\rangle \langle nm|\right), \tag{3.3}$$

where the oscillator wavefunctions $|m\rangle$ are arranged in order of increasing kinetic energy. Thus, the effects of antisymmetry are not correctly taken into account. Kallio and Day [13] and McCarthy [14] have developed techniques for solving the Bethe-Goldstone equation with the approximation (3.3). Earlier calculations have used a still more approximate Q due to Eden and Emery [20],

$$Q^{EE} = 1 - \sum_{N=1}^{N_{\max}} |nm\rangle \langle nm| ,$$

where N is the total energy quantum number of the product wavefunction $|nm\rangle$, and $N_{\max}$ is suitably chosen.

With the approximation (3.3), it is natural to go one step further and approximate the $E_{i'} + E_{j'}$ in the denominator of the Bethe-Goldstone equation, with an energy assigned to the oscillator product state $|nm\rangle$ in which $|\phi_{i'}\phi_{j'}\rangle$ has been expanded, instead of with the eigenvalues of $\phi_{i'}$ and $\phi_{j'}$:

$$E_{i'} + E_{j'} \simeq E_n^{\mathrm{HO}} + E_m^{\mathrm{HO}}. \tag{3.4}$$

The E_n^{HO} are then chosen, like $E_{i'}$ and $E_{j'}$, with a view to making $\mathscr{E}^{(1)}$ a good approximation to the total energy. No one really knows how to do this, so several choices of E_n^{HO} have been tried. One is simply to take the harmonic-oscillator energy eigenvalue; another is to subtract a constant energy, or the order of $\hbar\omega$, from this eigenvalue. A third choice is to take simply the kinetic energy of the oscillator state $|n\rangle$, as suggested by the investigations in nuclear matter. Results of these three 'guesses' for how to approximate U have been computed by Davies *et al.* [15], using the Reid potential and some further, minor approximations. They call the three approximations pure HO, shifted HO, and QTQ, respectively. The results for the three cases are quite different, as we might expect. The pure HO approximation gives much less binding energy than the other two; ^{208}Pb is scarcely bound in the pure HO approximation. The other two approximations also fall short in total binding energy by about 4 MeV per nucleon; this is not too surprising since in nuclear matter the same result is obtained with the analog of the QTQ prescription. The nuclear radii are also quite similar in QTQ and shifted HO, being about 7% too small for ^{40}Ca and about 13% too small for ^{208}Pb. The pure HO radii are about 5% larger than the others, but still not large enough. Other computations using different E_n^{HO} give a similar range of results [16].

The most puzzling result of the computations available to date is the difficulty in obtaining simultaneously the correct nuclear size *and* binding energy. This failing is not just due to the use of the Reid potential; on the contrary, other parametrizations of the two-nucleon force do as badly or even worse. Of course, it must be possible to find a 'recipe' for the energies E_i such that $\mathscr{E}^{(1)}$ has the observed properties. But it appears that this 'recipe' must play an active rôle in determining the nuclear size. It would be nice to understand this rôle in terms of higher-order correlation effects and/or corrections for the approximations (3.3) and (3.4). But, as yet, that understanding is lacking.

3.2. Plane wave representation and the local density approximation

The other set of orthogonal functions which permits factorization of a product of single-particle wavefunctions into functions of relative and center-of-mass coordinates is the plane-wave representation, whose basis vectors we will denote by $|k\rangle$. Computations which utilize the plane-wave representation can take advantage of two important facts:

(1) The complementary representation to momentum-space is ordinary coordinate space. This means that it is possible to work with single-particle wavefunctions $|\phi_i\rangle$ directly in coordinate representation, which is familiar and easy to visualize, and where we know they should be quite simple objects (unfortunately, the same cannot be said of the potential U).

(2) In large nuclei, the average field is constant throughout much of the nucleus. This means that, in this region, the single-particle wavefunctions $|\phi_i\rangle$ should be rather nicely represented as plane waves. So the plane-wave representation is especially suited to large nuclei, just where the oscillator representation becomes less convenient.

Not surprisingly, though, it is necessary to make certain approximations to take advantage of the plane-wave representation. These approximations are referred to as the Local Density Approximation, or LDA (ref. [17]).

As in the computations in harmonic-oscillator representation, the main approximation of the LDA is in the operator Q, which must be modified to take advantage of the properties of plane waves.

First, the product $|\phi_i\phi_j\rangle$ is expanded in a series of harmonic-oscillator eigenfunctions, each term being rewritten in relative and center-of-mass coordinates:

$$\phi_i(r_1)\,\phi_j(r_2) = \sum_N \sum_n c_{Nn}\,|N(R)\rangle\,|n(r)\rangle\,, \tag{3.5}$$

where $R=(r_1+r_2)/2$, $r=r_1-r_2$. Then, for each value of R, the matrix element of g in relative coordinates is calculated with a modified Q which depends on R

$$\langle\phi_{i'}\phi_{j'}|\,g\,|\phi_i\phi_j\rangle = \sum_N \sum_{N'} \sum_n \sum_{n'} c_{Nn}c_{N'n'}\,\langle N'|\,\langle n'|\,g(R)\,|n\rangle\,|N\rangle\,, \tag{3.6}$$

where the operator Q in the Bethe-Goldstone equation for g is approximated by Q that would be used in uniform nuclear matter of the density found in the nucleus at R:

$$Q(R) = Q(\varrho(R)) = \Bigl(1 - \sum_{k<k_F} |k\rangle\,\langle k|\Bigr)\Bigl(1 - \sum_{k'<k_F} |k'\rangle\,\langle k'|\Bigr) \tag{3.7}$$

k_F being the fermi momentum of nuclear matter at density $\varrho(R)$. The argument in favour of this approximation is that, since the correlations in g are of short range, they should mainly be sensitive to the other nucleon's behaviour in the region where the two nucleons are interacting. This argument is perhaps dubious, since it is exactly Q which keeps the correlations short-range: the action of Q is not to induce short-range correlations, but rather to suppress long-range correlations. Nevertheless, the argument is made, and the computations proceed.

Further approximations are introduced which greatly simplify the numerical work, but are probably not important in principle. The interaction g is averaged to make it a local function of the relative coordinates, in each angular-momentum projection. It is also averaged over spin orientations. The dependence of $g(R)$ on $\varrho(R)$ is approximated as two multiplicative factors, one for the short-range part of g, and one for the long-range part. All these approximations are in the nature of numerical tricks and can be shown to be unimportant. A further approximation is to replace the contribution to the average potential U of the incomplete occupation probabilities $P(i)$,

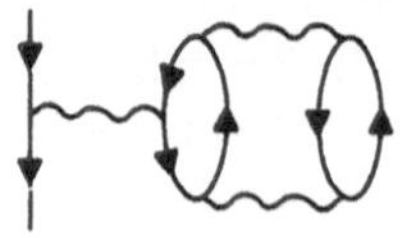

by a term proportional to $\partial \dot{g}/\partial \varrho$. This amounts to evaluating it for the last occupied level ϕ_i, again in the plane-wave approximation [9]. The main effect of this approximation is a change of a few MeV in the lowest few E_i.

Finally, we come to the question of the definition of U for states of large kinetic energy. Negele chose $U=0$ for all his intermediate states in the Bethe-Goldstone equation. As we know from nuclear matter computations, this does not make $\mathscr{E}^{(1)}$ sufficiently attractive. To compensate for this, he adds some attraction to g at short distances, in such a way that the binding energy of nuclear matter matches the observed -16 MeV, and that the radius of ^{40}Ca is correctly predicted. Within Negele's approximations, this is completely equivalent to adjusting the potential energy U of single-particle states of high kinetic energy, in order to get the 'correct' binding energy and saturation density ($k_F=1.36$) for nuclear matter. With this adjustment, Negele gets excellent results for the binding energies, single-nucleon removal energies, and density distributions of the nuclei ^{16}O, ^{40}Ca, ^{90}Zr, and ^{208}Pb. (ref. [17]).

Recently, Negele and Vautherin [18] have developed a method for greatly simplifying computations in the LDA. Their method revolves around an approximation to the density matrix of the wavefunction $|\psi_1\rangle$. They show that, to a good approximation and summing overspins,

$$\varrho_1\left(R+\frac{s}{2},\ R-\frac{s}{2}\right)=\varrho_1(R)\frac{3}{sk_F}j_1(sk_F)+ \\ +\frac{35}{2sk_F^3}j_3(sk_F)\left[\tfrac{1}{4}\nabla^2\varrho_1(R)-\tau(R)+\tfrac{3}{5}k_F^2\varrho_1(R)\right], \tag{3.8}$$

where k_F is the fermi momentum associated with $\varrho_1(R)$, j_1 and j_3 are spherical bessel functions, and

$$\tau(R)=\sum|\nabla\phi_i(R)|^2.$$

The approximation (3.8) is a terrific computational simplification in working the Hartree-Fock problem. Instead of a function of two three-dimensional variables, it is only necessary for the computer to store two functions of one three-dimensional variable, the functions ϱ and τ. Furthermore, instead of performing many integrals of the effective interaction g over the relative coordinate, these integrals can be performed once and tabulated. That also saves a lot of computing time. To see how great a simplification is involved, look at the expression Negele and Vautherin obtain for the total nuclear energy, using the same approximations as in Negele's LDA, plus (3.8):

$$\mathscr{E}^{(1)}=\int d^3R\,H(R) \tag{3.9}$$

$$H=\frac{h^2}{2m}\left[\tau_n+\tau_P\right]+A(\varrho_P,\varrho_n)+B(\varrho_P,\varrho_n)\tau_P+B(\varrho_n,\varrho_P)\tau_n+ \\ +C(\varrho_P,\varrho_n)|\nabla\varrho_P|^2+C(\varrho_n,\varrho_P)|\nabla\varrho_n|^2+D(\varrho_P,\varrho_n)\nabla\varrho_n\cdot\nabla\varrho_P, \tag{3.10}$$

where the subscripts n and P refer to neutrons and protons, and the coefficients A, B, C, and D are particular integrals of the effective interaction. Varying $\mathscr{E}^{(1)}$ leads to a local Schrödinger equation with a position-dependent effective mass. Negele and Vautherin also find simple ways to include effects of the spin-orbit force, as well as the relation between the effective interaction and the single-nucleon eigenvalues E_i. When they perform computations in this approximation, they get practically the same results as Negele got in his LDA – using a fraction of the computer space and time.

The computational virtues of an approximation like (3.10) was realized long ago by Skyrme [19], who proposed a similar theory, but without showing the connection to many-body theory. Now that the connection is better understood, this theory should make it possible to investigate many phenomena in a reliable microscopic model.

It is interesting to note that the mixed density also tells us the operator Q in a position-space representation:

$$\langle r_4 r_3 | Q | r_2 r_1 \rangle = (1 - \varrho(r_1, r_3))(1 - \varrho(r_2, r_4)).$$

Perhaps Equation (3.8) would thus make a better way of treating Q as a function of R. The LDA uses only the first term of (3.8). This hasn't been investigated but gives one suggestion of what can be done in this theory.

References

1. Brown, G. E. and Durso, J. W.: *Phys. Letters* **35B**, 120 (1971).
2. Brown, G. E. and Green, A. M.: *Nucl. Phys.* **A137**, 1 (1969).
3. Talmi, I. and Federman, P.: *Phys. Letters* **19**, 480 (1965); Averbach, N.: *Nucl. Phys.* **76**, 321 (1966).
4. Elias, J. E., Friedman, J. I., Hartman, G. C., Kendall, H. W., Kirk, P. N., Sogard, M. R., Van Speybroeck, L. P., and de Pagter, J. K.: *Phys. Rev.* **177**, 2075 (1969).
5. Gomes, L. C., Walecka, J. D., and Weisskopf, V. F.: *Ann. Phys. N.Y.* **3**, 241 (1958).
6. Jastrow, R.: *Phys. Rev.* **98**, 1479 (1955); Clark, J. W. and Westhaus, P.: *Phys. Rev.* **141**, 833 (1966).
7. Bressel, C. N., Kerman, A. K., and Rouben, B.: *Nucl. Phys.* **A124**, 624 (1969).
8. Day, B. D.: *Rev. Mod. Phys.* **39**, 719 (1967).
9. Bethe, H. A.: *Ann. Rev. Nucl. Sci.* **21**, 93 (1971).
10. Baranger, M.: *Nucl. Phys.* **A149**, 225 (1970).
11. Brandow, B. H.: *Phys. Rev.* **152**, 863 (1966); *Rev. Mod. Phys.* **39**, 771 (1967).
12. Day, B. D.: *Phys. Rev.* **151**, 826 (1966); *Phys. Rev.* **187**, 1269 (1969); Dahlblom, T. K.: *Acta Acad. Aboensis* **B29**, 6 (1969).
13. Kallio, A. and Day, B. D.: *Nucl. Phys.* **A124**, 177 (1969).
14. McCarthy, R. J.: *Nucl. Phys.* **A130**, 305 (1969).
15. Davies, K. T. R., McCarthy, R. J., and Saver, P. U.: to be published.
16. Davies, K. T. R. and McCarthy, R. J.: *Phys. Rev.* **C4**, 81 (1971).
17. Negele, J. W.: *Phys. Rev.* **C1**, 1260 (1970).
18. Negele, J. W. and Vautherin, D.: *Phys. Rev.* **C5**, 1472 (1972).
19. Skyrme, T. H. R.: *Phil. Mag.* **1**, 1043 (1956).
20. Eden, R. J. and Emery, V. J.: *Proc. Roy. Soc.* **A248**, 266 (1958).

PART D

ASTROPHYSICS

AN INTRODUCTION TO ASTROPHYSICS

V. CANUTO

Institute for Space Studies, Goddard Space Flight Center, NASA, New York, U.S.A.

1. Star Formation – Main Sequence – Nuclear Reactions [1, 2, 3, 4]

In order to have a panoramic view of the role played by neutron stars, the main objective of our lecture, one has to understand how stars are born, evolve and finally die. Of these three periods of evolution we must confess that the last one is probably the one that has attracted the majority of those physicists who have decided to enter the field of astrophysics. It is also true that by now evolutionary tracks have suffered a massive attack from heavily equipped numerical programs that are able to include the finest refinements that nuclear physicists and atomic physicists will provide. A general statement can be made: evolution of stars is fairly well understood. Less understood are the transitions between the various phases of evolution. No detailed understanding has yet been achieved of what happens after a star has long left the main sequence, i.e., the region where it spends most of its life. Phrased differently, we can ask how a star goes into its cemetery, where only two kinds of possibilities are presently known: white dwarfs and neutron stars. We still don't know for sure. The formation of a star is also an extremely complicated problem. We can, however, describe it as follows. Consider a huge gaseous mass. The only force likely to act is the omni-present gravitational force that attracts the various components (whatever they are) towards each other. The total energy of this proto-star is

$$E = -\, q\,(GM^2/R)\,.$$

To be exact this equation should be written as

$$E = -\, G \int_0^M \frac{M(r)\,\mathrm{d}M(r)}{r} \equiv -\, q\,\frac{GM^2}{R}$$

$$q = \int_0^1 \frac{M(r/R)/M}{r/R}\,\frac{\mathrm{d}M(r/R)}{M}$$

where q is of the order of unity. Stars for which the pressure is related to the density by a polynomial law of the type

$$P = P_0 \varrho^{(n+1)/n}$$

(also called polytropes of index n) have a value of q given by $3/(5-n)$. When the proto-star is dispersed at infinity $R \to \infty$, $E=0$. By contracting (R smaller) $E<0$. Therefore in going from a dissipated gas cloud to a finite system, the energy has gone from zero

Abecassis de Laredo and Jurisic (eds.), Selected Topics in Phys. Astrophys. and Biophys. 327–378.

to a negative value. *This loss of energy has appeared as radiation and it represents the total energy radiated by the star in shrinking* from an infinitely dispersed system to a finite value R. Consequently if the gravitation is the only source of energy then the luminosity L is given by

$$L \equiv -\frac{\mathrm{d}E}{\mathrm{d}t} = -q\frac{GM^2}{R^2}\frac{\mathrm{d}R}{\mathrm{d}t} \quad (L > 0).$$

It follows that the time required for a star to contract from infinity to R is given by (no mass loss)

$$t = -qGM^2 \int_{-\infty}^{R} \frac{\mathrm{d}R'}{R'^2 L(R')} = qGM^2 \int_{R}^{\infty} \frac{\mathrm{d}R'}{R'^2 L(R')}.$$

For constant L, we obtain ($G = 6.67 \times 10^{-8}$ dyne cm^2 gr^{-2})

$$t = q\frac{GM^2}{RL} \simeq \frac{E_{GR}}{L}.$$

For the Sun, $M_\odot = 2.10^{33}$ gr, $R = 6.95 \times 10^{10}$ cm, $L_\odot = 3.9 \times 10^{33}$ erg s^{-1}, $t \simeq 2 \times 10^7$ yr. The age of the Sun is not 20×10^6 yr but 4.5×10^9 yr as based on geological evidence. *Fossil algae* have been found by geologists that are older than a billion years. We can safely conclude that *thermal and gravitational energies are not sufficient to cover the surface losses for the whole life of the star.*

In the last century it was thought that the release of gravitational energy was the only source of luminosity for stars and the discrepancy with the age of the Sun was not easy to understand.* We now know the answer: the gravitational contraction does not go on forever: it is the mechanism through which the temperature rises until nuclear reactions can take place, in particular fusion of light elements to form heavy ones. At that moment the star enters in the longest period of its life, the main sequence region where our sun is located. *The gravitational shrinkage ceases and the luminosity of the star is entirely provided by the energy released by burning nuclear fuel.* How long does this burning process go on for? Suppose we transform 4 protons into a helium nucleus, i.e., we built up helium out of protons via some nuclear process. In short notation

$$4\mathrm{p} \rightarrow \mathrm{He}^4.$$

The total initial energy is simply the sum of the rest masses.

$$E_\mathrm{i} \equiv M_\mathrm{i}c^2 = 4m_\mathrm{p}c^2.$$

The final energy is however

$$E_\mathrm{f} \equiv M_\mathrm{f}c^2 = 4m_\mathrm{p}c^2 + T + V,$$

* A famous dispute arose between Cuvier and Kelvin about the age of fossils and the age of the Sun computed on gravitational energy release only [2].

where $T+V$ indicates the kinetic and potential energy of the helium nucleus. Since such a nucleus is bound,

$$T + V = - |B| \cong - 28 \text{ MeV},$$

where $|B|$ is the binding energy. Therefore we gain

$$E_{\text{i}} - E_{\text{f}} \cong 28 \text{ MeV or } 7 \text{ MeV per particle}.$$

How many particles do we have in the Sun? The density is known to be about 1 gr cm^{-3}, the radius 6.9×10^{10} cm and since

$$\varrho = 1 \text{ gr cm}^{-3} = M/\frac{4\pi}{3} R^3 \equiv \frac{m_{\text{p}}N}{4\pi/3R^3}$$

we get $N \simeq 10^{57}$. The total gain in energy is therefore

$$7 \times 10^{57} \text{ MeV} \simeq 7 \times 10^{+57} \times 10^{-6} \text{ erg} \simeq 7 \times 10^{51} \text{ erg}.$$

The Sun, however, shines every second 2 erg per gr, i.e., the so-called solar constant is known to be

$$\varepsilon = 2 \frac{\text{erg}}{\text{gr s}}$$

These are 10^{33} gr in the Sun so that every second the Sun emits

$$\eta = M\varepsilon = 2 \times 10^{33} \text{ erg s}^{-1}.$$

Consequently the Sun can live at the expenses of nuclear energy for as long as

$$\frac{7 \times 10^{51} \text{ erg}}{2 \times 10^{33} \text{ erg s}^{-1}} = 3 \times 10^{18} \text{ s} \cong 10^{11} \text{ yr}$$

quite along time! Even supposing that not all the material is ready for burning but only some percent of it, we still get a life long compared with the geological findings.

This simply means that the energy losses at the surface as measured by the luminosity of the star are most likely compensated by the energy release from nuclear processes in the interior, i.e., we shall demand that the luminosity of a star should be accounted for by the nuclear energy release, i.e.,.

$$L = \int_0^R 4\pi r^2 \, \mathrm{d}r\varrho\varepsilon$$

Here ε is the energy per gram, per sec released by nuclear reactions and it has to be computed from first principles. This is one of the fundamental equations of stellar structure.

What kind of nuclear reactions actually do take place? Evidently one starts by burning the lightest elements, protons, into something heavier; then this last product into something else and so on. At any time one product is depleted, say hydrogen is depleted, the nuclear fuel is temporarily out of business since the temperature is not high enough to trigger the next element's fusion. At that moment the gravitational contraction that has been jobless for a while takes over i.e., the star will collapse a little bit, the internal energy will go up, the temperature will increase and the situation is again ready for the next element to be burnt. At each step the chemical composition changes due to the depletion of certain elements; this we do observe through the spectroscopic study of the surface. How do we go about to compute the nuclear reactions rates occurring in the star? First of all, we presently know about few hundred stable isotopes. All of them in principle can react with one another. If so, there is really little hope that we can come up with 'the right nuclear reaction' that we claim is the responsible for the Sun to shine. Fortunately there exists a lucky circumstance that will help us out. The average temperature in the interior of a star is about 10^7 deg. In fact, the assumption of hydrostatic equilibrium implies that

$$NkT \simeq GM^2/R$$

i.e. the thermal energy content of the star must equal the gravitational binding energy, otherwise the star would collapse. Since $M = m_p N$ we have

$$T = \frac{Gm_p^2N^2}{NkR} \simeq \left(\frac{GM}{R}\right)\frac{m_p}{k} \simeq 10^7 \text{ K}$$

for the Sun. More exact computation gives $T_{\text{central}} = 1.57 \times 10^7$ K, $\varrho_c = 158 \text{ gr cm}^{-3}$*. From this we deduce that the average kinetic energy per particle is

$$E_{\text{Th}} = \tfrac{3}{2}kT \simeq 1 \text{ keV}.$$

How big is the Coulomb Barrier (C.B.) between two nuclei of charge Z_1 and Z_2? For $Z_1 = Z_2 = 1$ and $R = R_N \simeq 1$ fermi $= 10^{-13}$ cm, we have

$$E_{\text{C.B.}} = \frac{e^2 Z_1 Z_2}{R} \simeq 10^3 \text{ keV} \simeq 1 \text{ MeV}.$$

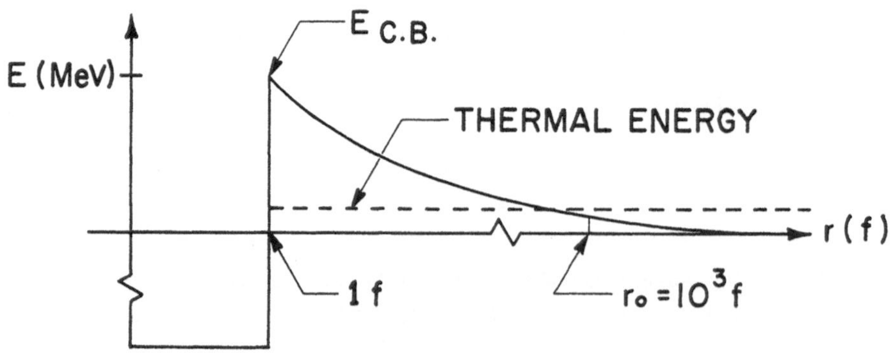

Fig. 1.

* The surface temperature is known to be about 6000 K.

Before feeling the benefit of the attractive nuclear forces, a particle must overcome the Coulomb Barrier. Without quantum mechanics we would be lost: the particle wouldn't be able to penetrate inside and there would be no nuclear fusion at all. In effect the distance of closest approach r_0 is

$$\frac{Z_1 Z_2 e^2}{r_0} = E_{\text{Th}} = 1 \text{ keV}$$

$$r_0 = Z_1 Z_2 e^2 / E_{\text{Th}} \simeq 10^3 \text{ fermi !}$$

about 1000 times larger than the nuclear radius. Fortunately there is always some tunneling due to quantum effects. i.e., there is a finite probability for a particle with $E_{\text{Th}} < E_{\text{C.B.}}$ to penetrate the barrier. Such a probability is proportional to ($E_{\text{Th}} \equiv E$)

$$P \simeq \exp - \left[\int_R^{r_0} \sqrt{U_l(r) - E}\, \mathrm{d}r \right],$$

where R is the nuclear radius and r_0 is the distance of closest approach defined by

$$U_l(r_0) \equiv E$$

$U_l(r)$ is in general

$$U_l(r) = \underset{\begin{pmatrix}\text{Coulomb}\\ \text{barrier}\end{pmatrix}}{\frac{Z_1 Z_2 e^2}{r}} + \underset{\begin{pmatrix}\text{Centrifugal}\\ \text{barrier}\end{pmatrix}}{\frac{\hbar^2}{2m} \frac{l(l+1)}{r^2}}$$

Evidently the penetration probability decreases very rapidly with decreasing energy. We could think of increasing the energy: this however does not work because when $E \gg$ thermal average, the number of particles decreases very rapidly as dictated by Maxwell's distribution

$$N(E) \propto e^{-E/kT}.$$

Therefore there is a delicate balance between the energy and the number of particles to make the process go. The battle between these two factors, however, gives rise to the right energy output.

For particles of higher Z, the Coulomb Barrier becomes prohibitively high compared with the average kinetic energy and there is just no hope to make anything go through. We therefore can forget about high value of Z; this will in turn reduce the few hundred nuclei to the first ten or so, giving an all new dimension to the problem.

Let us now first see what physical quantity we can compute and what we get out of it. The cross-section of any reaction of the type

$$a + X \rightarrow b + Y$$

goes with the Fermi '$1/v$' power i.e.

$$\sigma(a, b) \simeq \sigma_0 E^{-1/2} P(E),$$

where P is the Gamow probability factor. From W.K.B. approximation in quantum mechanics we know that

$$P_l(E) = \frac{K_l(R)}{k} \exp\left\{-2\int_R^{r_0} K_l(r)\,\mathrm{d}r\right\}$$

$$K_l^2(r) = \frac{2m}{\hbar^2} U_l(r) - k^2$$

$$U_l(r) = V(r) + \frac{\hbar^2}{2m}\frac{l(l+1)}{r^2}$$

$$V(r) = \frac{ZaZx}{r} e^2 .$$

Taking $l=0$ and performing the integrations one gets

$$P_{l=0}(E) = \text{const} \times E^{-1/2} e^{-2\pi\eta}, \qquad \eta \equiv \frac{2\pi Z^2 e^2}{\hbar v} .$$

In turn

$$\sigma = \sigma_0(E)\, E^{-1} e^{-2\pi\eta},$$

where $\sigma_0(E)$ is a very slightly varying function of E.

Now that we have the cross-section we can compute the astrophysically important quantity, i.e. the RATE r for a given process defined as *the number of reactions of such a type per cubic centimeter per second, i.e.*,

$$r \equiv \int\int \mathrm{d}N_1(v_1)\,\mathrm{d}N_2(v_2)\,\sigma v$$

with

$$\mathrm{d}N_i(v_i) = N_i \left(\frac{v_i}{2\pi mkT}\right)^{3/2} e^{-m_i v_i^2/2kT}\,\mathrm{d}^3 v_i .$$

Performing first the center of mass integration we obtain

$$r = N_1 N_2 \langle\sigma v\rangle$$

$$\langle\sigma v\rangle = \frac{2}{\sqrt{\pi}} (kT)^{-3/2} \int_0^\infty e^{-E/kt} E^{1/2}\,\mathrm{d}E v\sigma(v) .$$

In Figure 2 below we have indicated how the various terms in the last integral behave with energy

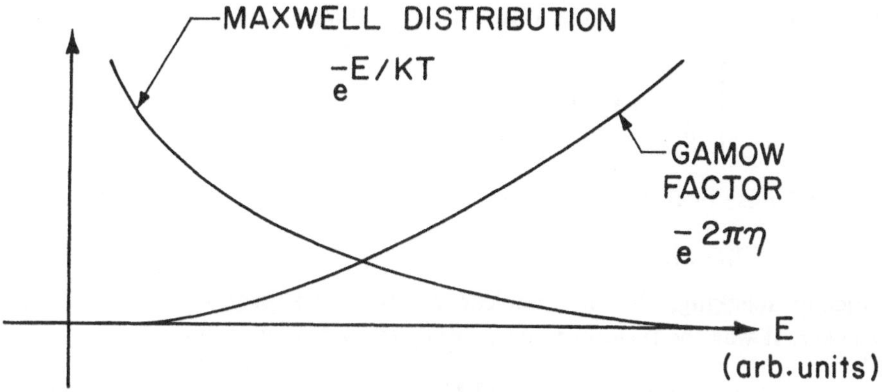

Fig. 2.

Using the above expression for $\sigma(v)$ we get after some algebra

$$\langle \sigma v \rangle = C\tau^2 \exp(-\tau)$$
$$\tau = 42.487\,(Z_a^2 Z_x^2 A)^{1/3} T_6^{-1/3}$$
$$A = \frac{A_a A_x}{A_a + A_x}$$
$$C \equiv \frac{7.207 \times 10^{-19}}{Z_a Z_x A}\, S(E_{\max}).$$

Finally

$$r \equiv N_1 N_2 C \tau^2 e^{-\tau}.$$

What do we do with r? If Q is the amount of energy released in a single reaction of the type $X(a, b)\,Y$, then the amount of energy per second and per gram, ε, is just

$$\varepsilon = (Q/\varrho)\cdot r = \frac{1}{\varrho} N_1 N_2 Q \langle \sigma v \rangle.$$

How strongly does $\langle \sigma v \rangle$ depend on temperature? Given a function of the form

$$f(x) = x^n$$

the exponent n is simply obtained after taking

$$\frac{\partial \lg f(x)}{\partial \lg x} = n.$$

Similarly if we suppose that $\langle \sigma v \rangle \simeq T^n$, i.e., if we parametize our cross-section with a polynomial expression and we apply the previous simple formula we find that

$$n = \frac{\tau - 2}{3}$$

We have for instance (for temperature corresponding to the Sun)

$$\tau/3 = \begin{cases} 4.7 & (p, p) \\ 19 & (p, C^{12}) \\ 25 & (p, N^{14}) \end{cases}$$

We will make use of this result in a moment.

1.1. p–p AND CNO CHAIN [1, 2]

Historically speaking, the first nuclear reaction was proposed in 1937 by Von Weizsäcker. It was the fusion of two protons to form deuterium, i.e.,

$$\begin{aligned} &p + p \to D + e^+ + \nu + 1.18 \text{ MeV} \\ &p + D \to He^3 + \gamma + 5.49 \text{ MeV} \\ &He^3 + He^3 \to He^4 + 2p + 12.86 \text{ MeV}. \end{aligned} \tag{1}$$

In total we get 26.20 MeV by transmuting 4p into He^4. We could also symbolically write

$$4p \to He^4 + 2\nu + Q_T (\equiv 26.20 \text{ MeV}).$$

We must note that this fundamental processes cannot be observed in laboratory and probably never will be because of its extremely small cross-section. The smallness is essentially due to the β-decay that must occur during the encounter of the two protons.

Until 1958 it was believed that his was *the* nuclear reaction. In that year, however, Johnston and Holgren measured the cross-section for the reaction

$$\alpha + He^3 \to Be^7 + \gamma + 1.6 \text{ MeV}$$

and found that it was 2500 times bigger than the previously adopted value. Therefore there can be two branches once He^3 is formed, i.e., the one given before and the following two

(2)	Q (MeV)	(3)	Q (MeV)
$p + p \to D + e^+ + \nu$	1.18	$p + p \to D + e^+ + \nu$	1.18
$p + D \to He^3 + \gamma$	5.49	$p + D \to He^3 + \gamma$	5.49
$\alpha + He^3 \to Be^7 + \gamma$	1.59	$\alpha + He^3 \to Be^7 + \gamma$	1.59
$e^- + Be^7 \to Li^7 + \nu$	0.06	$p + Be^7 \to B^8 + \gamma$	0.13
$p + Li^7 \to \alpha + He^4$	7.35	$B^8 \to Be^8 + e^+ + \nu$	10.78
		$Be^8 \to 2He^4$	0.09

Evidently, if the amount of He^4 (α-particle) is extremely low, then the two branches may contribute nothing at all. In the exponent $e^{-\tau}$ we have the reduced mass $\mathscr{A}$; now

$$\underset{(1.71)}{\mathscr{A}(He^3, He^4)} > \underset{(1.5)}{\mathscr{A}(He^3, He^3)}$$

This means that if He^4 abundance is enough to make the reaction go, then the last two branches would be more important at high temperature than the first branch, since the reaction $He^3 + He^4$ is a more rapidly increasing function of the temperature.

It is estimated that (2) and (3) dominate over (1) at $T_6 \gtrsim 20$. Energetically these reactions are perfectly acceptable. Their temperature dependence is however not very strong, $n \cong 4$, i.e., it goes like T^4. Now, observationally it so happens that the temperature changes very little from the sun to more massive stars. We would therefore get a very little change in energy production. However in Figure 3, we see that the luninosity changes a great deal by changing the mass. Therefore there must be some other mechanism, i.e., some other nuclear reaction that proceeds faster than T^4.

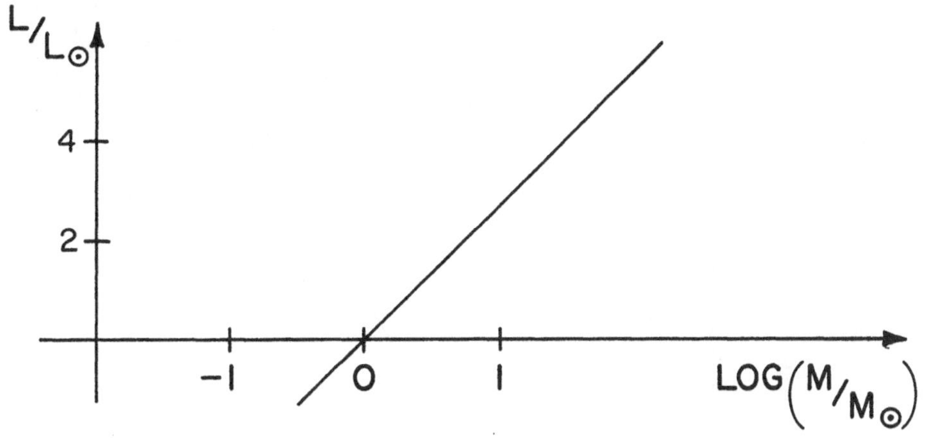

Fig. 3.

1.2. CNO CYCLE [2]

Another possibility is offered by the so-called CNO cycle given by the following series of transformations

	Q (MeV)
$\rightarrow C^{12} + p \rightarrow N^{13} + \gamma$	1.95
$N^{13} \rightarrow C^{13} + e^+ + \nu$	2.22
$C^{13} + p \rightarrow N^{14} + \gamma$	7.54
$N^{14} + p \rightarrow O^{15} + \gamma$	7.35
$O^{15} \rightarrow N^{15} + e^+ + \nu$	2.71
$N^{15} + p \rightarrow C^{12} + He^4$	4.96

Again it can be written as

$$4p \rightarrow He^4 + 2\nu + Q_T (\equiv 25.02 \text{ MeV}).$$

From the numbers computed before the temperature dependence of these reactions is like $T^{18.4}$ quite high to explain the luminosity variation. The energy produced by the two processes is represented in the Figure 4.

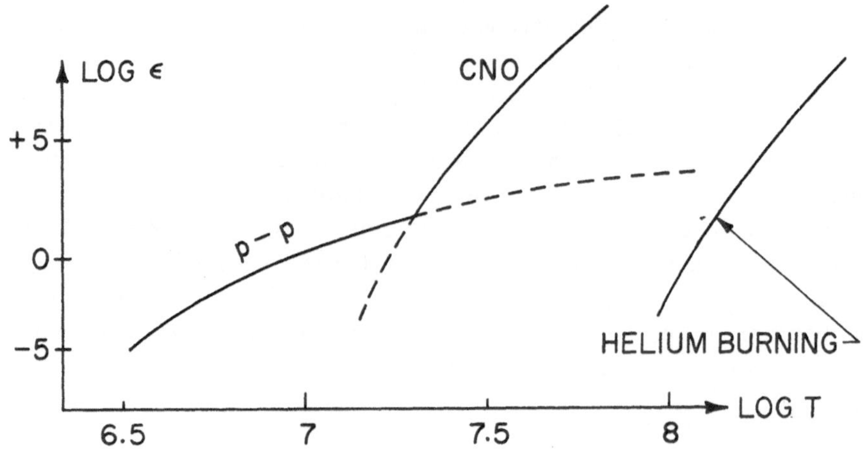
LOG ϵ
CNO
+5
0
-5
p-p
HELIUM BURNING
6.5
7
7.5
8
LOG T

Fig. 4.

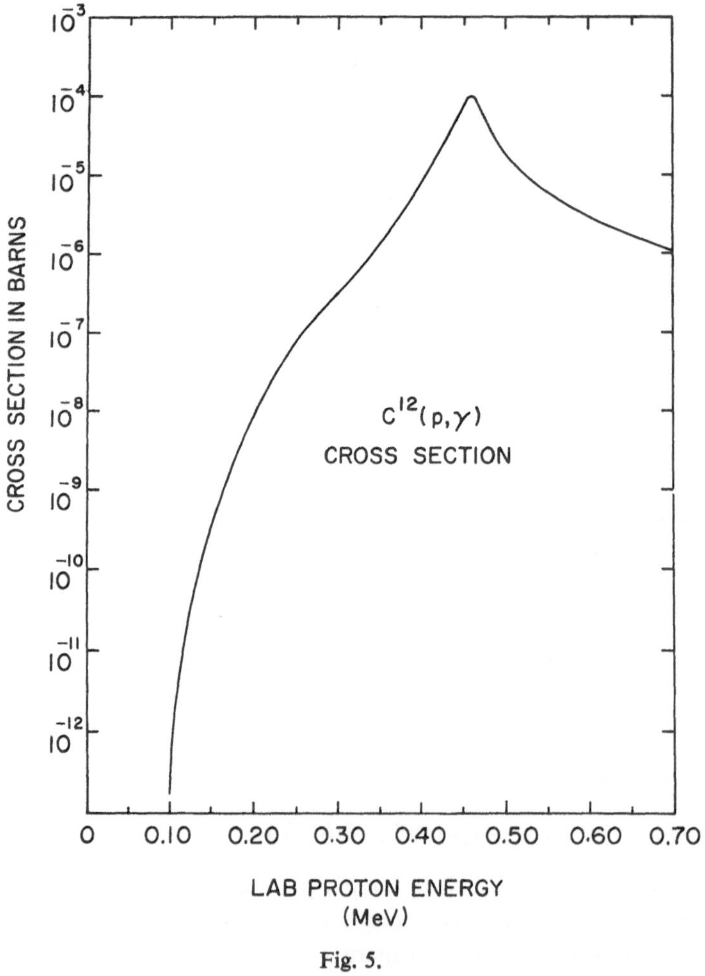
CROSS SECTION IN BARNS
10^{-3}
10^{-4}
10^{-5}
10^{-6}
10^{-7}
10^{-8}
10^{-9}
10^{-10}
10^{-11}
10^{-12}
$C^{12}(p,\gamma)$
CROSS SECTION
0
0.10
0.20
0.30
0.40
0.50
0.60
0.70
LAB PROTON ENERGY
(MeV)

Fig. 5.

As expected, the pp reaction dominates over CNO at low temperatures. This is directly a consequence of the Coulomb Barrier. As we can see, our Sun sits just before the onset of the CNO cycle. This series of nuclear reactions is all very nice in theory, but it is worthwhile to ask what is the experimental evidence that supports such reactions [4]. In Figure 5 we show the cross-section for $p + C^{12} \to N^{13} + \gamma$ vs. p_L energy. Some comments are in order. First of all, the values of σ vary over a factor of 10^6 in magnitude from 10^{-4} down to 10^{-10}. There is a resonance at 460 keV. The theoretical curve is impressively good. Analogous results exist for the other reaction of the CNO cycle. In the case of the Sun, the Gamow's peak occur at

$$E_0 = (\tfrac{1}{3})\,\tau kT \cong 20\,kT\,, \qquad \tau/3 \simeq 19\,.$$

For $T = 1.3 \times 10^7$ we have $kT = 1.1$ keV.

$$E_0 \cong 20 \text{ keV !}$$

This is too low an energy to observe σ experimentally. We have to extrapolate it all the way down, hoping that there are no resonances near $E = 0$. For instance it was once found that the reaction $N^{14} + p$ could have a resonance at 20 keV. However it had $J = \frac{7}{2}$. Now N^{14} has $J = 1$ and the proton $J = \frac{1}{2}$; in order to get $J = \frac{7}{2}$ one needs an orbital angular momentum $l = 3$. Such a high angular momentum would reduce σ by a factor of about 10^4. Therefore it can be disregarded.

1.3. Advanced Stage of Evolution [4]

After about 10^{10} yr, the depletion of hydrogen is almost complete and the star finds itself in a situation in which the core is almost purely formed of helium. In the outer layers of the star hydrogen is still abundant. Since the temperature is not enough to burn He^4, the interior does not have enough energy to support the external layers and the center collapses. More properly, since there is no nuclear reaction taking place there is no energy source capable of withstanding the outer layers and the 'standing by 'gravity enters into play. By contracting the star, T goes up, the core heats up and so does the envelope which expands and increases the surface area. This increased area means that in order to maintain the same luminosity the temperature lowers, i.e., the surface reddens in color. This is because $4\pi R^2 = L/\sigma T_e^4$. That is why these stars are called '*Red Giants*'. Now the problem is how does one make two helium nuclei interact? We have a problem here. All nuclei with masses between $A = 5$ and $A = 8$ are unstable and there seems to be no way to get through the gap. We have to wait until the reaction

$$3\alpha \to C^{12} + \text{something}$$

happens. It is hard to think that this is the right way to go about the problem since a three-body collision is not exactly an every day event. In 1952 Salpeter [5] came up with the idea that the building up of C^{12} does indeed take place, but via an intermediate step

$$He^4 + He^4 \to Be^8\,(10^{-7}\text{ s}) + \gamma - 95 \text{ keV} \tag{1.1}$$

$$Be^8 + He^4 \to C^*_{12} \to C_{12} + \gamma + 7.4 \text{ MeV}\,. \tag{1.2}$$

If Be^8 is formed and a small amount of it reacts with another He^4, then a stable C^{12} could be formed. The amount of Be^8 is actually so very small as deduced from Saha equation

$$N(Be^8) \propto N^2(He^4)\, e^{-X/kT} T^{-3/2}$$

(of the order of 1 in 10 billion of α-particles) that if any second reaction (with He^4) has to occur it better be fast! In nuclear physics jargon one would say that (1.2) has to be a resonant reaction.

After C^{12} is formed, the building up of heavier elements continues via a host of reactions like

$$\begin{aligned} C^{12} + He^4 &\to O^{16} + \gamma \\ O^{16} + He^4 &\to Ne^{20} + \gamma \\ Ne^{20} + He^4 &\to Mg^{24} + \gamma \\ &\text{etc.} \end{aligned}$$

At even higher temperature $T_6 = 600–900$ we have the so-called *carbon burning period*,

$$\begin{aligned} C^{12} + C^{12} \to\ & Ne^{20} + \alpha \\ & Na^{23} + H \\ & Mg^{24} + \gamma \\ & Mg^{23} + n\,. \\ & \text{etc.} \end{aligned}$$

and many others. In summary, we can say that

(1) It is up to main sequence stars to burn hydrogen to produce He^4.
(2) It is up to Red giant stars to burn He^4 to produce C^{12}.
(3) It is up to Red supergiants to produce heavier nuclei from the ashes of C^{12} into Ne, Na, Al and Sr^{28}.

Now the question is: does the fusion process keep going indefinitely? It's clearly impossible to make it work when Z_1 and Z_2 become 10 times the charge of the proton. Fusion becomes a very ineffective way to build up nuclei. Another important channel opens up at temperatures of the order 5×10^9 K. Intense photon fluxes can photodisintegrate a number of nuclei into α-particles. For instance, Si^{28} can be photodisintegrated into 7 α's; some other Si^{28} nucleus that escaped the breaking up can capture the 7α's system to form a radioactive nickel Ni^{58}; later on Ni^{58} decays inot Co^{56} and finally to Fe^{56}. Many other nuclei around the ironpeak are thought to be built up in this way. *These processes are known as α-processes.* The building up of higher nuclei after iron is thought to be achieved by neutron capture, a slow but secure mechanism for the synthesis of heavier and heavier nuclei. Neutrons interact with a $1/v$ law, i.e neutron processes are more rapid at low energy than at high ones, in total contrast with the charged particles mechanism. It is worth noticing that Gamow's idea of synthesizing everything from an original 'neutron ball' is incorporated in this picture of the nucleosynthesis at midway [5, 6, 7].

Where do we get the neutrons to start the process? Mainly through (α, n) reactions that are known in the laboratory to be exoergetic. Examples are

$$C^{13}(\alpha, n)\,O^{16}$$
$$Mg^{22}(\alpha, n)\,Mg^{25}$$
$$Mg^{25}(\alpha, n)\,Si^{28}$$
$$Mg^{26}(\alpha, n)\,Si^{29}(\alpha, n)\,S^{32}\text{ etc.}$$

We have already seen that it's possible to have even more direct reactions like

$$C^{12} + C^{12} \rightarrow Mg^{23} + n \quad Q = 2.603$$

during carbon burning or else

$$O^{16} + O^{16} \rightarrow Si^{31} + n$$

Many of the previous nuclear reactions are largely endothermic and they keep eating up energy out of the core. *In this stage nuclear processes tend to consume rather* than *release energy*. They do not oppose gravitational contraction, rather they go along with it. Just as the first hydrogen burning contracted the star and led to an expansion of the outer layers, here too, nuclear reactions and gravity together cause a more fantastic process. The outer layer gets expanded at such high rate that it is actually an explosion. Hoyle and Fowler have suggested that this could be the mechanism for supernova explosion. The ejected material most likely contains those heavy elements that were formed in the interior of heavy stars. The heavy elements go into the interstellar space out of which a new star will soon be born.

Note

Does the chain of thermo-nuclear reactions described before occur in any star? The answer is no; it depends on temperature and mass. Let us devise the way it enters. The interior of a typical star is highly non-degenerate, or what is the same quantum mechanics does not play any role. If that is so, it simply means that the mean separation distance is greater than the de Broglie wavelength, which usually triggers the entering of quantum mechanics. Therefore

$$\langle r \rangle \gg \lambda_{DB}\,.$$

But

$$\langle r \rangle = \left(\frac{m_p}{\varrho}\right)^{1/3}$$

$$\lambda_{DB} = \frac{\hbar}{m_e v} = \frac{\hbar}{\sqrt{2m_e kT}}$$

$$\left(\frac{m_p}{\varrho}\right)^{1/3} > \frac{\hbar}{\sqrt{2m_e kT}}$$

or else

$$1 > \frac{\varrho^{1/3}\hbar}{m_p^{1/3}\sqrt{2m_e kT}}.$$

Using

$$\varrho = M/\frac{4\pi}{3}R^3, \qquad \frac{P}{R} = \frac{GM\varrho}{R^2} \rightarrow P = \frac{GM\varrho}{R}$$

and since

$$P = kT\varrho/\mu m_{\mathrm{H}} \quad \text{we get} \quad kT = \frac{GM}{R}\mu m_{\mathrm{H}}$$

so that

$$1 > \frac{M^{1/3}\hbar}{Rm_p^{1/3}}\sqrt{\frac{R}{2m_e GM\mu m_{\mathrm{H}}}}$$

$$1 > 0.0914\,\mu^{-1/2}\left(\frac{M_\odot}{M}\right)^{1/6}\left(\frac{R_\odot}{R}\right)^{1/2}$$

or finally

$$\left(\frac{R}{R_\odot}\right) > \frac{8.36\times 10^{-3}}{\mu}\left(\frac{M_\odot}{M}\right)^{1/3}.$$

On the other hand, we know that the necessary central temperature for hydrogen burning is $T = 10^7$ K so that

$$T \equiv \frac{1}{5}\frac{\mu m_{\mathrm{H}}}{k}\frac{GM}{R} > 10^7$$

This implies that

$$\mu\left(\frac{M}{M_\odot}\right)\left(\frac{R_\odot}{R}\right) \geqslant 2.16.$$

Putting the two requirements together we finally obtain,
Hydrogen burning:

$$\mu^{3/2}\frac{M}{M_\odot} \geqslant 0.05, \quad T = 10^7\,K.$$

Similar computation gives the following conditions
Helium burning:

$$\mu^{3/2}\frac{M}{M_\odot} \geqslant 0.28, \quad T = 10^8\,K.$$

Carbon burning:

$$\mu^{3/2}\frac{M}{M_\odot} \geqslant 1.2, \qquad T = 7.0^8\,K.$$

Oxygen burning:

$$\mu^{3/2}\frac{M}{M_\odot} \geqslant 1.9 \qquad T = 1.3\times 10^9\ \mathrm{K}.$$

2. General Theorems About Cold Stars

The Onset of Quantum Effects

About 1850 two binary stars were found that had peculiar features. They were found to have

$M/M_\odot$	$L/L_\odot$	$R/R_\odot$	$\varrho\,(\mathrm{gr\,cm^{-3}})$	
0.7	10^{-3}	10^{-2}	10^5	*Procyon B*
1.0	3.10^{-3}	10^{-2}	10^5	*Sirius B*

Their existence faced astrophysiscists with the problem of understanding the existence of stars with *the same mass as the Sun, the radius of a planet and much lower luminosity*! One thing is for sure: they do not belong to the main sequence. They are not burning any nuclear fuel, otherwise they would be brighter. Two common explanations are usually given:

(1) They never turned on the nuclear processes.

(2) They have completely exhausted the nuclear fuel and in a very uneventful way they are just simply dying.

Let us consider what happens to a star when its nuclear fuel is completely exhausted. Since the star has no energy fuel any longer the only thing left to do is to contract. Let us *test* the assumption that the pressure is still given by the perfect gas law,

$$P = \frac{kT\varrho}{\mu m_\mathrm{H}}$$

From the virial theorem (or else the balance between gravitation and thermal pressure) we have

$$\int P\mathrm{d}V \equiv \int \frac{P}{\varrho}\,\mathrm{d}M \equiv \frac{k\bar{T}}{\mu m_\mathrm{H}}\,M = -\,W_g \equiv q\,\frac{GM^2}{R}$$

so that

$$\frac{\bar{T}M}{\mu m_\mathrm{H}} = q\,\frac{GM^2}{R},\ T \equiv q\,\frac{GM}{R}\cdot\left(\frac{m_\mathrm{H}\mu}{k}\right) \sim M/R$$

By decreasing R the temperature goes up like $1/R$. By contracting the system, the temperature increases (only) with the first power of the dimension of the system. However, the density increases more steeply than the temperature, so there has to be a point at which the perfect gas law fails. Let us compute Δp and Δx [8].

$$\Delta p \simeq (mkT)^{1/2}$$
$$\Delta x \simeq \mu_e(m/\varrho)^{1/3}$$

so that

$$(\Delta p)^3 (\Delta x)^3 \simeq \left[m^{5/6} \left\{ q \frac{GM}{R} \frac{\mu m_H}{k} \right\}^{1/2} \mu_e \left(\frac{4\pi R^3}{3M} \right)^{1/3} \right]^3$$

$$\simeq \left[m^{5/6} \mu (G \mu m_H R_\odot)^{1/2} \left(\frac{M}{M_\odot} \right)^{1/6} (16 M_\odot)^{1/6} \left(\frac{R}{R_\odot} \right)^{1/2} \right]^3$$

$$\simeq \left\{ 6.3 \times 10^{-26} \left(\frac{M}{M_\odot} \right)^{1/6} \left(\frac{R}{R_\odot} \right)^{1/2} \text{gr cm}^2 \text{ s}^{-1} \right\}^3$$

so that

$$(\Delta p)^3 (\Delta x)^3 \simeq 10^3 h^3 \left(\frac{M}{M_\odot} \right)^{1/2} \left(\frac{R}{R_\odot} \right)^{3/2} .$$

When a star collapses down to $10^{-2} R_\odot$, *the volume occupied by an electron in this space is approximately* h^3.

R. H. Fowler, in 1920, first pointed out that Pauli's principle becomes important and one is forced to use Fermi-Dirac statistics. Electron degeneracy seems to enter exactly at the point where white dwarfs are observed. If we do the same exercise for neutron stars we would find that neutron degeneracy enters right where pulsars are observed.

Chandrasekhar [9] made the first detailed study of the white dwarf mass limit by integrating the hydrostatic equilibrium equation

$$\frac{dP}{dr} = -\varrho \frac{GM(r)}{r^2} \tag{2.1}$$

$$M = \int_0^R 4\pi r^2 \varrho(r)\, dr = \int_0^R M(r)\, dr \tag{2.2}$$

with the equation of state corresponding to a cold, degenerate electron gas

$$P = \frac{\pi m^4 c^5}{3h^3} f(x) = 6 \times 10^{22} f(x) \text{ (dyne cm}^{-2}), \quad x = p_F/mc$$

$$f(x) = x(2x^2 - 3)(1 + x^2)^{1/2} + 3 \ln (x + \sqrt{1 + x^2}) \quad \begin{cases} 2x^4, & x \gg 1 \\ \frac{8}{5} x^5, & x \ll 1 \end{cases}$$

$$n = \frac{8\pi}{3} \frac{p_F^3}{h^3}, \; \varrho \equiv mn = 10^6 \mu x^3 \text{ (gr cm}^{-3}).$$

Eliminating p_F in favor of ϱ, one gets $P = P(\varrho)$, i.e. the equation of state.

Integrating Equations (2.1) and (2.2) one obtains a one-parameter family of solutions with respect to ϱ_c the prescribed central density. The results obtained by Chandrasekhar were remarkable; no stable configuration was found with $M \geqslant 1.44\, M_\odot$, the so called Chandrasekhar limit (Figure 6).

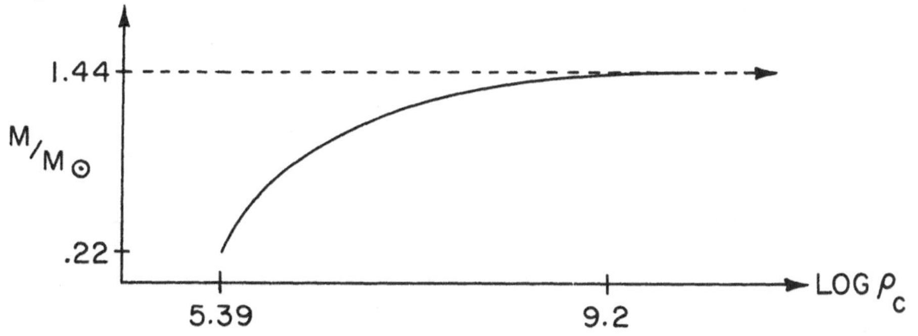

Fig. 6.

Better results have been produced recently by including all sorts of refinements and embellishments such as screening, chemical composition, etc. etc. The main feature is persistent. There is a maximum limit to the mass of the order of lower than 1.44 $M_\odot$.

What happens now if we increase ϱ_c? First of all, let us write down the equation of state for $x \gg 1$, i.e. high density limit. We have

$$\begin{aligned} P &= 6 \times 10^{22} \times 2 \times x^4 \\ &= 1.2 \times 10^{15} \mu^{-4/3} \varrho^{4/3} \\ P &= P_0 \varrho^{4/3}. \end{aligned}$$

Let us now compute Γ, the so-called adiabatic limit defined as

$$\Gamma = \frac{\varrho}{P} \frac{\mathrm{d}P}{\mathrm{d}\varrho}.$$

We obtain

$$\Gamma = \tfrac{4}{3}.$$

We will immediately show that when $\Gamma \geqslant \frac{4}{3}$ a star is unstable against collapse.

2.1. Hydrostatic Instability [10]

Let us consider a spherically symmetric star in hydrostatic equilibrium. The total energy of the star is given by

$$E = \int \varrho \left[\varepsilon - \frac{GM(r)}{r} \right] \mathrm{d}V,$$

where

r: distance from center

ϱ: density; v: the specific volume $= 1/\varrho$

$m(r)$: mass inside radius r

$\mathrm{d}V$: $4\pi r^2 \mathrm{d}r$

ε: internal energy density of matter.

By equation of state of matter we will mean a relation

$$\varepsilon = \varepsilon(\varrho, P),$$

where P and ϱ are the pressure and density.
From the definition given before we have

$$\Gamma = \frac{\varrho}{P}\frac{\mathrm{d}P}{\mathrm{d}\varrho} = -\frac{v}{P}\frac{\mathrm{d}P}{\mathrm{d}v}$$

since

$$P = -\frac{\partial \varepsilon}{\partial v}$$

we get

$$\Gamma = \frac{v}{P}\frac{\partial^2 \varepsilon}{\partial v^2}.$$

Suppose now that we perform a uniform expansion (or contraction) such that the radius r goes into r^* defined as

$$r \to r^* = (1+\alpha)\, r$$

We will have

$$\begin{aligned} v &\to v^* = (1+\alpha)^3\, v \\ \mathrm{d}V &\to \mathrm{d}V^* = (1+\alpha)^3\, \mathrm{d}V \\ \varrho &\to \varrho^* = (1+\alpha)^{-3}\, \varrho \\ m &\to m^* = m \\ \varepsilon &\to \varepsilon^* = \varepsilon + \left(\frac{\partial \varepsilon}{\partial v}\right)(\Delta v) + \tfrac{1}{2}\left(\frac{\partial^2 \varepsilon}{\partial v^2}\right)(\Delta v)^2 \\ &\qquad = \varepsilon - \mathrm{p}\left[3\alpha + 3\alpha^2\right] v + \tfrac{1}{2}\Gamma \frac{\varrho}{v}(3\alpha)^2\, v^2 . \end{aligned}$$

Therefore the total energy E changes to

$$E \to E^* = \int \varrho \mathrm{d}V \left[\left(\varepsilon - 3pv(\alpha + \alpha^2) + \tfrac{9}{2}\Gamma \varrho \alpha^2 v\right) - \frac{Gm(r)}{r}(1+\alpha)^{-1}\right]$$

or

$$\Delta E = \int \varrho \mathrm{d}V \left[\alpha\left(-3pv + \frac{Gm(r)}{r}\right) + \alpha^2\left(-3pv + \tfrac{9}{2}\Gamma pv - \frac{Gm}{r}\right)\right]$$

$$E(\alpha) = E(0) + \alpha E_1 + \alpha^2 E_2 + \cdots .$$

For the star to be in any kind of equilibrium (stable or unstable) the coefficient of α must be zero, i.e.

$$E_1 \equiv \frac{\partial E(\alpha)}{\partial \alpha} = 0 \to -\int 3p\mathrm{d}V + \int \frac{G\varrho m(r)}{r}\,\mathrm{d}V = 0$$

or

$$-3\int p\mathrm{d}V = Wg$$

which is just the already known *virial theorem.*
For the equilibrium to be stable, the coefficient of α^2 must be positive, i.e.

$$E_2 \equiv \frac{\partial^2 E}{\partial\alpha^2} \geqslant 0 \rightarrow \int\left[-3p + \tfrac{9}{2}\Gamma p - \frac{Gm\varrho}{r}\right]\mathrm{d}V \geqslant 0.$$

Using the virial theorem

$$\int\left[-3p + \tfrac{9}{2}\Gamma p - 3p\right]\mathrm{d}V \geqslant 0$$

$$\int\left(\Gamma - \tfrac{4}{3}\right)p\mathrm{d}V \geqslant 0.$$

Thus, the value $\Gamma = \frac{4}{3}$ is critical for stability. When a star happens to be left with a mass higher than the one allowed by this theory it will continue to shrink *until a new source of pressure is formed* which will sustain and stop the collapsing object.

2.2. About M vs ϱ [11]

The problem of the stability of a star will be treated here by using general arguments. Even though the results are not numerically exact, they contain all the necessary information that more detailed calculations later reveal. Let us first approximate the entire star by a mean density ϱ and mean energy per gram of matter ε. The total energy of a star of mass M is then as before

$$E = \int \varepsilon\varrho\mathrm{d}V - G\int\frac{m\varrho(r)}{r}\mathrm{d}V$$

m is the mass inside a radius r, i.e., $m = m(r)$.
Using average values we get

$$E = \varepsilon M - \lambda\frac{M^2 G}{R},$$

where λ is a coefficient of the order of unity. Let us now eliminate R in favor of ϱ and M through

$$\varrho = M/\frac{4\pi R^3}{3}, \quad R = (3M/4\pi\varrho)^{1/3}$$

we get

$$E = \varepsilon M - \lambda_1 G M^{5/3}\varrho^{1/3}.$$

Once we have specified the nature of the star and we have found the equation of state, we end up with a relation of the form

$$\varepsilon = \lambda_2 \varrho^{\Gamma - 1}$$

so that

$$E = \lambda_2 M \varrho^{\Gamma - 1} - \lambda_1 G M^{5/3} \varrho^{1/3} = \varrho^{1/3} M \left[\lambda_2 M \varrho^{\Gamma - 4/3} - \lambda_1 G M^{2/3}\right].$$

For $\Gamma = \frac{4}{3}$

$$E = \varrho^{1/3} \left(\lambda_2 - \lambda_1 G M^{2/3}\right) M,$$

the energy does not possess a minimum with respect to the density. For $\Gamma < \frac{4}{3}$, $E(\varrho)$ cannot have a minimum and the star has no stable equilibrium state.

For $\Gamma > \frac{4}{3}$, $E(\varrho)$ does have a minimum. It corresponds to the position of the stable equilibrium of the star. The relation between M and ϱ can be found by looking at the condition

$$\frac{\partial E}{\partial \varrho} = 0 \rightarrow \lambda_2 M (\Gamma - 1) \frac{1}{\varrho} \varrho^{\Gamma - 1} - \tfrac{1}{3} \lambda_1 G M^{5/3} \varrho^{1/3} \frac{1}{\varrho} = 0$$

or

$$M = \text{const.}\, \varrho^{(3/2)(\Gamma - 4/3)}$$

i.e., the sign of M vs. ϱ is the same as the sign of $\Gamma - \frac{4}{3}$. *Since only $\Gamma > \frac{4}{3}$ is allowed, it follows that curves with positive* $\mathrm{d}M/\mathrm{d}\varrho$ *are the only acceptable configurations.* The criterion is physically ovbious: an accretion of mass ($dM > 0$) must increase the density ($\mathrm{d}\varrho > 0$). We therefore expect curves of the type shown in Figure 7.

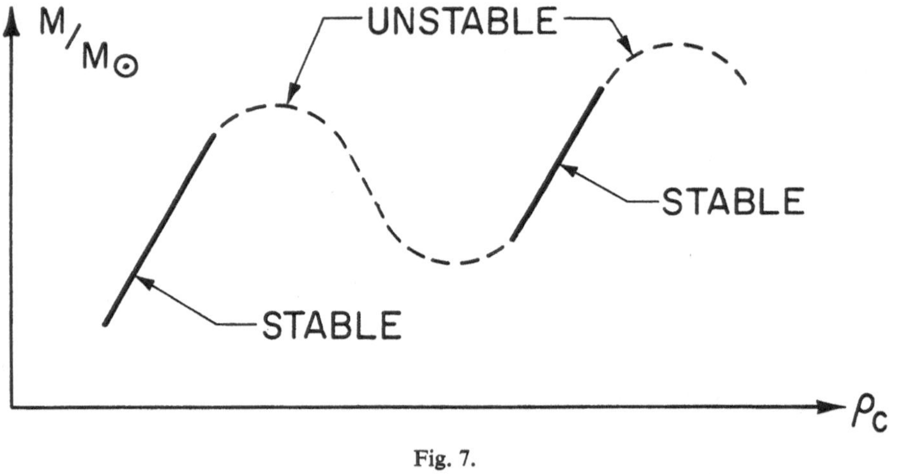

Fig. 7.

2.3. About radii of cold stars [12]

Consider a star for which kT is small compared with the average kinetic energy of a particle. For instance, for a white dwarf the pressure is supplied by degenerate electrons, whereas in a neutron star is supplied by degenerate neutrons. How do we go about having an equilibrium? The kinetic energy can be written as

$$\frac{\hbar^2}{2m}\frac{1}{r_0^2}N \qquad m \begin{cases} \text{electrons (White dwarfs)} \\ \text{protons (Neutron stars)} \end{cases}$$

where r_0 is an average separation between two particles and N is the number of particles. For the star to be bound, this positive energy must be balanced exactly by the gravitational binding energy of the nucleons, i.e.,

$$\frac{\hbar}{2mr_0^2}N = q\frac{GM^2}{R}$$

with $M = m_p N$, m_p being the mass of the proton. Since $R = r_0 N^{1/3}$ we obtain for r_0

$$r_0 = \frac{\hbar}{mc}\cdot\frac{1}{Gm_p^2/\hbar c}\cdot\frac{1}{N^{2/3}} = \frac{\hbar}{mc}\left(\frac{N_0}{N}\right)^{2/3},$$

where

$$N_0^{2/3} = \frac{\hbar c}{Gm_p^2} = \frac{1}{\alpha_G}.$$

The constant α_G is known as the gravitational coupling constant

$$\alpha_G = \frac{Gm_p^2}{\hbar c} = 5.902 \times 10^{-39}.$$

The numerical value of N_0 is

$$N_0 = \left(\frac{10^{+39}}{5.902}\right)^{3/2} = 2.10^{57}$$

which is about twice the number of nucleons in the Sun. For white dwarfs, $m = m_e$ and therefore the radius R is

$$R = r_0 N^{1/3} = \left(\frac{\hbar}{m_e c}\right)\left(\frac{N_0}{N}\right)^{1/3} N_0^{1/3} = 10^{-2} R_\odot (N_0/N)^{1/3}.$$

For neutron stars $m = m_p$, so that

$$R = r_0 N^{1/3} = \left(\frac{\hbar}{m_p c}\right)\left(\frac{N_0}{N}\right)^{1/3} N_0^{1/3} = 10 \text{ km } (N_0/N)^{1/3}.$$

Finally the ratio of the two radii is given by

$$\frac{R_{\text{W.D.}}}{R_{\text{N.S.}}} = \frac{m_p}{m_e}$$

i.e., the ratio of the radii of white dwarfs to neutron stars equals the ratio of the neutron mass to the electron mass.

2.4. Limiting Mass

Let us try to understand more closely the meaning of r_0. When the density is high enough that $r_0 \cong \lambda\!\!\!^{-}_c = \hbar/mc$ we are forced to abandon the non-relativistic approximation and use the relativistic form of the kinetic energy. This is easily achieved by using

$$E = \sqrt{(mc^2)^2 + p^2c^2} - mc^2 .$$

We have as before

$$mc^2 N \left\{ \left(1 + \frac{p^2c^2}{m^2c^4}\right)^{1/2} - 1 \right\} = q \frac{Gm_p^2 N^2}{r_0 N^{1/3}} .$$

Substitute now $p \cong \hbar/r_0$. We get

$$mc^2 N \left\{ \left(1 + \left[\frac{\hbar/mc}{r_0}\right]^2\right)^{1/2} - 1 \right\} = q \frac{Gm_p^2 N^{5/3}}{r_0} .$$

For small r_0's (high density)

$$\left(\frac{N}{N_0}\right)^{2/3} \simeq 1 - \frac{r_0}{\hbar/mc} + \cdots .$$

As $r_0 \to 0$, i.e., th density increases $N \to N_0$, so that

$$N \to N_0 = 2 \times \frac{M_\odot}{m_p}$$

represent the maximum number of nucleons that can be assembled into an *hydrostatically stable configuration. Since m_pN is the mass of the star, we also obtain that the maximum star is about* $2M_\odot$, twice the solar mass. Note that N_0 is independent of the nature of the particle supplying the pressure, electrons or nucleons. The reason is simply that at high density $E \to cp$ irrespectively of the kind of particles. Evidently the rule $M \to 2M_\odot$ is not exact as we saw before since other factors enter into play changing the picture slightly. It is however very instructive to see that a model independent calculation leads to a result very close to what much more refined computation will produce.

2.5. About Pulsations

The fundamental pulsation period of a star is related to the size and internal properties by the relation

$$P \equiv \frac{\text{Radius}}{\text{Sound Velocity}} = \frac{R}{V_s} = \frac{r_0 N^{1/3}}{V_s} .$$

By definition V_s is given in terms of pressure and density via the relation

$$V_s = \sqrt{P/\varrho} ,$$

where

$$P = -\frac{\mathrm{d}}{\mathrm{d}V} E = -\frac{\mathrm{d}}{\mathrm{d}V}\left[\frac{\hbar^2 N}{2mr_0^2}\right] = \frac{1}{3}\frac{\hbar^2}{mr_0^5}$$

and

$$\varrho = \frac{M}{V} = \frac{m_\mathrm{p} N}{r_0^3 N} = \frac{m_\mathrm{p}}{r_0^3}.$$

The previous definition of the pulsation period states the fact that P ought to be of the order of the time required for sound waves to propagate through the diameter of the star. Therefore

$$V_\mathrm{s}^2 = \frac{1}{3}\frac{\hbar^2}{mr_0^5}\cdot\frac{r_0^3}{m_\mathrm{p}} = \frac{1}{3}\frac{\hbar^2}{mm_\mathrm{p}}\cdot\frac{1}{r_0^2}.$$

Substituting r_0 from before (see 2.3) we have

$$V_\mathrm{s}^2 = \frac{1}{3}\frac{\hbar^2}{mm_\mathrm{p}}\frac{m^2c^2}{\hbar^2}\left(\frac{N}{N_0}\right)^{4/3}$$

$$V_\mathrm{s} = c\left(\frac{m}{m_\mathrm{p}}\right)^{1/2}\left(\frac{N}{N_0}\right)^{2/3}.$$

Finally we obtain the period P as

$$P = \frac{r_0 N^{1/3}}{c\left(\frac{m}{m_p}\right)^{1/2}\left(\frac{N}{N_0}\right)^{2/3}} = \left(\frac{\hbar}{mc^2}\right)\cdot\left(\frac{m_\mathrm{p}}{m}\right)^{1/2}\cdot N_0^{1/3}\cdot\left(\frac{N_0}{N}\right).$$

For white dwarfs ($m = m_\mathrm{e}$)

$$P_{\mathrm{W.D.}} = 0.7\left(\frac{N_0}{N}\right)\,\mathrm{s}.$$

For neutron stars

$$P_{\mathrm{N.S.}} = 10^{-5}\frac{N_0}{N}\,\mathrm{s}.$$

The ratio of the two pulsation periods goes like the $\frac{3}{2}$ power of the ratio of $m_\mathrm{p}/m_\mathrm{e}$, i.e.,

$$\frac{P_{\mathrm{W.D.}}}{P_{\mathrm{N.S.}}} = \left(\frac{m_\mathrm{p}}{m_\mathrm{e}}\right)^{3/2}.$$

The timing device of pulsars (0.033 s) cannot be either pulsating white dwarfs (0.7 s) or neutron star (10^{-5} s). Pulsations are out of question.

2.6. About maximum rotation rates

Consider a rotating body. A particle at the equator is bounded to the body only if

$$\frac{V_{\mathrm{surf.}}^2}{R} < \frac{GM}{R^2}$$

With the usual substitution

$$P_{\text{rot.}} = R/V_{\text{surf.}}$$

we have

$$\frac{R^2}{RP_{\text{rot.}}} < \frac{GM}{R^2}$$

or

$$P_{\text{rot.}} > R^{3/2}\cdot\frac{1}{(Gm_pN)^{1/2}} > \frac{r_0^{3/2}N^{1/2}}{(Gm_pN)^{1/2}} = \frac{r_0^{3/2}}{\sqrt{(Gm_p^2/\hbar c)}\cdot\sqrt{(\hbar c/m_p)}}$$

$$> \left(\frac{\hbar}{mc}\right)^{3/2}\cdot\left(\frac{N_0}{N}\right)\cdot N_0^{1/3}\sqrt{\frac{m_p}{\hbar c}} > \left(\frac{\hbar}{mc^2}\right)\left(\frac{m_p}{m}\right)^{1/2} N_0^{1/3}\left(\frac{N_0}{N}\right)$$

$$P_{\text{rot.}} > P_{\text{puls.}}$$

i.e., rotation is likely to be the right mechanism of timing in pulsars. We will see later that indeed rotation is the only likely mechanism to think of as a source of the pulsar mechanism.

3. Neutron Stars*

3.1. The Atmosphere

In order to have an idea about the atmosphere of a neutron star, let us compute a characteristic parameter, i.e., the scale height. A scale height 'h' is meant to represent a physical distance upon which there is a considerable variation of physical quantities as temperature and pressure. For instance for the atmosphere of the sun $h \simeq 300$ km, which is meant to indicate that T and P vary by a factor of two over that distance. This is surely true if we consider a simple law like

$$P = P_0 e^{-z/h}$$

which indicates the change of P with height. For instance, the scale height of the gas in this room is about 50 km.

We will find that for the gas at the surface of a neutron star $h \cong 1$ cm, i.e., over the distance of 1 cm the physical parameters do change significantly.

Consider a particle of mass m in hydrostatic equilibrium outside the star. The balance between the gravitational pressure pulling it down $(-g dm)$ and the pressure of the gas pushing it outwards establishes a differential equation

$$\mathrm{d}P = -g dm = -g\varrho\, \mathrm{d}r$$

$$\frac{\mathrm{d}P}{\mathrm{d}r} = -g\varrho\,.$$

If, as usual

$$P = \frac{k\varrho T}{\mu m_{\mathrm{H}}} \quad (\mu\text{: mean molecular weight})$$

* The main reference for this section is F. Dyson's paper [10].

we obtain

$$\frac{k}{\mu m_{\mathrm{H}}} \frac{\mathrm{d}T}{\mathrm{d}r} = - g \,.$$

The scale height is defined as

$$h = kT/\mu m_{\mathrm{H}} g$$

i.e., the ratio between the thermal energy (pushing the particle out) and the attraction (pulling it down).

For a neutron star ($G = 6.67 \times 10^{-8}$ dyne cm^2 gr^{-2})

$$kT = 10^6 \text{ K} = 10^{-10} \text{ erg}$$
$$\mu m_{\mathrm{H}} = \tfrac{1}{2} m_{\mathrm{H}} = 3 \times 10^{-24} \text{ gr}$$
$$g = \frac{GM}{R^2} = 3 \times 10^{13} \text{ cm s}^{-2} [\text{for the Earth } g = 10^3 \text{ cm s}^{-2}]$$

or

$$h \simeq 1 \text{ cm} \quad \text{for} \quad M = 10^{33} \text{ gr}, \quad R \simeq 10 \text{ km}.$$

In this picture we haven't considered the huge magnetic field present and this can change the situation. Regretfully no detailed study has yet been published of the changes brought by a huge magnetic field.

3.2. The surface

After the neutron star has formed out of a supernova explosion numerous physical processes enter into play that cool the star very quickly. Prominent among many of them are neutrinos created by many different processes, the most important in the early times being the URCA process

$$\mathrm{n} \rightarrow \mathrm{p} + \mathrm{e}^- + \nu$$
$$\mathrm{e}^- + p \rightarrow \mathrm{n} + \bar{\nu}$$

that takes place in heavy nucei. The energy loss by any process is defined as follows

$$\frac{\mathrm{d}U}{\mathrm{d}t} = \sum_{\mathrm{i}} \sum_{\mathrm{f}} W(i \rightarrow f) \, F_{\mathrm{i}} F_{\mathrm{f}} U_{\mathrm{f}} \,, \tag{3.1}$$

where $W(i \rightarrow f)$ is the probability that the transition $i \rightarrow f$ takes place, U_f is the energy carried away by neutrinos and F_{i}, F_{f} are the distribution functions of the initial and final state.

Given a specific process, one has to compute W from field theory and then insert it into Equation (3.1) to compute the energy loss. Here we will do an order of magnitude computation. We have

$$\frac{1}{U} \frac{\mathrm{d}U}{\mathrm{d}t} \simeq W \simeq \frac{1}{\tau} .$$

For muon decay (as a typical weak interaction) we know that

$$\frac{1}{\tau} = \frac{1}{192\pi^3} \frac{G_\mu^2}{c\hbar^7} \left[1 + \left(\frac{me}{m_\mu}\right)^2\right] (m_\mu c)^5$$

or

$$\tau = 10^{-6}\,\text{s} \left(\frac{E\,(\text{eV})}{10^8}\right)^{-5},$$

where we have substituted the rest mass with a more general expression, i.e., the initial energy.

We now have the following estimate for the cooling time

$$\frac{1}{t} \simeq \frac{1}{U} \frac{\mathrm{d}U}{\mathrm{d}t} \simeq \frac{1}{\tau} \simeq 10^6 \left(\frac{T}{10^{12}}\right)^5 \quad (\text{s}^{-1})$$

i.e.,

$$t \simeq 10^{-6} (T/10^{12})^{-5}\,(\text{s})^-$$

more accurate computations give

$$t \simeq 10^{-4} (T/10^{12})^{-4}\,(\text{s}).$$

Given any initial temperature T_0, we will reach

$$\begin{cases} T = 10^{10}\ \text{K} \\ T = 10^{9}\ \text{K} \\ T = 10^{8}\ \text{K} \end{cases} \text{after} \begin{cases} t = 10^4\ \text{s} \simeq 3\ \text{h} \\ t = 10^8\ \text{s} \simeq 3\ \text{yr} \\ t = 10^{12}\ \text{s} = 3 \times 10^4\ \text{yr} \end{cases}$$

More exact computation have been carried out by Tsuruta *et al.* [13] and confirm our results. The effect of a strong magnetic field will be considered later. At any rate, remembering that for the Crab pulsar $t \simeq 900$ yr, we obtain a temperature of about 10^7 K or less.

With the help of this number we can make an important deduction. We have said before that the density of the crust is of the order of 10^8 gr cm^{-3}. The electrons are highly relativistic and degenerate. This means they are not attached to the nuclei any longer but are quite free to move around. In this circumstance they do not screen the nuclear Coulomb fields; the nuclei see each other's bare charge and the repulsion is now

$$\mathscr{E}_c = Z^2 e^2 / a$$

instead of

$$\mathscr{E}_c = e^2 / a$$

a being an average internuclear distance. The repulsion is greater and the only way to minimize the energy is to think that the nuclei arrange themselves in a lattice structure, so as to keep a fixed distance between any two of them. The fact that most likely they arrange in a lattice structure doesn't yet mean that the resulting structure is a real solid, since we haven't computed the melting temperature, say T_m. If this happens to be greater than the previously computed surface temperature, then we are in business, the crust is an actual solid. Melting is a difficult process to understand but there are empirical rules that work very well even though sometimes it is not too easy to find a complete theoretical justification for them. Lindemann rule for instance states that a solid melts when the atoms vibrates with an amplitude comparable with the lattice site; in formula

$$\delta k T_m = \frac{Z^2 e^2}{r_i} = Z^2 e^2 N_z^{1/3}$$

or else when the kinetic energy is a sizeable fraction δ of the potential energy. Notice that this is the only formula allowed on dimensional considerations. δ is an unknown dimensionless parameter at this point. By Lindemann rule δ^{-1} is predicted to be 1/20. A computer simulation of the solid-liquid transition (with 32 particles) gives $\delta^{-1} \simeq 1/80$. We will take $\delta^{-1} \simeq 1/50$. In the table below we quote some of the numbers obtained by inserting N and Z appropriate for neutron stars surfaces.

$\varrho\,(\mathrm{g\,cm^{-3}})$	10^7	10^8	10^9
Z	26	26	28
$N\,(\mathrm{cm^{-3}})$	10^{29}	10^{30}	10^{31}
$T_m\,(\mathrm{K})$	10^8	2×10^8	6×10^8
$C_s\,(\mathrm{cm\,s^{-1}})$	2×10^7	3×10^7	4×10^7

Also given is the shear waves velocity to which we will return in the future. At any rate, the main point has been to establish that the surface temperature is lower than the melting temperature T_m, giving support to the idea that the crust is indeed a BCC solid of nuclei with charge Z.

Before going into the interior of the star let us consider the elastic properties of the crust. The energy of an undeformed BCC lattice is, say (classically)

$$\mathscr{E}(0) = \tfrac{1}{2} \sum \Phi(R_0),$$

where R_0 is the lattice distance from any given point. If we shear the crystal to assume a deformation δ, the new energy is

$$\mathscr{E}(\delta) = \tfrac{1}{2} \sum \{\Phi + \delta D\Phi + \tfrac{1}{2}\delta^2 D^2 \Phi + \cdots\}$$

where we have written

$$\mathscr{E}(\delta) = \tfrac{1}{2} \sum \Phi(\sqrt{R_0^2 + \delta})$$

for the deformed lattice and then expanded the square root into a power series in δ, with

$$\frac{\mathrm{d}}{\mathrm{d}\delta} \Phi(\sqrt{R_0^2 + \delta}) = \left[\frac{1}{r}\frac{\mathrm{d}}{\mathrm{d}r} \Phi(r)\right] \equiv (D\Phi)$$

with

$$r^2 = R^2 + \delta$$

The first derivative $D\Phi$ is zero if we are in equilibrium and therefore

$$\mathscr{E}(\delta) = \mathscr{E}(0) + \text{Quadratic form},$$

where

$$\text{Quadr. form} = \tfrac{1}{8}C_{11}(e_{xx}^2 + e_{yy}^2 + e_{zz}^2) + C_{12}(e_{yy}e_{zz} + e_{xx}e_{zz} + e_{xx}e_{yy}) + \\ + \tfrac{1}{2}C_{44}(e_{yz}^2 + e_{zx}^2 + e_{xy}^2)$$

with

$$C_{11} = \tfrac{1}{2}\cdot\frac{r_0}{\gamma}\cdot\sum D^2\Phi l_1^2$$

$$C_{12} = C_{44} = \tfrac{1}{2}\cdot\frac{r_0}{\gamma}\cdot\sum D^2\Phi l_1^2 l_2^2 .$$

The direction cosines 1_1, 1_2, 1_3 enter because by definition δ is given by

$$\delta = r_0^2 \sum_{\alpha\beta} e_{\alpha\beta} l_\alpha l_\beta$$

$e_{\alpha\beta}$ being the strain-components; r_0 is the site of the undeformed cube, whose various points at a distance R_0 can be read by using the definition

$$\mathbf{R}_0 = (l_1 r_0, l_2 r_0, l_3 r_0).$$

When the cube is deformed, the various points of the parallelepiped are given by

$$\mathbf{R}_\delta = (l_1 r_1, l_2 r_2, l_3 r_3).$$

For small deformations

$$R_\delta^2 = R_0^2 + \delta,$$

where δ is given before, and

$$e_{11} = \frac{r_1^2 - r_0^2}{r_0^2}, \quad e_{22} = \frac{r_2^2 - r_0^2}{r_0^2} \quad \text{etc.}$$

$C_{\alpha\beta}$ are called the elastic constants of the medium and have dimensions of E/V or dyne cm^{-2}. A simple computation with

$$\Phi = \frac{e^2 Z^2}{r}$$

gives

$$C_{44} = \frac{Z^2 e^2}{a^4}.$$

The so-called yield stress, S, is usually of the order of $10^{-2}\, C_{44}$, so that

$$S = 10^{-2} C_{44} = 10^{12} \varrho^{4/3} \quad (\text{dyne cm}^{-2}).$$

At this point it is interesting to compute the height of the highest mountain that could withstand the gravitational pull of a neutron star. We have

$$h = \frac{S}{\varrho g} = \frac{1}{30} \varrho^{1/3} \quad (\text{cm}).$$

For $\varrho = 10^8\,(\text{g cm}^{-3})$, $h \cong 20$ cm *or for* $\varrho = 3 \times 10^{10}\,(\text{gr cm}^{-3})$ *which would include deeper densities*, $h = 1$ m.

Dyson has pointed out that the absence of big mountains means that the star cannot have a quadruple moment Q transverse to its rotational axis greater than $10^{-5}\, I$, where I is the moment of inertia,

$$\frac{D_I}{I} \simeq \sqrt{\frac{a-b}{a}} \simeq 1 - \frac{1}{2}\frac{b}{a} \sim 1 - 10^{-5},$$

where a and b are the axis of the ellipsoid.

This in turn means that gravitational radiation is probably not the primary cause of slowing down of pulsars.

Taking the usual form for the gravitation radiation

$$\frac{\mathrm{d}E_G}{\mathrm{d}t} = \frac{G}{45 c^5} (\dddot{D})^2 = \frac{G}{45} \frac{D_I^2 \Omega^6}{c^5} = \frac{6G}{c^5} I^2 \varepsilon^2 \Omega^6,$$

where I is the moment of inertia, Ω the angular velocity and $\varepsilon = (a-b)/a$ the ellipticity. Using $M = 1 M_{\odot} = 10^{33}$ gr, $\Omega \simeq 200\ \text{s}^{-1}$ (Crab pulsar) we obtain

$$L_g \simeq 10^{45} \varepsilon^2 \quad (\text{erg s}^{-1}).$$

With an ε enormously generous of 10^{-3} we would get 10^{39} erg s^{-1}; such a gravitational radiation would reach the Earth with a flux given by ($R = 2 \times 10^3$ parsec)

$$\phi = L_g / 4\pi R^2 \simeq 10^{-8}\ \text{erg cm}^{-2}\,\text{s}^{-1},$$

whereas the existing receivers cannot go any lower than 10^4 erg cm^{-2} s^{-1}. If we think of registering gravitational waves by the methods so far devised we should think of cooling a receiver of several tons at a temperature of the order of 10^{-2} K.

3.3. The Interval Layer [14]

As we have already mentioned, the crust is made up of nuclei forming a BCC lattice. As the density increases, the electrons are being eaten up by the nuclei via inverse beta decay; we approach the region of neutron rich nuclei: the neutronization has started the deeper we go, the more we find that neutrons are less bound inside the nuclei until a moment in which they drip out and we are in the presence of a neutron gas floating on a (still) solid structure of nuclei. The nuclei disappear almost completely at $\varrho \simeq 2.4 \times 10^{14}$ gr cm^{-3} and from that moment on we have essentially neutrons, presumably in a form of a liquid, with a few protons present.

At this distance from the surface of the star, we apparently have the appropriate condition to be in a superfluid state as far as the neutrons are concerned.

The usual BCS theory gives the following formula for a gap Δ

$$\Delta = \frac{\hbar p_{\mathrm{F}}}{mb} e^{-1/N(0)V}$$

with p_{F}=Fermi momentum, b=range of the interaction, $N(0)$=density of electron states at the Fermi surface and V the strength of the interaction. For a typical neutron star we have

$$N(0) \simeq \frac{8\pi p^2 \, \mathrm{d}p}{h^3 \, \mathrm{d}E} \simeq \frac{8\pi}{h^3} m p_{\mathrm{F}} \simeq 4 \times 10^{42} \, (\mathrm{erg\ cm}^3)^{-1}$$

$$V = \int V(x) \, \mathrm{d}^3 x \simeq \frac{4\pi}{3} (10^{-13} \, \mathrm{cm})^3 \, (20 \, \mathrm{MeV}) = 10^{-43} \, \mathrm{erg\ cm}^3$$

$$N(0)\, V \simeq 0.4 \quad \exp(-{}^1/0.4) \simeq 0.1$$

$$\hbar p_{\mathrm{F}}/mb \simeq 20 \, \mathrm{MeV}$$

so that

$$\Delta = 2 \, \mathrm{MeV} \simeq kT_{\mathrm{c}}$$

$$T_{\mathrm{c}} \simeq 2 \times 10^{10} \, \mathrm{K}.$$

A detailed study of Δ was made [15] and the result is that

$$kT_{\mathrm{c}} = \Delta = E_{\mathrm{f}} \exp(-\pi/2 \operatorname{cotg} \delta m/m^*),$$

where δ is the phase shift and m^* the effective mass. For superfluidity to occur tg $\delta > 0$, i.e. the forces must be attractive. This formula is valid for any angular momen-

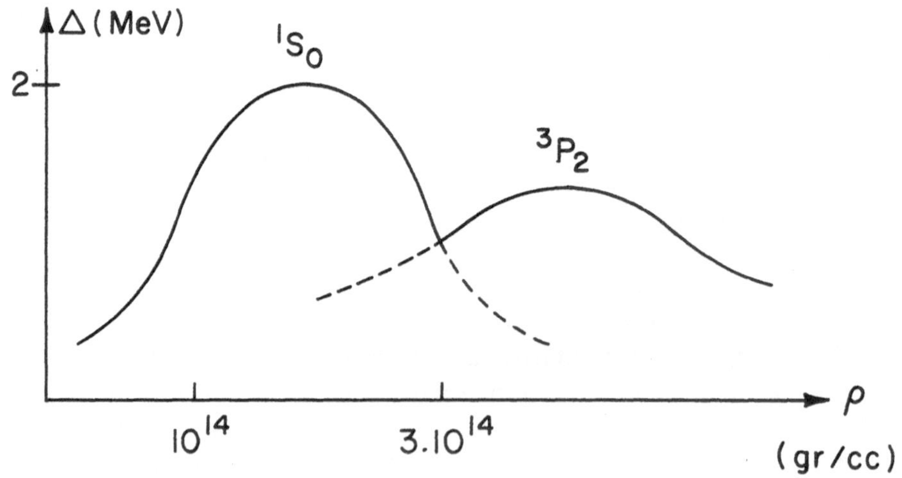

Fig. 8.

tum state. Now for $^1S_0$ state δ becomes negative above 210 MeV, or equivalently 3×10^{14} g cm^{-3} $\delta(^1S_0)<0$ and the gap cannot exist. However, $^3P_2$ is positive at that density and therefore the superfluid behavior can continue up to very high density (Fig. 8).

We therefore seem to have two different regions: up to 2×10^{14} gr cm^{-3} a superfluid layer of neutrons followed by a transition region which is probably not superfluid and then again a superfluid region of neutrons in $^3P_2$ states.

A very interesing aspect of having superfluids is that it is experimentally known that if you have a rotating bucket of superfluid substance, the angular momentum is carried by an array of vortex lines which form an exagonal lattice. The circulation around each line is

$$l = \int v \, \mathrm{d}s = h/2M .$$

Now the circulation about a large radius R is given by

$$L = \pi R^2 n l ,$$

where n is the density of vortex lines. Since L is also equal to $2\pi R v = 2\pi R \cdot \Omega R$ where Ω is the angular velocity, we have

$$\Omega = \tfrac{1}{2} n l = nh/4m .$$

Using $\Omega = 200$ s^{-1} (Crab Nebula) we obtain

$$n = 2 \times 10^5 \text{ cm}^{-2}$$

$$a = 2.5 \times 10^{-3} \text{ cm} ,$$

where

$$n = \frac{1}{(\sqrt{\frac{3}{2}})\, a^2} .$$

The spacing among vortex lines is very large compared to their dimension $\simeq 10^{-13}$ cm.

The most interesting thing about such an array of vortex lines is the existence of *sound waves* propagating perpendicular to the vortex lines at a very slow velocity V_T given by

$$V_T = \sqrt{\frac{\hbar \Omega}{8m}} \simeq 0.13 \text{ cm s}^{-1} .$$

These waves are known as *Tkachenko waves* [10]. As one can see these waves are independent of the density or other properties of the liquid. These waves are normal modes of the superfluid array of vortex lines. Their non-existence in laboratory experiment is essentially due to the lack of any highly rotating bucket of superfluid helium.

Ruderman has suggested that the observed 'wobble' of the Crab pulsar period is associated with a standing Tkachenko wave in the core. Using

$$P = 2R/V_T$$

and taking $R=5$ km, $V_T \simeq 0.13$ cm s^{-1}, P turns out to be $P=3$ months. The observed oscillation of the Crab pulsar also has a period of about 3 months.

Tkachenko waves are not exactly understood in their intimate nature; the derivation is not very transparent and the only existing simple derivati is the one that is based on an analogy with the *geostrophic waves.*

Take the equation of classical incompressible hydrodynamics for a fluid rotating with angular velocity Ω

$$\dot{v} + 2(\Omega \times v) + 1/\varrho\,[\text{grad}\, p - \tfrac{1}{2}\Omega^2 r_\perp^2] = 0\,.$$

This equation is written in a system of coordinate which rotate with the fluid. Take the curl of it,

$$\text{curl}\,\dot{v} = 2(\Omega\cdot\text{grad})\,v\,,$$
$$\text{curl curl}\,\ddot{v} = 4(\Omega\cdot\text{grad})^2\,v\,,$$
$$-\nabla^2\ddot{v} = 4(\Omega\cdot\text{grad})^2\,v\,.$$

Taking now a single plane-wave solution of the type

$$v = v_0 \exp(ikr - i\omega t)$$

we obtain the dispersion relation (θ is the angle between k and Ω)

$$k^2\omega^2 = 4k^2\Omega^2\cos^2\theta$$
$$\omega = 2\Omega\cos\theta\,.$$

Now the argument is that Tkachenko waves are the quantum mechanical analogue of the geostrophic waves.

Now

$$\omega = 2\Omega\cos\theta \simeq 2J_\perp/mR^2\,,$$

where $J_\perp$ is the component of the angular momentum $\perp$ to the direction of quantization Ω. We have ($J \simeq mR^2\,\Omega$)

$$\omega \simeq \frac{1}{mR^2}\sqrt{\langle J_\perp^2\rangle} \simeq \sqrt{\frac{\hbar\Omega}{mR^2}}\,.$$

Since

$$\langle J_\perp^2\rangle = \langle J^2\rangle - \langle J\rangle^2 \cong \hbar\,|J|$$

we deduce the velocity ($k \sim R^{-1}$)

$$V_T = \frac{\omega}{k} = \sqrt{\frac{\hbar\Omega}{m}}$$

which is the Tkachenko formula. In this way, without pretending of having gone into any totally satisfactory theoretical explanation, we believe we have been able to narrow down the origin of these 3-months oscillations superimposed on the regular slowing-down curve. Observationally we must add that the amplitudes are very small, even though 3 complete cycles have already been observed.

4. Pulsars

4.1. Historical background [16, 17]

The discovery of pulsars was reported on February 9, 1968 in *Nature*, by Hewish, Bell, Pilkington, Scott and Collins of Cambridge University. The discovery of pulsars is [18]: the best example of 'serendipity' defined as the faculty of making happy and unexpected discoveries by accident.*

The most intriguing feature of pulsars is that they emit short pulses of energy at exceedingly well maintained intervals**, typically of the order of 1 s. In the very first moments of the discovery, a certain flavour of mystery was thought to be inherent in those messages at the extreme that they were considered to be signals of an extra-terrestrial civilization. The rapid variation of pulse amplitude was suggestive of a code but it was not possible to separate any recognizable pattern. The almost simultaneous discovery of 3 more pulsars put an end to these speculations. By October, 1968, 10 pulsars were found. In the same month the eleventh pulsar was announced: PSR 0833 with two peculiar properties:

(1) It had the shortest period known up to that moment $P = 0.089$ s.

(2) It was situated at a position of a suspected supernova remnant. At the beginning of December 1968, 20 pulsars were found and in particular the now famous PSR 0532 that showed an even smaller period: 3 millisecond, $P = 0.033$ s. This pulsar lies in a very well known nebulosity, the Crab Nebula. By April 1969, 35 pulsars were found. By now they are about 70, to be compared with the number of theoretical papers, about 300!

4.2. Observational data [16, 17]

(A) The pulses were found to be repeating approximately every 1 s with a pulse width of the order of 20 μs (msec).

(B) The pulses do not always have the same shape and a detailed study has shown that three main categories can be devised.

(a) S pulsars: have simple pulse profiles; polarization can be small or large.

(b) D pulsars: have drifting subpulses; weakly polarized.

(c) C pulsars: have complex pulse shape: large degree of linear polarization.

(C) The mean flux density is of the order of a fraction of a flux unit ($\Phi U = 10^{-26}$ W m^{-2} Hz$^{-1} = 10^{-23}$ erg cm^{-2} Hz^{-1}). Individual pulses can have fluxes as high as 10^2 ΦU.

* Horace Walpole: *The Three Princes of Serendip*, Ceylon. [18]

** The first pulsar was recorded on November 28, 1967 and the code name used was CP 1919, Cambridge pulsar 19^h19^m.

(D) Pulsars have been studied at radio frequencies ranging from 40 Mhz to about 5 GHz and for the Crab Nebula data are available in optical, X-Ray and γ-Ray part of the spectrum.

(E) *Distance*. The determination of pulsar distance is based on the following argument. The interstellar medium is a plasma with an electron density that varies between

$$10^{-2} \leqslant N_e \leqslant 10\,(\text{particles cm}^{-3})$$

Any wave of frequency ω traveling in such a plasma satisfies a dispersion relation of the form

$$\omega^2 = c^2k^2 + \omega_p^2 \tag{4.1}$$

where ω_p, the so-called plasma frequency, is defined as

$$\omega_p = \sqrt{\frac{4\pi N_e e^2}{m}} = 5.64 \times 10^4 N_e^{1/2}\ \text{Hz}.$$

Equation (4.1) is equivalent to saying that the photon has acquired a certain mass, $m_\gamma c^2 \simeq \omega_p^2$. A simple derivation of Equation (4.1) is as follows:
Consider Maxwell's equations:

(1) $$\text{curl}\, H - \frac{1}{c}\dot{E} + \frac{4\pi}{c} j \equiv \frac{4\pi}{c}(-en\mathbf{v})$$

(2) $$\text{curl}\, E + \frac{1}{c}\dot{H} = 0$$

(3) $$\dot{v} = -\frac{e}{m}E.$$

Take now the curl of (1)

$$\text{curl curl}\, H - \frac{1}{c}\,\text{curl}\,\dot{E} = -\frac{4\pi}{c}\, en\, \text{curl}\, v.$$

On the other hand

$$\text{curl}\,\dot{E} = -\frac{1}{c}H$$

so that (curl curl $H = -\nabla^2 H$)

$$-\nabla^2 H + \frac{1}{c^2}H = -\frac{4\pi en}{c}\,\text{curl}\, v.$$

Take now the curl of (3)

$$\frac{d}{dt}\,\text{curl}\, v = -\frac{e}{m}\,\text{curl}\, E = +\frac{e}{mc}\dot{H} = \frac{e}{mc}\frac{d}{dt}H$$

$$\text{curl}\, v = \frac{e}{mc}H.$$

Finally substituting curl v into the previous equation for H we obtain

$$\ddot{H} - c^2\nabla^2 H + \omega_p^2 H = 0\,; \quad \omega_p^2 = \frac{4\pi e^2 n}{m}.$$

Considering a plane wave solution of the form

$$H = H_0 e^{-i\omega t + ikr}$$

we obtain

$$-\omega^2 + c^2k^2 + \omega_p^2 = 0$$

i.e.

$$\omega^2 = c^2k^2 + \omega_p^2$$

which is the desired relation.

The group velocity is now given by ($10^2 \leqslant \omega \leqslant 10^4$ MHz)

$$v \equiv \frac{d\omega}{dk} \cong c\left[1 - \omega_p^2/2\omega^2\right], \quad \omega^2 \gg \omega_p^2.$$

The approximation $\omega \gg \omega_p$ is good in the radio frequency range. If a narrow signal passes through a medium with variable electron density, the time delay between the arrival of a signal at frequencies ω_1 and ω_2 is

$$\Delta t \equiv t_1 - t_2 = \frac{2\pi e^2}{mc}\left[\frac{1}{\omega_1^2} - \frac{1}{\omega_2^2}\right]\int N_e \, dl,$$

where the integral is taken along the line of sight. $\int N_e dl$ is called the dispersion measure (DM) and is given in pc per cm^3. (1pc$=3l.y.$:or 3×10^{18} cm).

Measurements of Δt for two ω's give an estimate of l after N_e has been substituted. Canonical values of N_e range in the order of 0.1 electron cm^{-3}. The reported results are

	l (pc)
PSR 1919	125
PSR 0834	130
PSR 0950	30
PSR 1133	60

Many people have questioned the use of the universal value 0.1 for the electron density on the ground that cosmic rays can produce more ionization than the one expected and therefore influence the amount of electrons. No definite agreement has yet been achieved on this topic since our knowledge of the interstellar medium is rapidly changing. However this may turn out to be, the numbers presented before given us a very good idea of the distance of pulsars.

4.3. Spatial Distribution

Once we know the distance of pulsars we can consider their spatial distribution. First we note that out of 54 pulsars 40 of them lie within 1 kpc from the Sun.

This is clearly a selection effect, since pulsars are such weak emitters. Statistical analysis has shown that if we accept that we have observed *all* pulsars up to a DM of 20 pc m^{-3}, the statistics up to 500 pc suggest that there are probably 100 pulsar/(kpc)^3 or else 10^4-10^5 pulsars in our Galaxy. Evidently one must remember that many pulsars can be missing due to the beaming effect of the radiation. If we just happen not to be in the right line of sight we could automatically miss many of them. This would automatically affect our statistics.

From this kind of statistics, people have deduced that at least 10% of all supernovae outbursts produce pulsars. Too many factors enter into play in this problem. For the time being the belief is that they are mostly distributed in the galactic disc. There is apparently a tendency to cluster slightly below the galactic plane (few degrees).

4.4. Periods

An interesting feature is displayed by the following hystogram

Periods (sec)	$(\frac{1}{32}-\frac{1}{16})$	$(\frac{1}{16}-\frac{1}{8})$	$(\frac{1}{8}-\frac{1}{4})$	$(\frac{1}{4}-\frac{1}{2})$	$(\frac{1}{2}-1)$	$(1-2)$	$(2-4)$
# of Pulsars	1	1	2	11	23	14	3

from which we note the preponderance of periods between $\frac{1}{2}$ and 2 s.

For 20 pulsars we know the rate of change of P. For instance, PSR 1919 has a dP/dt of 10^{-15}. This means that in three years that had been observed ($1\text{ yr}\equiv 3\times 10^7$ s) its period has changed by 1 part in 10^7. A more exact clock could hardly be conceived!

4.5. Polarization

After the discovery of pulsars it was almost immediately realized that pulses were polarized.

For the pulsar PSR 0833 (Vela pulsar) at *1720 MHz, it was found that pulses are practically 100% linearly polarized.* Circular polarization is almost completely absent.

4.6. Optical and X-ray pulsars

PSR 0532 was found to pulsate not only in the radio range but also: (1) *optically*; (2) *in the X-ray region* ($\simeq 1-10$ keV); and (3) *in the γ-ray part of the spectrum up to 150 MeV.* Moreover there are x-ray pulsars that do not radiate in the radio range.

One main feature of the optical range is that there is a main pulse and then a quite remarkable interpulse 13.37 ms later (for Crab), whereas such a precursor appears only 1.64 ms before the main pulse in the radio region.

Since PSR 0532 has been observed in a wide part of the spectrum, it was possible to determine $P=P(t)$ quite accurately. It was found that the period can be fitted by a law of the form

$$P = a + bt + ct^2 \pm 2.09\text{ ns}$$

with (t in days, to $\equiv$November 15, 1968)

$$a = 0.3390 \times 10^8, \qquad b = 36.477 \pm 0.32, \qquad c = -(0.60 \pm 0.63)\, 10^{-4}$$

This tells us that pulsars are slowing down or that the period is lengthening. This important fact is common to all the pulsars observed and is a strong indication of what a theorist should look for. Evidently one should be very careful in extending these data to any other pulsar since no other optically emitting object has been yet found.

Besides the Crab, two other X-emitters were recently found

(A) *Cygnus X-1* (B) *Centaurus X-3*

(1) Cy X-1: (a) has a period of about 0.073 s.
(b) (20–25)‰ of the total X-ray luminosity (3×10^{36} erg s^{-1}) between 1 and 100 KeV is in form of periodic pulsations.
(c) If confirmed, Cy X-1 is the second fastest X-pulsar.

(2) Cen X: (a) has a period of 5 s.
(b) 70‰ of the total X-ray intensity is pulsating.

4.7. What is a Pulsar?

After having summarized the main properties of pulsars we have to ask ourselves: what is a pulsar?

Given the great angular velocity Ω, an object of density ϱ does not fly apart only if the acceleration due to gravity exceeds the centrifugal acceleration, i.e., if

$$GM/R^2 > v^2/R.$$

Substituting $v = \Omega R$ we obtain

$$GM/R^2 > \Omega^2 R$$

or

$$\sqrt{GM/R^3} > \Omega.$$

Since

$$\varrho = \frac{M}{V} = \frac{M}{\frac{4}{3}\pi R^3}\varrho$$

and

$$P = 2\pi/\Omega$$

the stars stay together only if

$$\varrho > \frac{3\pi}{G} P^{-2}.$$

For $P \lesssim 1$ s, $\varrho \gtrsim 10^{8-9}$ gr cm^{-3}. If $P = 10^{-3}$ s, ϱ goes up to 10^{14} gr cm^{-3}.

No other body except neutron stars at $T = 0$ is known to have a density as high as that. The parameters of a neutron star i.e., mass, radius and central density depend very sensibly upon the exact form of the equation of state $P = P(\varrho)$ employed.

The first equation of state employed by Oppenheimer [10] was that corresponding to a gas of free degenerate neutrons. The results were:

$$\begin{aligned} \varrho_c &= 2 \times 10^{14} \text{ gr cm}^{-3} \\ M &= 2 \times 10^{33} \text{ gr} = 1 \text{ M}_\odot \\ R &= 2 \times 10^{6} \text{ cm} = 20 \text{ km} \\ W &= GM^2/R = 10^{53} \text{ erg (grav. binding energy)} \\ \Omega &= \sqrt{G\varrho} = 3000 \text{ rad s}^{-1} \\ P &= 2\pi/\Omega = 2 \times 10^{-3} \text{ s}. \end{aligned}$$

It therefore follows the dogma that as far as obtaining the right period of revolution *neutron stars are the only plausible object that we presently know.*

How about general relativistic effect? It is worthwhile to remember that the Schwarzschild radius

$$R^* = 2GM/c^2 = 3 \times 10^5 (M/M_\odot) \text{ cm}$$

is not all that small compared with the radius of the star itself and consequently general relativistic effects are likely to come into play very early. (For the sun $R^* = 3 \times 10^5$ cm, to be compared with the radius 10^6 km!). One can also think that huge magnetic field could possibly occur and this would radically change the situation. This can be so only if the magnetic energy is comparable with the gravitational energy i.e. when

$$\frac{H^2}{8\pi} R^3 = \frac{GM^2}{R}.$$

For $M \simeq M_\odot$

$$H \simeq 10^{10} R^{-2}$$

i.e.

$$H = 10^{16} \text{ G} \quad \text{at} \quad R = 10^7 \text{ cm}.$$

At such a field intensity, the Larmor radius of an electron would be [20]

$$r_L = \frac{mc^2}{eH} = 4 \times 10^{-3} (\hbar/mc) = 4 \times 10^{-3} \lambda\!\!\!^{-}_c$$

i.e., the gyration radius would be about a thousand times smaller than the Compton wavelength of the particle itself, i.e. the particle would be localized well within its own Compton wavelength. At that stage we know quantum mechanics puts strong limitations on the interpretation of the electron as such.

Even less probable is that the equations of general relativity should be modified. Such modifications are expected to take place at characteristic length, times, and densities of the order of [19]

$$\begin{aligned} \lg &\simeq \sqrt{G\hbar^3/c^3} \simeq 10^{-33}\ \text{cm} \\ t &\simeq \lg/c \simeq 10^{-43}\ \text{s} \\ \varrho_g &\simeq c^5/\hbar G^2 \simeq 5 \times 10^{93}\ \text{gr cm}^{-3} \end{aligned}$$

How do we get to a neutron star phase during the evolution of a star? If we consider a NS as an object representing the final evolution of an ordinary star, then pure dimensional scaling will tell us a lot about them.

Consider the following features of the Sun

Mass	$M = 2 \times 10^{33}$ gr
Radius	$R = 10^6$ km
Ang. velocity	$\Omega = 10^{-6}$ rad s^{-1}
Mag. field	$B = 10^2$ G

The four main types of energies characterizing a star are

$$E_T = -\tfrac{1}{2}E_G = \frac{1}{2}\frac{GM^2}{R} \simeq 10^{48}\ \text{erg} \qquad \text{(GRAVITATIONAL)}$$

$$E_R = \tfrac{1}{2}I\Omega^2 = \tfrac{1}{10}MR^2\Omega^2 = 2 \times 10^{42}\ \text{erg} \quad \text{(ROTATIONAL)}$$

$$E_M = \frac{1}{8\pi}B^2\frac{4\pi}{3}R^3 = 10^{37}\ \text{erg} \qquad \text{(MAGNETIC)}$$

For an adiabatic collapse from $R=10^6$ km down to $R=10$ km we have that conserving the angular momentum $L=I\Omega=MR^2\Omega$ the several quantities scale down like

$$\begin{aligned} \Omega &\simeq R^{-2} \simeq 10^4\ \text{rad s}^{-1} \\ B &\simeq R^{-2} \simeq 10^{12}\ G \\ E_G &\simeq R^{-1} \simeq 10^{53}\ \text{erg} \\ E_R &\simeq R^{-2} \simeq 2 \times 10^{52}\ \text{erg} \\ E_M &\simeq R^{-1} \simeq 10^{42}\ \text{erg.} \end{aligned}$$

We see that with this simple minded model (if model at all!) we obtain values very close to the one proper of an object as a neutron star.

Lacking any more appropriate candidate we conclude that a pulsar is a *rapidly rotating magnetized neutron star*.

In principle there are no reasons whatsoever to believe that the dipole magnetic moment $\mathfrak{M}$ *should coincide with the revolution axis* Ω. Besides, if the rotation axis (or spin axis) coincides with the magnetic axis, there would be axial symmetry and we wouldn't see any pulse. It must therefore be an oblique rotation. Arguments have been given to indicate that for PSR 0532, $\alpha \simeq 45°$; other authors prefer 63°; still other observers seem to indicate 90°. In usual ordinary magnetic stars there seems to be a tendency to believe that $\alpha \simeq 90°$.

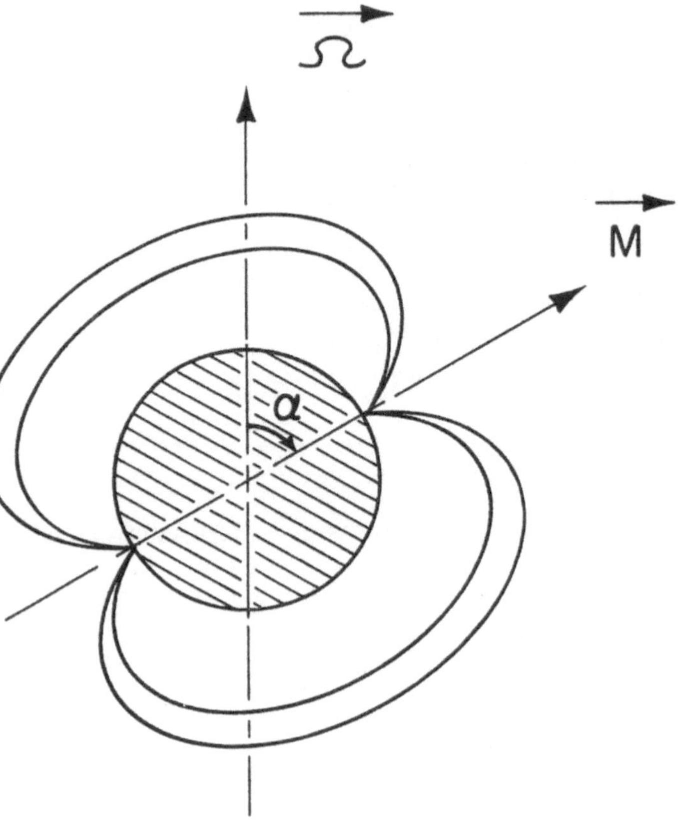

Fig. 9.

With this picture in mind we therefore try to understand the slowing down mechanism. The simplest thing to do is to imagine that the field of the star is a magnetic dipole. No physical reason exists to prefer this configuration to a more complicated one, except that of simplicity. The magnetic dipole is resolved into || and ⊥ component. Evidently the $\mathfrak{M}_{\|}$ does not vary with time and therefore is totally irrelevant.

The radiation from a dipole $\mathfrak{M}_{\perp}$ varying with angular frequency Ω can be easily seen from dimensional analysis [10]. The energy emitted (or more precisely) the rate of emission $\dot{E}_D$ can depend only on c, $\mathfrak{M}_{\perp}$ and Ω, i.e.

$$\dot{E}_D = k\Omega^{\alpha}\mathfrak{M}_{\perp}^{\beta}c^{\gamma}.$$

Since $\mathfrak{M}_\perp^2$ is an energy times a volume it follows that $\alpha=4$, $\gamma=-3$, $\beta=2$, i.e.

$$\dot{E}_D = k\,\frac{\mathfrak{M}_\perp^2 \Omega^4}{c^3}$$

$k=\frac{2}{3}$ would correspond to the magnetic dipole. Since $\mathfrak{M}_\perp = BR^3$ with $B=10^{12}$ G, $R=10^6$ cm we obtain

$$\mathfrak{M}_\perp = 10^{30}\ \mathrm{G\ cm^2}$$

therefore

$$\dot{E}_D = 10^{29}\Omega^4\,(\mathrm{erg\ s^{-1}}).$$

(1) *Take now* $\Omega \equiv 200$ (rad s^{-1}, Crab Nebula pulsar)

$$\dot{E}_D = 2\times 10^{38}\ \mathrm{erg\ s^{-1}} = 10^5 \cdot E_D^{(\mathrm{SUN})}.$$

This amount of energy agrees rather well with the *total energy* which we observe coming from the Crab Nebula. The energy is emitted from the pulsar and absorbed by the plasma in the surrounding debris. This can therefore be considered as the PRIMARY ENERGY SOURCE for the Crab.

(2) Let us now compute another quantity

$$\dot{E}_R = -\frac{\mathrm{d}}{\mathrm{d}t}\left(\tfrac{1}{2}I\Omega^2\right) = -I\Omega\dot{\Omega} = I\Omega^2\,\frac{\dot{P}}{P} = 2E_R\left(\frac{\dot{P}}{P}\right)$$

i.e. the rate of change of rotational energy of the pulsar.
Taking

$$I = \tfrac{1}{10}MR^2 = 10^{44}\ \mathrm{gr\ cm^2}$$

and the observed $\Omega=200$ rad s^{-1}, and $\dot{P}/P = 1.3\times 10^{-11}$ we obtain

$$\dot{E}_R = 5\times 10^{37}\ \mathrm{erg\ s^{-1}}$$

again of the order of the dipole emission.

(3) Another important number can be obtained by considering what happened just after the birth of a pulsar. Most likely it was spinning very fast say $\Omega \simeq 10^4\ \mathrm{s}^{-1}$. In this case

$$\dot{E}_R = 10^{45}\ \mathrm{erg\ s^{-1}}$$

very close to the peak of a supernova optical luminosity

$$\dot{E}_{SN} = 10^{43}\ \mathrm{erg\ s^{-1}}.$$

What does this mean? It can be interpreted as saying that the dipole radiation is enough to explain the supernova light as being absorbed by the surrounding stellar envelope and reemitted to an efficiency of 1/100.

The force due to the radiation pressure is simply obtained as

$$F = \dot{E}_R/C = 3 \times 10^{34}\ \text{dyne}.$$

This would accelerate a mass 10^{33} gr (solar mass) with an acceleration of

$$a = F/m = 30\ \text{cm s}^{-2}$$

or otherwise to a velocity of

$$v = 1000\ \text{km s}^{-1}$$

for a time as long as 30 days $= 3 \times 10^6$ s. This number is in good agreement with the observed velocity of the supernova debris.

(4) Let us finally go back to the slow-down process. Because of the almost identity of $\dot{E}_R$ with $\dot{E}_D$ we can think that this is actually the mechanism operating in the star i.e. we put [21]

$$\frac{\mathrm{d}}{\mathrm{d}t}\left(\tfrac{1}{2}I\Omega^2\right) = -\tfrac{2}{3}\mathfrak{M}_\perp^2 \frac{\Omega^4}{c^3}.$$

The solution of this differential equation is $(t<0)$

$$\Omega = \frac{\Omega_0}{\sqrt{1 + t/\tau}}$$

with

$$\tau = \frac{3c^3 I}{4\mathfrak{M}_\perp^2 \Omega_0^2} \simeq 5 \times 10^{11}\ \text{s} \simeq 2400\ \text{yr}.$$

At any time $(t<0)$ previous to the present $t=0$, the angular velocity Ω was greater than its present value Ω_0.
The present ratio is

$$\Omega_0/\dot{\Omega}_0 = P/\dot{P} = -\,2260\ \text{yr}.$$

The age of the Crab $t = -917$ yr. Putting this number into the previous equation we get

$$\Omega\,(\text{at } t = -\,917\ \text{yr}) = 2.3\Omega_0\,(t = \text{present time}).$$

This is not in agreement with what we supposed earlier about $\Omega \simeq 10^4\ \text{s}^{-1}$ i.e. hundred times greater than Ω_0. The only hope is to think in terms of gravitational radiation. If we start the problem all over again and add [22]

$$-\frac{D_\perp^2 \Omega^6 G}{45c^5}$$

that represent the energy loss per second by quadrupole gravitational radiation. We have

$$-\frac{\mathrm{d}}{\mathrm{d}t}\frac{1}{2}I\Omega^2 = -\frac{2m_\perp^2\Omega^4}{3c^3} + \frac{D_\perp^2 G\Omega^6}{45c^5}.$$

It is clear that due to the higher power of Ω, the gravitational radiation is important almost exclusively at times close to the birth of the star. Introducing the following dimensionless variables.

$$x = \frac{P}{P_0} = \frac{\Omega_0}{\Omega}, \quad \tau_m = \frac{3c^3 I}{2\mathfrak{M}_\perp^2\Omega_0^2}, \quad \tau_g = \frac{45c^5 I}{GD_\perp^2\Omega_0^4}, \quad \eta = \tau_g/\tau_m$$

we have

$$\frac{\mathrm{d}}{\mathrm{d}t}x^2 = \frac{2}{\tau_m}\left[1 + \frac{1}{\eta x^2}\right].$$

The implicit solution is

$$t = \tfrac{1}{2}\tau_m\left[x^2 - 1 - \frac{1}{\eta}\ln\left(\frac{1+\eta x^2}{1+\eta}\right)\right].$$

When the period was very short or equivalently when Ω was very large $x \ll 1$, we have

$$t_i = \tfrac{1}{2}\tau_m\left[-1 + \frac{1}{\eta}\ln(1+\eta)\right]$$

as initial time. Let us now define a characteristic time τ_0 measureable at present

$$\pi_0 = \left(\frac{1}{\frac{1}{P}\frac{\mathrm{d}P}{\mathrm{d}t}}\right)_{t=0}$$

We also have

$$\tau_0 = \left(\frac{1}{\tau_g} + \frac{1}{\tau_m}\right)^{-1} = \tau_m(1 + 1/\eta)^{-1}.$$

Eliminating τ_m we get for t_i

$$t_i = \tfrac{1}{2}\tau_0(1 + 1/\eta)\left[-1 + \frac{1}{\eta}\ln(1+\eta)\right]$$

and finally we can define the age of a pulsar as

$$-t_i = \frac{\tau_0}{2}(1 + 1/\eta)\left[1 - \frac{1}{\eta}\ln(1+\eta)\right]$$

Suppose now two cases:

(1) No grav. radiation $\eta \to \infty$, $\text{Age} = \tau_0/2$

(2) No mag. radiation $\eta \to 0$, $\text{Age} = \tau_0/4$

For the Crab, $\tau_0 = 2260$ yr and this implies that the age is either 1130 yr or 565 yr against the known age of 916 yr! If we require gravitational radiation to be of such an amount as to give the exact age, then η is determined ($\eta \simeq 5$), the ellipticity turns out to be 3×10^{-4} and at the present time one sixth of the lumisosity of the Crab is in the form 60 Hz quadrupole gravitational radiation. We have so far supposed that $\alpha = \pi/2$. The question is: Is there any tendency for the magnetic and rotation axes to align? If so, i.e., if $\alpha \to 0$, the magnetic dipole radiation will be turned off. The electromagnetic radiation emitted by a dipole exert a torque, say $\boldsymbol{\tau}$. Since a torque is by definition $\mathbf{r} \times \mathbf{F}$ we simply generalize it to a more general form and have

$$\tau_\alpha = \varepsilon_{\alpha\beta\gamma\delta} \int \chi^\beta T^{\gamma\delta} \, \mathrm{d}\Sigma_\delta ,$$

where $T^{\alpha\beta}$ is the electromagnetic energy-momentum tensor; $\varepsilon_{\alpha\beta\gamma\delta}$ is the usually totally anti-symmetrical tensor and $\mathrm{d}\Sigma_\delta$ is the surface on which we are integrating. By definition

$$T_{\alpha\beta} = \frac{1}{8\pi} \left[\left(- B_\alpha B_\beta + \tfrac{1}{2} B^2 \delta_{\alpha\beta} \right) + \left(- E_\alpha E_\beta + \tfrac{1}{2} E^2 \delta_{\alpha\beta} \right) \right].$$

For a point dipole we have

$$B_r = 2m \sin\alpha \left[\frac{1}{a^3} \cos\phi + \frac{\Omega \sin\phi}{a^2 c} \right] \sin\theta + \frac{2m}{a^3} \cos\theta \cos\theta$$

$$B_\theta = m \sin\alpha \left[\left(-\frac{1}{a^3} - \frac{\Omega^2}{ac^2} \right) \cos\phi - \frac{\Omega \sin\phi}{a^2 c} \right] \cos\theta + \frac{m}{a^3} \cos\alpha \sin\phi$$

$$B_\varphi = m \sin\alpha \left[-\frac{\Omega}{a^2 c} \cos\phi + \frac{1}{a^2} - \frac{\Omega}{ac^2} \right] \sin\varphi .$$

The components of the torque are easily computed to be

$$\tau_x = 0, \quad \tau_y = \frac{2m^2 \Omega^2}{3ac^2} \sin\alpha \cos\alpha, \quad \tau_z = -\frac{2m^2 \Omega^3}{3c^3} \sin^2\alpha .$$

Let us now write down the Euler equation for a rigid body

$$\frac{\mathrm{d}\mathbf{L}}{\mathrm{d}t} = \boldsymbol{\tau} .$$

We have in detail

$$I \frac{\mathrm{d}\Omega}{\mathrm{d}t} = \tau_z , \qquad I \frac{\Omega \mathrm{d}\alpha}{\mathrm{d}t} = -\tau_x , \qquad I \frac{\Omega \mathrm{d}\varphi \sin\alpha}{\mathrm{d}t} = \tau_y .$$

The first equation is nothing but the old energy balance already seen. Since $\tau_x=0$ there seems to be little hope from this equation that α will ever align. We have otherwise to warn about any hasty conclusion that the magnetic axis will never align. Plasma effect have been neglected. If instead of a rotating dipole we think in terms of a rigid, magnetized and perfectly conductive sphere in vacuum one obtains that τ_y and τ_z are unaltered, whereas τ_x changes to

$$\tau_x = \frac{m^2\Omega^3}{c^3} \sin 2\alpha$$

and this would give alignment in a time approximately equal to τ_m.

5. Electric Field [17]

So far we haven't considered two very important questions: the existence of electric fields and plasma effect.

(1) *Near a magnetized star rotating in vacuum there should exist an electric field* with magnitude

$$E = R\Omega H/c\,.$$

Evidently in the reference frame rotating with the star $E=0$. Putting numbers into it we easily find with $\Omega=200\ \text{s}^{-1}$, $R\simeq10^6$ cm

$$E = \frac{10^6\ \text{cm}\ 2\times10^2\ \text{s}}{\text{s}\ 3\times10^{10}\ \text{cm}}\ H = 10^{-2}H\,.$$

Even for H only 10^8, we get

$$E = 10^6\ \text{esu} = 10^8\ \text{V cm}^{-1}$$

or a potential difference

$$V = ER = 10^{14}\ \text{V!}$$

To begin with, the force played by the gravitational attraction in keeping the particles is practically negligible. In fact the ratio of the fwo forces is

$$\frac{\text{Electric force}}{\text{Grav. force}} = \frac{eR\Omega H/c}{m_\text{p}GM/R^2} \simeq 10^9$$

for the Crab. *The surface is dynamically unstable because the electrostatic forces will drive the charges into the vacuum!*

6. Plasma Effect [19]

So far we have shown that a reasonable overall picture can be achieved if the plasma outside the star is considered to be very tenuous or else if we work within the vacuum approximation.

The idea was related to the fact that the scale height turned out to be terribly small, ($\simeq 1$ cm). If so, after 1 cm the density has decreased by a factor of e, after 2 cm by a factor $2e$ and so on. Very soon we won't have any density and the majority of the external part of the star is just empty.

This, however, was shown not to be true, became as we have just said there could be a field as huge as 10^{14} V; if so, the computation of h, as performed before loses its meaning.

Suppose we now have a rotating neutron star. In the interior the electrical conductivity is supposed to be very high [23]. Let us see if this is true. A simple model for σ is obtained in the following way. Under the action of an electric field E the electrons move according to

$$m \frac{d\mathbf{v}}{dt} + m\nu\mathbf{v} = e\mathbf{E},$$

where ν is the collision frequency. Taking $E = E_0 e^{-i\omega t}$, i.e., neglecting the spatial dependence, we get

$$v_0 (-mi\omega + m\nu) = eE_0$$

If $v = v_0 \exp(-i\omega t)$

$$v_0 = \frac{eE_0}{m\nu - i\omega m} = \frac{eE}{m(\nu - i\omega)}$$

so that the conductivity is $(\mathbf{j} = \sigma\mathbf{E})$

$$\sigma = \frac{e^2 N}{m(\nu - i\omega)} \simeq \frac{e^2 N}{m\nu} \quad \text{for} \quad \omega \to 0.$$

Supposing that $\nu = c/l$ where l is a typical interparticle distance and considering $N \simeq 10^{33}$ cm^{-3} (i.e. 10^9 gr cm^{-3}) we obtain

$$\sigma = 10^{22}\ \mathrm{s}^{-1}$$

very close to the value 10^{23} that I got [23] after using the complicated machinery of Dirac spinors and relativistic Boltzman equations. To give you an idea, copper has $\sigma \simeq 10^{16}$ s^{-1}. To all practical purposes σ is infinite, the medium is infinitely conducting. If so, it follows that

$$\mathbf{E} + \frac{1}{c}(\boldsymbol{\Omega} \times \mathbf{r}) \times \mathbf{B} = 0$$

or else $\mathbf{E} \cdot \mathbf{B} = 0$. We have also assumed that the non-degenerate outside plasma is an excellent conductor. If the star is surrounded by vacuum, $\varrho = 0$, we can solve the Laplace equation for the external electrostatic potential which must be continuous at the stellar surface. The result is

$$\phi = -\frac{B\Omega R^5}{3cr^3} P_2(\cos\theta),$$

where B is the polar magnetic field. The surface charge density (Σ) computed from the discontinuity of the normal component of the electric field at the surface, is

$$\Sigma = -\frac{BR\Omega}{4\pi c} \cos^2\theta.$$

The Lorentz invariant $E \cdot B = 0$ vanishes in the interior as we have just seen. In the exterior would be

$$E \cdot B = -\frac{\Omega R}{c}\left(\frac{R}{r}\right)^7 B^2 \cos^3\theta \neq 0.$$

Within the surface charge layer the value of $E \cdot B$ must change continuously from zero to its external value. Thus, near the outer layer, the magnitude of E along B would exceed the gravitational force and the system would not be in equilibrium.

Now, if we have a plasma, the situation can change quite drastically. If such a plasma has a refractive index n, then it can be shown that the magnetic dipole power L changes by a factor n^3. We saw before that for isotropic (zero magnetic field) plasma

$$n^2 = 1 - \omega_p^2/\omega^2, \qquad \omega_p = 5.64 \times 10^4 \times N_e^{1/2}$$

and the inequality $n^2 < 0$ can easily be achieved.

In the case of magnetic stars the relation is more complicated and in any elementary plasma book you will find that in the case where $\Omega_H = eH/m_i c \gg \omega$, i.e., when the ion gyrofrequency is greater than the wave frequency then [19]

$$n^2 = 1 + \frac{4\pi\varrho c^2}{H^2} = 1 + c^2/v_A^2.$$

Where

$$v_A = H/\sqrt{4\pi\varrho}$$

is the so-called Alfvén velocity. *These low-frequency modes are called magnetohydrodynamic modes or Alfvén waves.* For instance, $n_e = 10^{14}$ cm^{-3} ($\varrho \equiv mn_e$), $\varrho > 10^{-10}$ gr cm^{-3}, $H < 10^6$ G, $n \sim 10$. This will change the output as a factor of more than 10^3! For the Crab pulsar, say $H \simeq 10^{-3}$ G, $\Omega_H \leqslant 10$ s^{-1}, so that the magnetohydrodynamic approximation to the dielectric constant is not valid any longer. What one does in fact, is to neglect the ion completely. Suppose now that the electron plasma frequency $\omega_p = (5.65 \times 10^4 \sqrt{N_e}) \gg \omega_H = eH/mc = 1.76 \times 10^7 H$. These electromagnetic waves are known as 'Whistlers' modes and their dispersion relation is

$$n = \frac{\omega_p}{\sqrt{\omega\omega_H \cos\theta}}$$

$$\omega_p \gg \omega_H, \qquad \omega_p \gg \omega, \qquad \omega_H \cos\theta \gg \omega.$$

Taking the following typical values

$$\omega_p \simeq 10^{5-6}, \quad N_e \simeq (1 - 10^3)/\text{cc}, \quad \omega_H = 10^4 \, (H = 10^{-3}\ \text{G})$$
$$\omega = \Omega = 200/\text{s}^{-1} \quad n \simeq 10^2 \sim 10^3$$

The radiation is decreased by a factor between $10^6 - 10^9$! These arguments do not have to be taken too seriously because we don't know how realistic we have been in all these estimates. However they point out to an enormously important problem.

7. Cooling of a Neutron Star [13, 24]

No matter how it is formed, a neutron star will cool by URCA process and by plasma neutrino emission to 10^{10} K within an hour. After that has occurred and the temperature is around 10^8 K, the huge magnetic field of the star enters into play and begins to accelerate the process by reducing the Compton scattering cross section and therefore increasing the mean free path. A magnetic field always acts in the direction $\perp$ to the electron velocity since

$$m\dot{v} = eE + (e/c)\, \mathbf{v} \times \mathbf{B}$$

If H $\parallel z$, then H will act very strongly in the v_x and v_y component of the electron velocity.

Suppose now that we have some radiation propagating along the magnetic field axis. Its electric field E will try to shake the electron in the y direction. In this direction, however, the strong magnetic field has already acted by forcing the particle to stay in a small orbit (helix). The action of E is therefore greatly reduced. Since the only dimensionless ratio you can construct is

$$(\omega/\omega_H)^2 \quad \omega \ll \omega_H,$$

One would expect that σ will be reduced by that same factor i.e.

$$\sigma_H = \sigma_0 \, (\omega/\omega_H)^2 \quad \text{for} \quad k \parallel H,$$

where σ_0 is the usual Thomson cross-section. With $h\omega \simeq \kappa T \simeq 10^3$ eV $h\omega_H = 10^{-8}$HeV (10^5 eV for $H = 10^{13}$) $\sigma_H = 10^{-4}\, \sigma_0$. The mean free path

$$l = n\sigma^{-1}$$

is increased by 10^4 ! The electromagnetic radiation does not scatter all that much, its getting to the surface is easier and therefore the cooling is faster. A detailed analysis was performed and the previous results were indeed confirmed.

8. Glitches [16, 17, 10]

Among many interesting features associated with pulsars we can't leave aside four radical changes occurred in the $P = P(t)$ function. As we said before, the period P is increasing with time i.e. Ω is decreasing with time. Pulsars are slowing down. Four times it happened that two pulsars, the Crab and Vela sped up as shown in Figure 10.

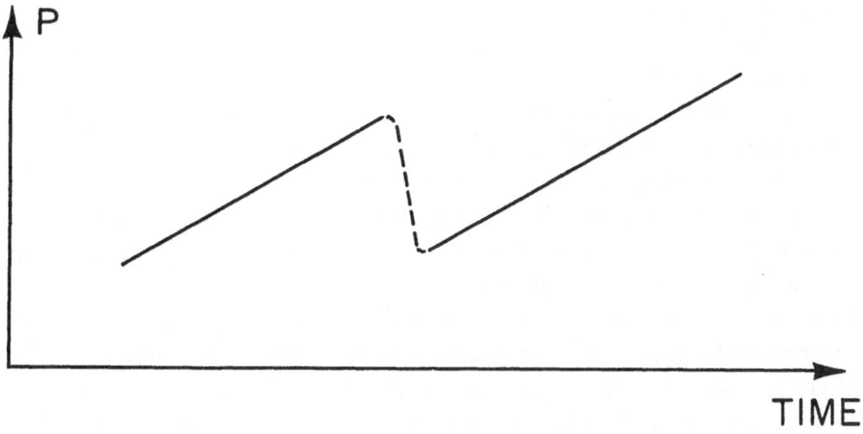

Fig. 10.

The data are as follows:

	CRAB		VELA	
	29 Sept., 1969	Oct., 1971	Feb. 1969	Aug., 1971
$\Delta\Omega/\Omega$	7×10^{-9}	2×10^{-9}	2.3×10^{-6}	2×10^{-6}
$\Delta\dot{P}/\dot{P}$	8×10^{-4}		10^{-2}	
τ	$4 \sim 7$ days		1.2 yr	
Q	0.9		0.145	

The post glitch behavior has been fitted with a curve of the form

$$\Omega = \Omega_0(t) + \Delta\Omega\left[Qe^{-t/\tau} + 1 + Q\right].$$

There was an apparent sudden change in period P by an amount

$$\Delta P \cong 208 \text{ nsec} \simeq 2 \times 10^{-6} P$$
$$\Delta P/P \simeq 2 \times 10^{-6}. \qquad \text{(Vela)}$$

After the change, the rate of lengthening of the period is increased by an amount

$$\Delta\dot{P} \cong 1.0 \times 10^{-15} \simeq 10^{-2}\dot{P}$$
$$\Delta\dot{P}/\dot{P} \sim 10^{-2}. \qquad \text{(Vela)}$$

Analogously for the Crab we have

$$\left.\begin{aligned}\Delta P &\simeq 0.12 \text{ nsec}\\ \Delta\dot{P} &\simeq 3 \times 10^{-6} \sim 10^{-4}\dot{P}\end{aligned}\right\} \qquad \text{(Crab)}$$

Many things can be said about these jumps. To begin with, their magnitude is no. the same. Vela has a much more pronounced jump than the Crab by a factor of 100t This puts already a great burden on theoretical astrophysicists.

There are two models for this phenomenon each with its drawback and advantages. Let us go through them.

The Crab Nebula is an amorphous mass of gas surrounding the pulsar. Apparently there have been (a) motions and (b) brightening of the wisp-like features just after the September, 1969 speed-up. Due to uncertainties in the distance one cannot place a better estimate than a *month as far as the response of the wisp to the glitches.*

It was also found that the dispersion measure (DM) of the Crab pulsar was increased by 2.5×10^{16} electron cm^{-2} during the two months following the September 1969 glitch. Conversely, a large DM increase occurred (4×10^{16} electron cm^{-2}) in April 1970 without any apparent glitch.

Unfortunately it so happened that the pulse shape or intensity was not changed during or after the glitch, whereas one would expect that these should have been the case if actually the magnetoplasma around the star was actively changed.

This cannot be claimed to be a model yet. It just points out some interesting coincidences.

A more detailed model has been proposed by Ruderman [17]. It is based on the idea that glitches are star-quakes analogous to earthquakes. Such seismic phenomenon are supposed to take place in the crust. As we know, Ω decreases with increasing time. The crust we have shown is a solid and it cannot change its form smoothly as the rotation slows down. The original oblateness was

$$\varepsilon \equiv \frac{25}{12} \frac{I\Omega^2 R}{GM^2} \cong \frac{\Omega^2}{G\varrho}$$

i.e. it was very high in the early times of the Crab (or Vela) when Ω was very high. Since Ω decreases, ε decreases. but as we said it cannot be a continuous process and the crust is straining itself. Electromagnetic radiation gives rise to a torque that can change the alignment between the axis of rotation and the original principle moments of inertia. This also strains the crust. All these are good reasons to believe that the crust sooner or later will 'crack', i.e. release the stresses that have been building up during the slow-down. Such a crack would *reduce the moment of inertia* and for conservation of angular momentum Ω would increase, i.e.,

$$L = I\Omega$$
$$\Delta L = 0 = \Delta I\Omega + I\Delta\Omega$$
$$\frac{\Delta\Omega}{\Omega} = \frac{\Delta I}{I} = \frac{\Delta R}{R}.$$

Since $I = MR^2$, $\Delta I = 2\,MR\,\Delta R$. For Vela $\Delta\Omega/\Omega = 2 \times 10^{-6}$ so that $\Delta I = 2 \times 10^{-6}$ I, or $\Delta R = 1$ cm ! i.e. a change of 1 cm occurred 500 parsec away is recorded today ! (1580 yr ago !). For the Crab ΔR is still more impressive $\Delta R \simeq 10^{-3}$ cm, i.e. an adjustment of the radius of 10^{-3} cm out of 10 km occurred at 2700 parsec, or else 6300 yr ago !

References

1. Schwarzschild, M.: *Structure and Evolution of the Stars*, Dover Publ. Inc., New York, 1965.
2. Reeves, H.: *Stellar Energy Sources from Stars and Stellar Systems*, Vol. VIII, University of Chicago Press, 1955.
3. Cox, J. P. and Giuli, R. T.: *Principles of Stellar Structure*, Vol. I, Gordon-Breach Publ., New York, 1968, p. 422.
4. Fowler, W. A.: *Nuclear Astrophysics*, Am. Phyl. Soc., Philadelphia, 1967.
5. Salpeter, E. E.: *Phys. Rev.* **88**, 547 (1952).
6. Burbidge, E. M., Burbidge, G. R., Fowler, W. A., and Hoyle, F.: *Rev. Mod. Phys.* **29**, 547 (1957).
7. Cameron, A. G. W.: *Astrophys. J.* **130**, 452 (1959).
8. Ostriker, J.: in H. Y. Chiu and A. Muriel (eds.), *Stellar Evolution*, MIT Press, 1972.
9. Chandrasekhar, S.: *Introduction to Stellar Structure*, Chicago Univ. Press, 1939.
10. Dyson, F.: *Neutron Stars and Pulsars*, Accademia Nazionale Dei Lincei, Roma, 1971.
11. Zeldovich, Ya. B. and Novikov, I. D.: *Stars and Relativity*, Univ. of Chicago Press, 1971.
12. Iben, I.: in H. Y. Chiu and A. Muriel (eds) *Stellar Evolution*, MIT Press, 1972.
13. Tsuruta, S., Canuto, V., Lodenquai, J., and Ruderman, M.: *Astrophys. J.* **176**, 739 (1972).
14. Canuto, V.: *Ann. Rev. Astr. Astrophys.* in press (1973).
15. Hoffberg, M., Glassgold, A., Richardson, R., and Ruderman, M.: *Phys. Rev. Letter* **24**, 775 (1970).
16. Hewish, A.: 'Pulsars' *Ann. Rev. Astron. Astrophys.* **8**, 265 (1970).
17. Ruderman, M.: *Pulsar: Structure and Dynamics*, Columbia Univ. preprint, 1972.
18. Ter Haar, D.: *Pulsars* Preprint, Univ. of Oxford, 1971.
19. Ginzburg, V. L.: *Sov. Phys. Uspekhii* **14**, 83 (1972).
20. Canuto, V. and Chiu, H. Y.: *Space Sci. Rev.* **12**, 3 (1971).
21. Pacini, F.: *Nature* **219**, 145 (1968).
22. Gunn, J. E. and Ostriker, J. P.: *Astrophys. J.* **157**, 1395 (1969).
23. Canuto, V.: *Astrophys. J.* **159**, 641 (1970).
24. Canuto, V.: *Astrophys. J.* **160**, L153 (1970).

PART E

BIOPHYSICS

THEORETICAL PHYSIOLOGY

B. C. GOODWIN*
School of Biological Sciences, University of Sussex, Brighton, England

1. Introduction

The subject of theoretical physiology does not exist, but we can quite easily create it for the purpose of these lectures. This is most easily done by a consideration of the time scales that identify different subjects, as follows:

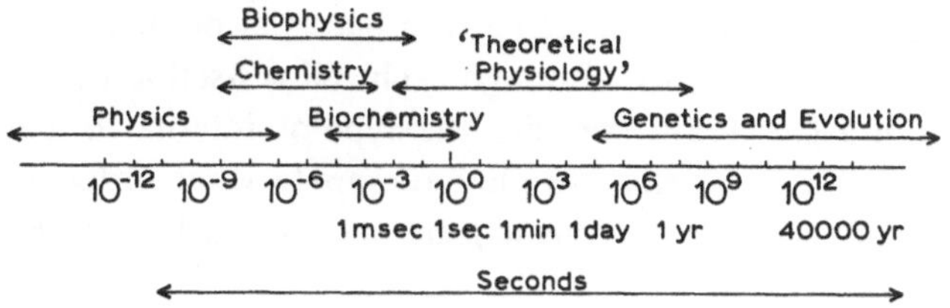

Fig. 1.

Defined in this form, the subject-matter of these lectures is concerned with processes whose rates of change vary over the range from about one millisecond to about 1 yr. This is ten orders of magnitude, and it will be necessary to introduce some criterion whereby different types of physiological process such as metabolic transformations, endocrine adaptations, wound healing, or embryological changes, may be distinguished from one another. In dynamical terms, which will be the major context for this study, the obvious criterion is in relaxation or characteristic times, the period over which a significant change can occur in the system under investigation. Fortunately biological processes lend themselves quite naturally to such an analysis, and different biological disciplines can quite easily be identified in these terms.

Associated with a distinction based upon the concept of characteristic times for different processes is a distinction with respect to functional organization, the particular ways in which biological processes are organized causally and structurally. For example, one type of functional organization is found at the level of metabolic transformations; another, though closely related, type of functional organization is found at the level of embryological change. Thus there emerges a hierarchical ordering of biological systems which is naturally associated with different characteristic times for various types of processes. Functional units may be identified for each hierarchical level, units which are defined by particular causal relationships between structural elements such as metabolites, proteins, membranes, nuclei, cells, gradients, etc. The analysis of a complex, organized system into units always does some violence to its totality, its unity and wholeness, and this analytical procedure must be balanced by

* On going to press, the publishers were unable to contact the author to check the final proofs.

Abecassis de Laredo and Jurisic (eds.), Selected Topics in Phys., Astrophys. and Biophys., 381–420.

holistic descriptions. But like the good butcher who does the least cutting, we may hope to identify the natural joints of the biological system by a functional analysis which exploits natural discontinuities of time scale.

1.1. States of organization of cells

Besides the use of relaxation time criteria to identify levels of organization of biological systems, I will exploit another analogy with physical systems. It is possible to consider three different states of order of cell populations in a manner somewhat similar to the three states of matter: gas, liquid, solid. In the first case, we may consider cells which have internal order, but no order in the population apart from that which emerges from averaging procedures. This is what we have in bacterial populations, where there are no interactions between the cells, although each cell is itself a highly-ordered system, as is a molecule in a gas. I will consider the type of internal organization which is found in such cells, in order that we may anticipate what order may emerge when there are interactions sufficiently strong to generate a simple kind of population regularity, namely temporal organization.

In cell populations ordered with respect to time, we will find behaviour analogous to that occurring in synchronized oscillators: clock-like behaviour with stable phase locking. This can occur as a result of interactions between intrinsically periodic systems, which cells are; or because of entrainment to an external periodicity, in which case we have the phenomenon of biological clocks. There are no necessary spatial constraints in such populations, so the temporal order is not accompanied by spatial ordering.

The third type of organization between cells emerges when there are interactions which constrain the relative movements of cells. Then the system can become organized both in space and in time, possessing as it were both a map and a clock. Embryos are examples of such organized systems, undergoing particular transformations of state at defined locations within their boundaries and at particular times, thus achieving quite extraordinary four-dimensional patterns of ordered change. Thus the embryo keeps time and constructs a map of itself. The brain is a system which possesses not only a clock and a map of the body in which it is located, but also a map of its environment, thus achieving a still higher level of spatial and temporal order than the embryo. However, I will not be concerned with the organization of the brain, which is beyond the subject matter of these lectures as well as beyond my own competence.

2. The Dynamic Organization of Single Cells

The most obvious characteristic that we encounter in the study of single cells such as the bacterium *Escherischia coli* is the sheer complexity of the system. Within the boundaries of such a cell, whose volume is only a few cubic microns, there are of the order of 10^4 different molecular species undergoing highly ordered transformations and interconversions. Thirty years of biochemistry was required before some coherent picture began to emerge regarding the interrelated pathways whereby one metabolic species is converted into another, a study which now goes under the name of intermediary metabolism. We may represent these transformations in a simplified form by means of branching reaction sequences as shown in Figure 2.

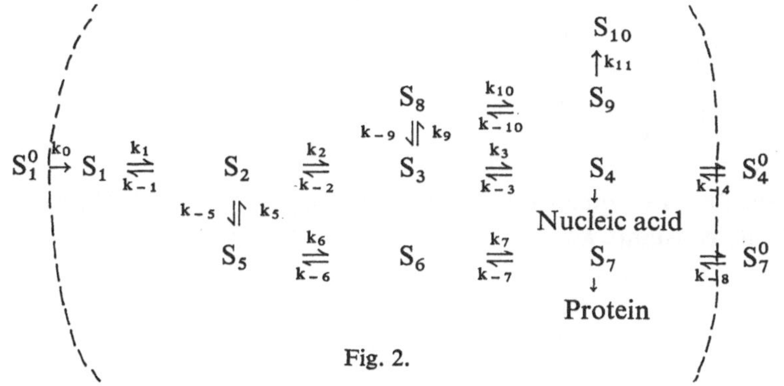

Fig. 2.

The dotted lines represent the boundaries of the cell, and the superscript '0' labels materials or nutrients from the outside which the cell takes up and uses for metabolism and growth. For a bacterium such nutrients may be glucose, amino acids, salts, etc. Metabolic pathways are in fact very much more complicated than this, and many reactions are bimolecular or multimolecular not unimolecular as shown. However, it was believed for many years that the basic principles governing reaction sequences in living cells were only those of mass action, the rates of reaction being determined by the concentrations of the reactants. Each reaction is catalyzed by a specific enzyme, so that enzyme concentrations are incorporated in the rate constants k_i. We are thus making the assumption that these enzyme concentrations are not changing significantly when we represent reaction sequences in this form. For relatively short times, of the order of seconds or a few minutes, this is a perfectly reasonable assumption.

We may now ask some elementary questions about the dynamic behaviour that a complex system of the type shown in Figure 2 could have. We could write down equations for the rates of reactions in the following form:

$$\frac{dS_1}{dt} = k_0 S_0 - k_1 S_1 + k_{-1} S_2$$

$$\frac{dS_2}{dt} = k_1 S_1 - (k_{-1} + k_2 + k_5)\, S_2 + k_{-2} S_3 + k_{-5} S_5$$

$$\vdots$$

$$\frac{dS_7}{dt} = k_7 S_6 - (k_{-7} + k_8)\, S_7 + k_{-8} S_7^0 .$$

Such a system of equations will have a unique solution for the values of the variable, at the steady state, defined by $dS_i/dt=0$. If we write the external nutrient termss assumed to be constant, as a vector

$$b = \begin{pmatrix} -k_0 S_0 \\ 0 \\ \vdots \\ -k_{-4} S_4^0 \\ \vdots \\ -k_{-8} S_7^0 \end{pmatrix}.$$

then the steady-state equation can be written in the form $b = KS$, where K is the matrix of rate constants K

$$K = \begin{pmatrix} -k_1 & k_{-1} & 0 & 0 & 0 \ldots & & & \\ k_1 & -(k_{-1}+k_2+k_5) & k_{-2} & k_{-5} & 0 \ldots & & & \\ \vdots & & & & & & & \\ 0 & & & & & & & \\ 0 & & & & & k_7 & -(k_{-7}+k_8) & k_{-8} \end{pmatrix}.$$

and S is the vector of steady-state values. We see that the solution is simply

$$S = K^{-1} b$$

so that each steady-state value is determined solely by the rate constants and the boundary conditions, the nutrient conditions. This is what one expects in the case of an open system: the values of the system variables at the steady state will be independent of their initial values.

It is possible to say something about the dynamics of such a system as well. It has been shown by Hearon [1] that a linear kinetic system of the type considered gives a characteristic equation whose roots are all real and non-positive. Thus the system is asymptotically stable at its steady state, and any perturbation from this state will be followed by a relaxation back to it. The relaxation time is of the order of seconds, since the rate constants are about $10^{-2}\,\mathrm{s}^{-1}$ (Appendix 1). The fact that many reactions are really second order, so that the equations have quadratic terms, does not greatly change this basic dynamic picture except to introduce the possibility of multiple steady states. However, what does change the picture of intermediary metabolism in a rather dramatic manner from the picture depicted in Figure 2 is a series of experimental studies, performed between 1950 and 1960, which radically altered our whole conception of the nature of the metabolic organization of single cells (see ref. [14]).

In the early 1950's investigations with bacterial mutants which were defective in their abilities to synthesize particular metabolites such as amino acids revealed a very interesting fact. In the defective enzyme occurred at the end of a reaction sequence, such as the steps from S_6 to S_7 in Figure 2, then the precursors S_5 and S_6, were found to be present in high concentrations. If now the final product of the sequence S_7 (let us assume it to be the amino acid isoleucine) is provided in the nutrient medium, so that S_7^0 is present, then it was found that S_5 and S_6 virtually disappear from the bacterium. If the system were governed by mass action, then we would expect these intermediates to occur at about the same concentration whether or not S_7^0 is present. But detailed studies showed that whenever an end-product such as S_7 is supplied externally, the whole of the reaction pathway leading to that product is somehow shut down and no intermediates are found. This is of course a very efficient way for the bacterium to behave, since it does not then waste materials on the internal production of metabolites provided gratis. But it was problematical how S_7 brought about the disappearance of S_5 and S_6. It was then discovered that S_7 exerts this effect by arresting or blocking the first step in the reaction pathway leading to it, viz., the step in which S_2 is converted to S_5. This block results from a very specific inhibition of the enzyme which catalyzes the transformation. Furthermore, it was found that *only* this step is blocked by S_7. This is clearly a sufficient condition for stopping the synthesis of the end-product, and so the logic of the system is impeccable.

After the initial discovery of this type of control, called feed-back inhibition of end-product inhibition, many other examples were discovered. Soon it was possible to generalize and to say that the first enzyme in a reaction pathway leading specifically to a unique end-product is subject to inhibition by the product. This observation was of very fundamental significance for the analysis of metabolic organization for it provided a logical foundation for the partitioning of complex metabolistic pathways into functional units with a degree of autonomy. Adding S_7^0 to the nutrient medium shuts down the reaction sequence leading to S_7, but has little effect on the rest of the system except to increase the materials available to it by an amount equal to the normal flux down the pathway to S_7.

Of equal importance was the realization that feed-back inhibition required the introduction of new postulates regarding enzyme structure and function. The enzyme bringing about the catalytic interconversion of S_2 and S_5 must have the property of reacting with S_7 in some very specific manner. Thus not only can the enzyme 'recognize' S_2 and bring about its conversion to S_5; it must react specifically with S_7 in such a manner that after reaction it is unable to act as an enzyme for S_2 conversion. Furthermore, S_2 and S_7 may be structurally totally dissimilar. An example is shown

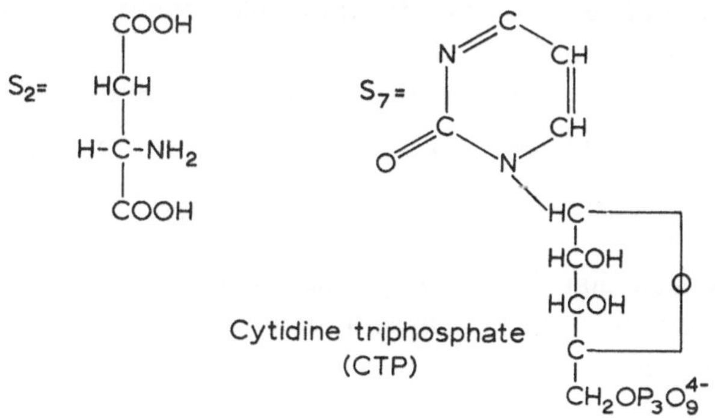

Fig. 3.

in Figure 3. The enzyme which must 'recognize' these two molecules is aspartate transcarbamylase, whose properties have been studied by Gerhardt and Schachman [2]. In this particular instance there are 6 reaction steps between aspartic acid and CTP rather than the 3 shown in Figure 3.

The capacity of enzymes to react specifically with molecules which have totally different steric properties was named by Monod 'allosteric behaviour'. It is envisaged in the following way. Suppose that an enzyme in its native state has the configuration shown in the left in Figure 4.

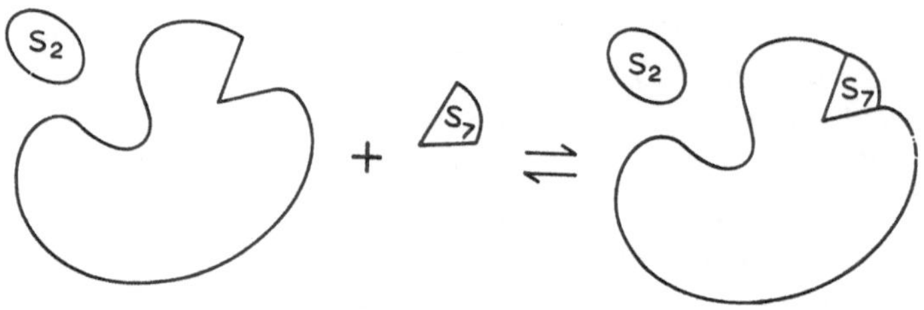

Fig. 4.

It can then react with S_2 by a complementary fit interaction. However, when a molecule of S_7 reacts with the enzyme, the shape of the site for combination with S_2 is altered and this reaction is no longer possible. Thus it is supposed that the enzyme has two specific sites, one of which is required for catalysis, the other for binding with the inhibitor.

These observations have been greatly extendedn in recent years, and of course the picture has become very much more complex, as it always does. Enzymes have been found which can react with as many as 6 different types of molecule, each one having a particular effect. It is now known that some molecules activate the enzyme; i.e., they increase its affinity for substrate rather than decrease it. Thus metabolic control involves a complex interplay of interacting molecules. However, there is an important generalization that emerges from these studies. The curves relating the activities of enzymes and activating or inhibiting molecules all have the same fundamental characteristics, because they all arise from a similar basic process of interaction. Although there is as yet no fundamental theory for these interactions, a biological theory is as follows.

Consider the reaction between an enzyme, E, and an end-product, P, to form an inactive complex, EP, as in Figure 4. We may write

$$E + P \underset{l_{-1}}{\overset{l_1}{\rightleftharpoons}} EP\,. \tag{2.1}$$

It may be assumed that there is very much more P in the system than E, since enzymes are present in catalytic concentrations. So we may write a conservation relation for total enzyme:

$$[E_0] = [E] + [EP]\,. \tag{2.2}$$

Here $[E_0]$ is the total initial enzyme concentration, $[E]$ is free enzyme, and $[EP]$ the inactive complex (square brackets refer to concentrations). At equilibrium, we can write from (2.1) the relation

$$l_1 [E] [P] = l_{-1} [EP]$$

so that

$$[EP] = \frac{l_1}{l_{-1}} [E] [P] = L [E] [P]$$

Substituting this in (2.2) and solving for $[E]$, we find

$$[E] = [E_0]/1 + L[P]. \tag{2.3}$$

This gives a curve of the form labelled (1) in Figure 5.

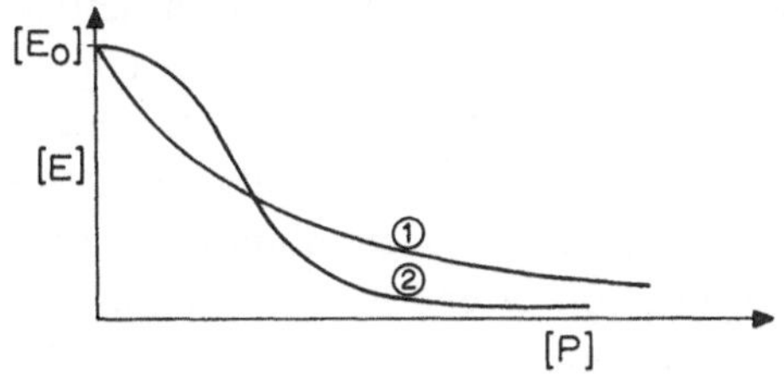

Fig. 5.

If it were assumed that each molecule of enzyme could combine with $n > 1$ molecules of P, so that there are n sites of this particular kind, then we would find

$$[E] = \frac{[E_0]}{1 + L[P]^n} \tag{2.4}$$

and the curve would have zero slope at $[P]=0$ and a point of inflection, as seen in curve (2) of Figure 5.

This is generally found to be the case. For example, aspartate transcarbamylase has 8 sites on every enzyme molecule for reaction with CTP, so $n=4$. In this instance the enzyme has 4 binding sites for reaction with its substrate, aspatic acid. Thus it is evident that enzymes may be highly complex molecules indeed. I should emphasize at this point that there is no basic theory which explains the specificity of reaction between enzymes, or proteins generally, and small metabolites, despite their fundamental importance in metabolic processes. A protein is a very large and very complex molecule, so it is hardly surprising that there is nothing analogous to a molecular orbital theory to explain its behaviour. Probably what is required is a phenomenological approach to complex polymers in order to achieve some quantitative treatment of protein behaviour in general and enzyme reactions in particular. So far all that we have in biology is control functions of the type described by Equations (2.3) and (2.4).

However, even with those we can do something in the way of a dynamical analysis, which extends in an important way the kinetic treatment described in the linear equations.

2.1. Enzyme adaptation

So far we have regarded enzyme concentrations as constants in our equations, a procedure which is perfectly legitimate so long as relatively short times are being considered. The variables used have been metabolic intermediates and end-products, whose concentrations can change significance in seconds or a few minutes. That is the time-scale for metabolic processes, and we can refer to that part of the cell which responds to stimuli and perturbations in seconds or a few minutes as the metabolic system. One observes such responses when the concentration of an external metabolite is halved or doubled, when the O_2 concentration is increased or decreased, or when the temperature is changed by a few degrees. The concentrations of metabolites in the cell then adjust to new levels and the metabolic system settles to a steady state within seconds or minutes of the change in environmental conditions. We refer to such behaviour as homeostasis, for the adjustments of state which occur within the system are always such that essential variables are maintained at relatively constant values. For example, the phenomenon of feed-back inhibition described above has the effect of maintaining the concentration of end-product, S_7, at a relatively constant value independently of its concentration outside the cell. This is an inmediate consequence of the curve in Figure 5. The rate at which S_7 is made by the enzyme catalyzing the interconversion of S_2 and S_5 is dependent on the concentration of free, uninhibited enzyme, $[E]$. The more end-product there is, the smaller will $[E]$ be (P is now identified as S_7 in the metabolic network of Figure 3), and vice versa. Thus the system is sensitive to the concentrations of particular variables, the ones that are essential to the maintenance of the living state and which are generally referred to as end-products of metabolic sequences, and those are maintained at relatively constant values despite variations in the environmental conditions. We may say that the metabolic system is organized dynamically in such a manner as to achieve this constancy or homeostasis with the greatest possible efficiency. Its time scale of operation is seconds to minutes, a rate which is determined primarili by the turnover number of enzymes; i.e., the number of molecules of metabolite that one enzyme can transform in one second. This number is of the order 10^2–10^4 mol s^{-1}. This, together with enzyme concentrations which normally fall within the range 10^{-6}–10^{-8} molar, gives the above estimate for relaxation times in the metabolic system (see Appendix 1).

However, what behaviour do we observe in cells when they are required to adapt to a much more extensive perturbation than those considered above, one which requires a change not simply of metabolite concentration, but the synthesis of new enzymes? It has been known for over 100 yr that bacteria are capable of adapting to very different environments, but only recently has it been possible to discover in some detail what is involved in this process. The phenomenon is known as enzyme adaptation, and its elucidation forms one of the most interesting and dramatic stories of contemporary biology. In fact, we may say that this analysis forms one of the cornerstones of molecular biology.

The key to the analysis of enzyme adaptation in bacteria was the discovery of sexuality in Escherichia coli by Lederberg in 1947. This allowed microbiologists to use genetic techniques to study mutant forms, and in the hands of Jacob and Monod [3] these techniques led to an unravelling of one of the most basic of molecular control processes in single cells. It is commonly observed that whereas bacterial sexuality was discovered in America, it was the French who knew how to put it to practical use.

The basic phenomenon underlying bacterial adaptation is the response of a population to a mixed carbon source. Suppose we have some Escherichia coli growing on a nutrient medium which contains both glucose and lactose. It is found that the bacterial growth curve is as shown in Figure 6.

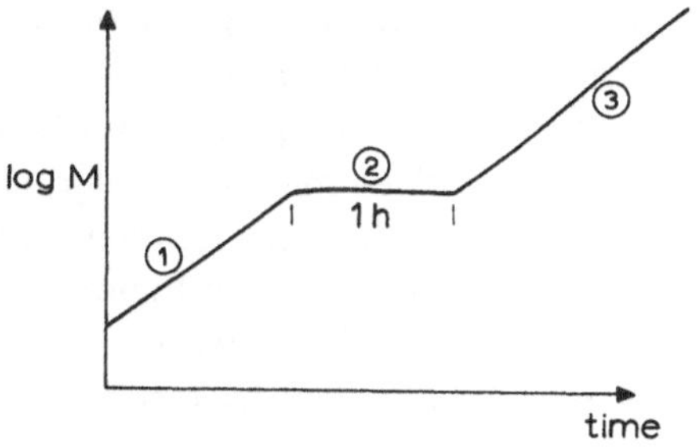

Fig. 6.

The ordinate is the log of bacterial mass per unit volume of medium, and the abscissa, time. The bacteria grow exponentially for some time as shown in that part of the curve labelled (1). Then there is a period of about 1 h during which there is no growth, after which the bacteria start to grow again at about the same rate as before. Analyses had shown that during phase (1) the bacteria use glucose exclusively for growth until it is exhausted. Then they stop growing, and when growth resumes, they use lactose. If during phase (3) glucose is added again, the bacteria immediately switch to glucose as the preferred carbon source without any lag phase such as (2). After a period of growth on glucose, the population will again go into a lag phase if the glucose is exhausted, resuming growth on lactose only after a period of 'adaptation'.

It was known that during phase (2) the bacteria synthesize some enzymes that are essential for the utilization of lactose as a carbon source, and that these enzymes are absent from cells growing in the presence of glucose (whether lactose is present or not). Thus a necessary condition for the synthesis of these enzymes is the absence of glucose. Presence of lactose and absence of glucose constitute sufficient conditons of induction of the lactose-utilizing enzymes in normal or wild type bacteria. Jacob and Monod succeeded in selecting mutants which made the enzymes all the time, irrespective of the presence of glucose or lactose. They were called constitutive mutants and were labelled i^- because they were no longer inducible by lactose. Normal cells were labelled i^+.

In studying the behaviour of these mutants by using the genetic techniques of recombination (combining different pairs of genes in one single cell, such as i^+ and i^- to observe their interactions), it was discovered that the constitutive or non-inducible mutants fell into two classes. In the first class, a cell with the pair i^+ i^- behaved as if it was normal, so it was dominant to i^-. However, in the second class, no such dominance occurred. The mutant gene continued to behave as a constitutive, producing the enzymes for lactose metabolism, while the wild-type gene behaved normally: there was no interaction. So this second class of constitutives was given a different label, O^-, while the corresponding wild-type was labelled O^+.

On the basis of these and a few other observations, Jacob and Monod were able to construct a self-consistent explanation of their results in molecular terms. This explanation is presented in a classical paper [3]. I will give only a schematic description because the model is so well known and widely described. Furthermore, I will introduce my own symbols, in anticipation of further analysis. I will use circles to designate macromolecules which carry out particular assembly or transformation processes, and squares for the 'pools' of metabolic intermediates. Arrows indicate causal relations. With these conventions we may represent the control scheme for the enzyme involved in the metabolism of lactose as in Figure 7. G_i is the gene referred to as i^+ prev iously in the wild-type or normal bacterium. It produces mRNA required for the synthesis of a specific protein, P_i, called by Jacob and Monod a repressor. This repressor has the natural property of combining with a site controlling the gene set, G_l, which produces messenger RNA required for the synthesis of the enzymes for

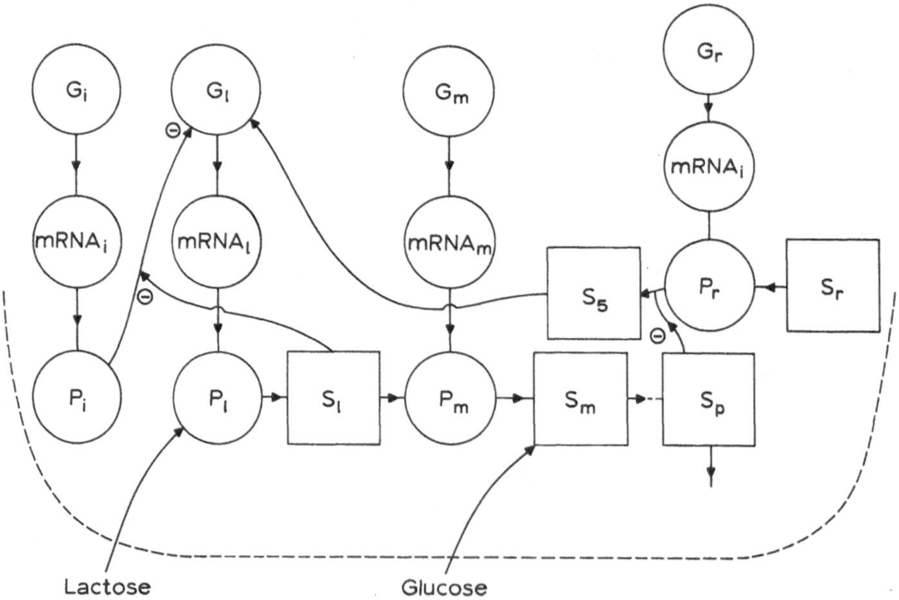

Fig. 7.

utilization of lactose, referred to collectively as P_l. The first product of lactose metabolism, S_l, can combine with the protein P_i and change its properties so that it no longer interacts with G_l. The process of interaction between S_l and P_i is the same as that described for the interaction of an end-product with one enzyme, resulting in inhibition. So once again we encounter the phenomenon of allosteric behaviour as an essential feature of intracellular control processes. Evidently if lactose is present, S_l will be produced and so P_i is inhibited. Then G_l is released from repression and so more P_l is made: the cell thus adapts to the presence of lactose by synthesizing the enzymes required for its metabolism. Evidently it is necessary for some P_l to be present in order for this adaptation or induction process to be initiated. Since cellular control processes are not perfect, there is always some P_l present in any cell, but it is undetectable unless one uses very sensitive techniques of measurement. We may say that cells make use of noise to initiate adaptive responses. A lack of perfection is evidently an advantage in the real world.

The glucose effect is not yet fully understood in detail, but the general picture is as shown. Since lactose is a disaccharide made up of glucose and galactose, its metabolism results in the production of glucose. The diagram shows glucose entering this metabolic pathway at the box S_m, with a 'final' box S_p at some later point in the sequence. This metabolic species, still unidentified, is referred to as the 'catabolite repressor', and it is responsible for the repression of G_l when glucose is present. S_p appears to act by inhibiting the production of another metabolite, S_5, currently believed to be cyclic AMP, which is necessary for the production of mRNA by G_l. So if S_p is present in high concentration, as it will be when glucose is available, G_l will be repressed.

We see that G_l is subject to two types of control, positive and negative, just as enzymes are. It is the negative or feedback control via S_p that produces stability in the system, and homeostatic performance. Thus if glucose is absent and lactose is present, the activity of G_l will be determined primarily by the concentration of S_p, which will be regulated in the same way as end-products are regulated by feed-back inhibition.

Now we may ask something about the rates of change expected in this control system. The important rate constants are those governing monomolecular synthesis. Whereas enzyme turnover numbers are in the range 10^2–10^4 mol s^{-1}, it requires several seconds for the synthesis of a single mRNA molecule or a single protein molecule. So we may expect that enzyme adaptation will be considerably slower than metabolic adaptation, as can be observed in any case from Figure 6. The adaptation period there was one hour, and this is typically the order of time required for significant changes to occur in enzyme concentrations in bacteria. In cells of higher organisms, adaptive responses may be larger, up to 12 h. So relaxation times in the system we are now considering, which I have called the epigenetic system, are of the order of hours.

I have said nothing so far about the constitutive mutant classes described earlier which have Jacob and Monod the key wherewith to unlock the door to an understanding of how gene activities are controlled in bacteria. They deduced that the mutant class i^- must produce a defective repressor molecule, P_{i-}, which was unable to repress G_l and so left it permanently on, i.e., in the constitutive state. This deduction followed from the observation that if a normal i^+ gene was present in a cell, it would overcome the deficiency of an i^-. Thus the i^+ gene must produce something which can interact with another gene in the cell. This repressor protein, P_i, has now been isolated and studied, and it behaves essentially as was predicted.

The other class of constitutive mutants are those defective in the reception of the repressor signal, Jacob and Monod argued; i.e., in terms of the diagram, they do not respond to P_i. Thus they are unaffected by genes producing P_i, and since they themselves produce no substance but only fail to respond, they cannot affect other genes. These were called operator genes, and such genes have also been extensively studied and their existence confirmed by many other investigations. So the essential characteristics of epigenetic control were revealed in these studies. During the last ten years a great deal of research has gone into the analysis of these control processes, and it has emerged that the most universal aspect of the system shown in Figure 7 is the negative feed-back loop, via S_p. Many control circuits have been shown to behave in such a manner, with an end-product such as S_p repressing the genes involved in the synthesis of the enzymes required for S_p production. However in this case *all* the genes are repressed, unlike the situation arising with feed-back inhibition, where only one enzyme was affected: the first one. The reason for this is again quite clear. Returning to Figure 3 for a moment it is evident that if the product S_7 is available to the cell, then it ought not to produce any of the enzymes involved in its production, since otherwise it would be wasting precious resources. Thus at the level of epigenetic control, the unit of function is controlled by coordinate repression of all enzymes involved in the sequence. We can draw this picture diagrammatically as shown in Figure 8.

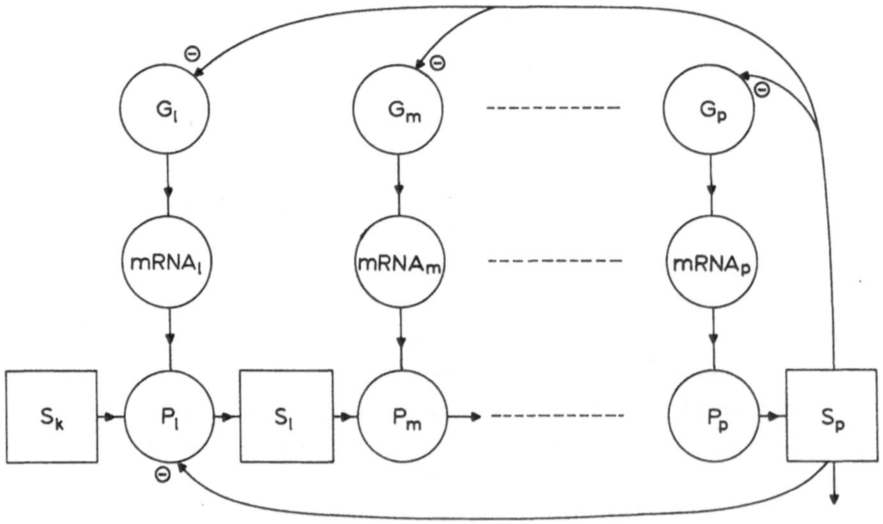

Fig. 8.

In this diagram both feed-back inhibition and feed-back repression have been included to show the difference between them. Many details have been left out, and in particular the fact that S_p must combine with a repressor protein in order to act as an active repressor. The point I would like to emphasize is that the metabolic and the epigenetic systems are not separate systems, but are related to one another in a hierarchical manner, the metabolic system being embedded in the epigenetic. Effectively, the parameters of the metabolic system contain the variables of the epigenetic system.

We may continue this hierarchical analysis by observing that the genes, G_i, are also subject to change over periods of time which are long compared with the relaxation time of the epigenetic system, i.e., days in the case of bacteria and months, years, or centuries in the case of higher organisms. But at the level of genetic change we know very little about the constraints which operate, and how to describe evolutionary processes apart from the elementary postulates of random mutation and natural selection. The concept of random mutation is clear and unambiguous. But the concept of natural selection is not, for we do not know what properties of organisms are selected by the evolutionary process. We do not, I believe, have any satisfactory theory of evolution of biological systems because we do not yet have any adequate definition of what a biological system is. I believe such a definition must be very intimately connected with the property of hierarchical organization that I have been describing, but we seem to be still some way from any formal description capable of capturing the essence of biological order.

2.2. The dynamics of the epigenetic system

Let me return now to the problem of writing down some equations for intracellular control dynamics which might give us some insight into the qualitative behaviour we might expect from this system. For this purpose I am going to simplify Figure 8 still further, and consider the whole unit as a single feed-back loop of the type shown in Figure 9.

Now introduce variables X_i and Y_i to represent the concentrations of mRNA and enzyme in the system, and we wish to derive equations for their rates of change in time. Let us make the assumption that all reactions are irreversible. This is certainly true for the production of X_i and Y_i since these are essentially polymerization reac-

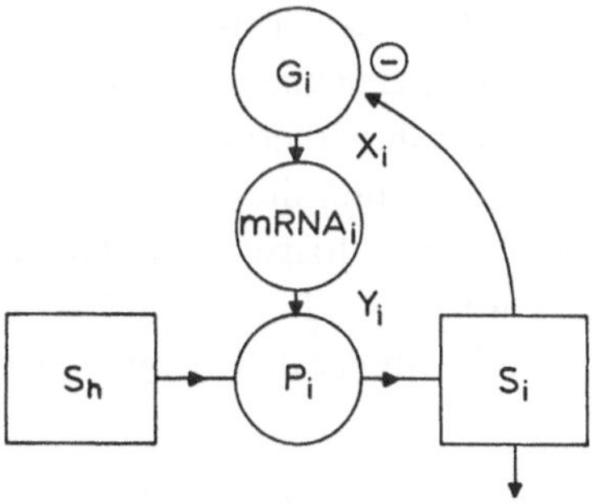

Fig. 9.

tions. We shall see that as regards S_i, the assumption has virtually no effect on the argument. I will also assume that each variable acts as a rate-limiting factor for the next step in the causal chain, which it therefore controls. Thus X_i determines the rate of protein synthesis; Y_i determines the rate of S_i synthesis; and S_i determines the rate of X_i synthesis. The last relationship is one that involves an interaction between S_i and a macromolecule of the type described by Figure 9, so we can write the activity of the gene G_i in the form

$$\frac{a_i}{1 + L_i S_i^n}. \tag{2.5}$$

If we assume that mRNA is degraded at a fixed rate, b_i, then we can write

$$\frac{dX_i}{dt} = \frac{a_i}{1 + LS_i^n} - b_i. \tag{2.6}$$

With X_i acting as the rate-limiting factor for protein synthesis, and assuming that protein is degraded at a constant rate, d_i, we can write

$$\frac{dY_i}{dt} = c_i X_i - d_i. \tag{2.7}$$

The metabolite S_i will be governed by an equation of the type considered previously, but with Y_i as the rate-limiting factor, so we have

$$\frac{dS_i}{dt} = l_i Y_i - k_i S_i. \tag{2.8}$$

At this point we can use the argument about different relaxation times for the metabolic and epigenetic systems to eliminate the variable S_i. The argument means that $dS_i/dt \simeq 0$ over time intervals which are of relevance for epigenetic changes. Thus we can write $S_i = l_i Y_i / k_i$ and substitute in Equation (2.6), giving an equation of the form

$$\frac{dX_i}{dt} = \frac{a_i}{1 + k_i Y_i^n} - b_i. \tag{2.9}$$

The pair of Equations (2.3) and (2.5) can then be taken to describe the behaviour of a simplified control circuit under particular conditions. There is a non-linearity in the system which arises from the control term, and one of the questions we want to ask is what the consequences of this non-linearity may be. Since I am proceeding in the simplest possible manner in order to get some feeling about the qualitative behaviour of such a system, I will take $n=1$.

Equation (2.7) and (2.9) can be integrated to give a first integral of the motion, since obviously

$$(c_i X_i - d_i)\frac{dX_i}{dt} - \left(\frac{a_i}{1 + k_i Y_i} - b_i\right)\frac{dY_i}{dt} = 0.$$

Some simplification is attained by changing variables to their steady-state values, so that we take

$$x_i = X_i - X_i^*$$
$$y_i = \frac{b_i k_i}{a_i}(Y_i - Y_i^*),$$

where X_i^* and Y_i^* are the values of the variables when $\mathrm{d}X_i/\theta t = \mathrm{d}Y_i/\theta t = 0$. This transformation puts the equations into the form

$$\frac{\mathrm{d}x_i}{\mathrm{d}t} = b_i\left(\frac{1}{1 + y_i} - 1\right)$$
$$\frac{\mathrm{d}y_i}{\mathrm{d}t} = \alpha_i x_i$$

where

$$\alpha_i = \frac{b_i c_i k_i}{a_i}.$$

Now the integration required for a first integral of the motion is

$$\alpha_i x_i \frac{\mathrm{d}x_i}{\mathrm{d}t} + b_i\left(1 - \frac{1}{1 + y_i}\right)\frac{\mathrm{d}y_i}{\mathrm{d}t} = 0$$

giving

$$\frac{\alpha_i x_i^2}{2} + b_i\left[y_i - \log(1 + y_i)\right] = \text{const} \equiv G(x_i, y_i) \tag{2.10}$$

What is it that is conserved in such a system? It is evident that the equations of motion can be recovered from the function $G(x_i, y_i)$ by the canonical equations of dynamics, viz.

$$\dot{x}_i = -\frac{\partial G}{\partial y_i}$$
$$\dot{y}_i = \frac{\partial G}{\partial x_i}. \tag{2.11}$$

The surface described by $G(x_i, y_i)$ is egg-shaped, which seems like a promising beginning for a biological theory. The phase-plane description is as shown in Figure 10.

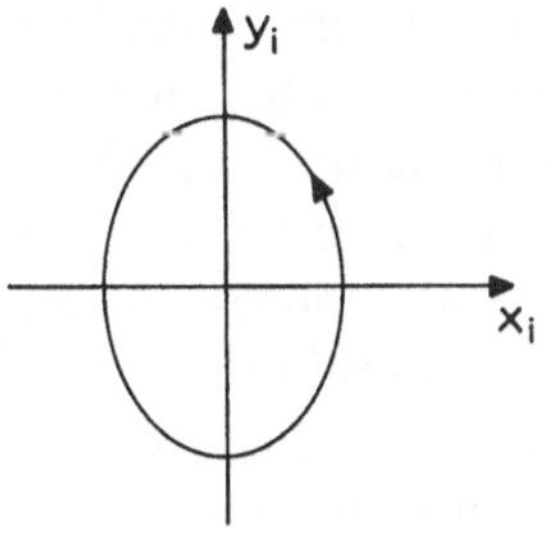

Fig. 10.

In terms of the original variables, X_i and Y_i, the picture is simply translated into the positive quadrant, with the centre of the closed curve at the steady-state value X_i^*, Y_i^* (Figure 11).

Thus this system executes continuous non-linear oscillations about the steady-state. This analysis has shown that one possible mode of behaviour of a control circuit of the type described in Figure 9 is continuous oscillations. This is of some interest, because phenomenologically it is well known that cells have rhythmic modes of behaviour. Indeed, it could be argued that rhythmic change is the basic dynamical mode of cell behaviour, since cell reproduction is an inherently cyclic process. Also there is the phenomenon of the biological clock, another widespread mode of cell behaviour.

Thus it is possible that rhythmic behaviour of cells arises from a very fundamental feature of their organization as regulatory systems: the existence of negative feed-back control processes. This conjecture is by no means a new one. However, it is very difficult to prove despite its apparent plausibility. It is nevertheless of some interest to see where arguments of the type presented so far may lead us. It turns out that one is not led anywhere very interesting biologically speaking, but the reasons for this are themselves instructive. In seeing where and why the analysis fails, we may learn something more about biological organization.

2.3. Statistical Mechanics and Biological Process

The canonical Equations (2.7) and the invariant $G(x_i, y_i)$ invite the construction of a statistical mechanics for cellular control processes. The control circuit of Figure 9 is a condensed description of a unit of epigenetic control in single cells modeled on the bacterial system. There is no detailed evidence that such a circuit forms the basis of gene control in the cells of higher organisms. However, it is known that nuclear and cytoplasmic states interact in defined ways, and the evidence available does suggest that the regulatory stability of, for example, a liver cell, depends upon control processes of the general type depicted in Figure 9. The details are of course very much more complex than this, and there are additional causal relationships. For example, metabolites such as S_i can affect the rate of synthesis of protein, Y_i, thus exerting control at the translational level. So anything we might deduce from the analysis presented can only be of an extremely tentative nature. However, the exercise is very instructive in certain respects.

Any cell will have many control circuits of the general type shown in Figure 8 and simplified to Figure 9. This number will be a few hundred. Thus the subscript i will run over values from 1 to n, n being of order 10^2. Each of these control circuits is operating in a common 'physiological space', since the synthesis of mRNA and protein depends upon common pools of nucleotides and amino acids, respectively, and, of course, energy is needed. So these units are really in interaction. We may regard this interaction as 'weak' in the same sense that the interaction between molecules of a gas is weak: it results in the equipartitioning of energy throughout all degrees of freedom in the gas. For control circuits, such partitioning means that their oscillatory behaviour is spread evenly over all elements; i.e., they excite one another and transfer oscillatory energy. Using such a postulate, we can write a constant of the motion of the whole system in the form

$$G(\mathbf{x}, \mathbf{y}) = \sum_{i=1}^{n} G_i(x_i, y_i) = \sum_{i=1}^{n} G_i.$$

We can then construct, formally, a statistical mechanics for this system, taking as the appropriate probability distribution that for the canonical ensemble, viz.

$$\rho = e^{(\psi - G)/\theta}$$

where ψ is the analogue of the Gibbs free energy function and θ the analogue of kT. Assuming ergodicity, we can then find phase averages, for various functions and explore the meaning of 'temperature' in such a system.

Since the main dynamical behaviour is oscillation or rhythmic variation, all 'thermodynamic' variables will describe some aspect of rhythmic activity in the cell. For example, the 'temperature', θ, turns out to be a measure of the mean deviation of the variables X_i and Y_i from their steady-state values, viz.:

$$\theta = \overline{\alpha_i X_i (X_i - X_i^*)} = \overline{\frac{\gamma_i Y_i (Y_i - Y_i^*)}{1 + k_i Y_i}} \quad \text{all } i,$$

where the bars refer to phase averages and γ_i is a constant. The general picture is that at 'thermodynamic equilibrium' the oscillatory energy is distributed evenly over all the control circuits in such a cell. This and many other expressions are given in my book on this subject ([4]).

The question now arises, can we use this general picture of rhythmic behaviour in cells to understand and explore the rhythms observed in real cells? Periodic phenomena are of very basic importance in biological dynamics, the cycle of cell growth and division and biological clocks being perhaps the most widely-known examples. Let us consider first the phenomenology of the biological clock. Many single-celled organisms have the capacity of ordering their physiological processes in time such that over the 24-hours of the solar day there is a well-defined relationships between difficult activities. For example, in photosynthetic organisms, those enzymes, required

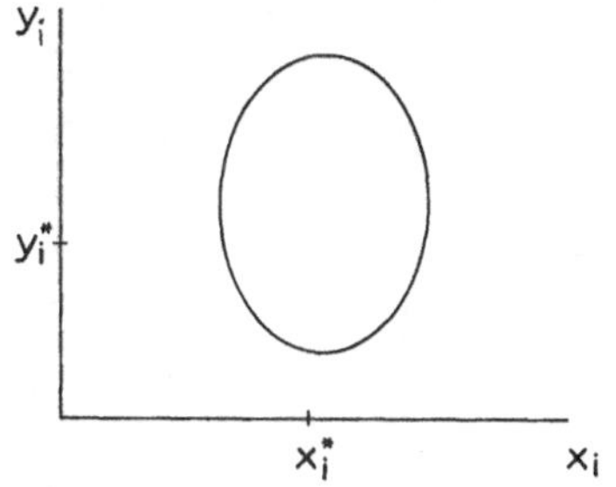

Fig. 11.

for photosynthesis are made during the day but not at night, whereas light-sensitive processes (those that are inhibited by light) occur at night. Cell division is also timed by this clock, which is found to continue in many such organisms independently of an environmental day-night cycle. Thus even under constant external conditions, many organisms continue to organize themselves temporarily by some intrinsic 'clock' or rhythmic time-base. The period is always in the vicinity of 24 h.

What is the period of the oscillations shown in Figure 11? Using the statistical mechanics we can get an estimate of the frequency for relatively large θ which is

$$\omega(y_i) \simeq \frac{b_i}{2} \sqrt{\frac{\alpha_i}{2\pi\theta}}.$$

With the appropriate values for the rate constants, this turns out to be in the range 2–12 h. Since the system is conservative, the amplitude of the oscillations depends upon the initial conditions, and so there is, mathematically speaking, no limit to the amplitude and hence to the frequency (since the oscillations are non-linear, frequency decreases as amplitude increases). However, there is a biological limit determined by the steady-state values, since the trajectory must not cross either the X_i or the Y_i axis (negative concentrations do not exist). Thus there is a plausible limit, though not an absolute one; i.e., there could be one or a few control circuits with periods of 24 h, but they could not all have such long periods and correspondingly large amplitudes.

Is there any way in which periods of 2–12 h could be extended to 24 h by some fairly natural means? Since the oscillations are non-linear, it is possible for interactions between them to result in what are known as sub-harmonic frequencies, i.e. frequencies less than those of either interacting oscillator. Thus two oscillators, one with a period of 4 h and the other with a period of 5 h can give a frequency of 24 h. The interaction is quite easily introduced, and is biologically perfectly plausible. We simply say that two control circuits interact with each other by reciprocal feed-back as shown in Figure 12. The interactions can be positive or negative, but they are shown as negative in this instance.

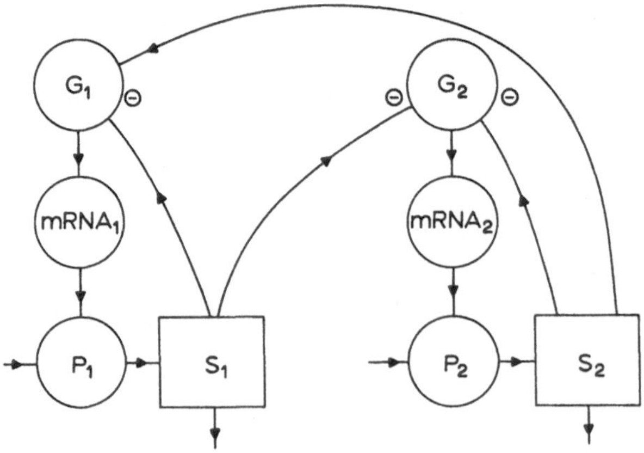

Fig. 12.

The equations of motion are then altered to the form

$$\frac{dX_1}{dt} = \frac{a_1}{1 + K_{11}Y_1 + K_{12}Y_2} - b_1$$

$$\frac{dX_2}{dt} = \frac{a_2}{1 + K_{21}Y_1 + K_{22}Y_2} - b_2 .$$

The equations for Y_1 and Y_2 are unaltered. These equations also have an integral, with a quadratic interaction term, and so the statistical mechanics can again be used. We then find that sub-harmonic resonance can occur, and the coupled pair of control circuits can be regarded as contributing to the 24 h organization of the cell. Indeed, we can consider that all or many of the circuits are in interaction in this way, with a selection of parameter values K_{ij} so that the 24 h periodicity is preferred.

What kind of stability will be shown by such a rhythmic system? And how does this compare with the stability of biological clocks? We can answer the first question very easily. Since the dynamical system is conservative, any particular oscillation will be orbitally stable. That is to say, it will stay on some orbit until it is perturbed, when it will move to another orbit and remain on it until disturbed again. Thus there is an infinite set of possible orbits, and hence an infinite set of frequencies which it can occupy. However, because of the influence of all the other oscillators with which any one is in weak interaction, there will be a mean trajectory and hence a mean frequency which any oscillator will have. This is also true of a coupled pair in sub-harmonic resonance. So the model shows weak or neutral stability.

It is very difficult to determine the exact stability properties of biological clocks, but the results of very elegant experiments performed by Winfree on the Drosophila eclosion rhythm suggest that they do not have strong asymptotic orbital stability; i.e., they do not return rapidly to a preferred orbit after they have been disturbed. Furthermore, the Drosophila eclosion clock can be stopped by a single critical stimulus. In terms of a two-variable model, this means that in the phase plane there is either a stable focus or a centre. The evidence taken as a whole is certainly in favour of a non-conservative oscillator with asymptotic stability for this and other clocks because of the fact that their free-running or autonomous frequencies are independent of perturbations which cause phase shifts. Such a perturbation results in a transient which may last for a few days, but eventually the system settles to a frequency which is the same for any given conditions of growth.

The phenomenon of synchronization or entrainment whereby an endogenous biological rhythm becomes locked with constant phase to an environmental periodicity, usually the day-night cycle, is a very basic aspect of clock behaviour, for it gives the organism the capacity to organize its physiological processes temporally in relation to environmental rhythms. For example, the unicellular dinoflogellate, *Gonyaulax polyedra,* has an endogenous rhythm of photosynthetic activity which normally coincides with the period of daylight. This rhythm persists under conditions of constant dim light, thus showing that it is not the light/dark cycle which causes the rhythm. However, when the organism is exposed to an environmental light cycle it enters a particular phase relationship with it such that photosynthetic processes coincide with the maximum light period, a relationship which is of obvious adaptive advantage to the organism.

A basic theoretical question that arises in relation to such a phenomenon is what the nature of the coupling might be between the endogenous rhythm and the environmental cycle. It has been generally assumed that the environmental stimulus acts as a forcing function, entering the equations of motion of the clock in the form of a periodic function $f(t)$ which is added to the right-hand side of either of the Equations (2.7) or (2.9), for example. If the clock is a conservative oscillator, then it is well known that this kind of forcing will result in resonance, so that the amplitude of the oscillation would be expected to increase greatly as the forcing frequency approaches the resonant frequency. There is no evidence from experimental studies on biological clocks that such an amplitude effect occurs, and this might encourage the conclusion that the oscillator cannot be conservative. However, such an inference is not warranted because it depends upon the assumption that the forcing function enters the system via a perturbation of one of the variables. There are no direct observations in support of such an assumption, and the alternative is equally possible: that the environmental periodicity affects one of the parameters of the 'clock'. Under parametric excitation there need not be any amplitude instability in a conservative oscillator. Suppose, for example, we assume that the light/dark cycle has its effect on gene activity through the parameter a_1 in Equation (2.9) so that a_i varies periodically. To see the effect of such a parametric variation on the oscillator, let us simplify Equations (2.3) and (2.5) by writing $z=1+K_iY_i$ in the case where $n=1$. Since from (2.3) we have

$$\dot{Y}_i = c_i\dot{X}_i,$$

assuming c_i to be constant, we find that

$$\ddot{z} = \alpha(\beta/z - 1), \tag{2.12}$$

where $\alpha=b_ic_ik_i$ and $\beta=a_i/b_i$. Thus β varies periodically if a_i does. For simplicity, let us suppose that this takes the form of a square wave as shown in Figure 13.

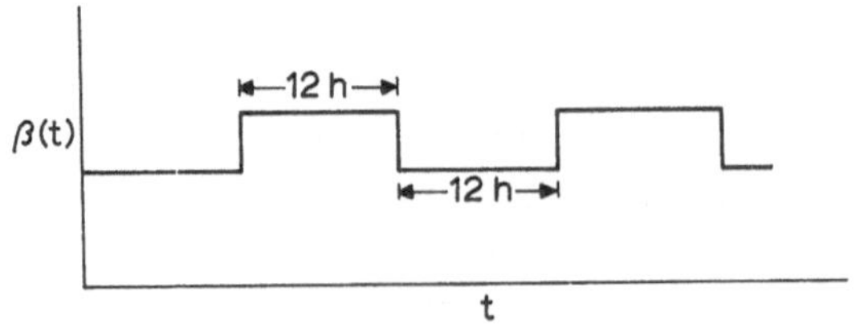

Fig. 13.

The motion of the system governed by Equation (2.12) may be interpreted as that of a particle of mass $1/\alpha$ moving in a potential field $V = z - \beta \ln z$, with total energy

$$E(z, \dot{z}) = \frac{\dot{z}^2}{2\alpha} + z - \beta \ln z .$$

The potential function has the shape shown in Figure 14. The motion of the particle is represented by the point P on the horizontal trajectory, the potential energy at any point on the motion being given by h. Periodic variation in β means that the shape of $V(z)$ changes periodically from curve (1) to curve (2). If the system is at P when β increases, then its position will suddenly change to the point P' whose position is

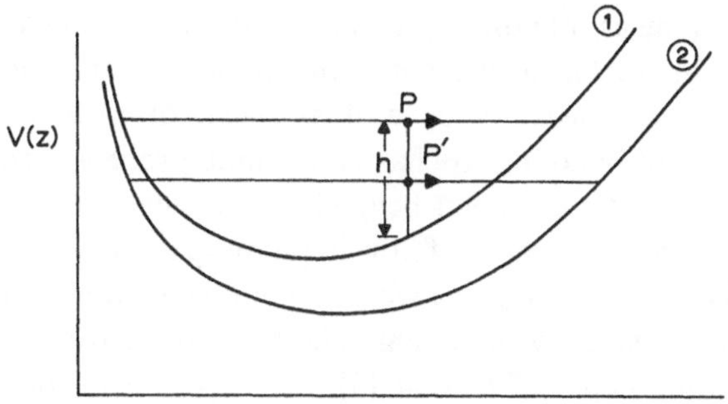

Fig. 14.

determined by the condition that the value of the potential, h, is unaltered. The system will then continue to move in the new potential field (2) until β changes stepwise and restores the original potential, when the position of the particle will again be determined by the fact that the value of the potential, h, is the same immediately, before and after the parametric change.

Such a parametrically excited system will entrain to the parameter frequency after a transient which will depend upon the initial phase relations of system variable and parameter, and upon the change in the steepness of the potential well with changes in β. The amplitude of the entrained oscillation will also depend upon both these factors, but the phenomenon of resonance will not occur and the amplitude will not, in general, vary greatly from the free-running to the parametrically driven mode. A detailed analysis of the system (2.12) with periodic β needs to be carried out and compared with the behaviour of a biological clock before any further conclusions can be drawn about the value of this conservative system as an approximate model of the process.*

2.4. Statistical mechanics for non-conservative systems

The question that naturally arises at this point is whether it is possible to use the general procedures of statistical mechanics for dynamical systems which have asymp-

* I am indebted to Professor K. Huang for the suggestion that the problem be cast in this form.

totic stability; i.e., whose trajectories converge on attractor sets in phase space. The measure of such attractor sets will, in gereral, be zero. We have seen that the experimental evidence for the dynamics of biological clocks is that they have asymptotic stability, and this is also the case for the dynamics of the bacterial cell cycle. Thus perturbation of a bacterial population from a steady generation time in defined conditions results in a return of the population mean to that same generation time after the disturbance has passed. This stability of frequency implies a stability of amplitude for the control system regulating initiation of DNA replication, since a non-linear system with two or more variables change of frequency implies change of amplitude, and *ipso facto* constancy of frequency implies constancy of ampitude. Thus it is evident that an adequate theory of both biological clocks and the cycle of cell growth and division requires the use of non-conservative, non-linear models. But it is necessary to have a statistical or stochastic element in the mode as well, since the experimental data provide fluctuation measures as well as mean or average values. Statistical mechanics is one way of handling stochastic dynamic processes, so it is of interest to see if its procedures can be extended beyond its classical domain of application to dynamical systems within invariance of flow in phase space, i.e., measure conservation.

I investigated this question in a paper which attempted to apply statistical mechanical procedures to a simple limit cycle system. The basic principles of the analysis came from a very interesting paper of Kerner [5] on this problem for dynamical systems with asymptotic point stability; i.e., whose attractor set consisted of a single point in phase space, the stable steady-state [6]. The first step in the procedure was to see if one could use the Lie-König theorem to obtain an invariant integral for the phase space with the limit cycle. Following the procedure described by Kerner it was possible to obtain what is known as a Last Multiplier of the dynamical equations, which had been derived in the form

$$\frac{\mathrm{d}r}{\mathrm{d}t} = a(1 - r^2),$$

$$\frac{\mathrm{d}\theta}{\mathrm{d}t} = \frac{b}{1 + r},$$

where the variables are polar coordinates. These equations define a system which is asymptotically stable to the limit cycle at $r=1$. The invariant integral was derived as

$$\begin{aligned} G(r, \theta) &= \frac{b}{4}\left\{\log \frac{1+r}{1-r} - \left(\frac{1-r}{1+r}\right)\right\} - a\theta \qquad (0 \leqslant r < 1) \\ &= \frac{b}{4}\left\{\log \frac{r+1}{r-1} + \left(\frac{r-1}{r+1}\right)\right\} - a\theta \qquad (1 < r \leqslant \infty). \end{aligned}$$

One can then formally construct a distribution function for some appropriate ensemble, say the canonical, and proceed on the assumption that the system when properly weighted for occupation of regions of phase space is ergodic so that time averages can be replaced by phase averages. This does yield values for such quantities as the mean generation rate, $(\bar{\theta})$, its variance, and so on, values whose qualitative behaviour fit experimental observations, for example with respect to skewness. However there are some rather fundamental objections to such a procedure. Probably the most serious one is that, whereas the existence of a Liouville theorem is a necessary condition for constructing statistical mechanics, it is by no means sufficient. Another condition is that the measure of the space over which the invariant integral is defined and which contains phase points which are occupied with finite probability should be finite. However, it turns out that the points of greatest interest in the present problem, those in the neighbourhood of the limit cycle, are just those points which produce a singularity in the weighting function on the last multiplier, which is $1/(1-r^2)$. So there is some question about convergence of the integrals, despite the results that can be obtained.

Another basic problem is that points on one side of the limit cycle can never cross over to the other side, despite the stochastic postulate of scattering and the maintenance of constant mean 'temperature'. This is quite readily found if one calculates the mean frequency of crossing of the limit cycle, $r=1$, which turn out to be zero. Obviously biological cells do not behave this way. If one starts a culture from two different states, let us say spores and stationary phase, then after a long period of growth on the same medium and under the same environmental conditions cultures started from these very different initial conditions would be indistinguishable. Furthermore, it is clear from the studies of Kubitschek [7] on the generation times of descendants of single cells that successive generation times vary with negative linear correlation about a mean. Thus one of the primary objectives of this study, to investigate the stability of the system, is frustrated by a basic topological constraint in the differential equations which is naturally preserved in statistical mechanics because of the fact that the stochastic or probabilistic component is introduced via assumptions about initial conditions. To get crossing of the limit cycle one would have to introduce additional assumptions. The clumsiness of this approach becomes more and more evident, and one is strongly inclined to look for other possible procedures.

One such is to use stochastic differential equations in place of deterministic ones, thus introducing variability in a somewhat more natural way (via parameter variation) than the initial condition assumptions of statistical mechanics. However, once again one quickly finds that non-linear equations become quite intractable by analytical procedures the moment they become sufficiently complex to show biologically interesting behaviour. It appears that the procedures that work in physics are either too fine or too coarse for biological processes considered in the global sense. Differential equations provide too detailed a description of system behaviour, and rapidly become intractable as one adds the non-linearities required for biological stability. Physical averaging procedures for systems described by differential equations necessarily and deliberately throw away much of the correlated behaviour between 'microscopic' variables, since the point is to get results from the laws of large numbers. So biological processes, which show stability and correlation of dynamical behaviour to generate macroscopic or global order, seem to require a somewhat different topology from that used in physics. We might say that Hilbert space is the appropriate space for normal mode analysis, but it is not an appropriate space for the study of organized behaviour. What space is? We do not yet know, but it appears that it must have a deep structure appropriate to biological processes if there is to be any natural elegance in the analytical handling of biological phenomena.

3. Correlations: Temporal and Spatial Order

The last sentence was a typical example of how to clothe ignorance in jargon. 'Deep structure' is an evocative phrase in biology these days and conjures up visions of constraints operating in some basically simple but globally very effective manner. It may be possible to say something rather more precise about this concept, but such precision as may be attained is far from the definition of a global analytical procedure. One can hope to make a connection with the theory of structural stability and catastrophes, say, or some other such context of general logical power and elegance, but only when there are good grounds for a marriage between empirically validated biological models and a higher-order logical context into which they fit naturally, like kinetic theory and thermodynamics. Having started those lectures with a search for a global theory of cellular processes and failed, for reasons which I hope are clear, I will now consider the problems that are posed by biological processes in which there is both temporal and spatial order, and suggest a global theory for such organization. However, you will see that the models proposed are no longer borrowed from physics. They come fairly directly from biology, and incorporated some basic non-linearities. These theories will fail as well, but may carry us further into biological processes than did the statistical mechanics, which was always somewhat superficial in its relation to biological phenomena. I will be concerned first with problems of coupling between 'microscopic' processes, and how a particular kind of non-linear behaviour can give rise to global organization.

The phenomenology with which I am concerned is that of experimental embryology. In this subject one observes the ordered unfolding of spatial and temporal patterns, and studies their properties by various perturbations. In general terms one can say that a system which organizes itself in four dimensions must be able to measure time and space. We have considered some aspects of temporal order in single cell and in cell populations, and what kind of dynamics could underly such time measurement.

Now let us consider the evidence for periodic processes in embryos, to see if we can use the same reasoning to explain how they keep time.

I will start with a very favourable case for the general proposition that I want to put forward. This is the development of the cellular slime mould *Dictyostelium*. What one can observe in these species is a beautifully clear and simple relationship emerying between temporal and spatial order, a relationship which I believe may hold throughout all species in its basic principles. *Dictyostelium* has a vegetative (non-reproductive) phase in its life cycle which consists of single amoebae that grow and divide on a surface with a nutrient source such as agar covered with bacteria. However, when the food resources have been depleted, the cells enter a new phase of their life cycle, one in which a multicellular organism is formed consisting of two cell types, structural cells and spores. The transformation from a population of single cells which has no global spatial or temporal order to an organism with a well-defined morphology is very instructive and has been used by many biologists as a prototype for embryological development. I am doing likewise.

The first globally visible evidence of this transformation of the cell population from a disordered to an ordered state cannot be seen by simply looking at the population. It is necessary to make a time-lapse film of the cells, with a frame interval of about 15–20 s. On such a film one observes waves of oriented amoeboid movement of cells propagating from centres which initially appear to have a random distribution over the cell population. Detailed study of single cell behaviour reveals that they rather suddenly begin to move toward a centre and then gradually this orientation disappears, motion becoming disoriented or random. Then again a cell will move towards the centre, the period between those directed movements being a few minutes (4–7), with the oriented motion lasting for about half of the period. All cells at a given distance from a centre orientate and move simultaneously. The result of this periodic motion is the aggregation of a large number of cells at the centre. Those cells then undergo further transformation spatially and temporally to produce the fruiting body by a series of changes which need not concern us. This development is described in many excellent monographs and books on development.

What processes are responsible for the aggregation phenomenon? It is now known that the cell of the centre initiates a wave by releasing a chemical which has in recent years attracted much attention: cyclic AMP. We encountered this metabolite earlier in the study of the control mechanism governing feed-back repression of the genes for lactose metabolism. These cyclic AMP acted as a necessary condition for message transcription from the genes, and when glucose was present, cyclic AMP was absent, presumably by the inhibition of the enzyme producing it, this inhibition being caused by a product of glucose metabolism, the postulated but as yet unidentified 'catabolite repressor'. In general it has been found that cyclic AMP is produced whenever cells experience a deficiency of nutrients, and Dictyostelium is no exception. Why certain cells should start producing cAMP before others is not clear, but this could arise simply from variations over the population with respect to their last meal (ingestion of bacteria). The interesting feature of these cells is that they produce cyclic AMP periodically, or a least release it periodically. Of course it is just such rhythmic behaviour that we would anticipate from a metabolic system with negative feedback control of the general type we have been considering. But what about the period of 4–7 min? This is not what we expect of epigenetic rhythms, those we have implicated in biological clock behaviour with periods of several hours. However, it is just the sort of period we would expect of a metabolic oscillator, where the relaxation times are of order of seconds to minutes. Furthermore, if developmental processes use a clock, then it must run considerably faster than one cycle per day, since many developmental systems undergo most of their transformations in a day or less. Dictyostelium takes about 18 h so a clock with a period of a few minutes would be in the right range for developmental processes. In fact a metabolic oscillator with a period of a few minutes has been discovered and extensively studied in the past few years. It is known as the glycolytic oscillator, and its existence was first demonstrated in yeast cells. Now it is possible to reconstruct this metabolic system in vitro and study the details of its dynamic behaviour (see the review by Hess and Boiteux [8]). The basic principles of its operation are exactly those which we would anticipate in terms of cellular control processes: there are one or two key rate-limiting enzymes which are under feed-back control by products of their activity, with the result that for particular values of the parameters a limit cycle oscillation in the concentration of the metabolites occurs, with periods of about a minute or more. Furthermore, this glycolytic process is a very fundamental part of cellular metabolism, so this oscillation is 'available' to any cell type. There is no evidence that this is the origin of the periodicity of cyclic AMP release in Dictyostelium, but since glycolysis and cAMP production are metabolically closely connected, a causal relationship is possible.

Given that cells produce cyclic AMP periodically, how do organized waves propagate from a centre? It appears that an amoeba which is stimulated by a pulse of cyclic AMP will itself release the same substance, so that the signal is transmitted from cell to cell. This transmission is possible only if the mean distance between cells is less than some critical value, for the amount of cyclic AMP released per cell is always about the same, and of course diffusion will attenuate the signal. But what prevents the wave from travelling backwards towards the centre or initiator cell, thus effectively destroying any spatial order in the system? At this point we encounter one of the really basic non-linearities of biological activities: after cells have been stimulated, they have in general a refractory period during which they are unresponsive. This is a very natural property of a system which uses energy to maintain metastable states and must pass through a transient after a perturbation. So it is this refractory property of the cells which results in one-way propagation of the signal.

At first many centres or initiator cells will be competing for amoebae, but in time it is the centres with highest frequency which will tend to dominate. There will also be phase differences between centres with about the same frequency which will result in refractory boundaries being established between domains dominated by different centres. So after an hour or more the population of amoebae will be fairly equally partitioned among the centres, and these sub-populations will then proceed to form developing 'embryos'.

Thus we see how the addition of one simple non-linear property, that of refractoriness, can result in the emergence of a primitive radial spatial order in a population of cells. In fact there are a number of other features of the aggregation phenomenon in Dictyostelium that are important and which I have simply assumed: the fact that the process takes place on a surface; that cells can move in response to the stimulus; that they respond metabolically by releasing cAMP after being stimulated by it. It is necessary to take account of these facts and to attempt to explain cellular responses in physiological and biochemical terms. This has been done very thoroughly by Cohen and Robertson [9, 10]. But for the moment I want to focus attention upon basic and simple cellular property which provides a mean whereby spatial order in cell populations can emerge from rhythmic behaviour of single cells. This property is simply the transmission of a metabolic signal from cell to cell by diffusion, with cells entering a refractory state for some period of time after being excited. We may then speak of a wave of excitation which propagates radially from a centre across a population of cells. Since the centre initiates the signal periodically, we can after to the process as periodic wave propagation, these being essentially non-linear excitation waves. Of course, the most familiar example of an excitation wave is the action potential that propagates along a nerve fibre. However, this may be regarded as a specialized example of a very general property of biological systems, which derives from their metabolic organization and the properties of membranes.

3.1. Breaking symmetry in biological systems

Is it possible to generalize from the behaviour of Dictyostelium and say that all developmental systems have this kind of organization? Although the cellular slime mould is often presented as a model of embryological development, it is important to realize that in many respects it is very difficult from 'normal' embryogenesis, the existence of many cells initially being most unusual. Most embryos start as a single cell which undergoes division to produce a population. So the aggregation process is peculiar to the cellular slime moulds, and much caution must be exercised in attempting to generalize from this picture. There are very few other examples known of developing systems which give clear evidence of periodic wave propagation. However, this may be because people have not looked. In our laboratory two developmental systems have been studied using time-lapse films, *Xenopus laevis* and *Tubularia*. No clear periodicities were observed in *Xenopus* during gastrulation, but the morphogenetic movements in this species during gastrulation are peculiar and would tend to mask periodic waves of movements of presumptive mesoderm. In *Tubularia* a very interesting periodicity was observed during hydranth regeneration, with a period of about 6–8 min. The very early chick embryo has also been observed to show periodicities like those in Dictyostelium.

However, we can ask a somewhat different and more fundamental question: can one use the properties of rhythmic metabolism and wave propagation to generate a stable axis of organization in an embryo, i.e., to break symmetry and introduce linear rather than radial spatial order? To give this question concrete relevance, we may turn to an example which serves as a prototype for this problem. This is the case of early development in the egg of the seaweed, *Fucus*. These eggs when released are symmerrical spheres about 70 μ in diameter. They are fertilized by free-swimming sperm, and some 6–8 h later the first evidence of asymmetry can be observed in the form of an outgrowth from one part of the egg. This outgrowth undergoes elongation and then the egg divides into two cells, the division plane being perpendicular to the long axis through the outgrowth, and asymmetric so that one cell is smaller than the other. This smaller cell becomes the root or rhizoid and the other the thallus [11].

It appears that under normal conditions the point of sperm entry determines where the outgrowth and hence the root will form, which is at the point diametrically opposite that of sperm penetration. However, many different stimuli can over-ride this initial influence of the sperm and establish a different axis. These include such stimuli as directed light from a source, a gradient of pH or KCl, or a gradient of a metabolic inhibitor, dinitrophenol. The membrane seems to be the receptor site for these signals, and the primary source of organizational response leading to an axis in the embryo. Once the axis has been established, it is stable and cannot be reversed.

I want now to generalize from the Dictyostelium model and to present a theory which can be applied to many different developmental systems in which symmetry is broken and polarity is established. This polarity must be labile initially, and then become stabilized. Furthermore, the model must apply to unicellular and acellular as well as to cellular systems, since the example of the fucoid egg shows that symme try is broken and polarity is established in the egg before cleavage. There are many other examples of polarity in single cells, and spatial order in syncytia (acellular systems), such as most insect embryos. The model, to be useful, must be applicable to a great diversity of forms.

3.2. The wave-broom model

Since our model must apply to systems which are not partitioned into cells, and it must generate stable spatial order, it is natural to assume that membranes are somehow involved. Furthermore, since one of the basic features of the model will be a propagating activity wave of some kind, our attention is again directed to membranes since these are excitable surfaces and can propagate activity waves, the best known example being the nerve action potential. However, this is a virtually universal property of biological membranes, although most rates of propagation of activity waves are much smaller than that of myelinated nerve axons, which are highly specialized for rapid conduction.

The general process I have in mind is as follows. Suppose that we have enzymes E_1 and E_2 located on a membrane surface, and that these enzymes catalyze the metabolic reaction sequence shown in Figure 15.

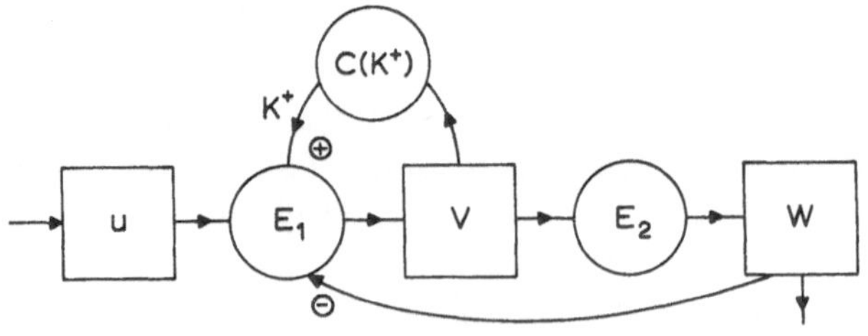

Fig. 15.

This is the familiar metabolic control circuit that incorporates the essential features of the glycolytic oscillator. E_1 catalyzes the conversion of u into V, and E_2 that of V into W. V has the effect of releasing some molecular species from a bound (or otherwise constrained) state, and this activates E_1. This is identified as potassium in the model, but this is only one of many possible candidates. W is a feed-back inhibitor of E_1. So this metabolic system is intrinsically capable of oscillatory behaviour.

Now let us consider the picture on the membrane. This is shown in Figure 16. It is known that many enzymes are localized in or on membranes, and so I assume that

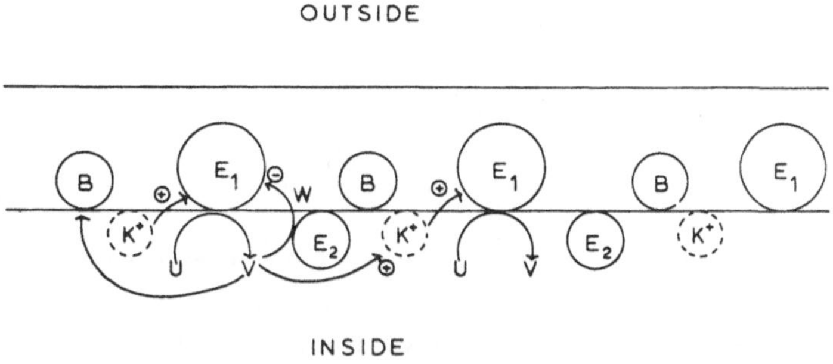

Fig. 16.

E_1 and E_2 are so situated. Only E_1 is actually held in the membrane, E_2 being either bound or soluble. I assume that potassium ion is normally adsorbed to sites on the membrane, and that the molecular species V can form a bound complex with a specific protein in the membrane, designated **B**.

An activity wave can be initiated by any event that causes a release of potassium ions, which will activate E_1 and start off the reaction sequence. Since V releases K^+, the wave of activation of E_1 spreads along the membrane in all directions from the point of initiation. Two consequences of this activity wave are desired: the first is the generation of a gradient in the spatial distribution of the substance V; and the second is periodic reinitiation of the wave from the point where it first occurred. These consequences may be realized by the following postulates. Assume that some time after the release of K^+ and the activation of E_1, the sites B for the binding of V become exposed (e.g., W could act as an activator of B as well as an inhibitor of E_1, but this detail may be left unspecified). These sites revert to an unavailable condition by relaxation some time after being activated if they do not combine with V. The metabolite V will then be picked up by the membrane-bond protein B and held in the form of an association complex behind the wave-front, so that there is a spatial asymmetry with respect to the sites of V-production by E_1 and its pick-up by B. The fraction of V picked up will then have moved by a pure diffusional process through a certain distance in a direction opposite to that of wave propagation. Thus bound V will have a maximum at the point of initiation of the wave. This process is shown schematically in Figure 17, where the propagating site of production of V is shown as a delta-function which decays symmetrically along the one-dimensional spatial axis. The pick-up func-

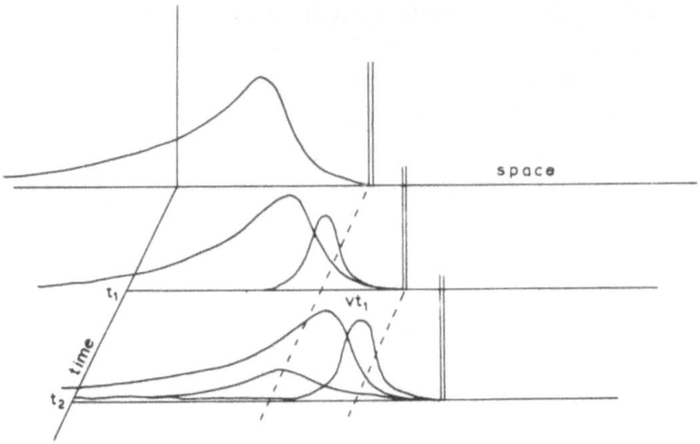

Fig. 17.

tion expressing the activation of the sites B and their decay is shown travelling along behind the wave-front.

Since we have assumed that V causes the release of K^+, it is a natural consequence of the accumulation of V in some region of the membrane that an initiation of the wave will recur in this region, since there will be a finite decay rate of the B-V complex. As the wave recurs and a gradient of V is built up, the recurrence of the wave will increase in frequency until we have a rhythmically recurring event generating a steady gradient in the bound material, V. In this way symmetry can be broken, and an initially uniform system, either with or without cells, may become spatially organized.

If we particularize the model to apply it to Dictyostelium, we could take U to be ATP, E_1 to be adenyl cyclase, V=cAMP, E_2=phosphodiesterase, and W=AMP. cAMP initiates the process by causing a local release of K^+. Amoebae then break symmetry and become polarized. The accumulation of cAMP could result in a periodic release of this substance from the amoeba when a threshold of depolarization is reached; or by some other release mechanism.

It is now possible to write down equations which describe such a process, and then to solve them under certain conditions to study the behaviour of the model and compare this with biological processes. Let $X(x, t)$ be the concentration of the diffusible form of the metabolite, and $Y(x, t)$ be the concentration of the bound form, where x is the spatial dimension and t is the time. The equations are of the form

$$\frac{\partial X(x, t)}{\partial t} = D \frac{\partial^2 X(x, t)}{\partial x^2} + \beta Y(x, t) - f(x, t) X(x, t) + P(x, t). \qquad (3.1)$$

$$\frac{\partial Y(x, t)}{\partial t} = - \beta Y + f(x, t) X(x, t). \qquad (3.2)$$

X is subject to diffusion, with a diffusion constant D. We assume that the bound form, Y, is subject to decay at the rate βY. The rate of formation of Y is taken to be the product of the concentration of X and the concentration of binding sites, the latter being described by the function $f(x, t)$, which also contains the rate constant for the process. $P(x, t)$ is a forcing function of any kind that is to be included in the problem.

Since $f(x, t)$ can be chosen to have plausible form, it is convenient analytically to use the function

$$f(x, t) = k\alpha^2 v (vt - x) e^{-\alpha(vt-x)}.$$

Here v is the velocity of the wave, k is the product of a rate-constant for the formation of the bound metabolite and a concentration parameter, α is a parameter that determines the shape of the function, and the term $\alpha^2 V$ is present to normalize the function.

Thus we have the relation

$$\alpha^2 v \int_{x/v}^{\infty} (vt - x) e^{-\alpha(vt-x)} \, dt = 1$$

which is readily verified.

For the case in which we consider the passage of a single wave from the origin, propagating at velocity v in both directions along a doubly-infinite one-dimensional space, $P(x, t)$ would be zero and there would be an initial condition on X in the form of a delta-function. This would propagate along the membrane in both directions.

I will now simplify the problem so that a simple analytical solution can be obtained and we can get some insight into the behaviour of the variables. Let us suppose that there is a pulse of X that is produced at the point $x=\rho$ on the membrane, and that this decays without propagating. Let us assume further that the amount of X that is picked up and becomes Y is small compared with the amount of X produced. Then we can ignore $f(x, t)$ $X(x, t)$ in the first equation. Let us also assume that Y is very stable so we can ignore $\beta Y(x, t)$ in both equations. Then we can write down the solution for X immediately which is

$$X(x, t) = \frac{M}{2\sqrt{\pi D t}} e^{-(\varrho - x)^2/4Dt},$$

where M is the concentration of X in the pulse. We suppose that the pick-up function $f(x, t)$ propagates in the normal way, so that $Y(x, t)$ is just the solution of the equation

$$\frac{\partial Y(x, t)}{\partial t} = \frac{K\alpha^2 v M (vt + \varrho - x)}{2\sqrt{\pi D t}} e^{-\alpha(vt+\varrho-x)} e^{-(\varrho-x)^2/4Dt}$$

i.e.,

$$Y(x, t) = \int_0^t \frac{K\alpha^2 v M (vt + \varrho - x)}{2\sqrt{\pi D t}} e^{-\alpha(vt+\varrho-x)} e^{-(\varrho-x)^2/4Dt} \, dt,$$

taking $Y(x, 0) = 0$.

The picture corresponding to this process is shown in Figure 18.

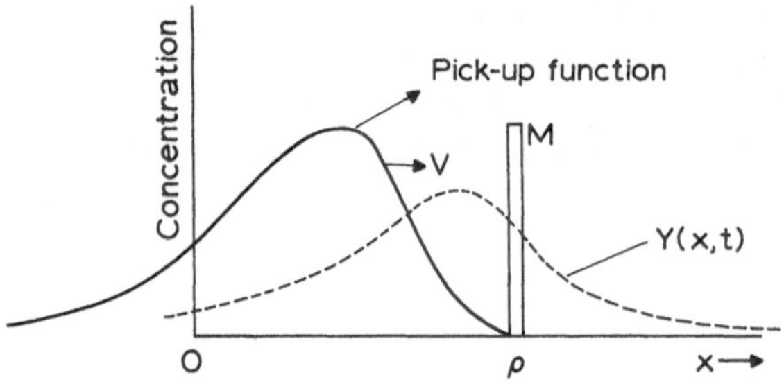

Fig. 18.

We may then ask questions such as how far the peak of Y is from the point ρ after the process is finished, how much Y there is in comparison with M, and how long the process takes to be essentially completed. Then we can look at the more general problem.

Writing $\beta=(\varrho-x)/V$, $\gamma=(\varrho-x)^2/4D$, the integral takes the form

$$Y(x,t)=\frac{K\alpha^2v^2M}{2\sqrt{\pi D}}\int_0^t(\tau+\beta)\,e^{-\alpha v(\tau+\beta)}e^{-\gamma/\tau}\tau^{-1/2}\,\mathrm{d}\tau\,.$$

Now we are interested in Y when t is large, so we can use the standard form

$$\int_0^\infty x^{\nu-1}e^{-(\beta/x+\gamma x)}\,\mathrm{d}x=2\left(\frac{\beta}{\gamma}\right)^{\nu/2}K_\nu(2\sqrt{\beta\gamma}),$$

where K_ν is a Bessel function of imaginary argument. Thus evidently we get the result:

$$Y(x,t)=\frac{K\alpha^2v^2M}{2\sqrt{\pi D}}\left\{2\left(\frac{\gamma}{\alpha v}\right)^{3/4}K_{3/2}(2\sqrt{\gamma\alpha v})+\right.$$

$$\left.+\,2\beta\left(\frac{\gamma}{\alpha v}\right)^{1/4}K_{1/2}(2\sqrt{\gamma\alpha v})\right\}e^{-\alpha v\beta}\,.$$

These Bessel functions have simple closed forms, viz.,

$$K_{1/2}(a)=\sqrt{\frac{\pi}{2x}}\,e^{-x}$$

$$K_{3/2}(x)=\sqrt{\frac{\pi}{2x}}\,e^{-x}\left(1+\frac{1}{x}\right).$$

It is somewhat simpler to write the solution in terms of a variable whose origin is the point $x=\varrho$, and to measure backwards from this point in relation to the diagram in Figure 16. So we take $\xi=\varrho-x$ as this new spatial variable, and then we find the solution

$$Y(\xi, \infty) = \frac{KM}{4}\left(\frac{\alpha v}{D}\right)^{1/2} e^{-(\alpha+\sqrt{\alpha v/D})\xi}\left[1+\xi\left(2\alpha+\sqrt{\frac{\alpha v}{D}}\right)\right].$$

The appropriate units to use are microns, μ, and seconds α has units $1/\mu$, v has units μ/sec, and D the units μ^2/sec. So $\alpha v/D$ has units $1/\mu^2$, and the units are all right. The shape of this distribution is essentially like the pick-up function, but it does not have the value 0 at $\xi=0$. It has the general shape shown in Figure 17.

Now we may ask some questions about distances, assigning some plausible values to the parameters. We take $v=4\ \mu/s^{-1}$, which is about the rate of propagation of the wave in Dictyostelium, $D=4\times10^2\ \mu^2/s^{-1}$, which corresponds to a molecule about the size of cyclic AMP; and we take $\alpha=1/16$, which means that the peak of the pick-up function occurs 16 μ behind the wave-front. With $v=4$, this means that the time for the sites to become receptive is about 4 s. This may be somewhat short, but it will do to give us some order-of-magnitude estimations.

The peak of $Y(\xi, \infty)$ occurs at a distance $1/(\alpha+\sqrt{\alpha v/D})$ from the point of origin of the pulses M. For the values above, this is about 11 μ. The amount of Y is found by integrating over the spatial variable. This turns out to be about 25% of the amount in the pulse, so our assumption about Y being small in relation to X is not a good one. For the value given to D, the decay time is a few seconds. Thus in 4 s, the pick-up function will have travelled past the origin of the pulse of M and this pulse will have decayed to about 1/150 of its original value. We could then say that the substance produced at the point ϱ will have been displaced through about 11 μ in 4 s. This gives us some idea of the 'transport' rate such a process is capable of.

Complicating the problem slightly, we can look for a solution of $X(x, t)$ in the case of a propagating delta-function, still making the assumption that X is considerably longer than Y.

In this case the form is

$$X(x, t) = M\int_0^{\pm vt} \frac{\pm\, d\xi}{2\sqrt{\pi D(t\mp\xi/v)}}\exp\left[\frac{-(x-\xi)^2}{4D(t\mp\xi/v)}\right].$$

This integral is taken over the spatial variable ξ from the origin up to the wave-front niboth directions. Writing $\eta=t-\xi/v$, $\sigma_1=(x+vt)^2/4D$, $\sigma_2=(x-vt)/4D$, and $\varrho= =v^2/4D$, this integral becomes

$$\frac{M}{2\sqrt{\pi D}}\left\{e^{2v(x+vt)/4D}\int_0^t \eta^{-1/2}e^{-(\sigma_1/\eta+\varrho\eta)}\,d\eta + \right.$$

$$\left. + e^{-2v(x-vt)/4D}\int_0^t \eta^{-1/2}e^{-(\varrho\eta+\sigma_2/\eta)}d\eta\right\}.$$

This leads to the solution

$$X(x,t) = \frac{M_v}{2\sqrt{\pi D}} \left\{ e^{v(x+vt)/2D} \sum_{n=0}^{\infty} \frac{(-\varrho)^n}{n!} \sigma_1^{n+1/2} \Gamma\left(-(n+\tfrac{1}{2}), \frac{\sigma_1}{t}\right) + \right.$$

$$\left. + e^{-v(x-vt)/2D} \sum_{n=0}^{\infty} \frac{(-\varrho)^n}{n!} \sigma_2^{n+1/2} \Gamma\left(-(n+\tfrac{1}{2}), \frac{\sigma_2}{t}\right) \right\}.$$

For large σ_i/t, this is approximately

$$X(x,t) \simeq \frac{M_v}{2\sqrt{\pi D}} \{\sigma_1^{-1} e^{-(x+vt)/4Dt} t^{3/2} + \sigma_2^{-1} e^{-(x-vt)^2/4Dt} t^{3/2}\} =$$

$$= 2M_v \sqrt{\frac{D}{\pi}}\, t^{3/2} \left\{ \frac{e^{-(x+vt)^2/4Dt}}{(x+vt)^2} + \frac{e^{-(x-vt)^2/4Dt}}{(x-vt)^2} \right\}.$$

This shows us how the solution decreases away from the wave-front. The graph looks qualitatively as shown in Figure 19.

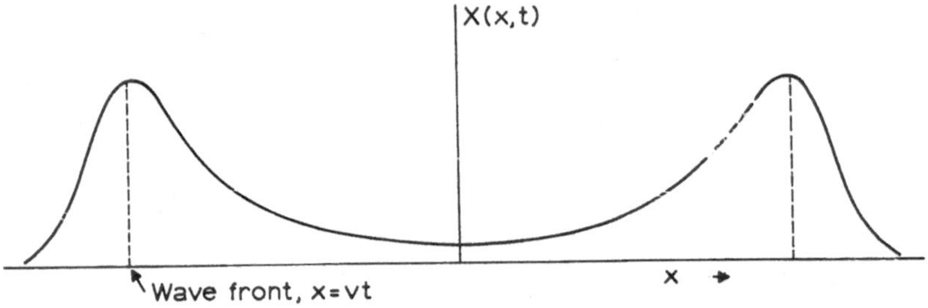

Fig. 19.

To find $Y(x,t)$ we must perform the integration

$$Y(x,t) = \alpha^2 v \int_{x/v}^{t} (v\tau - x)\, e^{-\alpha(v\tau - x)}\, X(x,\tau)\, d\tau .$$

The pick-up starts when the wave-front reaches the point x, which is at time x/v, and proceeds to time t. This integration is not an easy one, but it is considerably simplified by writing $x = pvt$ where $0 \leqslant p \leqslant 1$, and then expanding $\Gamma(\sigma_i/\tau, -(n+\frac{1}{2}))$ to the first order in the variable $\sigma_i/\tau = v^2[(1\pm\varrho)/4D]\,\tau$. For the parameter values used above, this in less than $2\times 10^{-2}\,\tau$, so an expansion to first order will give a good approximation for $\tau \leqslant 20$ s, say. In 20 s the wave will have travelled 80 μ, which would be the whole length of a fucus egg, and a substantial part of an amphibian embryo. So this approximation will give useful information. The general shape of Yx, t as obtained by this method is as shown in Figure 20.

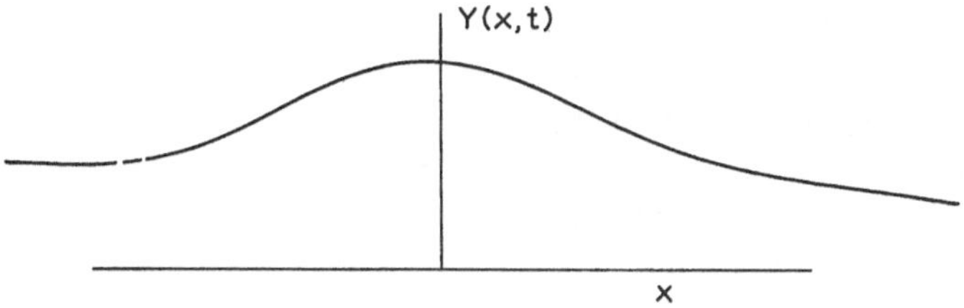

Fig. 20.

This is the result of one wave propagation over a finite distance, with the simplifying assumptions made above. The system with periodic repetition of the wave from the origin, decay of the bound form at the rate β, and no restrictions on the relative magnitudes of X and Y can only be solved by computation, which is in progress. We can, however, say something about the expected solution.

Evidently the process described behaves as a pump which 'transports' X towards the origin, the gradient being in the form of bound metabolite, Y. The origin will then act as if it were a source of substance because of the decay, and under particular parametric conditions we may expect a steady-state to be established, with a monotonic gradient of Y over the cell or the embryo. If we are considering an embryo which is undergoing cleavage into cells, then we need to suppose that every cell is in communication with every other cell so that wave propagation can occur and substances can diffuse from one cell to the next. Embryonic cells have been shown to be in electrical communication with one another, and substances of the size of cyclic AMP can certainly diffuse from cell to cell. So there is no difficulty in applying the model to either cellular or acellular systems.

There are several very interesting questions which need to be answered regarding the behaviour of the wave-broom model. There will be some optimal relationship between v, D, α, and the period of the repeated wave which generates a gradient which peaks well at the origin and extends substantially from the origin. If the frequency of the periodic wave is very high, then effectively the concentration of X will remain the same everywhere and no gradient in Y can result. However, for D values about $10^2 \mu^2 s^{-1}$, which is the range for molecules of the type we are interested in, the frequency of the wave and its velocity would both have to be implausibly high to result in a stationary, flat distribution of X everywhere.

Another very important question is the dimensionality of the problem. Most membranes can be regarded as 2-dimensional surfaces, so the problem should be presented in two spatial dimensions. This does not add significantly to the analysis, but again those questions will finally have to be resolved by computation.

3.3. THE ESTABLISHMENT OF A METRIC

So far we have considered only the problem of polarity, breaking symmetry and establishing an origin for a spatial axis. It is now necessary to consider the next step in the developmental process, which is the establishment of some metric on this axis so that cells or regions of the embryo know where they are in relation to the whole and so can behave accordingly. Furthermore, this metric must be adjustable to the size of the whole, since large or small embryos of any particular species are perfectly scaled in relation to their size. Thus embryos are pattern invariant, size independent systems, a property referred to as regulation. Not only do embryos adjust to variations in size; they also adjust to variations in rate. Development takes different times at different temperatures, but the final organism is always the same. So both space and time are scaled.

At this point we have a choice of paths to follow regarding scaled metrics in the embryo (or the unicellular organism, since ciliate protozoa behave exactly as embryos in regard to space and time regulation). The classical procedure is to regard the gradient of substance, such as Y, as the carrier of information. In this case it is the concentration of Y at any point which determines how that part of the embryo will behave. Then it is necessary that this gradient have positive slope throughout the dimensions of the embryo, and that the maximum and minimum values be independent of size. The wave-broom model is one which will tend to have these properties, since the amount of material transported with every wave will be very largely independent of the length of the axis, once a certain minimum has been exceeded. However, it is by no means evident that the model as described will produce a gradient with positive slope over a variety of axial lengths. It seems likely that modifications would have to be introdueed to achieve this result.

One such modification which is suggested by the stoichiometry of most allosteric binding processes is that the protein in the membrane which binds X has multiple sites, say $n > 1$, with cooperative behaviour. Then the concentration of Y as a function of X will have sigmoid characteristics of the type we have considered previously in relation to feed-back inhibition, as shown in Figure 21.

Such a non-linearity would have the effect of exaggerating whatever gradient is generated by the linear model, described by Equations (3.1), (3.2). Then in place of the terms $f(x, t)\, X(x, t)$ we would have

$$\frac{f(x, t)\, X^n(x, t)}{1 + LX^n(x, t)},$$

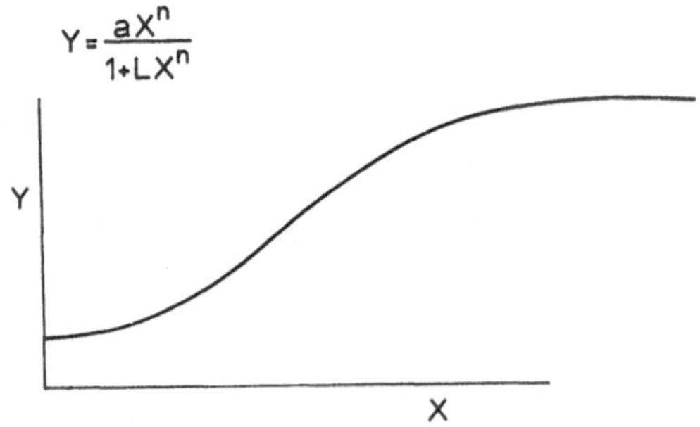

Fig. 21.

making the equations non-linear. The decay term of Y would give a term $n\beta Y(x, t)$, since for every molecule of protein that relaxes, n molecules of X would be produced. The variable Y now refers to the concentration of n-meric proteins, which could be referred to as 'morphogens'; i.e., those substances responsible for generating changes of epigenetic state in the embryo. Such equations could only be studied by computer. However, this would be of interest for two reasons. The first is to see if regulation of a gradient occurs 'naturally' in such a model, without further complication. The other is to test the multimeric model against some well-established experimental results.

The experiments I am referring to are those performed primarily by Locke on the insect curticle. In this system one can see in a very elegant manner certain consequences of perturbing a gradient system in defined ways. A theoretical model has been proposed by Lawrence *et al.* [12] to account for these observations on the basis of stable homeostatic properties of cells; i.e., the cell is the unit in their model, and it stabilizes the gradient by virtue of homeostatic performance. The behaviour of their model under perturbation is consistent with the experimental observations. However, the basic assumption of their theory, that cells are capable of maintaining any one of a number of different concentrations of a single metabolite, the particular value being 'set' by a concentration gradient, is not very plausible biochemically speaking. Thus it is of some value to investigate the stability properties of an alternative model in which the non-linearity is of the type shown in Figure 21. This model makes some predictions about the expected shape of the gradient itself, which can also be tested against experiment. These studies are in progress.

3.4. The phase-shift model

The alternative path that we can pursue in studying the establishment of a metric in the embryo after an axis has been generated is to suppose that the embryonic cell or region measures, not a concentration of a substance, but some other distance-measuring variable. One possibility is that once the metabolite M has accumulated sufficiently at the origin, it initiates not simply the wave of activity which is associated with the production of M, but another wave of essentially similar type involving different enzymes and metabolites. If we assume that this wave propagates at a somewhat lower velocity v_2 than the first wave (velocity v_1) so that $v_2 < v_1$, then the time interval between the occurrence of these two waves at any point will be a measure of the distance of that point from the origin. The principle involved is the same as that used to measure the distance of a thunder-bolt by timing the interval between the lighting and the thunder. But our waves are non-linear waves which do not travel in elastic media and do not reflect from boundaries. Since it is assumed that both these waves are initiated periodically from the same region, we may refer to the time interval between their arrival at any given point as a phase angle, 2π units of phase being the same as T units of time, where T is the period of the rhythmically recurring waves. This model has been developed in some detail in the publication by Goodwin and Cohen [16], and the theory is referred to as the phase-shift model. The gradient of spatial or positional information is then in the form of a phase gradient, not a gradient of substance.

There is a number of experimental implications of this theory which have been investigated in the original paper and in subsequent publications (Cooke and Goodwin [13], Goodwin [14, 15], Cohen and Robertson [9, 10]). None of these resolves the problem of the nature of the gradient of positional information, whether of substance or of phase, and indeed it is very difficult to design experiments which can distinguish clearly between them. It is much easier to study periodicities in development, and in this area some progress has been made. I think it can fairly be stated that periodic phenomena are now recognized as of possibly central importance in developmental processes. However, the wave-broom model shows that periodicities cannot be taken as evidence for a phase metric, since the periodic wave could simply be setting up a gradient of substance. So at the moment a direct study of the phase-shift model is rather difficult experimentally, and attention is being directed in our laboratory to the more general question of periodicities, polarity, and transport in developing systems.

3.5. Regulation

I have as yet said nothing about the problem of regulation in relation to the phase-shift model. A solution of this was given in the original paper of Goodwin and Cohen, but it uses rather expensive physiological machinery, and is hence rather implausible. My own feeling is that organisms achieve regulation by some simple and direct consequence of the basic process used to establish polar axes, and as yet we have not discovered this principle. A study of the wave-broom model may give some suggestions as to how regulation arises effortlessly in developmental processes. Temporal regulation is readily obtained from the phase-shift model, since every relevant process is a metabolic one. In diffusion-based theories the different temperature dependence of diffusion and energy-dependent metabolism creates problems, but is is possible to resolve those at the expense of additional regulatory processes. But neither type of theory as presently developed provides a simple and convincing explanation of spatial regulation, the adjustment of the metric according to overall dimensions.

Another problem that I have not considered is how axes are established in two and three dimensions. All higher animals have three dimensional order, and regulation occurs, of course, along all three axes. So far I have discussed only the establishment of one axis. In many organisms the egg has the information required for setting up axes in the form of differentiated regions of the egg, the differences being on the membrane or cortex. These are presumably generated by the position of the egg in the ovary, where spatial order can result from heterogeneities such as the position of blood vessels relative to the oocyte and its position in relation to the surface of the tissue. Whatever their origin, the 'boundary conditions' required for initiating a breaking of symmetry in animal cells are usually present in the egg. Thus the case of Fucus is rather exceptional. However, having provided a solution for this case, we can readily explain the others.

There are many other problems in developmental biology that I cannot go into here but which are very interesting and challenging. Perhaps the most dramatic of these is how higher organisms generate the extraordinary detail of neural connections in the brain. However, the principle whereby brains are formed are almost certainly basic and elementary in the sense that cells and membrane networks can quite easily undergo the spatial and temporal transformation required to generate remarkable four dimensional order. What these principles of biological organization are which underly such processes remain tantalizingly obscure. However, I hope I have suggested some avenues of approach which employ mathematical techniques of relevance and interest, which can help to guide further experimental research.

Appendix 1. Rate Constants and Relaxation Times of the Metabolic System

Bacterial volume is of order $4\mu^3 = 4\times 10^{-15}$l. About 10^6 protein molecules per bacterium, 10^3 different species. Therefore about 10^3 mol per enzyme species. This gives a concentration of enzyme in the bacterium of order

$$\frac{10^3}{4\times 10^{-15}\times 6\times 10^{24}} \simeq 4\times 10^{-8} \text{ molar}.$$

Turnover number for enzymes (i.e., number of molecules of substrate converted into product per second) is of order 10^3. With 10^3 mol of enzyme, 10^6 mol of metabolite (product) will be produced per second. This represents a concentration change in the bacterium of

$$\frac{10^6}{4\times 10^{-15}\times 6\times 10^{24}} \simeq 4\times 10^{-5} \text{ molar per second}.$$

Since the steady state value of S is about 10^{-3} M, it is evident that the above concentration represents a small perturbation and it occurs in a second or so.

With $k_i \simeq 10^{-2}$ per second, we find $k_i S_i \simeq 10^{-5}$ M sec^{-1}, agreeing with the above culation.

References

1. Hearon, J. Z.: *Bull. Math. Biophys.* **15**, 121 (1952).
2. Gerhardt, J. C. and Schachman, H. K.: *Biochemistry* **4**, 1054 (1965).
3. Jacob, F. and Monod, J.: *J. Mol. Biol.* **3**, 318 (1961).
4. Goodwin, B. C.: *Temporal Organization in Cells*, Academic Press, London and New York (1963).
5. Kerner, E. H.: *Bull. Math. Biophys.* **26**, 333 (1964).
6. Goodwin, B. C.: *J. Theoret. Biol.* **28**, 375.
7. Kubitscheck, H. E.: *Proc. Berkeley Symp. Math. Statist. Probab.* **4**, 549 (1967).
8. Hess, B. and Boiteux, A.: *Ann. Rev. Biochem.* **40**, 237 (1971).
9. Cohen, M. H. and Robertson, A.: *J. Theoret. Biol.* **31**, 101 (1971).
10. Cohen, M. H. and Robertson, A.: *J. Theoret. Biol.* **31**, 119 (1971).
11. Jaffe, L.: *Adv. Morphogen.* **7**, 295 (1968).
12. Lawrence, P. A., Crick, F. H. C., and Munro, M.: *J. Cell Sci.* **11**, 815 (1972).
13. Cooke, J. and Goodwin, B. C.: *Some Mathematical Questions in Biology* vol. II, p. 35 (1970). Ed. by J. D. Cowan, Amer. Math. Soc., Providence, R.I.
14. Goodwin, B. C.: *Some Mathematical Questions in Biology*, vol. III, p. 75 (1970). Ed. by J. D. Cowan, Amer. Math. Soc., Providence, R.I.
15. Goodwin, B. C.: *Symp. Soc. Exp. Biol.* **25**, 417 (1971).
16. Goodwin, B. C. and Cohen, M. H.: *J. Theoret. Biol.* **25**, 49 (1969).

GENERAL REFERENCES

I. *Biological Clocks and Cell Synchronization*

1. Aschoff, J. (ed.): *Circadian Clocks*, North-Holland Publ. Co., Amsterdam (1965).
2. Bünning, E.: *The Physiological Clock*, Academic Press, N.Y. (1964).
3. Biological Clocks, Cold Spring Harbor Symp. Quant. Biol. (1960), p. 25.
4. Mitchison, J. M.: *The Biology of the Cell Cycle*, Cambridge Univ. Press (1972).
5. Goodwin, B. C.: *Temporal Organization in Cells*, Academic Press, London (1963).
   Goodwin, B. C.: *Europ. J. Biochem.* **10**, 511 (1969).
   Goodwin, B. C.: *Europ. J. Biochem.* **10**, 515 (1969).
   Goodwin, B. C. *Symp. Soc. Gen. Microbiol.* **19**, 223 (1969).
6. Winfree, A. T.: in M. Gerstenharber (ed.), *Lectures on Mathematics in the Life Sciences*, vol. II p. 109, Amer. Math., Soc. Providence, R.I.

II. *Developmental Biology*

1. Bonner, J. T.: *The Cellular Slime Molds*, Princeton Univ. Press. (1959).
2. Berrill, N. J.: *Developmental Biology*, McGraw Hill Book Co., N.Y. (1971).
3. Trinkhaus, J. P.: *Cells into Organs*, Prentice-Hall, New Jersey (1969).
4. Waddington, C. H.: *Principles of Embryology*, Allen and Unwin, London (1956).

III. *Theoretical Biology*

1. *Towards a Theoretical Biology* (ed. by C. H. Waddington), 4 vols., (1968, 1969, 1970, 1972). Edinburgh University Press.
2. *Some Mathematical Questions in Biology* (ed. by M. Gerstenhaber), vols. I and II (1969, 1970). Ed. by J. D. Cowan, vols. III and IV (1970).